Handbook of Pressure Sensitive Adhesive Technology

Handbook of Pressure Sensitive Adhesive Technology

Second Edition

Edited by

Donatas Satas

Springer Science+Business Media, LLC

ISBN 978-1-4757-0868-4 ISBN 978-1-4757-0866-0 (eBook)
DOI 10.1007/978-1-4757-0866-0

Originally published by Van Nostrand Reinhold in 1989

Library of Congress Catalog Card Number 88-21107

Library of Congress Cataloging in Publication Data

Handbook of pressure sensitive adhesives/edited by Donatas Satas.—
2nd ed.
p. cm.
Rev. ed. of: Handbook of pressure-sensitive adhesive technology.
c 1982.
Includes indexes.

1. Adhesives. I. Satas, Donatas. II. Handbook of pressure sensitive adhesive technology.
TP968.H36 1989
668'.3—dc 19 88-21107

Contents

Preface

The first edition of this book was well accepted by the pressure sensitive adhesive industry in the United States, as well as internationally. This confirmed our belief that a book covering various aspects of this technology was needed and we appreciate having had the opportunity to fill this need.

This second, extensively revised edition is the continuation of the effort. We have included a considerable amount of new material. Some of the new material covers omissions in the first edition. New material also reflects some of the changes that have taken place in this technology since 1982, the publication year of the first edition. Although the technological changes are not as frequent and pervasive as some science popularizers try to make us believe, some important new directions are noticeable in pressure sensitive adhesive technology. The use of aqueous emulsions has been increasing as replacement for solventborne products, prompting the development of an increased number of emulsion polymers and aqueous tackifying resin dispersions. Counteracting the replacement of solventborne adhesives is the increased use of drying in an inert gas atmosphere and solvent recovery by condensation. Dynamic testing of adhesive rheological properties has been used for a long time by some companies, but the wider recognition of the usefulness of this method is relatively recent. It is expected that this method will become an important tool in the design of adhesives. These subjects have been covered in additional chapters.

As before, we attempted to draw contributors from many nations. Our gratitude goes to the authors who devoted many hours to preparing various chapters in this book. We would also like to express our gratitude to all those researchers whose names appear only in the references, but whose published work really made this book possible.

Our appreciation goes to Susan Munger, Gail Nalven, Jim Woods, and Trumbull Rogers of Van Nostrand Reinhold Company, to my daughter A. Maria for her help in editing, and to my wife Saule for assistance in the index preparation.

D. Satas

Contributors

Charles Bartell. B.S.Ch.E. (1942) from Cooper Union Engineering, M.S. (1949) and Ph.D. (1951) from Ohio State University. Was associated with Rohm & Haas Company, Permacel, W. R. Grace Company, and served as research director at Mystik Tape. Currently consultant for Mystik Tape.

Erhard Braeunling. Doctorate in chemistry. Project leader at Beiersdorf AG, Hamburg, West Germany. Engaged in the development of electrical tapes for many years. Currently in quality control of tape products.

G. L. Butler. B.Sc. in chemistry and natural philosophy at University of St. Andrews, Scotland. Development Manager of DRG Sellotape Products, Borehamwood, England, where he is responsible for the development and quality control of pressure sensitive products.

Anne Z. Casey. B.S. degree in chemistry from Rensselaer Polytechnic Institute (1981). The technology described in her chapter was a part of her project responsibilities in the Adhesive Applications Laboratory, Hercules, Inc., Wilmington, Delaware. She is currently a senior development chemist with Hysol Aerospace and Industrial Products Division of The Dexter Corporation.

Sung Gun Chu. Ph.D. in physical polymer chemistry from the University of Texas (1978). His main interests are property–structure relations of polymeric materials. Currently he is a project leader in adhesives at Hercules. He has been involved in the development work of amorphous polyolefins and block copolymers for adhesive applications.

Carl A. Dahlquist. Now retired and part-time consultant, was a research scientist of 3M Company, St. Paul, Minnesota. Main field of interest is adhesion, rheology of adhesive materials, mechanical properties of polymers. Author of many papers and patents.

Norman P. DeBastiani. B.S. in chemistry from American International College (1960). Was associated with Plastic Coating Corporation. Currently in charge of development and quality control of photographic, graphic arts, and pressure sensitive products at Chartpak, Leeds, Massachusetts.

R. Dowbenko. Ph.D. in organic chemistry. Involved during his career in the synthetic aspects of coating materials. Currently manager of polymer research at PPG Industries, C7R Division, R&D Center, Allison Park, Pennsylvania.

James R. Erickson. Senior research chemist at Shell Development Company, Houston, Texas. Specializing in new polymer and adhesive development for cross-linkable adhesive systems.

Earle E. Ewins, Jr. Senior technical associate at Shell Development Company, Houston, Texas. Responsible for compounding principles, applications, and associated technologies for pressure sensitive adhesives.

Keiji Fukuzawa. Doctor of Engineering from Yokohama University, Japan (1954). Chemical consultant, Tokyo, Japan. Specializes in pressure sensitive adhesives, tapes, labels, surgical products. Member Japan Consulting Engineers' Association.

Whiteford D. Grimes. Chemical engineer (Virginia Polytechnic Institute). Was associated with DuPont, Sun Chemical, and Stauffer Chemical Company. Thirty years experience in film processing, laminating, and coating technology. Currently general manager at Avery Graphic Systems Division-Automotive.

Fred H. Hammond, Jr. Graduated from Colby College and studied physical chemistry at Massachusetts Institute of Technology. Early research involved studying the surface chemistry and polymer rheology of pressure sensitive adhesives and development of Polyken probe tack tester. He is currently information services specialist at The Kendall Company, Lexington, Massachusetts.

George M. Harris. B.S.Ch.E. from Illinois Institute of Technology (1950). Was associated with The Kendall Company in R&D, technical service, and marketing functions of corrosion protective tapes. Currently vice president at Tapecoat Company, Evanston, Illinois.

Andrew D. Hickman. B.S. in chemistry from University of Illinois (1974). He is associated with Dow Chemical Company, initially in technical service for latex products and organic chemicals, and more recently as an opportunity analyst for latex products.

John J. Higgins. B.S. from Iona College and M.S. in chemistry from Fordham University. Currently sales representative with Exxon Chemical Company USA, Florham Park, New Jersey.

Frank C. Jagisch. B.E. and M.S. from Stevens Institute of Technology. Currently a product application and customer technical service specialist with Exxon Chemical Company, Baton Rouge, Louisiana.

J. Darrell Jones. B.Sc. (Hons.) in chemistry from the University of Wales-Aberystwyth. Associate Development Scientist at Dow Corning Corporation in the release coating/pressure sensitive field.

David N. Kendall. B.A. and M.A. in chemistry from Wesleyan University (1938 and 1939) and Ph.D. in chemical physics from Johns Hopkins University (1943). After research in chemistry, physics, and analytical chemistry at National Lead and American Cyanamid Company, he founded Kendall Infrared Laboratories (KIRL) in 1953. In addition to solving problems where unknown materials are involved, KIRL analyzes and consults on cases of product liability, patent infringement, and arson.

William H. Korcz. Staff research chemist at Shell Development Company, Houston, Texas. Currently involved in venture development projects. Primary duties are focused in development of opportunities and technologies related to the packaging industry.

Richard M. Krawiec. B.S. in chemistry from University of Connecticut and M.S. in chemistry from Drexel University. Was engaged previously by Occidental Chemical in the area of emulsion polymerization. Currently he is employed by Hercules, Inc., as an adhesives application specialist.

Roland R. Milker. Doctor of Science, University of Bonn, for research on phosphorus compounds. Was associated with Kimberly Clark/Koblenz as quality supervisor and process engineer. Since 1980 he has been R&D manager of adhesive tapes and nonwoven fabrics at Lohmann GmbH & Co. KG, Neuwied, West Germany. Specializes in acrylic polymerization, acrylic pressure sensitive adhesives, and coating technology.

Walter Moehren. Chemical engineer, University of Aachen (1977). Conducted research on polyurethane chemistry. Was engaged in management of silicone specialty product sales. Since 1982 head of Tape Application Department at Lohmann GmbH & Co. KG, Neuwied, West Germany.

Helmut W. J. Mueller. Graduated as a chemical engineer, Berlin (1958). Currently in charge of the pressure sensitive adhesive work team at BASF, Ludwigshafen, West Germany. Previous experience in automotive industry.

Lee A. Mushel. B.S., University of Wisconsin, Platteville (1965). Chemist. Currently specializing in computer controls for converting industry. Previous experience in product development and manufacturing of flexible packaging.

Yolanda A. Peters. B.S. in chemistry from Wayne State University and M.S. in chemistry from Michigan State University. She is a senior technical specialist at Dow Corning Corporation in the release coating/pressure sensitive field.

Valentinas Rajeckas. Doctor of Technical Sciences. Professor and head of the Department of Leather and Textile Products Technology at Kaunas Polytechnic Institute, Kaunas, Lithuania, USSR. Specializes in theoretical principles and instrumental methods in prediction of adhesive joint strength.

Richard K. Rathmell. Chemical engineer with an extensive industrial experience mainly with Du Pont Co. Mr. Rathmell was vice president of Wolverine Corporation and is currently associated with Waste Management Corporation of North America in Wakefield, Massachusetts. He holds a dozen U.S. patents, including one on a solvent recovery system.

John Rienau. B.S. in mechanical engineering from New Jersey Institute of Technology (1958). He has forty years experience in the area of slitting and rewinding, and is business development manager at John Dusenbery Company, Inc.

David J. St. Clair. Staff research chemist at Shell Development Company, Houston, Texas. Currently specializing in new polymers and their development for adhesives, sealants, and coatings applications.

A. Maria Satas. B.S.Ch.E., University of Illinois (1983). Chemical engineer. Currently marketing researcher at Satas & Associates. Warwick, Rhode Island.

Donatas Satas. B.S.Ch.E., Illinois Institute of Technology (1953). Chemical engineer. Independent consultant since 1975 (Satas & Associates, 99 Shenandoah Rd., Warwick, Rhode Island). Specializes in adhesives, coating, and laminating technology.

James A. Schlademan. B.S. in chemistry from Carleton College and Ph.D. in organic chemistry from Clarkson University. In 1970 he started at Neville Chemical Company and was in charge of adhesives and rubber application research. In 1979 he joined ARCO Chemical Company and worked on the synthesis and adhesive applications of thermoplastic elastomers. In 1984 Dr. Schlademan joined The Goodyear Tire & Rubber Company, and is currently head of the Wingtack and Specialty Polymer Section.

Loretta A. Sobieski. B.S. in chemistry from Bowling Green State University (1976). Technical Specialist at Dow Corning Corporation. Experience in the formulation and use of silicone elastomers, fluids, and silanes, including paper coatings and adhesives.

N. Eugene Stucker. B.S.Ch.E., University of Kansas. Currently a polymerization specialist in the plant technical department with Exxon Chemical Company U.S.A., Baytown, Texas.

Thomas J. Tangney. B.S. in chemistry from Saginaw Valley State College (1981). Technical representative at Dow Corning Corporation. Areas of interest include surface science, siloxane polymer chemistry, and silicone pressure sensitive adhesives. Several patents and publications in these areas.

Alexander Zettl. Awarded the degree of Dipl.-Ing. (FH) in applied chemistry in Nuernberg, West Germany, (1972). Since then he has been employed as an application technologist in the adhesives sector in the Department for Marketing Dispersions in BASF AG, Ludwigshafen, West Germany. His main activities are concerned with pressure sensitive adhesives.

1. Pressure Sensitive Adhesives and Adhesive Products in the United States

Donatas Satas
Satas & Associates
Warwick, Rhode Island

Pressure sensitive adhesives are used for many different applications. Adhesive coated products are subdivided into tapes, labels, and miscellaneous products. The latter category consists of a large number of diverse products; it shows the highest growth rate and presents the largest opportunity for new product development. In addition to these products, pressure sensitive adhesives are also used in the lamination of various films, foils, and papers, especially to difficult-to-adhere surfaces.

Pressure sensitive adhesive products consist of two components: backing and adhesive. Usually the backing is the most important component; the purpose of the adhesive is to secure the backing to another surface in order to perform some function. Thus the adhesive secures a label that carries printed information, electrical vinyl film tape is secured to bare wires to provide insulation, or packaging tape is applied to a carton to keep it closed. In all of these cases, the adhesive provides only an easy way to apply the flexible backing—the backing could be applied by other adhesives, or by other means completely. However, pressure sensitive adhesives provide a convenient and fast means of application that would be difficult to meet by other methods. An additional feature of pressure sensitive adhesives is their easy removability—the adhesive bond is relatively weak and can usually be broken without damage to the substrate. This feature of pressure sensitive adhesives is not always required; there are many uses where a permanent bond is preferred. While such a permanent bond is easily achieved with products on weak backings, such as paper labels, it is more difficult to design adhesives that cross-link on aging to yield strong and permanent bonds on strong backings.

While pressure sensitive adhesives are clearly a feature of secondary importance for most products, the successful application of these products and their position in the market would not be possible without pressure sensitive adhesives. It is difficult to visualize the large array of medical and first aid tapes and other products without pressure sensitive adhesives. A relatively new and successful product: easily removable note pad pages with a strip of pressure sensitive adhesive was made possible because of the adhesive. Numerous tapes serve a large number of applications, and it is difficult to find a product area where pressure sensitive adhesive products are not used.

HISTORY

The start of the pressure sensitive adhesive industry was in medical application of adhesive tapes and plasters. The use of adhesive masses and plasters can be traced from the beginnings of recorded medical industry.[1] These were not pressure sensitive adhesive products, and it is not entirely clear at which point these masses evolved into products sufficiently resembling the pressure sensitive adhesives of today. During the mid-nineteenth century, natural rubber was added to the adhesive formula that previously contained resins and beeswax. The invention of rubber pressure sensitive adhesives is attributed to Dr. Henry Day.[2] Such an adhesive consisted of India rubber, spirits of turpentine, turpentine extract of cayenne pepper, litharge, pine gum, and other ingredients. A U.S. patent was issued to Shecut and Day in 1845.[3] European developments followed, with a German patent issued in 1882 to a druggist named P. Beiersdorf for a plaster based on gutta-percha, and the establishment of a continuing business in pressure sensitive adhesive products. In 1899, a natural rubber-based zinc oxide containing adhesive that resembled modern adhesive formulations was developed. The large-scale manufacture of pressure sensitive adhesive coated surgical tapes started in 1874 when Robert Wood Johnson and George Seaburg opened a plant in East Orange, New Jersey. In 1886 Johnson separately founded Johnson & Johnson Company, which is still a major manufacturer of hospital tapes.[6]

For quite some time, pressure sensitive adhesive applications were limited to hospital and first aid uses. Only in the 1920s did the first industrial applications start to appear. The first electrical tapes were produced during the period 1920–1930.[4] This product was not strictly a pressure sensitive tape, but rather was a friction tape. The adhesive mass was designed to adhere to itself. Its cohesive strength was low and the adhesive failed cohesively during the unwind of the tape.

One of the first important industrial uses of pressure sensitive tapes was for masking applications in automobile painting. It is said that a manufacturer of hospital tapes noticed that the tape sales of one of his distributors was increasing at an unusually fast rate. A further investigation disclosed that the distributor was selling the tape to the automobile industry to hold paper sheets employed as paint masks. At about the same time, Richard G. Drew of the Minnesota Mining and Manufacturing Company (3M), attempted to develop a masking tape for the automotive aftermarket, and chose to use crimped paper as the backing. This lead to the development of saturated paper tapes, which became the most important tape category for many applications in addition to masking.[5] Drew started 3M's first tape laboratory in 1926, and he later developed cellophane tape for package closure application.[6] This was the first transparent film tape on the market that found many uses in offices and households.[7,8] The further development of various tapes escalated at a tremendous rate. Every conceivable flexible backing was used for pressure sensitive tape construction. Tapes found uses in every industry for a tremendous variety of purposes.

The development of pressure sensitive adhesive coated labels took place later, with the work being done at different companies, probably because labels are essentially a printed product. Knowledge of printing and die cutting were the most important operations, rather than adhesive compounding or the capability of adhesive coating. R. Stanton Avery in Los Angeles, California, started printing stickers and selling them by direct mail in the late 1930s. Kleen Stik was engaged in printable sheet stock manufacturing at the same time in an effort that was similar to Avery's business. Avery put the label business on a firm foundation during World War II, expanding later into label stock through Fasson. Avery International expanded through increased sales in the United States and also abroad through their numerous branches in Europe.[6]

ADHESIVES

Pressure sensitive adhesives may be classified according to the physical form in which the adhesive is supplied or according to the chemical composition, that is, the main elastomer used. The form in which the adhesive is supplied determines the application equipment required, which is not easily interchangeable.

Natural rubber adhesives dominated pressure sensitive applications for a long time, but the shortage of natural rubber during World War II prompted the search for substitutes. Polyisobutylene was among the first polymers to replace natural rubber.[9] Many other elastomers have been used for pressure sensitive adhesives, but the dominance of natural rubber has been threatened only recently by polyacrylates and by block copolymers, which are especially useful for hot melt adhesives.

The adhesive types and their percent distribution on a dry weight basis are listed below. It is estimated that total U.S. adhesive consumption on a dry basis amounts to 500 million pounds.

Solutions in solvents	38%
Hot melts	22%
Aqueous emulsions	21%
Calenderable	19%
Radiation curable	negligible
	100%

Adhesives are classified into the following categories according to the elastomer used. Percentages are based on the dry weight of the compounded adhesives. The amount of elastomers used with these adhesives varies: rubber-based adhesives are heavily compounded with tackifying resins and often might contain filler as well, while acrylic adhesives usually consist of a pure polymer.

Block copolymers	29%
Natural rubber	28%
Polyacrylates	23%
Butyl rubber and polyisobutylene	9%
Styrene-butadiene rubber (SBR) random polymer	8%
Other	3%
	100%

Solution adhesives are still the most important category, although they are being replaced by both aqueous and hot melt adhesives. The reasons for replacement are economical, and are caused by environmental regulations,which require the removal of solvent vapor from exhaust gases, thus increasing the investment in processing equipment, and by high and potentially unstable solvent prices. Solution adhesives are based on several elastomers as shown below:

Natural rubber	42%
Block copolymers	20%
Polyacrylates	16%
SBR random polymer	11%
Other	11%
	100%

Solution adhesives are the easiest to apply and are the most versatile in their properties. They have been used for a long time, so there is a considerable amount of accumulated compounding know-how. Many companies have invested in solvent recovery equipment and, in such cases, high solvent costs are of small importance. The recent development of inert gas drying allows the recovery of solvents by condensation and helps to remove the economical disadvantages of the adhesive solution coating process. If solvent costs are excluded, then rubber/resin solutions become the least expensive adhesives. Although the use of solution adhesives will continue to decrease slowly, it is not expected that they will be completely replaced by other adhesive types.

Hot melt adhesives are based almost exclusively on block copolymers tackified with resins. The advantages of hot melt application are high coating speeds and low floor space requirements; the disadvantages are a low degree of versatility in adhesive compounding and inherently poor shear resistance at higher temperatures. The main applications of hot melt adhesives are for oriented polypropylene film tapes and for label stock.

Aqueous adhesive emulsions are the fastest growing adhesive category. Acrylic copolymers dominate these adhesives; butadiene-styrene copolymers and natural rubber latex are also used in a more limited way. Aqueous adhesives have many advantages, but their growth is inhibited by two main shortcomings: aqueous emulsions are more difficult to coat, because they wet low-energy surfaces poorly and do not level well to yield smooth coatings, and aqueous adhesives often exhibit a poor resistance to water, because of the presence of surface active agents. These disadvantages are being successfully addressed: proper compounding can overcome the coating difficulties, and these polymers can be prepared without external emulsifiers, thus greatly improving their water resistance.

Calenderable adhesives are 100% solid compounded materials processable on calendering equipment. The importance of calendered tapes is decreasing. Coating thickness cannot be controlled at lower coating weight levels, and calendering equipment requires a heavy investment. Most calenderable adhesives are composed of traditional natural rubber, or natural rubber/SBR blends, compounded with tackifying resins. A significant amount of butyl rubber-based adhesives are calendered for corrosion-resistant tapes. Similar adhesives are also applied by the extrusion process.

Radiation curable adhesives are 100% solid reactive coatings, consisting of acrylic functionality containing oligomers and monomers. These adhesives have a sufficiently low viscosity so that they can be applied by conventional coating equipment. They cure upon ultraviolet (UV) or accelerated electron irradiation yielding a useful pressure sensitive adhesive. The process so far has not penetrated the manufacturing of pressure sensitive adhesive products to any appreciable extent. The main success of radiation technology was the limited acceptance of silicone release coatings. Radiation curable adhesives are of interest for applications where the adhesive coatings must be applied over a small area only and where other adhesives become difficult to use.

Adhesives were traditionally compounded in-house by tape and label manufacturers, and their formulations were jealously guarded. These adhesives first became available on the merchant market with the advent of acrylic solution polymers in the late 1950s. It is estimated that currently only 17% of all pressure sensitive adhesives are purchased, the remainder being compounded in-house.

MANUFACTURERS

Some time ago it was possible to subdivide the pressure sensitive adhesive products manufacturers into two main groups: tape and label producers. This distinction became less definitive as tape producers started to add label stock on film backings to their product line, and the label and label stock producers started to expand into tape products. In addition, acquisitions made the demarcation between these two main product groups even hazier.

The major U.S. tape manufacturers are listed below; they produce about 80% of all tapes. The remainder is manufactured by a larger number of small specialized companies.

3M Company
Permacel
Johnson & Johnson Company
Fasson (Division of Avery International)
Tuck Tape
The Kendall Company (Subsidiary of Colgate Corporation)
Anchor Continental
Shuford Mills, Inc.
American Tape Company
Nashua Corporation

The label business is subdivided into label stock manufacturing and label manufacturing. Label stock is produced by several larger companies; label production is primarily a converting operation consisting of printing and die cutting. Major manufacturers producing about 80% of label stock are:

Fasson (Division of Avery International)
3M Company
Flexcon Corporation
Morgan Adhesives (Subsidiary of Bemis Corporation)
Fitchburg Coated Products, Inc.
Brown Bridge Mills, Inc. (Subsidiary of Kimberly-Clark Corporation)
Dennison Manufacturing Company
S. D. Warren Company (Division of Scott Paper Company)

Some of these label stock manufacturers also produce labels, but label printing is much more diverse. In addition to a few larger companies, a large number of small manufacturers are engaged in this business. There is some in-house label stock manufacturing by the larger label printers, but in general, the label industry depends upon label stock produced by specialized manufacturers.

It is estimated that the annual volume of tapes amounts to $2.2 billion and that of labels to $1.3 billion. In addition, pressure sensitive adhesives are used on other products outside the categories of tapes and labels.

TAPES

Tapes are the most important of the pressure sensitive adhesive products. Most tapes are wound on themselves, that is, the adhesive is in contact with the tape backing. When the tape is unwound, the adhesive must remain on the same side of the backing to which it was applied. It should not split or transfer, even if aged at an elevated temperature or at other unfavorable conditions and under considerable pressure, which can be exerted on the adhesive in a tightly wound roll. The unwind force must be low, but not too low. If unwind is too easy, such a tape might show a tendency to flag when applied: the overlap might not adhere sufficiently firmly. To meet these requirements, release, backsize, and prime coatings may be necessary. Release coating of tape backings has evolved into a considerable and largely proprietory technology. For some tapes it might be necessary to use release coated liners, which detract from the easy application of the tape. Many foam and foil tapes are constructed with liners.

Many different tapes are on the market, with estimates ranging from 200 to 400 different varieties.[2] It all depends on the definition of what constitutes a different tape and what is merely a variation of the same product.

Tapes may be classified according to construction or according to use. Classification according to both criteria produces some repetition, but it also allows coverage of the various aspects of pressure sensitive tapes.

Tapes may be subdivided according to their construction into the following categories.

- Fabric tapes
- Paper tapes
- Film tapes
- Nonwoven fabric tapes
- Foil tapes
- Reinforced tapes
- Foam tapes
- Double-face tapes
- Transfer tapes

As is obvious from the preceding list, the backing, rather than the adhesive, provides the main distinction among various tapes, although classifications like acrylic adhesive tape or hot melt adhesive tape might be heard occasionally.

Tapes may be classified according to their function. Bemmels,[2] in describing various tapes, uses classifications such as holding or bundling, masking, sealing, protecting, reinforcing, splicing, stenciling, identifying, packaging, insulating, mounting. We will try to classify tapes according to a more general area of application, which has to do with both the function of the tape and the industry for which the tape is intended. Thus the following tapes will be discussed.

- Hospital and first aid tapes
- Office and graphic art tapes
- Packaging and surface protection
- Building industry products
- Electrical tapes
- Automotive industry products
- Shoe industry tapes
- Appliance industry products
- Splicing tapes
- Corrosion protective tapes
- Miscellaneous tapes and pressure sensitive products

Fabric Tapes

Woven fabric is the oldest backing material used for pressure sensitive adhesive tapes. Hospital tapes were, and many still are, made with a woven fabric backing. Fabric has some unique characteristics that are difficult to obtain with other backings, and therefore fabrics are still used despite their higher cost. The combination of high tensile strength with a high flexibility is not obtained in other materials. A low tear strength can be incorporated into a fabric backing without sacrificing the tensile strength. The tape could be made to tear easily across or lengthwise, eliminating the need for cutting. This property has been useful for hospital tapes and other applications.

Fabric backing can be heavily filled and coated with starch, nitrocellulose, vinyl, or acrylic coatings. Fabrics can also be used without any backcoating for tape backings. The rough surface of a woven cloth provides sufficient release if the adhesive does not flow into the fabric. While cotton and rayon are the most common fibers used, polyester fabric backings are used for many applications, including high tensile packaging uses. Nylon fabric tapes have found some uses in the shoe industry. Glass fiber fabrics are used for their high tensile strength and especially for their resistance to high temperatures. Such tapes with silicone adhesive are useful over a temperature range −54–232°C. Glass also does not support combustion.

Paper Tapes

Paper is the least expensive material available in web form, and its dominance has only recently been threatened by some low cost polymeric films.

Paint masking tape for the automotive industry, an important application for a low cost tape, prompted the development of a family of saturated paper tapes. These tapes proliferated beyond the use for paint masking and became general-purpose tapes serving many diverse applications. Two main types of saturated paper backings are used: creped paper and flatback paper. The saturation process and various tape products are discussed in detail by Bartell in a separate chapter.

In addition to saturated paper tapes, unsaturated kraft tapes have been developed more recently. They are especially popular in Japan. The main disadvantage is poor delamination resistance. Because of it, efficient silicone-based release coatings are used. These tapes exhibit poor adhesion to their backing and have a tendency to flag.

Various specialty papers are also used: rope fiber paper for higher tensile strength, and nylon high-temperature papers for increased temperature resistance in electrical applications.

Film Tapes

Polymeric films have found numerous applications as tape and label backings. The properties inherent in polymeric films (impermeability, thinness, smooth surface, good dielectric properties, inertness) are desired for many electrical, packaging, and decorative applications. Some films are also inexpensive on the basis of surface area. They are competing with paper for applications where price is the determining factor, and are replacing paper in many cases. Some properties of typical film tapes are shown in Table 1-1. The table lists only the most commonly used film tapes. There are many more varieties of film tapes produced.

Cellophane is the oldest transparent film tape. It was and still is widely used as an office and general-purpose household tape. Cellophane tape is being replaced by other film tapes. Matte cellulose acetate film tape replaced cellophane in some offices, paper mending, and general-purpose applications, despite the fact that cellulose acetate film is more expensive than cellophane. Cellophane is hygroscopic, becoming quite brittle when dry and quite soft in a humid atmosphere. These deficiences are absent in acetate tapes. Furthermore, a matte acetate film surface does not reflect light and is therefore inconspicuous. It also accepts writing.

Rigid or lightly plasticized vinyl film tapes were the predominant tapes

Table 1-1. Physical Properties of Some Film Tapes

FILM	TENSILE STRENGTH (kg/cm)	ELONGATION (%)	THICKNESS (mm)
Clear cellophane	5.7	19	0.07
Clear acetate	3.6	20	0.07
Etched acetate	3.6	25	0.08
Acetate fiber tape	8.6	3	0.15
Polyester transparent	4.5	110	0.05
Transparent vinyl	3.6	16	0.07
Carton brown vinyl	6.3	40	0.07
Polypropylene clear	3.2	14	0.06
Polypropylene clear	6.8	73	0.06

used in Europe for packaging and general-purpose applications. In the United States, saturated paper and polyester film tapes were used for such general-purpose applications. More recently, biaxially oriented polypropylene film tapes have taken over packaging applications. Vinyl tapes are still of some interest, because of their easy printability and easier, less noisy, unwind. Plasticized vinyl tapes are used for electrical insulation, protecting racks immersed in electroplating solutions, masking metal surfaces to be plated, sealing cans, protecting metal window frames, and fabricating corrosion protective tapes for underground pipes and printable tapes for general labeling. Colored vinyl tapes are used for identification and floor markings.

Polyester film tapes are used in electrical applications and variously where high tensile strength and high tear are important. Oriented polypropylene tapes have replaced polyester tapes in some applications because of their lower price.

Polyethylene film tapes have not been as successful for general use as some other films. It is used for special applications, some of which are large volume uses. It is used for some medical tapes, electrical insulating, duct insulating, and carpeting. One large application is corrosion-protective tapes for underground gas and oil lines.

Tetrafluoroethylene (TFE) tapes offer a high dimensional stability, resistance to elevated temperatures, chemical inertness, and a low coefficient of friction. Because of these properties, it has found many applications despite its high cost. In combination with a silicone adhesive, TFE tapes are good for continuous use at 250°C. TFE tapes are useful in many electrical applications, such as coil winding, transformers, cables, relays, condensers, and resistors. The tape is useful for wrapping various rollers to prevent sticking of adhesives and coatings, to line conveyors, chutes, guide rails,and hoppers in order to eliminate friction. It is useful as a dry lubricant to prevent abrasion, rubbing, and squeaking. It provides a nonstick surface in heat sealing equipment and glue dispensers.

Many tapes are also available with a TFE coating over a fabric, usually a glass fiber cloth.

Polyimide film tapes are used in specialty electrical applications for high-temperature insulating, flat cable construction, and masking in soldering operations.

Nonwoven Fabric Tapes

Nonwoven fabrics do not find many uses as tape backings. There is a successful hospital tape with a nonwoven fabric backing. These fabrics are used for some electrical tapes where the porous nonwoven fabric helps to hold the resinous impregnant.

Foil Tapes

Aluminum, lead, and some copper foil tapes are used in the industry. Aluminum is by far the most important foil tape, and it is used in applications where a moisture seal is required. Foil tapes are also used in electrical applications such as a static bleedoff on metallic structures, in heating insulation and air conditioning ducts as reflective heat shielding, in high-temperature masking operations (as experienced in the manufacture of printed circuit boards), in repairing sheet metal, in blocking radiation on X-ray plates, and other applications. Dead soft aluminum foil is most commonly used for tape backings. Aluminum foil laminated to cloth, foam, and other bulkier materials is used for vibrational and acoustical damping. Properties of some aluminum and lead foil tapes are given in Table 1-2.

Table 1-2. Properties of Foil Tapes

BACKING	ADHESIVE	TOTAL THICKNESS (mm)	TENSILE STRENGTH (kg/cm)	PEEL ADHESION (g/cm)	ELON-GATION (%)	USEFUL TEMPERATURE RANGE (°C)
Aluminum	Acrylic	0.09	3.6	730	5	−65–177
Aluminum	Acrylic	0.13	5.4	450	5	−65–177
Aluminum	Acrylic	0.13	5.4	730	5	−65–177
Aluminum	Acrylic	0.29	9.0	730	13	−65–177
Aluminum	Silicone	0.09	3.2	220	5	−65–316
Aluminum foil glass cloth	Silicone	0.15	13.4	—	—	−54–316
Lead	Acrylic	0.13	3.0	450	15	−65–177
Lead	Rubber	0.15	3.0	390	15	−60–106
Lead	Rubber	0.19	3.6	390	15	−60–106

Lead foil tape is used for its chemical resistance, especially in plating operations, for its high density as a radiation shield and acoustical dampening tape, and for its low water vapor transmission as a sealing tape. Many foil tapes, especially lead foil, are sold with a release liner, because of the difficulty in providing an adequate release or prime coating on lead.

Copper foil pressure sensitive adhesive tapes do not find many applications. In flat cables, which is a major application for copper foil, heat activated adhesives have found a greater acceptance than pressure sensitive ones.

Slitting and winding of foil tapes present special problems. Score slitting deforms the foil, leaving a bead at the edge. Also, the razor blades are dulled easily. Shear slitting might deform the edge, unless the knives are set very shallowly. Rider rolls are generally required to keep the foil smooth and to iron out any wrinkles or deformations.

Reinforced Tapes

Packaging operations such as the bundling of heavy materials, for example, steel rods, require high tensile tough tapes. The toughness, as measured by the area under a stress-strain curve, is perhaps a better index of the performance of these tapes in heavy-duty applications than just the tensile strength.

The first reinforced tapes were constructed by laminating acetate fiber rope to acetate film. These filaments are embedded in the adhesive mass. They can move and distribute the load evenly over the fibers, resulting in high tensile strength tapes. Some of these tapes are listed in Table 1-1. Other yarns, especially glass and rayon, and other films, polyester and polypropylene, are also used for the construction of reinforced tapes. High tensile strength polyester fiber and glass fiber fabrics have also been used as reinforcements.

Reinforced tapes are used for heavy-duty bundling, closing and reinforcing fiberboard cartons, palletizing, as a tear tape for cardboard packages and as an edging for photo albums and documents. Some reinforced tapes are printable, or the film is printed in reverse.

Foam Tapes

Foam tapes are used for several different functions and, therefore, products of several types are on the market. Foam tapes are used for sealing and gasketing. Lower density foams are generally used for these applications. Foam tapes made from various foam/foil and foam/film laminates are used for thermal and acoustical insulation. An important application for two-sided, adhesive coated foam tapes is in the mounting of various objects: reflectors for automobiles, trucks, trailers; side trim; mounting of balancing weights to automobile wheels;

bulletin boards; telephone pen caddies; bathroom fixtures to walls; medallions to wine and liquor bottles; and electrocardiogram terminals to patients and many other mounting applications. Foam tapes are suitable for such applications because the foam can conform to the uneven surfaces. Foam distributes the load and prevents stress concentration that causes failure in such applications.

Polyethylene, vinyl, urethane, and polychloroprene foams are used for tape applications. Closed cell radiation cross-linked polyethylene foams manufactured by Sekisui Chemical Company are becoming very important for tape construction. The foam is available in several formulations: cell size, bulk density, and thickness. Polyurethane foams have been important for mounting applications. High-density foams with a well-adhered skin are useful for this purpose. Low-density polyurethane foams are used for sealing and insulating purposes and also for miscellaneous general-purpose uses where the low cost of such foams becomes an important advantage. Vinyl foams are useful for general-purpose applications. Proper adhesives must be chosen that are not affected by plasticizer migration. Polychloroprene foams are used because of their solvent resistance in automotive and other applications.

The properties of mounting foams are illustrated in Table 1-3 and of some foam sealants in Table 1-4. Foam tapes are sold faced with silicone release coated paper.

Double-face Tapes

Double-face tapes can be subdivided into two groups: thin tapes, usually coated polyester or polypropylene film (0.013 mm thick or heavier) or lightweight paper, and heavy double-face tapes, usually coated foam or fabric. Heavy tapes are used for adhesion to a rough surface where the bulk is required for the conformability, or a heavier adhesive layer is required for distributing the force over a larger area. Many mounting applications require heavy double-face tapes, and foam tapes are the most popular. Holding down carpets and securing lithographic printing plates to the printing cylinder are two examples of applications of heavy double-face tapes.

Lightweight tapes are used for applications such as splicing, lamination, business forms, disposable sandpaper pads, nameplates, and tags. Table 1-5 lists various carrier webs used for double-face tapes, illustrating the variety of constructions that are needed. Most commonly used backings for double-face tapes are listed in Table 1-5.

Most double-face tapes require the use of a two-sided coated interliner, although a double-face tape has been produced that is wound on itself without an interliner. Synthetic adhesives used are sufficiently incompatible to prevent the diffusion of polymer from one side to the other, and the tape can be unwound quite easily.[10] Tapes with different adhesion levels on each side have been

Table 1-3. Properties of Pressure Sensitive Foam Mounting Tapes

FOAM	ADHESIVE	DENSITY (kg/m^3)	THICKNESS (mm)	TENSILE STRENGTH (kg/cm)	ELONGATION (%)	PEEL ADHESION (gm/cm)	TEMPERATURE LIMITS (°C)
PVC[a]	Acrylic	256	0.79	0.68	150	500	−18–71
PVC	Acrylic	256	1.59	1.16	150	500	−18–71
PVC	Acrylic	256	3.12	2.06	150	500	−18–71
PU[b]	Acrylic	400	0.79	0.50	350	710	−35–93
PU	Acrylic	400	1.59	1.34	350	710	−35–93
PU	Acrylic	400	3.12	2.06	350	710	−35–93
PU	Rubber	400	0.79	0.50	350	535	−35–93
PU	Rubber	400	1.59	1.34	350	535	−35–93
PU	Rubber	400	3.12	2.05	350	535	−35–93
PE[c]	Rubber	96	0.70	0.89	250	557	−29–82
PE	Rubber	96	1.59	1.25	250	557	−29–82
PE	Rubber	96	3.12	2.32	250	557	−29–82

[a] PVC. Polyvinyl chloride.
[b] PU. Polyurethane.
[c] PE. Polyethylene.

Table 1-4. Properties of Pressure Sensitive Adhesive Coated Foam Sealants

TYPE OF FOAM	DENSITY (kg/m^3)	MAXIMUM TEMPERATURE (°C)
PVC[a] closed cell	256	82
PVC closed cell	176	82
PVC closed cell	96	71
PU[b] open cell	32	135
PU[b] open cell	32	121
Polychloroprene closed cell	192	71

[a] PVC: Polyvinyl chloride.
[b] PU: Polyurethane.

Table 1-5. Backings Used for Double-face Tapes

BACKING	BACKING THICKNESS (mm)	TOTAL THICKNESS (mm)
Polyester film	0.013	0.157
Polyester film	—	0.127
Polyester film	0.025	0.097
Polyester film	0.025	0.178
Biaxially oriented polypropylene film	0.030	0.150
Polyethylene film	0.076	0.178
Vinyl film	0.076	0.216
Vinyl carpet tape	—	0.127
Semiopaque tissue paper	0.076	0.140
Tissue paper	—	0.114
Repulpable tissue paper	—	0.089
Rope tissue	0.076	0.140
Rope tissue	0.051	0.147
Cloth	—	0.381
Cloth, flame retardant	—	0.330

produced. A product with a permanent adhesive on one side and a removable one on the other is of interest for packaging applications.

Transfer Tapes

Transfer tapes consist of an unsupported adhesive coating applied over a silicone release coated paper. The release paper used must have a sufficient differential release in order to retain the adhesive on the side to which it was originally coated. The release level, however, must be sufficiently low to allow a good adhesive transfer.

Transfer tapes, like thin film double-face tapes, are used for various lamination purposes and for adhesion of nameplates. They are especially useful in providing the adhesive in the exact location for later use. Removal of the release paper exposes the adhesive and makes it available for lamination.

Most transfer tapes consist of unsupported adhesive coating. The adhesive used is often acrylic, but rubber-resin adhesives are also widely used. There is a product on the market that has free fibers dispersed in the adhesive as a reinforcement. The fibers increase the cohesive strength, tensile strength, and tear resistance of the adhesive coating[11] without sacrificing thinness or transparency. Silicone pressure sensitive adhesive tape on silicone coated release paper is also on the market. This product is used for its resistance to high temperature and for its good adhesion to low-energy surfaces, such as unetched TFE and polypropylene.

Most transfer tapes have the adhesive coating up to the edge. Some have a free, uncoated edge that is helpful in facilitating the removal of the release paper. The free edge tapes are made by pattern coating and slitting in the uncoated area. Slitting of transfer tapes presents some problems unique to these tapes. The adhesive may easily transfer to the equipment during handling. Also, the adhesive may be pushed by the knife over the edge of the release backing and adhere to the next adhesive layer. This causes legging during unwind and may cause lifting of the adhesive in the area of contact. The release paper and the slitting method should be such as not to produce free fibers at the slit edge. Fibers may adhere to the adhesive coating and also cause legging and lifting. Adhesive coatings that flow excessively are not suitable for transfer tapes where the adhesive extends to the edge. The release lining should be sufficiently thick to prevent contact between adjacent adhesive layers.

Hospital and First Aid Tapes

Hospital tapes were the first application of pressure sensitive products. This application has grown to a sizable business. In addition, the same products are consumed at physicians' offices and at home for first aid uses. Similar products are also used for foot care and for athletic protection. These products are covered in a separate chapter.

Office and Graphic Art Tapes

Clear cellophane film tape was the first product for this application. These applications have developed into a large business requiring a diverse assortment of tape and various coated film products.

Packaging and Surface Protection

The competition between paper tapes and film-backed tapes is most visible in the area of packaging. Packaging applications are quite sensitive to price, and film tapes are replacing more expensive saturated paper backings. While unplasticized vinyl films do not find a wide acceptance in the U.S. market, biaxially oriented polypropylene film tapes are becoming the most frequently used tapes for packaging. Saturated paper is still maintaining its dominant position in surface protective tape and sheet products. Packaging tapes are discussed in a separate chapter by Fukuzawa and Satas, and saturated paper is discussed in another chapter by Bartell.

Building Industry Products

Pressure sensitive tapes do not find many applications in the construction industry, although the volume of adhesives and sealants consumed by this industry is very large.

Tapes are used for temporary applications such as paint masking. Double-face foam tapes are used for temporary attachment of gypsum panels to interior surfaces until permanent adhesives develop their full strength. Tapes are used to keep buildings sealed during cold weather by securing plastic film sheets. Similarly, dust is kept out by film partitions held by tapes. Expensive finishes and plastic or metal surfaces may be protected by pressure sensitive coated sheets.

Permanent uses of pressure sensitive tapes include employing foam tapes as sealants, sealing of floor-to-wall joints to eliminate draft, and bridging narrow cracks using a lightweight tape that is overpainted. Double-face tapes of heavier construction are used to secure jacketing over pipe insulation. Pressure sensitive polyethylene coated fabric, aluminum foil, and other construction tapes are used for insulation of heating and air conditioning ducts. Water vapor transmission is important in many sealing applications. Table 1-6 shows the water vapor transmission rates of some of the tapes used in building construction. Polyvinyl fluoride film pressure sensitive adhesive tapes have been used for sealing corrugated cement-asbestos siding and for an experimental roofing system.[12]

Electrical work on the construction site requires large quantities of electricians' tape. Door frames and moldings may be manufactured from sheet metal, particleboard, or low-quality wood. These can be covered with woodgrain printed vinyl film with pressure sensitive adhesive coating.

Very lightly metallized polyester film with a pressure sensitive adhesive is used to control solar radiation by applying the film over window glass. Acrylic adhesives are used because of their clarity and stability in the presence

Table 1-6. Water Vapor Transmission of Some Tapes Used in Building Construction

TAPE	WATER VAPOR TRANSMISSION (g/m^2, 24 hr)
Foil tapes	0
Polyethylene film	2.3–3.9
Polyethylene coated fabric	11–14
Polyester film	19–23
Vinyl film	31–47
Vinyl coated fabric	23–47
Cotton fabric	31–93
Glass fabric	62–93

of light. The product must be of a high quality optically and it requires careful coating. Various other coated and embossed films are used to restrict light passage through the windows.

Printed vinyl film with a pressure sensitive adhesive is used for wall covering, shelf inserts, and similar decorative uses. The applications for wall covering are potentially very large. Pressure sensitive products have not made a large penetration, mainly because of the poor repositionability of adhesive coated wall coverings. Many ingenious ways have been suggested. One method involves placing glass beads on the surface of the adhesive, to prevent contact between the adhesive and the wall, until pressure is applied which either crushes or drives the glass beads into the adhesive coating. Incorporation of encapsulated activating agents has also been suggested, allowing the activation of the adhesive on application of pressure.

Adhesive coated floor tiles are successfully used for the fast installation of tiles. Hot melt adhesives based on block copolymers and on amorphous polypropylene are used. A heavy adhesive coating is required and hot melt application is especially suitable.

Applications of pressure sensitive products for building construction have been discussed by Wilson,[12] Wallace,[13] and Gerstel.[14]

Electrical Tapes

Electrical tapes can be subdivided into two categories: tapes intended for the original equipment manufacturers (OEMs) and electricians' tapes used for electrical insulation during installation. The electrical properties of such tapes and their testing is discussed by Braeunling in a separate chapter. The chemical purity, because of the required noncorrosiveness, and the resistance to deterioration at high temperatures are the important requirements of electrical tapes. Natural rubber and synthetic elastomer-resin pressure sensitive adhesives are used for continuous performance up to 130°C. Acrylic adhesives are considered to be suitable up to 155°C and silicone resin-elastomer adhesives for continuous use up to 180°C.

The electricians' black friction tape is not a pressure sensitive product; the adhesive is very soft and it splits on the unwind. This product was the prototype for the development of electricians' vinyl tapes and also for many other industrial tapes. The current electricians' tape has either a plasticized vinyl film backing, or a polyethylene film backing. These tapes are conformable and are required to meet nonflammability specifications.

A large variety of tapes is used for OEM applications. In addition to electrical insulation, the tapes are used to hold, reinforce, and protect electrical wires, to provide a matrix for varnish impregnation, to identify wires in the electrical circuitry, and for many other uses. Some tapes are used for temporary purposes to help in manufacturing or in assembly. For example, printed circuit board manufacturers use pressure sensitive tapes to protect the terminals during manufacturing.

Many different backings are used for OEM tapes, as shown in Table 1-7. Cloth tapes made from cotton, acetate taffeta, or glass fibers are used for electrical tapes. Cloth tapes offer more protection against physical damage because of their bulk. Their surface is rough and absorbent and better suited for varnish impregnation. Cloth tapes are heavier than film tapes and therefore not as suitable where space is limited.

In addition to vinyl film tapes, which are standard electricians' tapes, polyester film tapes are widely used for OEM applications. Films from 0.013 to 0.05 mm thick are used for tape backings. Some less common films such as polyimide or tetrafluoroethylene find uses for OEM applications. TFE tapes exhibit a high dielectric strength and good arc resistance. TFE vaporizes under sustained arcing, but does not leave a carbonaceous path. Properties of TFE coated glass cloth tapes with silicone pressure sensitive adhesive are shown in Table 1-8.

Electrical coils are often encapsulated in resins for protection from the environment: moisture, dirt, mechanical abuse, and thermal shock. Film tapes do not form a good bond with the impregnating varnish (usually epoxy resin). The bond formation has been improved by using polyester web-polyester film laminates. Polyester webs alone are also used as a matrix for varnish.

Paper tapes are used for OEM applications. Nonwoven fabrics and paper made from unusual fibers such as Nomex find uses for higher temperature applications.

Table 1-7. Properties of Some Electrical Tapes

BACKING	BACKING THICKNESS (mm)	TOTAL THICKNESS (mm)	TENSILE STRENGTH (kg/cm)	ELONGATION (%)	DIELECTRIC STRENGTH (volts)	INSULATION CLASS (°C)
Films						
Polyester	0.013	0.025	2.3	100	3800	130
Polyester	0.025	0.063	4.5	100	5000	130
Polyester	0.050	0.089	8.9	100	6500	130
Vinyl	—	0.178	3.6	200	9500	80
Vinyl	—	0.254	5.3	190	11,500	80
Vinyl	—	0.381	10.6	250	15,000	80
Polyimide	0.025	0.075	5.3	60	7500	180
Polyimide	0.050	0.100	9.8	80	11,500	180
Polypropylene	0.032	0.063	6.8	75	8000	85
Polyvinyl fluoride	0.050	0.089	2.5	45	6000	130
Fluorohalocarbon	0.050	0.089	2.3	50	8500	130
Cloth						
Cotton	—	0.267	8.9	5	3000	105
Acetate taffeta	—	0.203	8.0	10	3000	105
Glass (rubber adhesive)	—	0.191	30.3	6	2500	130
Glass (silicone adhesive)	—	0.191	32.1	7	3500	180
Paper						
Kraft, creped	—	0.254	3.6	10	2500	105
Kraft, microcreped	—	0.152	8.0	10	2500	105
Nomex, flatback	—	0.089	3.9	8	2800	155
Nomex, creped	—	0.381	5.3	50	2600	155
Laminates						
Aluminum foil, glass cloth	—	0.165	17.8	5	—	180
Polyester film, paper	—	0.127	6.2	3	4500	105
Polyester film, nonwoven	—	0.152	7.1	35	5000	130
Acetate film, acetate cloth	—	0.203	8.9	10	5500	105

Table 1-8. Properties of TFE Coated Glass Cloth Tape with Silicone Adhesive

BACKING THICKNESS (mm)	TOTAL THICKNESS (mm)	TENSILE STRENGTH (kg/cm)	ELONGATION (%)	DIELECTRIC STRENGTH (volts)	TEMPERATURE RANGE (°C)
0.076	0.127	8.9	<5	4000	−73 to 260
0.076	0.127	13.4	<5	3000	−73 to 260
0.127	0.178	17.9	<5	4000	−73 to 260
0.152	0.203	17.9	<5	5000	−73 to 260
0.254	0.305	38.8	<5	6500	−73 to 260

Automotive Industry Products

A large volume of electrical tapes are used in the electrical system of a car. These products are similar to the OEM tapes discussed before. Other automotive pressure sensitive products are discussed by Grimes in a separate chapter. A general discussion of such products has been published in *Chemical Week.*[15]

Shoe Industry Tapes

In the shoe industry, pressure sensitive tapes are used to cover the backseam. The tape reinforces the seam and reduces pressure spots. Most of the tapes for this application are cotton-fabric-backed, although some paper tapes and polyester and nylon fabric tapes are also used. Pressure sensitive tapes are also employed as binding tapes and reinforcing tapes in various shoe construction applications.

Appliance Industry Products

Decorative vinyl strips, especially woodgrain, backed with pressure sensitive adhesive are widely used in the appliance industry. Nameplates are attached by transfer tapes or double-face tapes. Foam gasketing with a pressure sensitive adhesive coating for easy installation is often used, and foam pads for sound insulation are often secured with pressure sensitive adhesives.

Splicing Tapes

Tapes are widely used to splice various webs during conversion or manufacturing operations. The paper industry requires the use of repulpable tapes, so that waste paper can be returned to the beater and the adhesive will not gum up the equipment. Various water-dispersable polymers have been compounded into pressure sensitive adhesives for this application. Double-face tapes and transfer tapes are used for lap splices; film and paper tapes are used for butt splicing. Splicing of silicone coated release paper requires silicone adhesive tapes.

Corrosion Protective Tapes

This is a very large application for pressure sensitive polyethylene tapes. These products are discussed in detail by Harris in a separate chapter.

LABELS

The pressure sensitive label industry is a separate business, different from tape manufacturing in several aspects. In the case of labels, the backing, its printability, flatness, and the ease of die cutting, and the properties of the release paper are important properties that often overshadow the pressure sensitive adhesive. Firms experienced in these areas have established themselves in the pressure sensitive label business, rather than tape manufacturers expanding into the label area.

A label stock manufacturer sells its product faced with a release paper either in rolls or in sheets for later printing and die cutting by a converter. The conversion operation can be a part of the same organization. The adhesive is usually applied over the release paper and dried. Then the label stock is laminated at the end of the line. Larger operations might also apply the silicone release coating in-line.

Labels are discussed in detail in a separate chapter.

OTHER PRODUCTS

Pressure sensitive adhesives are used in many applications that do not fit into the categories of tapes and of labels. For example, such important applications as pressure sensitive adhesive coated floor tiles, wall covering, automotive wood-grained films, and various decorative sheets are outside of these categories. This area of miscellaneous products is growing as pressure sensitive adhesives continue to find new applications.

REFERENCES

1. *Professional Uses of Adhesive Tape*. Third edition. Johnson & Johnson, New Brunswick, N.J., 1972.
2. Bemmels, C. W. Pressure-Sensitive Tapes and Labels. *Handbook of Adhesives*. (I. Skeist, ed.). Van Nostrand Reinhold, New York, 1977.
3. Shecut, W. H., and Day, H. H. U.S. Patent 3,965 (1845).
4. Levine, I. W. U.S. Patent 1,573,978 (1926).
5. Drew, R. G. U.S. Patent 1,760,820 (1930), Reissue 19,128 (1934) (assigned to Minnesota Mining and Manufacturing Co.).
6. *Paper Film and Foil Converter* **51**(10): 90–96 (1977).
7. Hendricks, O. J., and Dahlquist, C. A. Pressure-Sensitive Adhesive Tapes. *Adhesion and Adhesives,* Vol. 2, pp. 387–408 (R. Houwink and G. Salomon,). Elsevier, Amsterdam, 1967.
8. Drew, R. G. U.S. Patent 2,177,627 (1939).
9. Lambert, R. J., and Smith, A. G. U.S. Patent 3,205,088 (1965) (assigned to Minnesota Mining and Manufacturing Co.).
10. Kalleberg, M. O. U.S. Patent 2,889,038 (1959) (assigned to Minnesota Mining and Manufacturing Co.).
11. Kalleberg, M. O., and Turner, C. E. U.S. Patent 3,062,683 (1962). (Assigned to Minnesota Mining and Manufacturing Co.).
12. Wilson, P. H. in *Adhesives in Building*. Building Research Inst. Natl. Academy of Sci.—Natl. Res. Council. Publication No. 979. Washington, D.C. 1962.
13. Wallace, D. A., in *Adhesives in Building*. Building Research Inst. Natl. Academy of Sci.—Natl. Res. Council. Publication No. 979. Washington, D.C. 1962.
14. Gerstel, M. H., in *Adhesives and Sealants in Building*. Building Research Inst. Natl. Academy of Sci.—Natl. Res. Council. Publication No. 577. Washington, D.C. 1958.
15. *Chemical Week,* October 16, 1974.

2. Pressure Sensitive Adhesive Products in Western Europe

Donatas Satas
Satas & Associates
Warwick, Rhode Island

The development of pressure sensitive adhesive products in Western Europe parallels that in the United States. The first products were medical tapes. A German patent was issued to a druggist, P. Beiersdorf, in 1882 for a medical plaster based on gutta-percha. From this beginning, Beiersdorf A. G. has developed into the largest tape producer in Europe and the second largest in the world, after the 3M Company. Beiersdorf expanded from hospital tapes into industrial tapes and then into pressure sensitive labels as well. This tendency of tape manufacturers to leave label stock and label production to specialized companies was also true in the United States, and only recently have the two product areas begun to merge.

TAPES

European production volume of pressure sensitive tapes is roughly equal to that of U.S. production. It is estimated that 2.6 billion square meters of tapes are manufactured annually in Western Europe. Production is distributed among various countries as shown in Table 2-1.

Italy exports about 60% of their production to other European countries as well as to the United States. Italian tape production is situated mainly in northern Italy, except for the Manuli plant, which is located in the south.

Tapes in Europe are produced by some 50 manufacturers, of which ten produce 65% and three produce 30% of the total tape volume. Table 2-2 lists the largest European tape producers.

Beiersdorf A.G., with headquarters in Hamburg, is the largest tape manufac-

Table 2-1. Western European Tape Production by Country

COUNTRY	PERCENT OF THE TOTAL
Italy	40
Germany/Switzerland/Austria	26
United Kingdom	11
France	7.5
Benelux	7.5
Spain	6.5
Scandinavia	1.5
	100

turer and operates nine manufacturing facilities in Europe: four in the Federal Republic of Germany and one each in France, Italy, Spain, Denmark, and Austria. The company also owns NOPI, a major polypropylene film tape manufacturer. Beiersdorf markets their tapes under the trade name "tesa" all over the world. The company also produces adhesives for various applications, including pressure sensitive ones, which are produced for both their internal use and for sale on the market, and also sells specialty labels.

3M Group is the second largest tape manufacturer, with production facilities in the United Kingdom, West Germany, France, and Italy. 3M is an important

Table 2-2. Largest European Tape Producers

Beiersdorf A.G.	Hamburg, West Germany
3M Group	West Germany, U.K., France, Italy
Manuli Autoadesivi SpA	Milan, Italy
Lohmann KG	Neuwied, West Germany
Transatlantik	Schwarzenbach, West Germany
DRG Sellotape, Ltd.	Boreham Wood, Herts., U.K.
Supertape	Ettenleur, The Netherlands
Monta-Klebebandwerk GmbH	Immenstadt, West Germany
Comet SARA	Como, Italy
Boston SpA	Bolata, Italy
Cellux	Rorchschach, Switzerland
S.A.P.E.	Chenove Cedex, France
Novacel	Deville Les Rouen, France
N.A.R. SpA	Legnaro (Padova), Italy
Canadian Tape	Northern Ireland, U.K.
Advance Tape	Leicester, Leics., U.K.
Rotunda	Manchester, Lancs., U.K.
Nitto Belgium NV	Genk, Belgium
Autoadesivitalia	Bergamo, Italy
Stalar	Barcelona, Spain
Muroll	Frastanz, Austria

Table 2-3. Distribution of Industrial Tapes Among Various Applications

APPLICATION	PERCENT OF THE TOTAL
Packaging	63
Masking	11
Protective	4
Diaper	2
Double face	2
Electrical	2
Reinforced	1
Other	15
	100

supplier of industrial and office tapes. Manuli Autoadesivi SpA is the largest packaging tape manufacturer.

Industrial tapes are distributed among various applications as shown in Table 2-3.

Packaging tapes have mainly film backings: 60% biaxially oriented polypropylene, 35% polyvinylchloride, and 5% other backings. This distribution is different from that of the packaging tapes in the United States, and reflects a somewhat different path of tape development.

Saturated paper tapes, although developed primarily for masking applications, became the most important general-purpose tapes in the United States. Rigid vinyl film tapes had a similar position in Europe, and these tapes still retain a portion of their original broad use. In the United States, saturated paper tapes were gradually displaced from packaging and general-purpose uses by film tapes, especially by biaxially oriented polypropylene tapes. This inexpensive product has also gained wide acceptance in Europe, replacing vinyl film tapes for less demanding applications. It is expected that vinyl film tapes, because of their printability, low noise level during application, and other properties, will maintain their current volume level.

Glass fiber reinforced packaging tapes are used less often in Europe than in the United States, with Germany perhaps being an exception. In the United States polypropylene film is replacing polyester film for this product. In Europe, polypropylene film is replacing the vinyl film that was used in the construction of glass fiber reinforced tapes.

Heavier, oriented polypropylene film tapes are used for palletizing, bundling, and other applications requiring high tensile strength. These products compete with glass reinforced tapes and have a larger market share in Europe than in the United States.

LABELS

The annual consumption of labels in Europe is estimated at 900 square meters. Pressure sensitive labels were developed by Avery International in the United States, and this company dominates this business field in Europe as well. The adaptation of pressure sensitive labeling for primary labeling has advanced further in Europe. Pressure sensitive labels are more expensive, but the application equipment is cheaper, and for short production runs it is often more economical to use pressure sensitive than glue applied labels. The European market is more diversified—such conditions favor the use of pressure sensitive labels for primary label applications.

Label printing in Europe also took a different path from that in the United States. While flexographic printing of narrow material in-line with die cutting and removal of matrix dominates in U.S. label printing, letterpress printing has about the same market share as flexographic printing in Europe, with a smaller amount of labels printed by gravure, as shown in Table 2-4. Rotary letterpress is an especially important label printing process. It is generally accepted that flexographic printing, if properly used, can yield a product of equal quality to that obtained by letterpress. Letterpress, however, is more tolerant of poorer workmanship.

The consumption of pressure sensitive labels in various countries is shown in Table 2-5.

The consumption of pressure sensitive labels for various end uses is shown in Table 2-6.

Pressure sensitive label stock manufacturers are listed in Table 2-7.

Raw Materials

Solvent solutions of adhesives are still used predominantly in Europe. Solvent recovery is commonly employed. Neither aqueous dispersions nor hot melts

Table 2-4. Label Printing by Various Processes

PROCESS	PERCENT OF THE TOTAL
Letterpress	48
Flexography	41
Gravure	11
	100

Table 2-5. Consumption of Pressure Sensitive Labels in Various European Countries

COUNTRY	ANNUAL CONSUMPTION (m^2)	PERCENT OF THE TOTAL
West Germany	246	27
France	157	18
United Kingdom	150	17
Benelux	117	13
Italy	110	12
Other	120	13
	900	100

Table 2-6. Consumption of Pressure Sensitive Labels for Various Applications

APPLICATION	PERCENT OF THE TOTAL
Electronic data processing	29
Retail price marking	16
Food labeling	11
Alcoholic drinks	4
Soft drinks	1
Cosmetic product labeling	12
Pharmaceutical product labeling	11
Other nonfood products	16
	100

Table 2-7. Major Pressure Sensitive Label Stock Manufacturers

COMPANY	LOCATION
Fasson Adhesive Products, Ltd.	Leiden, the Netherlands
Jackstadt GmbH	Wuppertal, West Germany
Raflatac/Sterling	Tampere, Finland
Samuel Jones & Co., Ltd.	St. Neots, Hunts., U.K.
Heinrich Hermann	Stuttgart, West Germany
Novarode	Sint-Genesius-Rode, Belgium
Mactac	Soignies, Belgium
SA Pilot	Versailles, Yvelines, France
Torres Hostench	Gerona, Spain
Meto	Hirschhorn/Neckar, West Germany

have replaced solvent adhesives to the same extent they have in the United States. Calender applications of the adhesives are rarely used.

Table 2-8 shows the distribution of pressure sensitive adhesives between major technology groups. A somewhat faster growth of aqueous dispersions

Table 2-8. Major Adhesive Groups by Technology

ADHESIVE	ANNUAL CONSUMPTION (1000 tons)	PERCENT OF THE TOTAL
Rubber solution adhesives	104	59
Acrylic solution adhesives	19	11
Acrylic emulsions	27	15
Hot melts	26	15
	176	100

than of hot melts is expected. The growth is partly at the expense of solution adhesives.

Aqueous adhesives have penetrated label applications more than tape manufacturing, and have about 60% of the label market. Hot melts are used for several important tape products, such as OPP film tape, glass filament reinforced tapes, and double-sided tapes. Aqueous dispersions are mainly acrylic polymers, hot melts are based on block copolymers (S-I-S), and solution adhesives are dominated by natural rubber, with some usage of styrene-butadiene copolymers.

Acrylic adhesive development took a slightly different path in Europe. In the United States, adhesive properties are usually tailored during polymerization by the emulsion producer, while in Europe blending and compounding to adjust adhesive properties and machineability are more often done by the coater. BASF is perhaps mainly responsible for this trend. The most important supplier of acrylic adhesives, this company also performs a considerable amount of customer service consisting of blending and compounding their emulsion products to meet the buyer's specific product and machineability requirements.

Polyvinyl ethers were and still are used in Europe for specialty pressure sensitive adhesives, including the compounding of acrylic adhesives.

Table 2-9 shows adhesive consumption distribution according to end products.

In Europe, the tackifying resins employed are predominantly of synthetic origin, while in the United States rosin derivatives constitute the most important

Table 2-9. Consumption of Pressure Sensitive Adhesives by End Use

USE	ANNUAL CONSUMPTION (1000 tons)	PERCENT OF THE TOTAL
Tapes	131	75
Labels	27	15
Transfers	6	3
Other	12	7
	176	100

Table 2-10. Type of Tackifying Resins

RESIN TYPE	PERCENT OF TOTAL
Hydrocarbon	75
Rosin derivatives	15
Polyterpenes	5
Other	5
	100

group. Table 2-10 shows the distribution of resin types for pressure sensitive adhesive use.

The main resin producers in Europe are Esso Chimie and Hercules. Some resins are also sold by Bergvik (Sweden), Olon (Finland), DSM, and Neville.

Radiation curable silicone release coatings were first offered commercially by Goldschmidt A.G. of West Germany. The use of 100% active, thermally curable silicone release coatings has penetrated the industry further than in the United States. Silicone release coatings are produced in Europe by Dow Corning, Wacker Chemie GmbH, G.E. Plastics B.V., and Rhone-Poulenc.

European equipment firms have always produced excellent, although expensive coating equipment. German, Swiss, and Italian firms are especially active in coating equipment sales worldwide.

The author appreciates the assistance and suggestions for this chapter received from Mr. Helmut Mueller, BASF; Mr. G. L. Butler, DRG; Mr. Mario Magnini, Enichem Elastomeri SpA; Mr. A. M. Deurloo, Akzo; and Dr. Erhard Braeunling, BDF.

3. Pressure Sensitive Adhesive Technology and Markets in Japan and Asian Countries

Keiji Fukuzawa
Consulting Chemical Engineer
Tokyo, Japan

JAPAN

The pressure sensitive industry is divided into three main areas: tape, label stock, and surgical products manufacturing. The tape manufacturers are organized into the Japan Adhesive Tape Makers Association, which was established in 1961 and currently has 23 members. The total number of tape and label stock manufacturers is estimated at 100. The major pressure sensitive product manufacturers are listed in Table 3-1. The Association reports statistical data on the production of member companies annually. These data are shown in Figure 3-1. The product value data were converted by assuming that the U.S. dollar is equal to 170 yen. The data show that the pressure sensitive tape business rapidly increased until 1973, when the recession caused a sharp decrease in production. Recovery was complete by 1982.

Recent statistical data on various types of tapes, including the production of nonmember companies, are shown in Table 3-2. The most important tape product is kraft paper tape used for various packaging applications. The demand for this tape amounts to 400 million square meters annually. The largest manufacturer is OJI Tape Company, Ltd. Japanese paper tape is the product most often used for masking applications. Cellophane film tape is firmly entrenched and has not been replaced by matte acetate film tape. Oriented polypropylene film has an important part of the film tape market, but has not grown to the dominance of the packaging market it has achieved in the United States.

Table 3-1. Major Pressure Sensitive Adhesive Product Manufacturers in Japan

COMPANY	MAIN PRODUCTS
Nitto Industrial Co., Ltd.	Various industrial tapes, surgical products
Nichiban Co., Ltd.	Stationary tape, surgical products, industrial tapes
Sekisui Chemical Co., Ltd.	Cellophane tape, kraft paper tape, OPP tape
Sumitomo 3M Limited	Various industrial tapes, reflector sheet
FSK Co., Ltd.	Label stock, gummed tape, PSA sheet
Sohken Paper Manufacturing Co.	Label stock, kraft paper tape
Modern Plastic Industries Co., Ltd.	Label stock, double-face tapes
Kanzaki Paper Manufacturing Co.	Label stock
Hisamitsu Pharmaceutical Co., Ltd.	Plaster, surgical products
Taisho Pharmaceutical Co., Ltd.	Plaster, surgical products
Suzuki Nihon Do Co., Ltd.	Plaster
Ohishi Koseido Co., Ltd.	Plaster
Hitachi Chemical Co., Ltd.	Protective sheet, electrical tapes
Hitachi Kasei Polymer Co., Ltd.	Double-face tape
Dainippon Ink & Chemicals Inc.	Double-face tape, label stock
Kamoi Kakoshi Co., Ltd.	Japanese paper tape
Kikusui Tape Co., Ltd.	Kraft paper tape
Kyowa Limited	Cellophane tape, vinyl tape
OJI Kako Co., Ltd.	Kraft paper tape, label stock
Okamoto Co., Ltd.	Fabric tape
Rinrei Tape Co., Ltd.	Japanese paper tape, fabric tape
Sugawara Industrial Co., Ltd.	Fabric tape, reinforced tape
Sayama Kako Tack Paper Co., Ltd.	Label stock
Semedain Co., Ltd.	Polyester film tape
Sony Chemicals Corp.	Double-face tape
Teraoka Seisakusho Co., Ltd.	Fabric tape, OPP tape, electrical tapes
Toyo Chemical Co., Ltd.	Vinyl tape, OPP tape
Toyo Ink Manufacturing Co., Ltd.	Double-face tape
Uni Industry Co., Ltd.	Cellophane tape
Yazki Sogyo Co., Ltd.	Vinyl tape

Natural rubber is the most often used raw material for adhesives, although use of block-copolymer-based hot melt adhesives has grown considerably.

Production practices are also somewhat different in Asia. Instead of slitting the tapes into individual rolls, slicing of the log roll is preferred. Several tape companies have developed their own tape slicing machines.

The major tape manufacturers are Nitto Electric Industrial Company, Ltd., Nichiban Company, Ltd, Sugarawa Industrial Company, and OJI Tape Company.

The major label stock manufacturers are FSK Company, Ltd., OJI Kako Company, Ltd., Kanzaki Paper Manufacturing Company, Ltd., Sohken Paper Conversion Company, Ltd., and Nichiban Company, Ltd. The production volume of label stock and related products is shown in Table 3-3.

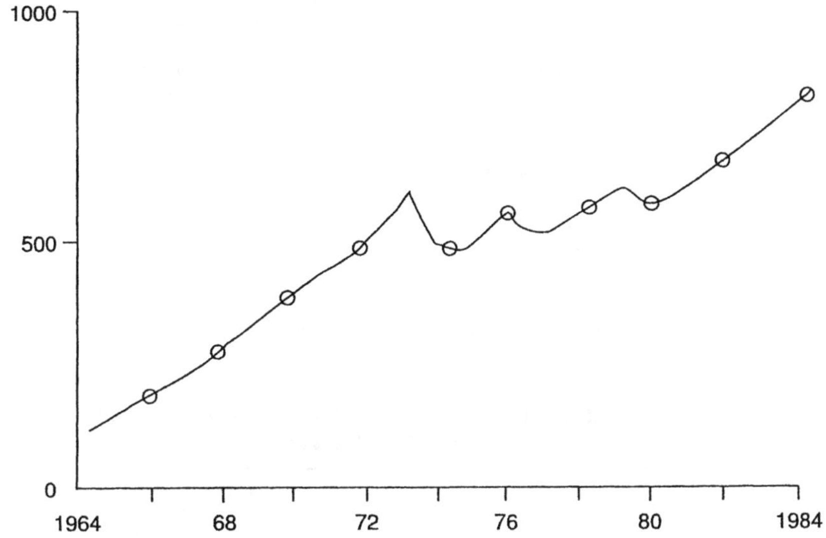

Fig. 3-1. Production of industrial pressure sensitive adhesive tapes. (Data by Japan Adhesive Tape Makers Association.)

Table 3-2. Production of Industrial Tapes in Japan

PRODUCT	MILLION m^2	US$ MILLION (1985)[a]
Paper Tapes		
Kraft paper	401.4	153
Japanese paper	66.1	50
Crepe paper	3.9	7.5
Others	8.2	10.3
Fabric Tapes		
Rayon fabric	79.6	104.2
Cotton fabric	0.2	1.0
Others	13.6	30.7
Film Tapes		
Cellophane	121.7	95.2
Vinyl	90.2	98.2
Polypropylene	78.2	49.2
Polyester	24.0	37.2
Acetate	1.4	3.9
Teflon	0.2	7.3
Other films	0.6	2.1
Foam tapes	0.7	3.3
Metal foil tapes	1.6	4.9
Rubber tapes	0.4	3.8

Table 3-2. (*cont.*)

PRODUCT	MILLION m^2	US$ MILLION (1985)[a]
Filament reinforced tapes	1.6	4.5
Double-face tapes		
Paper and nonwoven	37.0	107.4
Fabric	2.7	10.6
Plastic	0.9	3.8
Foam	0.9	11.0
Total	935.1	799.1

Note: Data by Sogo-hosoushupan Co., Ltd.
[a] US$ = 170 yen.

Table 3-3. Production of Pressure Sensitive Label Stock and Sheeting

PRODUCT		PRODUCTION LEVEL	
		MILLION m^2	US$ MILLION (1985)[a]
With release sheet			
Labels	paper	390	307
	film	95.2	174.7
Decorative labels		1.4	4.9
Reflective labels and signs		2.5	62.1
Without release sheet			
Protective sheeting		154.2	118
Total		643.3	666.7

Note: Data by Sogo-hosoushupan Co., Ltd.
[a] US$ = 170 yen.

Table 3-4. Pressure Sensitive Products for Hospital and Other Applications

PRODUCTS	US$ MILLION (1983)[a]
Plaster	100
Hospital tapes	90.6
First aid products	77.1
Transdermal delivery tapes	25.2
Salicylic acid plaster	29.4
Athletic tapes	17.6
Cockroach catching tape	88.2
Total	428.1

[a] US$ = 170 yen

Hospital pressure sensitive adhesive products are manufactured by the following major companies: Nitto Industrial Company, Ltd., Nichiban Company, Ltd., Taisho Pharmaceutical Company, Ltd., Hismitsu Pharmaceutical Company, Ltd., and Sumitomo 3M Company, Ltd. The production of various hospital products is shown in Table 3-4.

KOREA

There are about 60 pressure sensitive adhesive product manufacturers in Korea. The major companies are shown in Table 3-5. The largest tape manufacturer is Daeil Chemical Company, Ltd. Packaging tapes made with oriented polypropylene and cellophane film backings are the main products.

TAIWAN

Some 50 companies manufacture pressure sensitive tapes in Taiwan. The major manufacturers are listed in Table 3-6. Production volume was estimated at $250 million in 1985, of which 40% was exported to various countries, including the United States.

The distribution in Taiwan among various pressure sensitive adhesive products is estimated to be 40% industrial tapes, 30% label stock, and 10% surgical products. Taiwan excels in the production of plasticized vinyl film tapes and in OPP packaging tape.

CHINA

There are 40 surgical product and 30 industrial tape manufacturers in China. Medical plasters have been made in China for a long time. The Chinese tape

Table 3-5. Pressure Sensitive Adhesive Tape Manufacturers in Korea

COMPANY	MAIN PRODUCTS
Daeil Chemical Co., Ltd.	Industrial tapes, surgical products, gummed tape
STC Chemicals Corp.	Cellophane tape, OPP tape
Duck Sung Tape Co., Ltd.	OPP tape
Hae Sung Tape Industry Co., Ltd.	Cellophane tape, OPP tape, polyester tape
Lucky Tape Co., Ltd.	Vinyl tape
Sae Rih Tape Co., Ltd.	Label stock
Nan Mi Industrial Co., Ltd.	Label stock
Tong Sung Co., Ltd.	Label stock
Sinsin Pharmaceutical Co., Ltd.	Surgical products
LBL Trading Co., Ltd.	Fly, cockroach, vermin catcher tape

Table 3-6. Pressure Sensitive Adhesive Manufacturers in Taiwan

COMPANY	MAIN PRODUCTS
Four Pillars Enterprise Co., Ltd.	PVC film, kraft paper, OPP film, crepe paper, cellophane, double-face, and polyester film tapes, label stock
Asia Chemical Corp.	OPP Tape, PVC tape
Globe Industries Corp.	OPP, PVC, Cellophane, polyester, double-face tapes
K.K. Converter Co., Ltd.	Label stock
Chi-Ming Industries Corp.	PVC tape
Leaf Tapes Industrial Co., Ltd.	OPP tape
Taiwan Kikusui Tape Co., Ltd.	Kraft paper tape
Tung Tai Paper Corp.	Kraft, paper gummed tape
Poly Chemicals Co., Ltd.	PVC tape
San Tai Electrical & Machinery Co., Ltd.	Lamination, OPP tapes
Tsai Ho Yuan Industry Co., Ltd.	Cellophane tape
Kuo Chun Industrial Co., Ltd.	Kraft paper tape
Chuan Yi Industries Corp.	Polyester film, OPP, double-face tapes
Ho Fung Industry Co., Ltd.	Lamination, label stock, double-face tapes
Taiwan 3M Co.	Matte acetate tape
Taiwan Nitto Industrial Co., Ltd.	PVC tape

industry is attempting to modernize production. Twenty-three adhesion societies were established in 1982 in each of the prefectures.

SOUTH ASIA

There are about ten pressure sensitive product manufacturers in the Philippines. Major manufacturers and their main products are listed in Table 3-7. Tape

Table 3-7. Pressure Sensitive Adhesive Tape Manufacturers in the Philippines

COMPANY	MAIN PRODUCTS
Century Paper and Plastic Corp.	OPP tape, masking tape, label stock
Hope Adhesive Paper Products Inc.	Label stock
Globe Tape Inc.	OPP tape
United Cathay Industrial Corp.	OPP tape
Ridjo Paper Corp.	OPP tape
Industrial Tape Co.	OPP tape
Good Hope Tape Inc.	OPP tape, masking tape
Sibesco Manufacturing Co.	OPP tape, masking tape, plasticized PVC tape, label stock
3M Philippines Co.	OPP tape, masking tape, label stock
Sacoba Manufacturing Corp.	OPP tape

Table 3-8. Pressure Sensitive Adhesive Tape Manufacturers in South Asian Countries

COMPANY	MAIN PRODUCTS
Thailand	
Union Thai Nichiban Co., Ltd. (Bangkok)	Cellophane tape
Louis Adhesive Tapes Co., Ltd. (Bangkok)	Cellophane tape
Simkon (Thailand) Co., Ltd. (Bangkok)	Cellophane tape
Siam Printing Accessories, Ltd.	Label stock
Indonesia	
P.T. Sumilindo Tape Industry, Ltd. (Jakarta)	Plasticized PVC tape, cellophane tape
P.T. Rungkat Cahaya (Surabaya)	OPP tape
India	
The Bhor Industries, Ltd. (Bombay)	Plasticized PVC tape
Johnson & Johnson, Ltd. (Bombay)	Hospital and industrial tape
Singapore	
Four Pillars Manufacturing Industries Corp.	Plasticized PVC tape
Malaysia	
Loytape Berhad	Cellophane tape, OPP tape Plasticized PVC tape, gummed tape
Central Industrial Corp Sendirian Berhad (CIC)	Label stock, masking tape Kraft paper tape

manufacturers in other South Asian countries are shown in Table 3-8. Loytape Berhad of Malaysia is the largest pressure sensitive tape manufacturer in South Asia. It produces gummed tape, cellophane tape, and OPP tape. Central Industrial Corporation Sendirian Berhad (CIC) produces label stock, kraft paper, and masking tapes. Loytape and CIC are sister companies—both are members of the Ganda Group.

4. Tack

Fred H. Hammond, Jr.
Kendall Co.
Lexington, Massachusetts

Pressure sensitive tack is the adhesive property related to bond formation. It has been defined[1] as the property that enables an adhesive to form a bond with the surface of another material upon brief contact under light pressure. It is the dominant property that distinguishes the broad class of adhesives that are the subject of this book from other materials. Pressure sensitive tack is differentiated from the tire maker's tack of elastomer-to-elastomer bonding and from the printer's tack of resistance-to-separation by a viscous fluid between two surfaces. Implicit in definition of pressure sensitive tack of commercial adhesives is the assumption that they separate from adherends cleanly, that is, without leaving visible residue. The ASTM definition[2] of pressure sensitive tack requires that the bond established be of measurable strength. In this chapter, the word tack means pressure sensitive tack.

Tack has been called quick stick, initial adhesion, stickiness, quick grab, and other terms. It is generally conceived to be a property associated with, and to be measured at, low contact pressures and short contact times. Investigators into the science of tack would prefer to describe the extent of molecular contact between adhesive and adherend at any instant, but this is not yet possible. Instead, tack is described in terms of indirect phenomena observed when quickly formed adhesive bonds are also quickly broken in procedures hoped to be related to molecular contact or function of the adhesive.

Tack is not a simple material property, such as density or modulus, but is rather a composite response of the adhesive's surface chemical and bulk physical properties upon being brought into contact with another material. In practical terms, tack is defined by the test used to measure it. The merit of a given tack test is how well it reveals the rate at which a bond forms, relates to adhesive and adherend compositions, is interpretable in terms of basic adhesive chemical/physical properties, and how well it predicts an adhesive's function in proposed applications.

A person's basic impression and first measure of tack is the sensation experienced in bringing the thumb or finger lightly into contact for a short time with a pressure sensitive adhesive and then quickly withdrawing it. By varying the pressure and time of contact and noting the difficulty of pulling the thumb from the adhesive, one perceives how easily, quickly, and strongly the adhesive can form a bond. As a test for tack, the thumb test has the defects of being subjective, nonquantifiable, uncertain in differentiation among adhesives except for tests made at the same time, and indicative only of the adhesive/skin bonding, which may or may not relate to adhesive/other-adherend bonding. Even so, the thumb test made by a trained and experienced person is often as good an indication of tack as any instrument test.

Instrument tests for pressure sensitive tack can be classified into three groups:

1. Rolling ball, rolling cylinder, or rotating drum tests in which distance traveled, energy lost, or force observed in the interaction of these objects with an adhesive (tape) is measured.
2. Peel tests, 90 or 180 degree, involving low external contact pressure and short application time, in which the peel force is measured.
3. Probe tests, in which the tip of a probe is brought into contact under low pressure for a short time with the adhesive and the butt tensile strength of the bond formed between them is measured.

ROLLING BALL TACK TESTS

Rolling ball tests are probably the oldest and most widely used tack tests, having been used in one form or another for at least 60 years. In the 1940s, the Bauer and Black Company was measuring the tack of medical tapes by rolling a 12.5-mm (½-in.) -diameter stainless-steel ball from various elevations down a 30-degree incline out onto adhesive faceup tape and noting the distance traveled. In their earlier technical literature the Ashland Chemical Company[3] described a similar apparatus in which the simple incline was replaced by a tube. Douglas Aircraft Corporation[3] replaced the straight incline with a channel formed into a 300-mm (12-in.) radius, used 11-mm (7⁄16-in.) or, more commonly, 28.5-mm (1.125-in) -diameter steel balls and took the product of release height times the weight of the ball divided by the rollout distance as the measure of tack. Ping-Pong balls have been recommended for measuring low tack adhesives. To assure balls rolled without sliding on the incline so as to give more reproducible approach velocities and to smooth the transition from incline to adhesive surface

Kamagata, Saito, and Toyama[4] redesigned the ramp into a half sine wave form.

Rolling ball tack is the Pressure Sensitive Tape Council's Test Method No. 6.[5] In this procedure an 11-mm (7⁄16-in.) -diameter stainless-steel ball is rolled down an inclined track to come into contact at the bottom with the horizontal, upward-facing adhesive. The distance the ball travels out along the tape is taken as the measure of tack.

Ball rollout distance gives an inverse, compressed scale of tack; the greater the distance, the less tacky the adhesive, but not in proportion to the ratio of distances. The reciprocal of rollout distance has sometimes been taken as the tack value. Attempts have been made to create scaling of rolling ball tack values that would group combinations of release heights and rollout distances for equivalent tack levels. Table 4-1 was created years ago at Bauer and Black, then part of the Kendall Company. Such scaling tables have limited utility because of the high frequency of exceptions to them that are observed.

Table 4-1. Tack Level

ROLL OUT (in)	RELEASE HEIGHTS (in)					
	1	2	3	4	5	6
¼	—	—	—	—	—	7.0
½	—	—	—	—	—	6.9
¾	—	—	—	—	—	6.3
1	2.0	3.5	4.4	5.0	5.4	5.9
1¼	1.8	3.3	4.2	4.8	5.2	5.7
1½	1.5	3.0	3.9	4.5	5.0	5.4
1¾	1.3	2.8	3.7	4.3	4.7	5.2
2	1.0	2.5	3.4	4.0	4.5	4.9
3	0.8	2.0	2.8	3.2	3.9	4.3
4	0.5	1.6	2.5	3.0	3.4	3.9
5	—	1.2	2.1	2.5	3.1	3.5
6	—	1.0	1.8	2.3	2.9	3.2
7	—	0.9	1.5	2.1	2.6	3.1
8	—	0.8	1.4	2.0	2.4	2.9
9	—	0.7	1.2	1.8	2.3	2.6
10	—	0.6	1.1	1.6	2.1	2.5
11	—	0.5	1.0	1.5	2.0	2.4
12	—	0.5	1.0	1.3	1.8	2.3
13	—	—	0.9	1.2	1.7	2.2
14	—	—	0.8	1.1	1.6	2.1
15	—	—	0.8	1.0	1.5	2.0
16	—	—	0.7	1.0	1.4	1.9
17	—	—	0.7	0.9	1.3	1.8
18	—	—	0.6	0.8	1.3	1.7
19	—	—	0.6	0.7	1.2	1.6

In the 1950s Dow of Smith and Nephew, England[6] proposed an alternate procedure in which balls of increasing diameter were rolled onto the central 100-mm section of faceup adhesive tape on a 300-mm, 30-degree incline until one was just stopped by the adhesive. This ball's diameter in thirty-seconds of an inch was taken as the measure of tack. The method was cited in early Japanese adhesive tape patents but has not been widely used.

The ready availability of balls of different size, composition, surface texture, density, weight, moment of inertia, etc., invites the investigation of the effects of these parameters on apparent rolling ball tack. Gillespie[7] made an extensive study using stroboscopic photography to determine the time/position of the ball and thus its velocity and energy. Table 4-2 presents his analysis of the mechanics of the rolling ball test. Gillespie found instances where balls, while on the adhesive, slowed down as if they were acted on by a constant force and lost energy in proportion to distance traveled. In these instances, ball rollout distances are proportional to the $\frac{3}{2}$ power of the release height. While investigating filled compounded commercial adhesives, Gillespie found more instances where energy was lost in proportion to time of contact rather than in proportion to distance. He found some adhesives changed from one mode to the other as the release height was changed. Polytetrafluoroethylene balls tended to lose energy with time in the beginning portion of a rollout and with distance in the final portion. He was unable to summarize the data with a unified theory.

Gillespie found rollout distances and the area of contact between the ball and adhesive to vary with how well the tape was held down. He constructed a vacuum hold-down system that gave more consistent results than the common practice of hand tensioning of the backing.

Kamagata, Saito, and Toyama[4] reported stroboscopic rolling ball measurements in which the velocity of the ball, while rolling on the tape, decreased in direct proportion to time. Results are the same as Gillespie's constant-force findings. Kamagata *et al.* reported a direct linear relationship of rollout distance with ball diameter for seven adhesives, but with a different constant of proportionality for each adhesive. They also found rollout distance increases as temperature in the range of 5° to 35°C (centigrade) was lowered.

Mizumachi and Saito[8] have reported stroboscopic photography measurements of roll distances as a function of time for 6.4-, 8-, and 9.6-mm-diameter balls released at 60-, 115-, and 191-mm elevations on half sine wave incline tack testers at 6° and 20°C for seven commercial tapes of diverse backing and adhesive constructions. They interpret their data in terms of two rolling friction coefficients: one a constant, the other velocity dependent. From basic equations of motion of a rolling sphere and these rolling friction coefficients, they derived the equations in Table 4-3 for the velocity of the ball and the distance rolled at any given time, and the rollout distance as a function of the ball's initial approach velocity and release height. These equations were found to describe

Table 4-2. Rolling Ball Tack Test Mechanics Analysis

Prior to release: ball has potential energy (P.E.) $= gmh = gmd \cos \theta$		
On incline: ball's P.E. is converted to kinetic energy (K.E.) $= mv^2 + Iv^2/2r^2$		
	for solid balls $I = 2mr^{2/5}$	
	for thin shell balls $I = 2mr^{2/3}$	
At bottom of incline,		
ball has:	velocity $(V) = (gh/0.7)^{1/2}$	
While rolling on tape:	kinetic energy (K.E.) $= 0.7mv^2$	
	If ball loses energy	If ball loses energy
Deceleration:	proportional to distance	proportional to time
Velocity:	$-dv/dt = a$	$-dV/dt = b/v$
	$v = Vt - at$	
Distance at time t:	$v^2 = V^2 - 2ax$	$v^2 = V^2 - 2bt$
Time to stop:	$x = Vt + at^2/2$	$x = v^2(1 + (2bt/v^2 - 1)^{3/2})/3b$
Roll out distance	$T = V/a$	$T = V^2/2b$
	$X = v^2/2a$	$X = (gh)^{3/2}/1.76b$
where	$X = gh/1.4a$	
	g = acceleration due to gravity	I = moment of inertia
	m = mass of ball	V = velocity of ball at bottom
	h = height of release	T = time to stop
	d = position on incline	X = roll out distance
	θ = angle of incline	
	v = translational velocity	
	a = deceleration constant	
	b = deceleration constant	
	x = distance along tape	

the entire rollout of balls in individual tests at all conditions examined. However, the friction coefficients varied from one test to the next, so that averages of many test runs were thought to be needed for correlation with adhesive physical properties. How generally this theory of rolling friction summarizes rolling ball tack data and permits interpretation in terms of adhesive properties awaits the accumulation of more data on well-characterized adhesives.

The ultimate simplicity, low cost, ease of execution, and generally good reproducibility of the rolling ball tack test causes it to be used in adhesive tape quality control. It was developed when most pressure sensitive adhesives were compounded from natural rubber and tackifying resins for which there is an approximate relation between rolling ball value and expectation. For synthetic adhesives, or those adhesives compounded from synthetic elastomers, however, the correspondence of rolling ball tack to expectation is often less than good. Many functional synthetic acrylate adhesives do not stop the ball in the maximum prescribed length of tape and thus have no apparent rolling ball tack. The wide range of contact times in each test, multiplicity of contacts of the same ball area, dependence on degree of tape hold-down, lack of applicability to all

Table 4-3. Rolling Ball Friction Coefficients

If interaction of rolling ball and adhesive is described in terms of constant and velocity-dependent friction coefficients:

the velocity (v) at any time (t) =

$$v = v_o \exp\left(-\frac{5g\phi_1}{7R}t\right) - \frac{\phi_0}{\phi_1}\left(1 - \exp\left(-\frac{5g\phi_1}{7R}t\right)\right)$$

the distance traveled ($x-x$) at any time =

$$(x - x_0) = \frac{7R}{5g\phi_1}\left(v_0 + \frac{\phi_0}{\phi_1}\right)\left(1 - \exp\left(-\frac{5g\phi_1}{7R}t\right)\right) - \frac{\phi_0}{\phi_1}t$$

and the roll out distance =

$$(x - x_0)_{st} = \frac{7R}{5g\phi_1}\left\{\left(\frac{10}{7}gH\right)^{1/2} - \frac{\phi_0}{\phi_1}\log\left(1 + \frac{\phi_1}{\phi_0}\left(\frac{10}{7}gH\right)^{1/2}\right)\right\}$$

In terms of release height (H), the roll out distance =

$$(x - x_0)_{st} = \frac{7R}{5g\phi_1}\left\{v_0 - \frac{\phi_0}{\phi_1}\log\left(1 + \frac{\phi_1}{\phi_0}v_0\right)\right\}$$

where:

g = gravitational constant

R = radius of ball

v = velocity of ball at initial contact with adhesive

adhesives, inverse scaling, and emphasis on adhesive softness limits its use in theoretical studies and as a tape function test.

While tack measurement by rolling cylinders released from an elevation to roll out onto adhesives is a logical analogy to rolling ball measurements, there is no published literature describing such measurements. Hammond modified the rolling cylinder instrument of Voit and Geffken[9] for measuring the tack of printing inks to mount pressure sensitive adhesive tapes, but was unable to summarize data on commercial tapes in terms of adhesive physical properties or function. The conceptual advantages of cylinders over balls include the feasibility of lower contact pressures, more straightforward calculation of contact pressure, and the opportunity to limit contact of a cylinder's surface with adhesive to a single instance. Practical problems of assuring that cylinders roll exactly in the direction intended and the need to fabricate cylinders and track may be reasons why this method has not been examined more extensively.

A tack measuring instrument in which a lightweight aluminum wheel rolled over adhesive tape fastened face up on the flat rim of a larger diameter rotating drum was devised by Bull, Martin, and Vale.[10] It was simple to construct, convenient to use, scanned a large area of adhesive, and could be used in a chamber for tests at different humidities and temperatures. Contact times between adhesive and cylinder surface could be varied over wide limits even though they are coupled with the rate of bond making and breaking. Contact pressures could be adjusted from very low to any level that might be required to simulate end use conditions. Tack was measured as the magnitude of the restraining force required to hold back the freely turning wheel as it rolled over the controlled variable speed driven drum. They described the apparent increase of tack (force) with the rate at which a tape sample is removed from a roll and with adhesive coating weight; the decrease of tack upon Weather-Ometer aging and drum surface speed (decreasing contact time) and the dependence of tack on composition of natural rubber/polyterpene resin adhesive mixtures. There is less use of this procedure than it appears to warrant, possibly because of the inherent confounding interrelationship of contact time and rate of bond formation/rupture, and the lack of a commercially available instrument. The principle of this method is shown in Figure 4-1.

The pulling of a cylinder over the adhesive was proposed by Mizumachi[11] as an easier way of measuring tack-related rolling friction. He derived theoretical expressions predicting the dependence of rolling friction on translation velocity in terms of stress and strain failure criteria in combination with simple rheological models of the adhesive. The applicability of these theoretical equations can be expected to be examined in future work.

For the aluminum wheel system of Bull *et al.*, Druschke[12] substituted a freely turning, toothed, stainless-steel wheel suspended from the load cell of a standard tensile tester in such a way as to be in very light contact with adhesive tape mounted on the face of a rotating drum (Fig. 4-2). As each tooth on the wheel sequentially came in contact and was pulled from the adhesive, a number of individual tack measurements were made. Druschke made toothed wheel, initial peel adhesions, quick stick, probe tack, and Zosel tack measurements

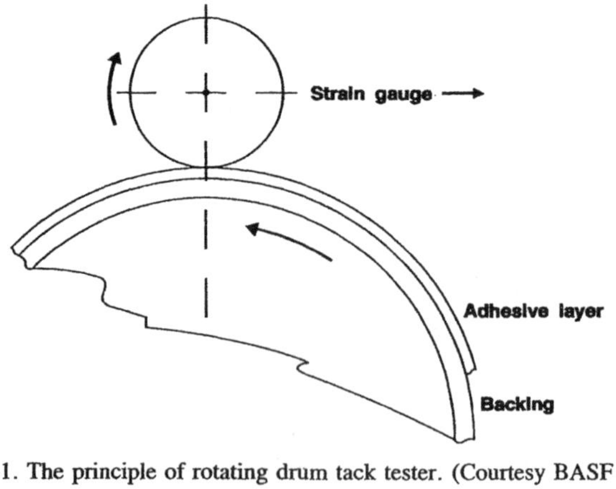

Fig. 4-1. The principle of rotating drum tack tester. (Courtesy BASF A.G.)

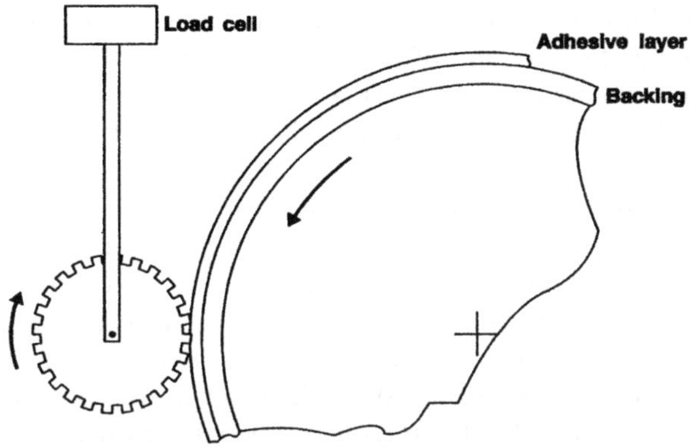

Fig. 4-2. The principle of toothed wheel tack tester. (Courtesy BASF A.G.)

on adhesives differing in chemical composition, component concentration, molecular weight, cross-link density, surface morphology, glass transition temperature, and modulus. He also examined tack dependence on adhesive thickness, backing stiffness, test separation rate, adherend surface roughness, and interface energy. His paper is of particular interest because of comparisons and discussions of differences and similarities among tack procedures. Further examination of toothed wheeled tack testing can be expected to be carried out in the future.

Instead of rolling cylinders over adhesives, the inverse operation of passing tape over freely turning cylinders has been used to measure tack. Johnston[3] mentioned a device consisting of two parallel rollers mounted one above the other in a constant rate of extension machine, in which tape in very light contact was drawn over them, that was used in the 1950s. Chang[13] writes of a single roll mounted in a tensile test machine onto which tape was placed with pressure due only to the weight of its backing and then peeled away at an approximately 90 degree angle. Both of these procedures have many of the characteristics of peel tests.

PEEL TESTS FOR TACK

An individual's method of assessing tack, if not by the thumb test, often involves forming a loop of tape—adhesive side out—bringing the loop into light contact for a short time with some solid surface, and then noting how difficult it is to remove the tape. Salespeople have used this procedure for years to demonstrate the relative quality of their products by lifting objects off the desk of a potential buyer. They also have used the gimmick of throwing a loop of tape against the wall or other surfaces to which tacky tapes stick and nontacky tapes do not. Johnston[3] describes how in the 1950s the loop procedure was adapted to the hinged metal box test for tack, in which the greatest weight that could be lifted was taken as the measure of tack. He pointed out this test is obviously cumbersome and almost the only one that does not separate the adhesive from the adherend. Also in the 1950s a procedure referred to as the Kreck test using a loop of adhesive tape lowered in a tensile test machine onto curved metal panels was proposed to the ASTM but was not accepted as a standard. Kamagata, Saito, and Toyama[4] developed a Self Pressure Tack Tester in which 25-mm (1-in.) by 300-mm (12-in.) tape samples formed into circular loops were suspended from a peg on the upper jaw of a tensile tester and subsequently brought at 300 mm/min (12 in./min) into contact with a horizontal flat plate mounted in the lower jaw. After some short period of time, the loop was peeled from the plate and the force required recorded as the measure of tack. A similar loop test[3,12] is one in which the free ends of the tape are brought together and clamped in the upper jaw of a tensile tester to form a pear or teardrop shape that is brought into contact with a flat, 25- or 50-mm (1- or 2-in.) -wide horizontal stainless-steel plate or other adherend.

Loop tack tests have the advantage of requiring only simple fixtures mounted on standard tensile testers, being quite easy to carry out, and being acceptably reproducible. Even in a single test, however, they are subject to variable peel angles and contact times. They also suffer from the large influence the stiffness or flexibility of the tape backing plays in the magnitude of the force measured. The cause of this influence is the compression zone that presses the adhesive into contact with the adherend just ahead of the zone where adhesive and tape is being peeled off and the manner in which compression magnitude depends on backing properties. Loop tack tests can very effectively assess the tack of adhesives on film or sheet (paper) backings for applications that are closely approximated by test conditions.

Peel tests for tack measurement made at short contact times and low pressure, presumably while the bond is less than complete, are logical adaptations of standard adhesion peel tests for bond strength. The force is usually assumed to be indicative of completeness of bond formation and taken directly as the tack measure. Tack has been suggested to be the limiting value of peel adhesion extrapolated to zero contact pressure and time. A rarely used tack measure is the ratio of peel force after a short contact time at low pressure to standard adhesion peel force at long time and high pressure.

In the Pressure Sensitive Tape Council's "Quick-Stick" test,[5] the tape is placed on the flat stainless-steel plate test surface with no contact pressure beyond that of its own weight. The tape is then peeled from the surface at 90 degrees at 300 mm/min (12 in/min) and the peel force taken as the tack value. An equivalent quick stick test is AFERA 4015.[12]

Peel tack is almost as much influenced by the backing on which the adhesive is coated as are loop tack tests. In peel tests, the same force couple consisting of the measured tension peel force acting on one side of the peel line and the resultant compression force acting on the opposite side, pressing the adhesive against the adherend surface, is present. Plastic film, stiff paper, or metal foil tape backings transfer the force couple across the line of peel into a compressive contact force, often causing an initially low peel force to cascade rapidly into a much higher force, and consequently to a higher tack value. On the other hand, cloth or soft paper backings cannot effectively transfer peel force torques to compressive forces, so the peel force and tack value remain at low levels.

Apparatus for carrying out peel tack tests are simple, commercially available fixtures for mounting on constant rate of extension tensile testers. As long as tape end use application conditions are closely similar to those in the test, tape peel tests offer a fairly good method of comparing tapes. Peel tack test contact times are relatively long and vary from start to finish in a single test. The more application conditions differ from the situation in which the tape is merely laid down on a substrate and removed right away, the less useful they are.

PROBE TACK TESTS

Tack tests have been tried in which the thumb or finger is used as a probe being brought into contact with, and pulled away from, adhesives mounted on a balance or other force measuring system.[4] While the mechanics of the systems have been imaginative, the procedure has not been widely adopted. The variability of surface chemical and physical texture from person to person and from time to time for the same person is too great to be acceptable.

In another tack test[4,14] balls have been freely suspended within a ring mounted in a tensile tester, lowered into contact with the adhesive, and the force required to pull them from the adhesive measured. The weight of the ball determined the contact pressure. Contact time was controlled by the mechanics of the tensile tester. Kamagata *et al.*[4] found some correspondence between the energy to pull off a ball and the energy absorbed by an equal contact area in a rolling ball test.

Mechanical probe tack tests are intended to be simulations of thumb or finger tack tests. In them, the tip of a probe is brought into contact with a supported adhesive under low contact pressures for a short time and then pulled away at a fixed rate, during which the peak force of separation is measured. Wetzel[15,16] pioneered the procedure using an Instron Tensile Tester and special fixtures. Other probe tack designs[14,17–20] are reported in the literature. Hammond[21] developed the self-contained instrument in which the individual factors involved in bond formation can be varied one at a time. It is available commercially as the Polyken Probe Tack Tester.

Probe tack can reveal how bond formation increases with increasing contact pressure. As shown in Figure 4-3, probe tack of smooth adhesives has some magnitude even at low contact pressures and rises quickly to an asymptotic value at high pressures, while probe tack of rough surface adhesives is low at low contact pressures and rises slowly as pressure is increased. The comparable increase of probe tack with pressure is observed with smooth adhesives on rough surfaces.

Probe tack can reveal how bond formation increases with increasing contact time. As shown in Figure 4-4, probe tack of smooth surface adhesives is appreciable even at short contact times and increases to a limit quickly; probe tack of rough surface adhesives is low at short contact times and increases gradually as contact time is increased. Again, comparable increase of probe tack with time is observed with smooth adhesives on rough surfaces.

Probe tack can reveal how initial bond strength tack depends on the chemical composition of the adherend.[22–24] Probe tack values for three commercial adhesive tapes as measured on stainless steel and various plastic probes using the Polyken instrument are shown in Figure 4-5. The three tapes decrease in tack with decreasing surface energy of the probe. It is also seen that the tack values and relative ranking of the adhesives are slightly different on different composition probes.

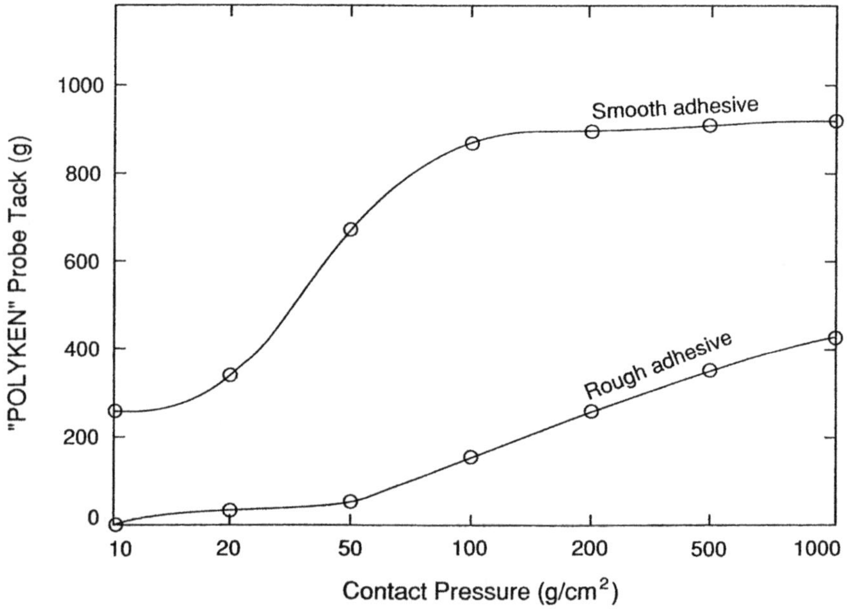

Fig. 4-3. Probe tack of smooth (film-backed) and rough (cloth-backed) pressure sensitive adhesive on a standard ½-cm-diameter stainless-steel probe, 1-sec contact time, 1-cm/sec separation rate at room temperature.

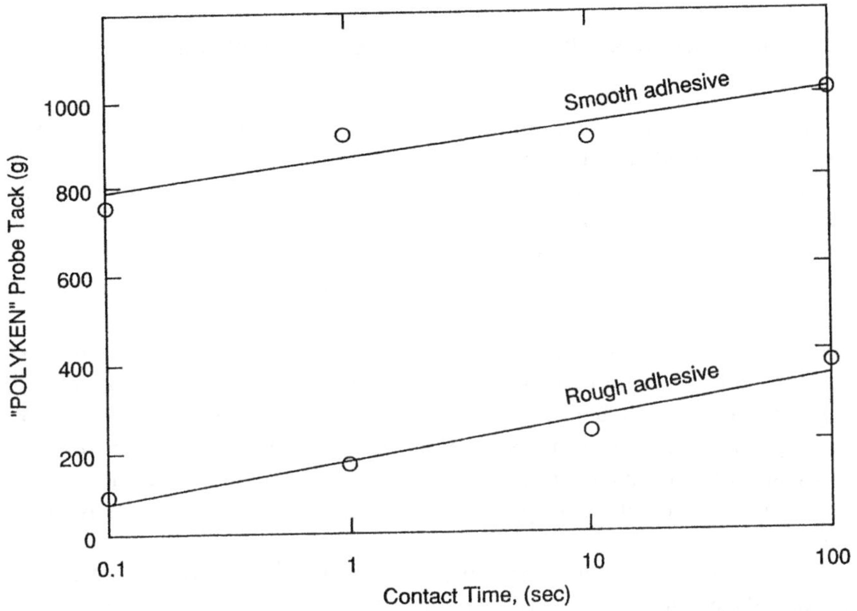

Fig. 4-4. Probe tack of smooth (film-backed) and rough (cloth-backed) pressure sensitive adhesive on a standard ½-cm-diameter stainless-steel probe, 100-g/cm^2 contact pressure, 1-cm/sec separation rate at room temperature.

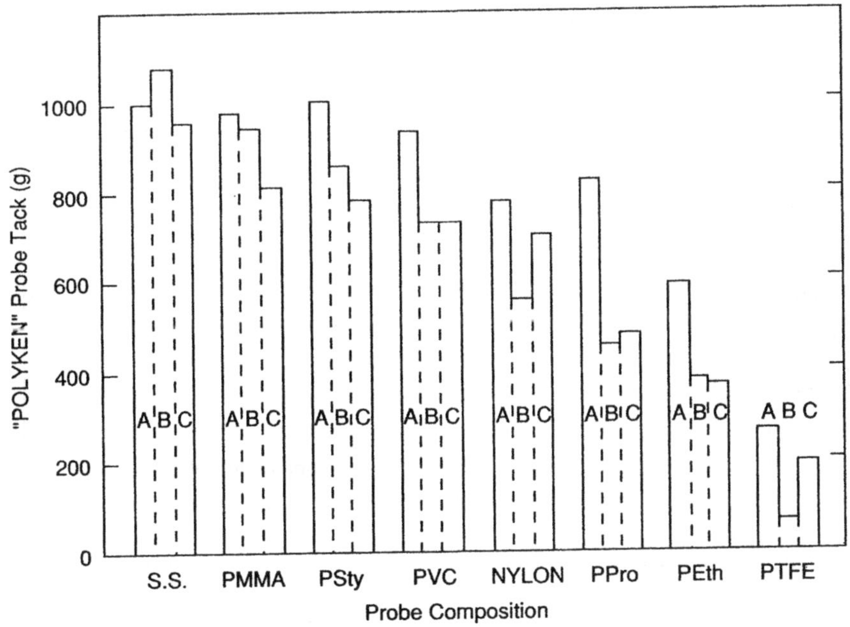

Fig. 4-5. Probe tack of three commercial adhesive tapes—A, B, and C—on ½-cm-diameter probe, 100-g/cm^2 contact pressure, 1-sec contact time at room temperature.

These three factors are the primary ones involved in bond formation. Two others are associated with breaking of the bond.

Probe tack can reveal how apparent bond strength increases with increasing rate of bond breaking. Probe tack values typically increase with the rate of separation of the probe from the adhesive, as illustrated in Figure 4-6.

While rate of bond formation can be expected to increase with increasing temperatures, probe tack values most often do not. As shown in Figure 4-7, the probe tack of a smooth surface adhesive decreases with increasing temperature. This decrease in probe tack is the consequence of the rate at which bond strength decreases with the temperature, being greater than the rate at which the nearly complete contact increases. Probe tack of the rough surface adhesive remains almost constant over this range of temperature, because the rate of increase in bond formation is approximately balanced by the decrease in bond strength.

No single tack value can characterize an adhesive because of the differences from one adhesive to another in rate of bond formation under different pressures, times, adherend surface roughness, and composition. Correspondingly, there is not just one probe tack value, since the several factors involved in formation can be varied to approximate those in a particular adhesive application. However,

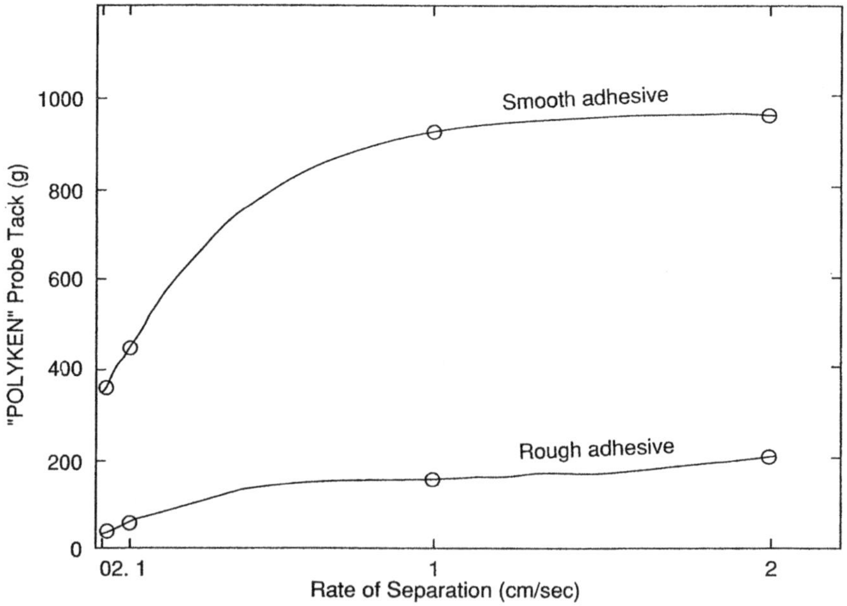

Fig. 4-6. Probe tack of smooth (film-backed) and rough (cloth-backed) pressure sensitive adhesive on a standard ½-cm-diameter stainless-steel probe, 100-g/cm^2 contact pressure, 1-sec contact time, at room temperature.

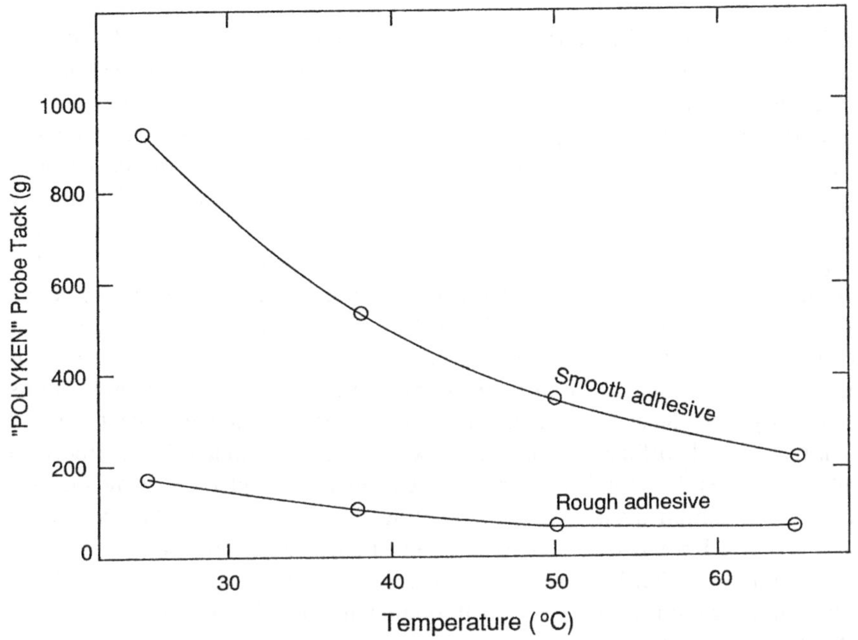

Fig. 4-7. Probe tack of smooth (film-backed) and rough (cloth-backed) pressure sensitive adhesive on a standard ½-cm-diameter stainless-steel probe, 100-g/cm^2 contact pressure, 1-sec contact time, 1-cm/sec separation rate.

there is an advantage in establishing a generally acceptable standard set of test conditions for comparing adhesives. The ASTM D14 "Pressure Sensitive Tack of Adhesives Using an Inverted Probe Machine"[1] prescribes a stainless-steel probe with a surface roughness of 2 μin. rms, a contact pressure of 100 g/cm^2, a contact time of 1 sec, and a rate of separation of 1 cm/sec as standard test conditions.

Irrespective of how pressure sensitive adhesive tack is defined or measured, it is ultimately the manifestation of the surface chemical and bulk physical properties of the adhesive interacting with those of an adherend. With a substantial degree of understanding, the rolling ball, peel, and especially probe tack tests can be interpreted in terms of surface energy, contact angle, wetting, modulus, loss tangent, etc.

SURFACE CHEMISTRY AND TACK

Pressure sensitive tack, and bond formation in general, ultimately involve molecular interactions at the adhesive/adherend interface. The major driving force toward bond formation is the surface chemist's "work of adhesion," which can be related to surface energies of the adhesive and adherend. Zisman's[25] concept of critical surface tension for a solid provides a basis for estimating surface energies and also understanding why low surface energy adhesives are desirable. Small drops of liquids placed on solid surfaces make characteristic angles of contact with the solid that depend on the liquid's surface tension. Plots of contact angles made by different liquids versus that liquid's surface tension, especially in the region of low contact angles, are most frequently straight lines. Zisman suggested, and it has been borne out since, that the surface tension values obtained by extrapolating these lines to zero contact angle are characteristic of the solid's surface. The critical surface energies of pressure sensitive adhesives can also be measured following the methods of Zisman, but care must be used in selecting test liquids to avoid complications of solubility, specific chemical reactions, and time dependency of contact angle.[26] Often the critical surface tension can be well enough estimated by reference to published tables[26–28] for low-energy plastic surfaces, and then using the one for the composition most closely resembling the adhesive or adherend. Table 4-4 contains a brief listing of polymers and their critical tensions as found in the literature. Gardon[29] showed how critical surface tensions may be related to the solid's chemical composition through Hildebrand's solubility parameter and cohesive energy density.

A simple liquid with a surface tension less than a solid's critical surface tension will make a zero contact angle with, spontaneously wet, and spread quickly over the solid suface. Adhesives with surface tensions less than the adherend's critical surface tension can also be expected to wet and spread onto a solid, even though this may require some period of time because of the adhesive's viscosity. Sharpe and Schonhorn[30,31] first recognized the significance of strong bond formation of the adhesive having a lower surface energy than the adherend. Pressure sensitive adhesives of hydrocarbon composition, because of their low surface energy, generally exhibit good tack to the widest variety of adherends.

Table 4-4. Critical Surface Tensions of Various Polymers

POLYMER	CRITICAL SURFACE TENSION (dyne/cm)
Polytetrafluoroethylene	18
Polypropylene	29
Polyethylene	31
Polystyrene	33
Polyvinylacetate	37
Polyvinyl chloride	39
Polyethylene terephthalate	43
Polyhexamethyleneadipamide	46

The surface chemistry that gives the greatest tack may not always have a critical surface tension less than the adherend but may be some higher value. Zisman pointed out that the maximum "work of adhesion" between simple liquids and solids occurred with liquids having surface tensions greater than the solid's critical surface tension. This circumstance should be true of adhesives and adherends as well. Huntsberger[23,32–34] has shown that the complete wetting of solid surfaces does not require zero contact angle, but can occur for liquids making contact angles up to 90 degrees. Huntsberger also reasoned that the greatest rate of wetting would be achieved by adhesives having surface tensions somewhat greater than the solid's critical surface tension. Toyama *et al.*[35] have observed experimentally a maximum in probe tack when the surface energy of the adhesive was slightly greater than that of the adherend.

RHEOLOGY AND TACK

While surface chemistries and molecular interaction energetics provide the driving potential, the adhesive's physical properties generally determine the rate and extent of contact (and bond strength).[36] Pressure sensitive adhesives differ from other materials, and among each other, primarily because of differences in their rheological properties. Rheological properties are the major factors in the phenomenon of pressure sensitive tack.

To have pressure sensitive tack, adhesives have to be easily deformed in the time scale allowed for bond formation, usually on the order of a fraction of a second. Dahlquist[37–39] found the 1-sec compliance of a typical pressure sensitive adhesive with good probe tack to be 10^{-6} cm^2/dyne. Toyama *et al.*,[18] and Kraus and others[40–42] found comparable values. Adhesives with good rolling ball tack need to be easily deformed in a time scale of about 0.001 sec. Tackifier resins, plasticizers, solvents, and temperature all contribute to compounded elastomer adhesive tack by increasing the ease of short time scale deformation.

For good tack, the adhesive deformation in the bonding stage has to be in large part viscous. It is essential to relax the stress created in the adhesive when it is made to conform to the topological irregularities of the adherend. If bonding deformation were only elastic and recoverable, the stored stress would assist external stress in the rupture of the bond. Pressure sensitive adhesives have a high ratio of viscous-to-elastic properties at time scales comparable to bonding times. In other words, a pressure sensitives adhesive's loss tangent should be maximum at bonding time scales. Consistent with this, pressure sensitive adhesives are characteristically very good absorbers of mechanical energy at audio frequencies.

Because tack is measured as the force or energy to break the bond, adhesives need to have a high modulus at the strain rates and elongation magnitudes imposed on them during rupturing of the bond. Tackifiers function by raising the glass transition temperature of elastomers so that the adhesive mixture has a high modulus at high strain rates and normal ambient temperatures. The difference between a plasticizer and a tackifier is that, whereas a plasticizer decreases modulus at all temperatures, times, and frequencies, a tackifier decreases modulus at high temperatures, long times, and low frequencies, but increases modulus at low temperature, short times, and high frequencies. Elastomers with large increases in modulus at some particular elongation also tend to make high tack adhesives. Natural rubber is the best example of such an elastomer.

Apparent tack and bond strength are greater when an adhesive has the ability to distribute stress throughout its volume, so that more of the adhesive is actively resisting bond rupture at any instant. Higher tack is associated with adhesives that have higher elongations to break in simple tensile strength tests. Tackier adhesives form longer "legs" during bond strength tests. High molecular weight and flexible backbone polymers provide this stress-distributing property.

COMPOUNDING AND TACK

The art of compounding or synthesizing pressure sensitive adhesives involves adjusting chemical and physical properties to give a balance of tack, peel adhesion, and cohesive strength (shear creep resistance). While it is not at all necessary to measure the rheological properties of the mixture or composition, such measurements can be extremely helpful. Empirical relations between tack and composition are readily established in simple "make it and test it" adhesive experiments. Probe tack tests are particularly well suited to empirical adhesive composition studies.

The probe tack of mixtures of an elastomer and a tackifying resin is simply related to composition. Wetzel[15] showed this in his original publication. Hammond,[21] Toyama and Ito,[24] and others using the Polyken Probe Tack Tester have reported this to be generally true. The probe tack of rubber-resin mixtures as a function of percent resin concentration gives the smooth curves in Figure 4-8. Hammond found that plotting probe tack as a function of resin content expressed as parts per hundred gave curves with two straight line sections as in Figure 4-9. In the region of low resin concentration, tack is unchanged from that of the elastomer. Above a characteristic point, tack increases with resin concentration, the rate of increase depending on the chemical type and melting point of the resin. At some critical concentration, also dependent on resin chemical type and melting point, tack values become erratic and abruptly drop to zero. Hock[43,44] observed that phase changes in the mixtures paralleled changes in tack. Marginal compatibility between elastomer and resin was postulated as the cause of tack.[4,45] However, fully compatible systems having tack

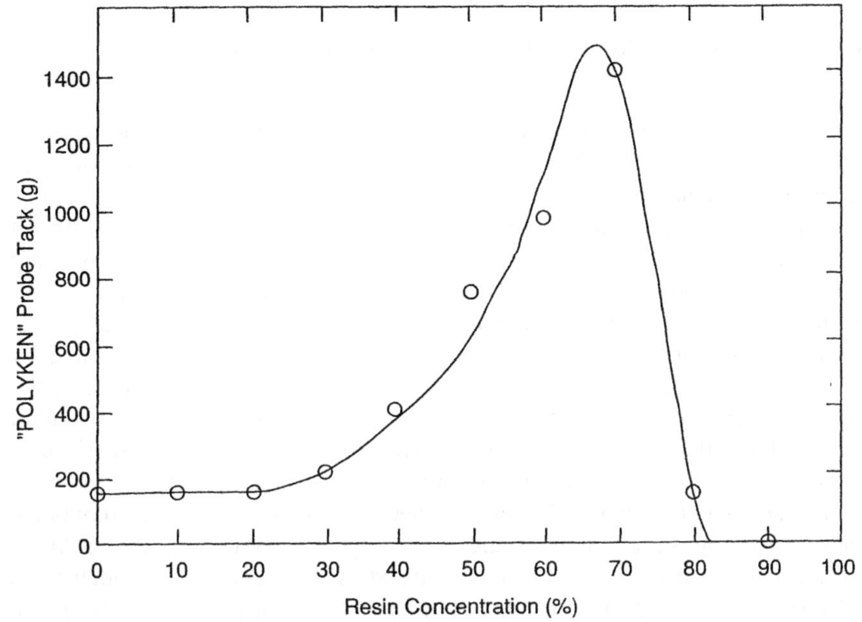

Fig. 4-8. Probe tack of natural rubber/Staybelite Ester No. 10 adhesives on a standard ½-cm-diameter probe, 100-g/cm^2 contact pressure, 1-sec contact time, 1-cm/sec separation rate at room temperature.

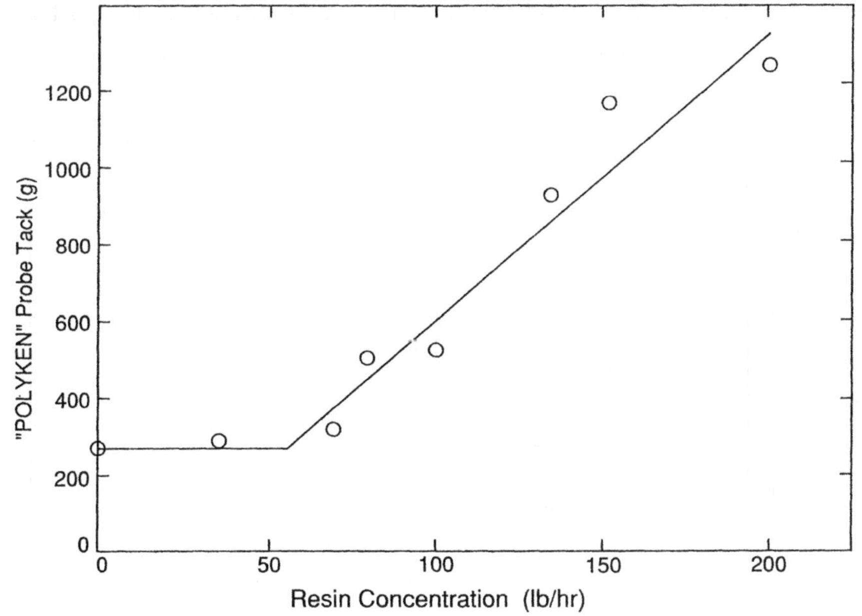

Fig. 4-9. Probe tack of natural rubber/Staybelite Ester No. 10 adhesives on a standard ½-cm-diameter probe, 100-g/cm^2 contact pressure, 1-sec contact time, 1-cm/sec separation rate at room temperature.

have since been observed. This postulate therefore applies only to some systems. More fundamental to tack than phase changes are the changes in the rheological properties of the adhesive.

The form of the relation between probe tack and composition illustrated in Figure 4-9 is quite typical of two-component mixtures of elastomer and resin.[46,47] Fortunately, tack/resin concentration relations determined in two-component systems carry over to three-component systems.[48] As one example, if an adhesive is made from two elastomers and a resin, both are tackified by the resin, and the tack of three-component mixtures will be the average of the two-component tacks at the same resin concentration. Such a system is natural rubber, SBR-1011, and Staybelite Ester No. 10, whose tack is summarized in Figure 4-10. As a second example, if an adhesive is made from two elastomers and a resin that tackifies only one of them, the tack of a three-component adhesive will be simply that of the two-component system of the tackified elastomer at the resin concentration calculated as if the second elastomer were not present. Such a system is natural rubber, SBR-1011, and a polyterpene resin with a high melting point, for which tack as a function of concentration is shown in Figure 4-11.

Tack sometimes can be disproportionately affected by small amounts of certain materials.[48] Stearic acid, lanolin, or paraffin wax added to a natural

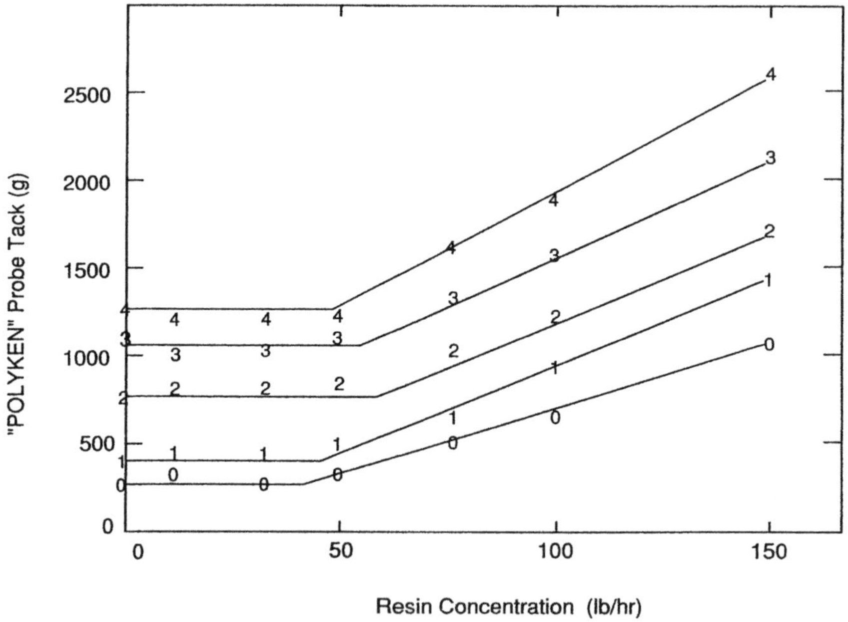

Fig. 4-10. Probe tack of natural rubber/SBR-1011/Staybelite Ester No. 10 adhesives. (0) NR only; (1) NR/SBR = 3/1; (2) NR/SBR = 1/1; (3) NR/SBR = 1/3; (4) SBR only.

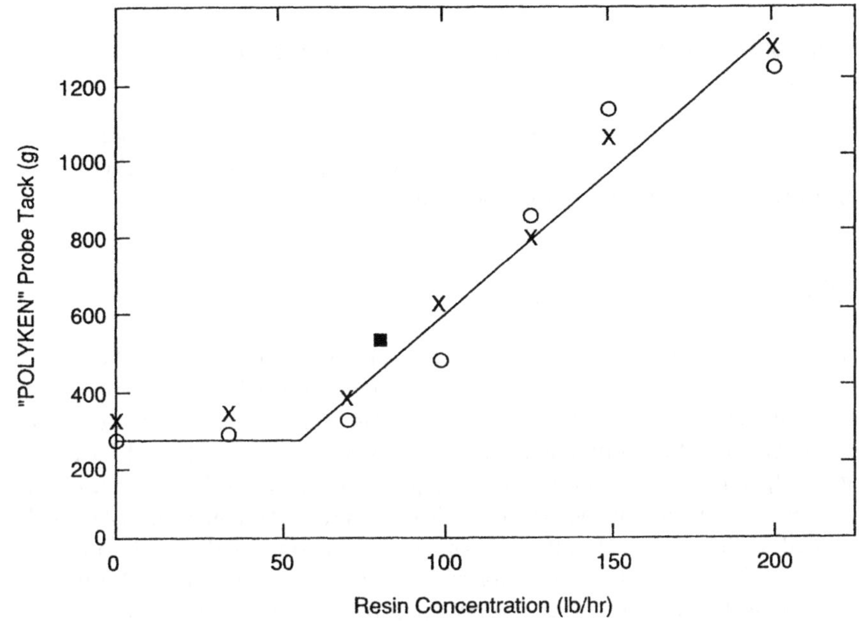

Fig. 4-11. Probe tack of natural rubber/SBR-1011/Piccolyte S115 adhesives. (O) Natural Rubber only; (x) NR/SBR = 1/1.

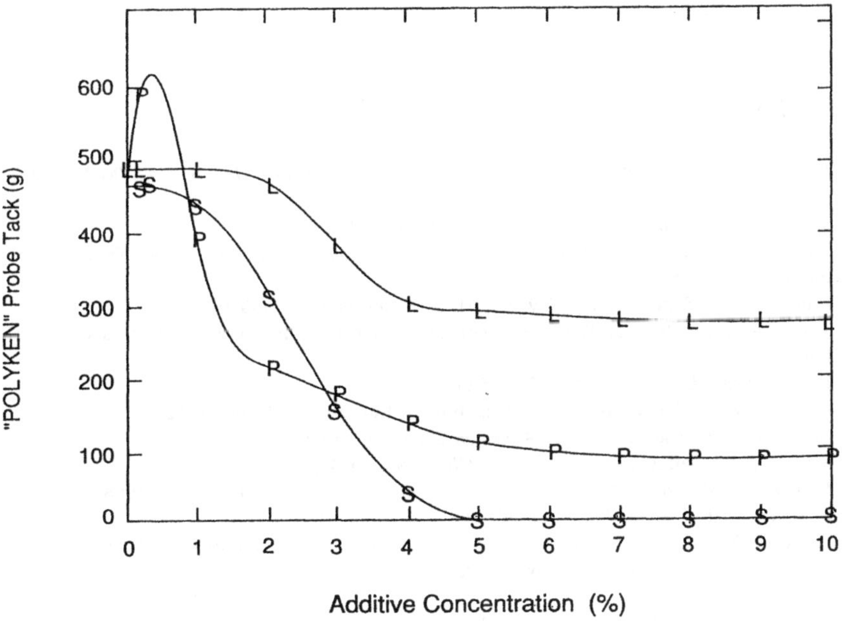

Fig. 4-12. Probe tack of natural rubber/Piccolyte S115 adhesives (100 lb/hr P S115) with (S) stearic acid; (P) paraffin (sealing wax); (L) lanolin.

rubber/rosin ester adhesive decreases the probe tack as shown in Figure 4-12. These small quantities of additives do not change to any extent the rheological properties of the adhesives; they modify the surface.

CONCLUSION

In the three decades from 1950 to 1980, pressure sensitive tack became a science of surface chemistry polymer rheology. The adhesive scientist now knows the chemistry that enables a bond to form and the viscoelastic properties that allow the bond to form quickly and yet be strong. Tack, and bond formation in general, is favored by the surface energy of the adhesive being comparable to, or lower than, that of the adherend. Tack requires the adhesive to be easily deformed at the time scale allowed for bond formation, and this deformation to be, to a large extent, energy absorbing and viscous in nature. Moreover, since appraisal of tack involves the breaking of the bond, tack is enhanced by a higher, very short time adhesive modulus and greater ultimate elongation to break. The formulator of commercial pressure sensitive adhesives still must look at a number of compositions, but he or she has insight into what materials to use, how much, and why.

REFERENCES

1. *Pressure Sensitive Tack of Adhesives Using an Inverted Probe Machine,* ASTM D2979–71 (1982).
2. *Standard Terminology of Adhesives,* ASTM D 907–82 (1985).
3. Johnston, J. Tack, in *Proc. Industry Technical Seminar.* Pressure-Sensitive Tape Council, Itasca, Ill. April 5–7, 1983.
4. Kamagata, K., Saito, T., and Toyama, M. *J. Adhesion* **2:** 279 (1970).
5. *Test Methods for Pressure-Sensitive Adhesives.* Sixth edition. Pressure-Sensitive Tape Council, Itasca, Ill.
6. Dow, J. *Proc. Inst. Rubber Ind.* **1:** 105 (1954).
7. Gillespie, R. H. *Rolling Ball Tack.* Unpublished Kendall Co. Report (1962).
8. Mizumachi, H., and Saito, T. *J. Adhesion* **20:** 83–97 (1986).
9. Voit, A., and Geffken, C. F. *Ind. Eng. Chem.* **43,** 1614 (1951).
10. Bull, R. F., Martin, C. N., and Vale, R. L. *Adhesives Age* **11:** 20–24 (May 1968).
11. Mizumachi, H. *J. Appl. Poly. Sci.* **30:** 2675–2686 (1985).
12. Druschke, W. Adhesion and Tack of Pressure-Sensitive Adhesives. Presented at the Association des Fabricants Europeens de Rubans Auto-adhesifs Meeting, Edinburgh, Scotland. October 1–4, 1986.
13. Chang, F. S. C. *Rubber Chem. Tech.* **20:** 847–853 (1957).
14. Brunt, N. *Rheol. Acta* **2:** 242–277 (1958).
15. Wetzel, F. Characterization of Pressure-Sensitive Adhesives. *ASTM Bulletin No. 221,* 64–68 (1957).

16. Wetzel, F. Function of Resin Derivatives in Pressure-Sensitive Adhesives. Presented at the *Am. Chem. Soc. Div. Paint, Plastics Ink Chem.*, Miami Meeting No. 1. 225–232 (1957).
17. Beatty, J. R. *Rubber Chem. Tech.* **42:** 1040–1053 (1969).
18. Kambe, H., and Kamagata, K. *J. Appl. Polym. Sci.* **13:** 493–504 (1969).
19. Midgely, A. Carboxylated Styrene Butadiene Rubbers in Water Based Pressure-Sensitive Adhesive Systems. *Pressure-Sensitive Tape Council Proc.*, Itasca, Ill. May 7–9, 1986.
20. Parry, S. A., and Ritchie, P. F. *Adhesives Age* **9:** 28–33 (November 1966).
21. Hammond, F. H. Jr. Polyken Probe Tack Tester. *ASTM Spec. Publ. 360* (1963).
22. Bates, R. *J. Appl. Polm. Sci.* **20:** 2941–2954 (1972).
23. Huntsberger, J. R. *J. Paint Tech.* **39:** 507 (April 1967).
24. Toyama, M., and Ito, T. *Polymer-Plast. Tech. Eng.* **2**(2): 161–229 (1973).
25. Zisman, W. A. in *Contact Angle, Wettability and Adhesion.* ACS Advances in Chemistry Series No. 1 (1964).
26. Dann, J. R. *J. Colloid Interface Sci.* **32**(2): 302–331 (February 1970).
27. Crocker, G. *J. Rubber Chem. Tech.* **42:** 30–69 (1969).
28. Toyama, M., Ito, T., Nukatsuka H., and Tikeda, M. *J. Appl. Polym. Sci.* **17:** 3495–3502 (1973).
29. Gardon, J. L. *J. Phys. Chem.* **67:** 1935 (1962).
30. Sharpe, L. H., and Schonhorn, H. *Int. Sci. Technol.*, 26 (April 1964).
31. Sharpe, L. H. and Schonhorn, H., in *Contact Angle, Wettability and Adhesion.* ACS Advances in Chemistry Series No. 43 (1964).
32. Huntsberger, J. R. *Chem. Eng. News,* 82–86 (November 1964).
33. Huntsberger, J. R. *J. Adhesion* **7:** 289–299 (1976).
34. Huntsberger, J. R. *Adhesives Age* (December 1978).
35. Toyama, M., Ito, T., and Moriguchi, H. *J. Appl. Polym. Sci.*, 2039–2048 (1970).
36. Kaelble, D. H. *Rubber Chem. Tech.* **45:** 1604–1622 (1972).
37. Dahlquist, C. A. Tack in *Adhesion, Fundamentals and Practice.* Ministry of Technology, Maclaren, London, 142 (1969).
38. Dahlquist, C. A. Tack, in *Proc. Nottingham Conference on Adhesives,* Sept. 22, 1966.
39. Dahlquist, C. A., in *Aspects of Adhesion,* Vol. **5,** 183 (D. J. Alner, ed.), Univ. of London Press, 1969.
40. Kraus, G., and Rollmann, K. W. *J. Appl. Polym. Sci.* **21:** 3311–3318 (1977).
41. Kraus, G., Jones, F. B., Marrs, O. L., and Rollmann, K. W. *J. Adhesion* **8:** 235–258 (1977).
42. Kraus, G., Rollmann, K. W., and Gray, R. A. *J. Adhesion* **10:** 221–236 (1979).
43. Hock, C. W. *J. Polym. Sci. Part C* **3:** 139–149 (1963).
44. Hock, C. W. *Adhesives Age* **7**(3): 21–25 (1964).
45. Kamagata, K., Kosaka, H., Hino, K., and Toyama, M. *J. Appl. Polym. Sci.* **15:** 483–500 (1971).
46. DeWalt, C. *Adhesives Age,* 38–45 (March 1970).
47. Wake, W. C. *J. Inst. Rubber Ind.* (April 1972).
48. Hammond, F. H. Jr. *Pressure-Sensitive Tack Measurements.* Unpublished Kendall Co. Report (1971).
49. Bauer, R. F. *J. Polm. Sci.* **A-2**(10): 541–48 (1972).
50. Bikerman, J. J. *J. Coll. Sci.* **2:** 163–175 (1947).
51. Chang, F. S. C. *Adhesives Age* **1:** 32–39 (November 1958).
52. Forbes, W. G., and McLeod, L. A. *Trans. Inst. Rubber Ind.* **35**(4): 154–174 (August 1958); *Rubber Age* **81:** 97 (1957); *Rubber World* **136:** 7676 (1957); *Rubber Chem. Tech.* **32:** 48 (1959).
53. Iyengar, Y., and Erickson D. E. *J. Appl. Polym. Sci.* **11:** 2311–2324 (1967).
54. Kambe, H., and Kamagata, K. *Adhesion Soc. Japan* **4** (4) (1968).

55. Kobe, H., and Kamagata, K. *J. Appl. Poly. Sci.* **13:** 493 (1969).
56. Sherriff, M., Knibbs, R. W., and Langley P. G. *J. Applied Polym. Sci.* **17:** 3423–3438 (1973).
57. Schonhorn, H. *J. Appl. Phys.* **37:** 4967 (1966).
58. Toyama, M., and Moriguchi, H. *J. Appl. Polym. Sci.* **14:** 2295–2303 (1970).
59. Uffner, M. W. *Adhesives Age,* 30–33 (December 1967).
60. Zosel. A. *Col. Poly. Sci.* **263,** 541 (1985).
61. Zosel, A. *Adhesion,* No. 3, 14 (1986).

5. Peel

Donatas Satas
Satas & Associates
Warwick, Rhode Island

One of the often measured characteristics of pressure sensitive adhesives is the resistance to peel that is determined by measuring the force required to peel away a strip of tape from a rigid surface. Resistance to peel is also called peel adhesion and peel force. The term peel adhesion is probably used most frequently, although objection could be made to the accuracy of such a term. Adhesion denotes a state in which two surfaces are held together, rather than the force required to break them apart. The peel force, as measured by the test, is a function of many factors, not only of the interfacial adhesive bond.

Peel resistance data can yield more information about the adhesive character and its expected performance than other commonly used tests of pressure sensitive products: tack and resistance to creep. However, peel resistance does not necessarily relate to the adhesive performance, and it should not be assumed that the peel resistance and the strength of the adhesive bond are synonymous. Peel is a special type of failure that might not relate to many other ways of failure that might take place in use.

MODE OF FAILURE

The standard peel tests are carried out at a constant peel rate, and it is expected that a pressure sensitive adhesive tape will strip cleanly from the adherend, leaving no visually noticeable residue. This type of failure is called adhesive failure, and it occurs at or near the adhesive–adherend interface.

Some adhesives may also fail cohesively, leaving adhesive residue on the test panel. If the adhesive has not firmly adhered to the backing, it may transfer to the test panel, leaving no adhesive on the backing. The magnitude of the peel force depends on the mode of failure, and it should be clearly stated if

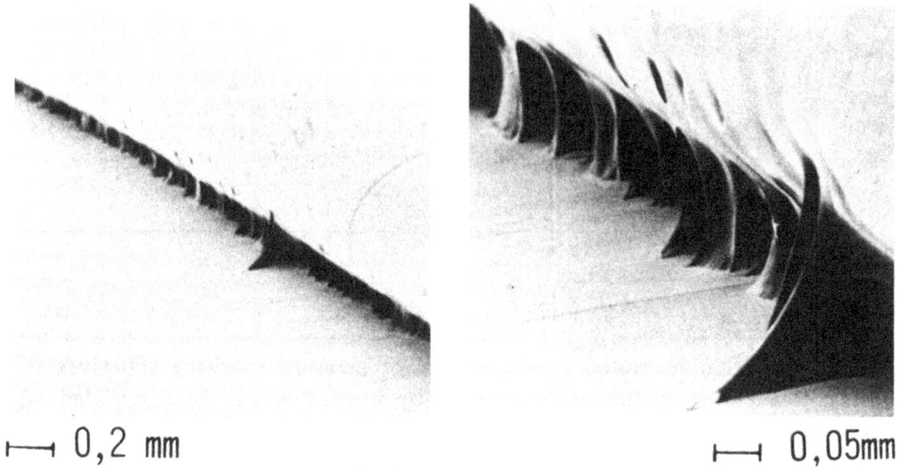

Fig. 5-1. Adhesive failure. (Courtesy of BASF A.G.)

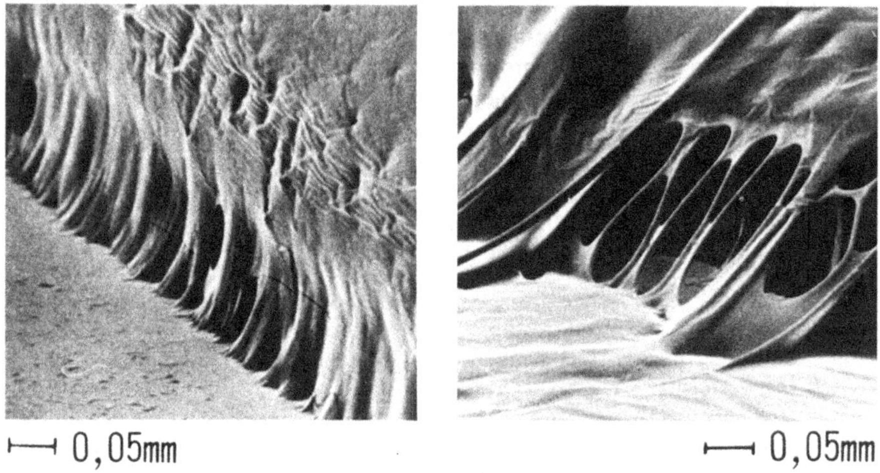

Fig. 5-2. Cohesive failure. (Courtesy of BASF A.G.)

the mode of failure is other than adhesive. Usually, failure other than adhesive is considered as a sign of a faulty product design. Figures 5-1, 5-2, and 5-3 show the adhesive, cohesive, and adhesive transfer modes of failure.

TESTING

There are several geometrically different combinations to test the resistance to peel of pressure sensitive adhesive tapes. The most common arrangements are

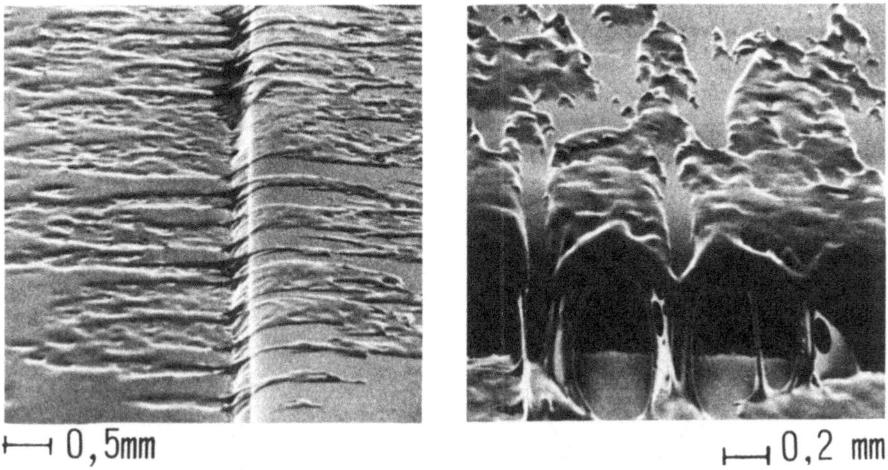

Fig. 5-3. Adhesive transfer mode of failure. (Courtesy of BASF A.G.)

shown in Figure 5-4. The test shown in Figure 5-4a, the 180-degree peel adhesion test, is the most frequently used to determine the peel force of pressure sensitive adhesive tapes. It is easy to carry out and yields fairly consistent results. The backing properties, however, have a larger effect on the test results than in other tests. Some tapes, such as glass filament reinforced tapes, may fracture when bent at a sharp angle of 180 degrees.

The 90-degree peel test is shown in Figure 5-4b. This test is used less frequently because it requires a special fixture, such as shown in Figure 5-5, to maintain the peel angle constant at 90 degrees throughout the test. The 90-degree peel test is preferred for theoretical analysis because it provides the best understood geometry. Buckling of the flexible member, sometimes experienced during the 180-degree peel test, may be eliminated in a 90-degree test. In both of these tests the tape is applied over a rigid steel panel.

The drum test (Fig. 5-4c) is rarely used, but if the drum is replaced by a roll of tape, then the test measures the unwind force, which is an important property of a pressure sensitive tape.

The T-peel test (Fig. 5-4d) measures the force required to separate an adhesive-to-adhesive bond. While it is of interest for some specialty tapes, this test is not used for the common characterization of the tapes. It is a useful quick test to check the anchorage of the adhesive mass to the backing. The T-peel test can also be used to measure the force to separate the tape from a flexible adherend. The test has been found useful for research studies on autoadhesion[1] and stress relaxation.[2]

The basis for most standard tests of pressure sensitive tapes is ASTM D1000–66 or later versions.[3] These tests have been specifically designed for

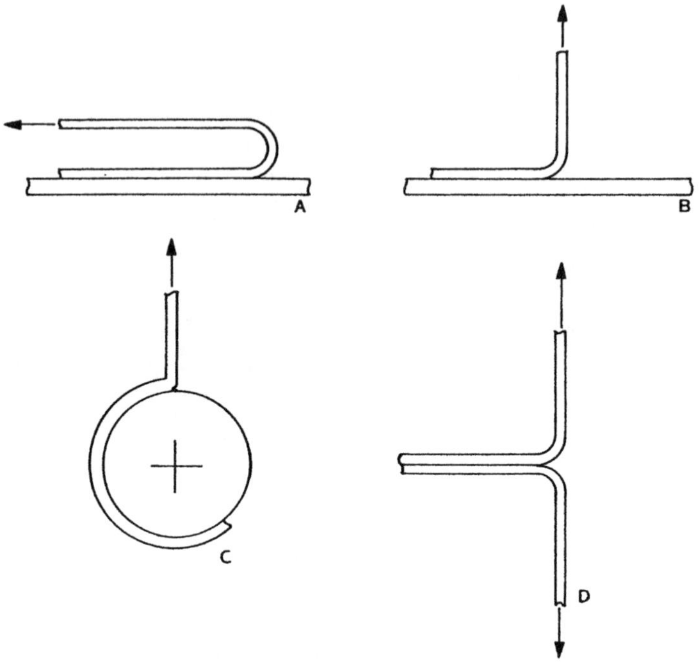

Fig. 5-4. Schematic diagram of various peel tests: (a) 180-degree peel; (b) 90-degree peel; (c) drum peel (tape unwind); (d) T-peel.

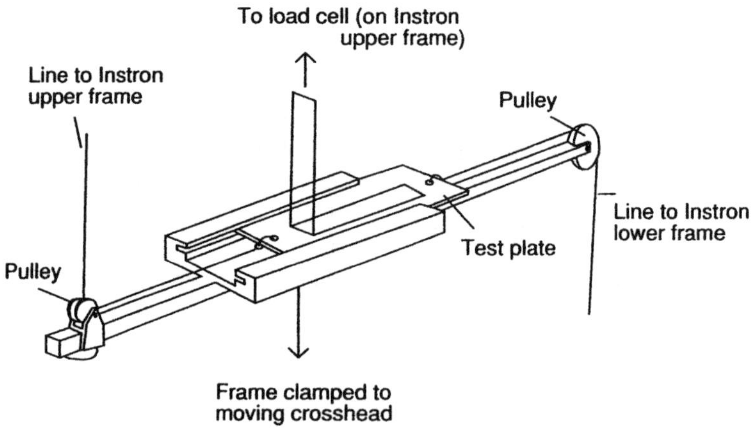

Fig. 5-5. Main features of 90-degree angle peel fixture.

electrical tapes, but most of them are also suitable for other tapes. Method D1000 contains 19 test methods, and about half are of interest for all pressure sensitive tapes. The procedures of interest for peel testing include sample preparation, adhesion to steel plate (180-degree peel test), drum adhesion (90-degree peel test), and other tests. The testing is specified to be performed at 0.5 cm/sec (12 in./min). While the driven jaw is moving at 0.5 cm/sec, the actual peel rate is, because of the geometry in the 180-degree peel test, 0.25 cm/sec. Stainless-steel plates finished to a surface roughness of 0.25–0.50 μm are used. The test results are reported in grams per centimeter or in ounces per inch width.

The T-peel test is described by ASTM D1876–61T and is used to check adhesive-to-adhesive bonds.[4]

The Pressure Sensitive Tape Council (PSTC), a manufacturers' trade association, has several tests for peeling: Peel Adhesion for Single Coated Tapes at 180° Angle (PSTC-1), Peel Adhesion for Double Coated Tapes at 180° Angle (PSTC-3), Adhesion to Liner of Pressure Sensitive Tapes at 180° Angle (PSTC-4), Unwind Adhesion (PSTC-8), High Speed Unwind Adhesion (PSTC-13), and Adhesion to Fiberboard at 90° Angle and Constant Stress (PSTC-14).[5] Some of these tests are similar to ASTM tests and have been adopted from ASTM procedures. Unwind adhesion (at 12 in./min) and high-speed unwind adhesion (at 200 ft/min) are important to determine the behavior of the tape in roll form. Adhesion to fiberboard is designed mainly for evaluation of the performance of carton closure tape. A constant force is applied in this test and the time to failure is measured.

Federal Test Method Standards 101 (Preservation, Packaging and Packaging Materials) and 147 (Tapes, Pressure-Sensitive and Gummed, Methods of Inspection, Sampling and Testing) are tests that involve the measurement of peel force. The National Electrical Manufacturers' Association (NEMA) has an interest in pressure sensitive tape testing, mainly for tape applications in original equipment. In addition, the Tag and Label Manufacturers' Institute (TLMI) has tests for 180-degree and 90-degree peel adhesion similar to the ones described.

European organizations A.F.E.R.A. (Association des Fabricants Européens de Rubans Auto-adhésifs) and FINAT (Fédération Internationale des Fabricants et Transformateurs d'Adhésifs et Thermocollants sur Papiers et autres Supports) also have 180-degree peel tests. They differ mainly in the time elapsed before the test is carried out. While the PSTC test allows a maximum of 1 min between the tape application and the start of the test, A.F.E.R.A. requires a maximum of 10 min and FINAT has two tests with time periods of 20 min and 24 hr. In addition the FINAT test is carried out employing a glass surface instead of the stainless-steel surface used in other tests.

Table 5-1. Peel Adhesion as a Function of Contact Time

CONTACT TIME	PEEL ADHESION (*N*/2 cm)
10 min	4.0
30 min	4.6
1 hr	5.3
3 hr	5.5
24 hr	8.0

The contact time has a significant influence on the peel test results as shown in Table 5-1.[6]

Peel tests are fairly reliable and the data scattering is less than in the case of many other tests used in pressure sensitive adhesive evaluation. The Table 5-2 shows the statistical evaluation of various tests.[6]

The peel force value reported as a single number at standard testing conditions does not carry as much information as the peel data presented as a function of peel rate or peel temperature in a form of a curve.[7]

Various special tests have been devised. A peel test employing a dead load is sometimes used to evaluate pressure sensitive label stock, and a schematic diagram of such test is shown in Figure 5-6. A 1.25-cm (½-in.) -wide strip is adhered to the surface of a square bar. The bar is placed in a holder that keeps it inclined to give a peel angle of 70 degrees. A 200-g load is applied and the time to failure is recorded.

Diefenbach[8] describes an apparatus used to measure the peel force of low adhesion pressure sensitive tapes, such as is used for securing the postage stamp coils. A peel force of about 30 g/cm is expected of such tapes.

An interesting way to measure the peel force has been described by Šalkauskas and Paulavičius.[9] A triangular cut is peeled at 90 degrees by applying a dead load. The peel rate decreases because the width of the peeled tape or

Table 5-2. Statistical Evaluation of Adhesion Tests

TEST	NUMBER OF SAMPLES	MEAN VALUE	STANDARD DEVIATION	COEFFICIENT OF VARIATION (%)
Peel adhesion after 10 min	25	3.8 *N*/2 cm	0.31 *N*/2 cm	8
Peel adhesion after 24 hr	42	9.3 *N*/2 cm	0.54 *N*/2 cm	5
Quick stick	50	10.2 *N*/2 cm	0.34 *N*/2 cm	3
Probe tack	43	11.3 N/cm^2	1.24 N/cm^2	11
Toothed wheel tack	39	0.7 *N*/cm	0.140 *N*/cm	18
Rolling ball tack	50	3.9 cm	1.46 cm	38
Zosel tack	25	13.9 J/m^2	4.67 J/m^2	34

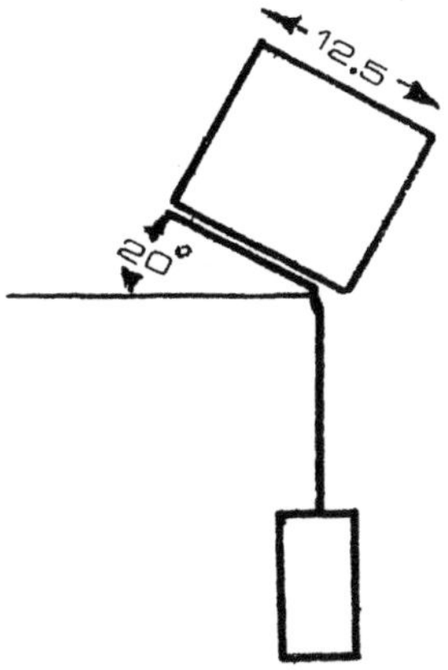

Fig. 5-6. Peel test for label stock.

coating increases until equilibrium is reached and the peeling ceases. The bond strength is measured by dividing the dead load applied by the width of the tape at equilibrium point. The values so obtained are somewhat lower than the values obtained by a regular 90-degree peel test.

The hot peel test is an important performance test that indicates the ability of tape, especially masking tape, to hold the masking aprons when baked in a hot oven. The test is performed by peeling the tape from a stainless-steel panel at a rate of 10 cm/min while a blast of hot air at 160°C impinges upon the tape.[10]

RESPONSE TO STRESS

Pressure sensitive adhesives are viscoelastic materials exhibiting both viscous and elastic response to stress. The nature of the response depends on the rate of force application and on temperature. This variation of adhesive response to stress is illustrated in the generalized peel-force–peel-rate curve shown in Figure 5-7. The dotted line at low peel rates represents the cohesive failure. The adhesive breaks up into legs (Fig. 5-2) that elongate and fail cohesively. The viscous flow response dominates in this region. The stress at the adhesive–adherend interface is insufficiently high to cause an adhesive failure. The energy is expanded instead in deforming the adhesive. A relatively large volume of the adhesive is under stress and this accounts for high peel resistance values in the region of cohesive failure. In the cases of very high molecular weight or cross-linked adhesives, the cohesive failure might not be observed at all.

The solid line in Figure 5-7 represents the adhesive failure at the adhesive–

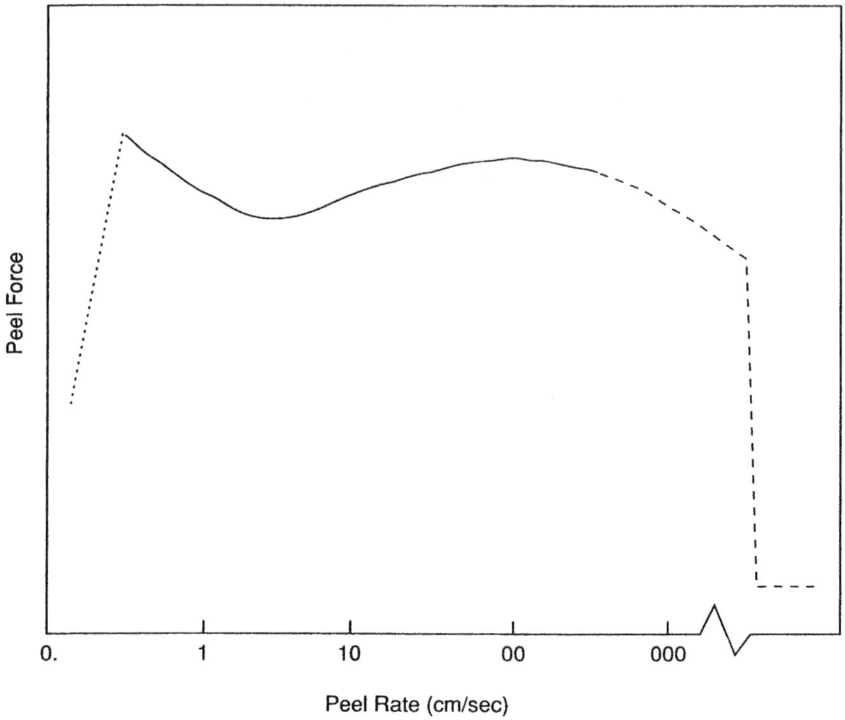

Fig. 5-7. Generalized peel force vs. peel rate curve.

adherend interface. At normal peel rates, this type of failure is expected from pressure sensitive tapes and labels.

At high peel rates adhesives exhibit so-called "stick-slip" failure. The peel force oscillates in a periodical manner between a higher and a lower peel-force value. This failure is denoted by the striped portion of the curve in Figure 5-7. The transition from a smooth adhesive failure to a raspy and jerky failure has been observed and called a yield point by Hendricks and Dahlquist.[11] At very high peel rates the peel resistance drops and the response is that of a glassy, high modulus material. Very high peel rates are required to induce a glassy failure to pressure sensitive adhesives at room temperature; however, this type of failure is easily demonstrated at low temperatures.

Various mechanical models have been suggested to describe the behavior of pressure sensitive adhesives in peel. A single Voigt element and two or multiple Maxwell elements have been proposed by Hata, a single Maxwell element by Fukuzawa,[12] and a more complex Burger's model by Aubrey.[13] (See Fig. 5-8.) The dashpot represents a viscous response to stress, the spring an elastic response, and the hook in the Burger's model the interfacial adhesive

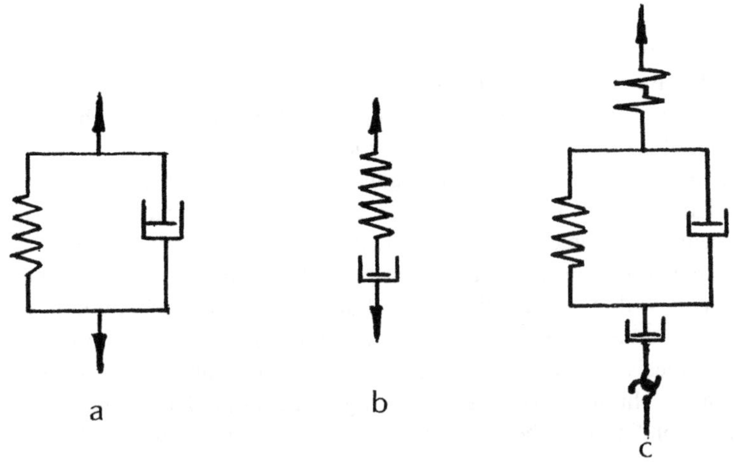

Fig. 5-8. Models of pressure sensitive adhesive tape in peel: (a) Voigt element; (b) Maxwell element; (c) Burger's model.

bond. Many different elements contribute to the peel strength: energy to deform the backing, both elastic and viscous responses in the bulk adhesive, and the interfacial forces between the adhesive and the adherend. There is no easy way to separate the contribution by each of the elements, and the peel force, as measured by the tests, is the result of the combined effect of these factors. It is erroneous to assume that the peel adhesion is mainly the function of the interfacial bond between the adhesive and adherend.

TRANSITION FROM COHESIVE TO ADHESIVE FAILURE

The transition from cohesive to adhesive failure in pressure sensitive adhesives and other polymeric materials has been discussed by many investigators. McLaren and Seiler[14] have shown that low molecular weight polyvinyl acetate adhesives fail cohesively in a peel test, while higher molecular weight materials exhibit adhesive failure. Similar behavior was also observed at elevated temperatures. The peel force showed a continuous increase with increasing temperature until the mode of failure changed from interfacial to cohesive. At that point, the peel force started to decrease with increasing temperature. Huntsberger[15] has shown similar behavior for poly(*n*-butyl methacrylate) adhesive. When the temperature decreases, the relaxation rate becomes slower than the rate of force application, and the stress concentration occurs at the edges, resulting in an interfacial failure. Further lowering of temperature causes a decrease in peel force because of a higher stress concentration.

This transition in pressure sensitive adhesives was first discussed by Bright.[16] Figure 5-9 shows some of these data. Segment C of the curve represents clean adhesive failure. Segment B represents the transitional period that takes place in a narrow temperature range, and segment A represents a cohesive failure within the adhesive mass. The effect of the molecular weight of acrylic pressure sensitive adhesives on the cohesive–adhesive failure transition point has been discussed by Satas[17] and the effect of cross-linking by Satas and Mihalik.[18] More recently this topic has been interestingly discussed by Aubrey.[13,19–21]

It is generally agreed that the transition from cohesive to adhesive failure is associated with the transition from liquidlike viscous to rubberlike elastic behavior,[13,19–23] although there have been other mechanisms proposed as well.

The adhesive bond is difficult to break at or near this transition region: the adhesive exhibits considerable legging and a large deformation is required to cause debonding or cohesive failure. Such pressure sensitive adhesives serve

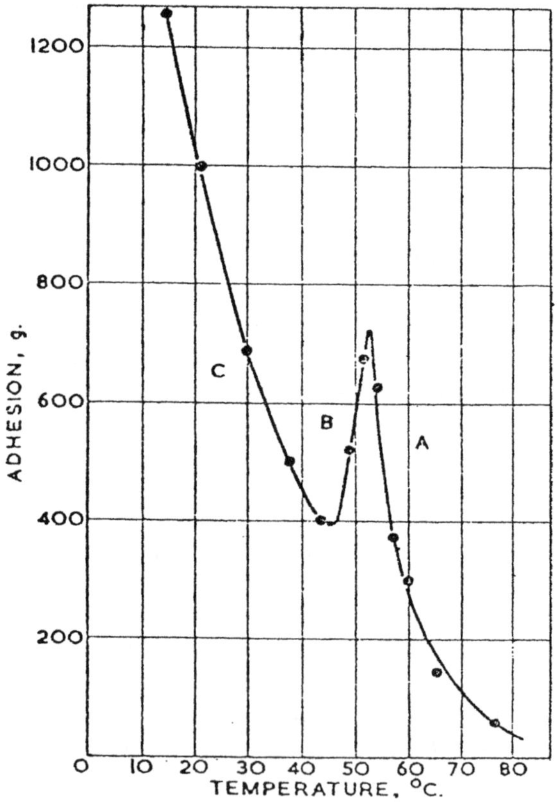

Fig. 5-9. The dependence of peel force on temperature showing the transition from adhesive to cohesive failure.

well for medical applications: they can extend and conform to the skin movement without debonding[24] (see also Chapter 25), they serve well as laminating adhesives for flexible materials, as adhesives that are applied to vibrating or flexing surfaces, and as tapes where the end flagging must be prevented. Gent and Petrich[23] have also pointed out that high adhesive strength is obtained from the materials that have large deformation energies within the stress limit set by the interfacial bond strength.

Gent and Petrich[23] have discussed the quantitative treatment of peel at low peel rates. The peel force P is given approximately by the sum of tensile stresses f in the elements of the adhesive layer up to the point of failure.

$$P = \int_0^{y_m} f\,dy + h\int_0^{e_m} f\,de \tag{1}$$

where

y = distance through which an element of the adhesive is stretched
e = corresponding fractional extension
y_m and e_m = maximum values obtained
h = thickness of the adhesive layer.

Equation (1) is based on two main assumptions: the substrate film is completely flexible, and the elements of the adhesive can be treated as independent.

Equation (1) can be rewritten in terms of stress limits

$$P = h\int_0^{f_m} f\,de \tag{2}$$

where f_m = maximum stress attained.

Equation (2) predicts that the peel strength depends on the deformation energy of the polymer up to the point at which the tensile stress is large enough either to break the adhesive bond or to rupture the polymer.

TRANSITION TO OSCILLATING FAILURE

At high peel rates, usually above 100 cm/sec, a second transition of the peel force is observed from a smooth adhesive failure to a failure in which the peel force is oscillating in a nonrandom manner. The peel-force versus peel-rate curve has a negative slope in that region, as shown by the striped line in Figure 5-7.

This behavior has been observed and discussed by many researchers. In pressure sensitive adhesives it has been discussed by Hendricks and Dahlquist[11] and more recently by Aubrey.[13,20,21] Greensmith[25,26] in the studies of tear proper-

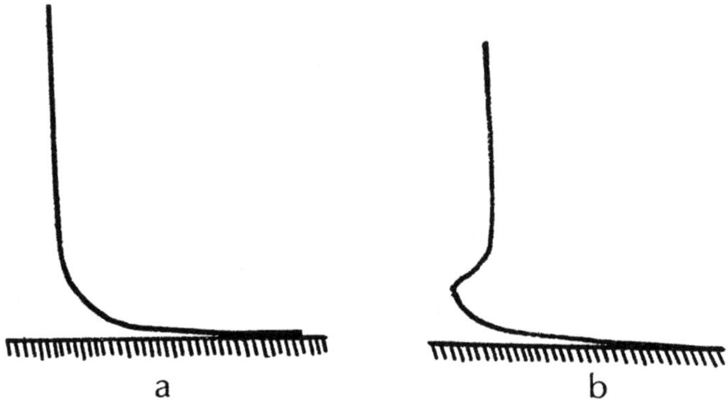

Fig. 5-10. The configuration during stick-slip unwind (according to Duke[30]).

ties of rubber has reported steady and also nonrandomly oscillating stick-slip forces. Similar behavior in peel tests of adhesive coatings has been reported by Dannenberg,[27] Gardon,[28] Lannus and Zdanowski,[29] and Duke.[30] This transition has been attributed to a change from elastic to glassy response, or to localized plastic yielding.

Such peel behavior at high speed rates is often observed during the unwind of pressure sensitive tape from a master roll. The unwind is accompanied by a noise, which, according to Duke,[30] is caused by a flapping of free tape at the point of separation. The tape configuration alternates between that shown in Figures 5-10a and 5-10b. The frequency of the sound is approximately that expected from the determined frequency of kinks found in the peeled tape.

The tape unwind has been studied on the equipment shown in Figure 5-11. The tape is unwound at a high speed at essentially a 90-degree angle. The point of failure is observed to fluctuate between points x_0 and x_1. When the point of failure travels from x_0 to x_1, the tape is peeled at a slower rate and

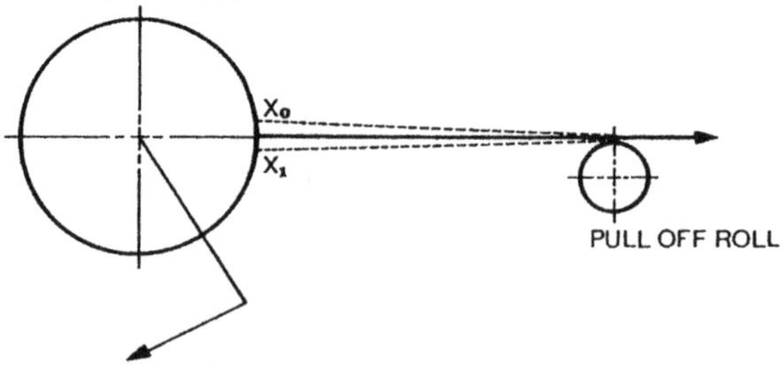

Fig. 5-11. Schematic diagram of tape unwind test.

part of the energy is absorbed elastically by the adhesive. When the point of failure reaches position x_1, the stored energy is released and the point of failure jumps quickly from x_1 back to x_0, developing a loose loop as shown in Figure 5-10b.

The pattern of peel-force variation depends on the mode of failure. In the range of cohesive failure, the peel force varies greatly in a random manner around some mean value (Fig. 5-12a). The magnitude of these variations decreases after the transition to adhesive failure. The more legging adhesive exhibits, the larger the variations. Peel-force variation for an adhesive exhibiting considerable legging is shown in Figure 5-12b. As the peel rate increases, the legging decreases, and the random variations of the peel force become much lower (Fig. 5-12c). In the region of stick-slip failure of the peel, force starts to fluctuate periodically, as shown in Figure 5-12d. The wavelength and the amplitude of

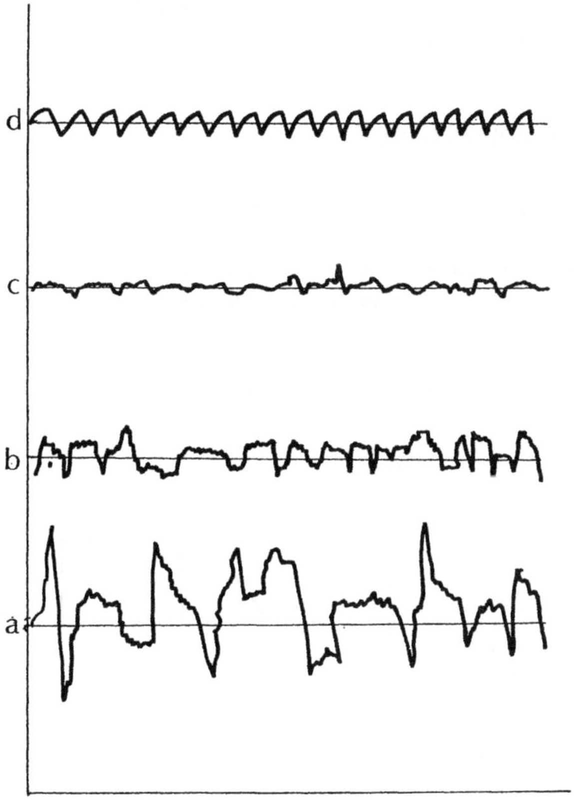

Fig. 5-12. Various peel-force variations observed during the tape reel and unwind: (a) cohesive failure range; (b) leggy adhesion; (c) rubbery adhesion; (d) stick-slip failure range.

Table 5-3. Electrostatic Charges Developed During Tape Unwind

UNWIND RATE (cm/sec)	PEEL FORCE (g/cm width)	TYPE OF OSCILLATION	ELECTROSTATIC CHARGE (V)
8	106	Random	0
13	98	Random	500
21	89	Random	1000
32	80	Nonrandom	0–1000
43	89	Nonrandom	0–2000

such variations increases with increasing peel rate. Oscillations of the peel force correspond to the oscillations of the fracture locus that travels between points x_0 and x_1 shown in Figure 5-11.

It is well known that an electrostatic charge is generated during tape unwind. It was observed that during the steady unwind the electrostatic charge developed is maintained at a constant level; during a nonrandomly oscillating force period, the electrostatic charge also oscillates. Table 5-3 shows the observed data.[31]

EFFECT OF TEMPERATURE

The effect of temperature on peel force is equivalent to that of peel rate. An increase in temperature shifts the peel curve (Fig. 5-7) to the right; the same effect could be achieved by reducing the peel rate. This is well illustrated in Figure 5-13, which shows the adhesion of butadiene-styrene copolymer to polyester film.

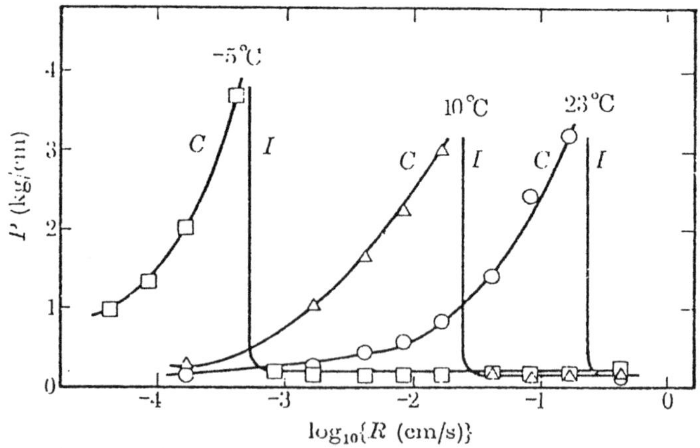

Fig. 5-13. Average peel force P against rate of peel R for butadiene-styrene copolymer against polyester film (Gent and Petrich[23]).

Peel adhesion data obtained at various temperatures can be reduced to a reference temperature according to the well-known WLF equation.[32] The measured peel force P is reduced by the standard convention[33] to a reduced peel force PT_0/T, and the peel rate r to a reduced rate ra_T, where $T_o = 296$ K and a_T is defined as shown below:

$$\log a_T = 9.8 - \frac{17.44\,(T - T_g)}{51.6 + T - T_g} \tag{3}$$

where $T_g = 230$ K.

The results of this reduction of data obtained by Kaelble[22] are shown in Figure 5-14. Such curves when plotted on normal coordinates log P versus log r at constant temperature yield curves similar to the peel-force–peel-rate plot shown in Figure 5-7.[34] The effect of temperature on the peel has been discussed by many other researchers: Huntsberger,[15] Hendricks and Dahlquist,[10] and Aubrey and Sherriff.[21]

The investigation of peel behavior as a function of temperature is an interesting tool to obtain information about pressure sensitive adhesives. Such tests

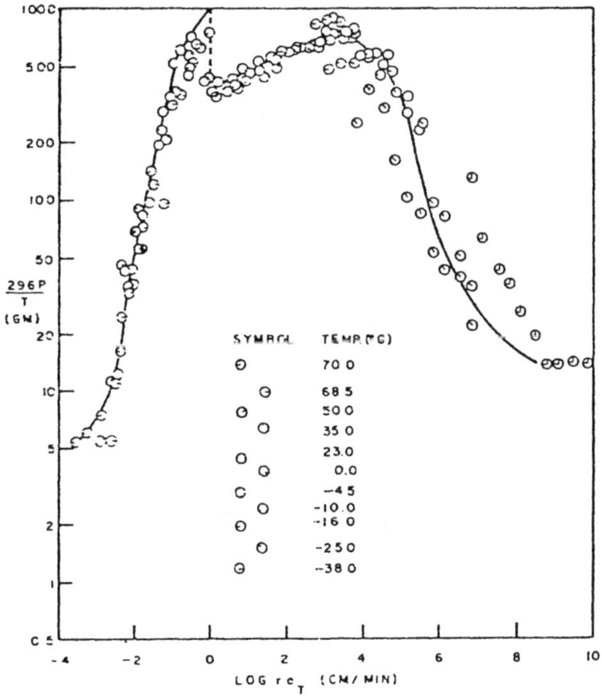

Fig. 5-14. Temperature-reduced peel adhesion to polystyrene (Kaelble[22]).

can be carried out by employing a bar over which a temperature gradient is established. One end of the bar is heated and the other is cooled. Heating is accomplished by insertion of a heater into a hole drilled through the bar. Cooling is done by circulating cold water through a hole provided for that purpose at the other end of the bar. Stainless-steel plates used for peel tests are attached to the bar by clamps. The peel test is started at the cooler portion of the bar and continued toward the increasing temperature. A schematic diagram of such a bar is shown in Figure 5-15. The temperature at which the transfer from adhesive to cohesive failure takes place is called the splitting temperature, and it is a useful measure of the adhesive. Table 15-2 in Chapter 15 shows a group of pressure sensitive acrylic polymers of various molecular weights. The splitting temperature clearly follows the molecular weight variations as well as the creep resistance of the adhesives.

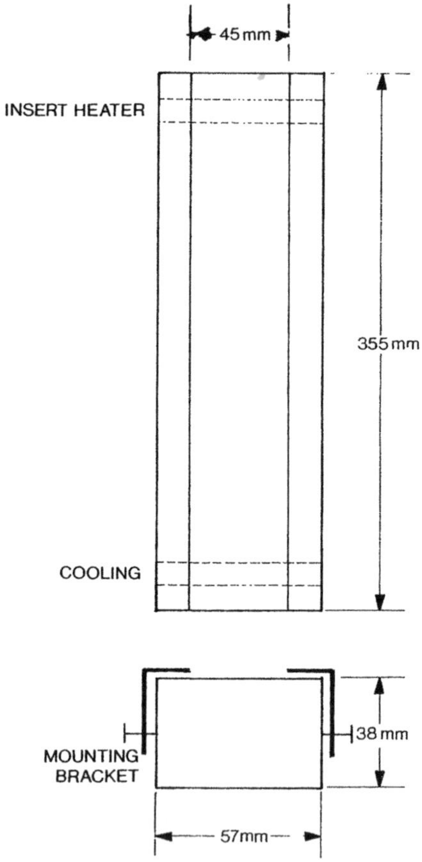

Fig. 5-15. Schematic diagram of temperature gradient bar.

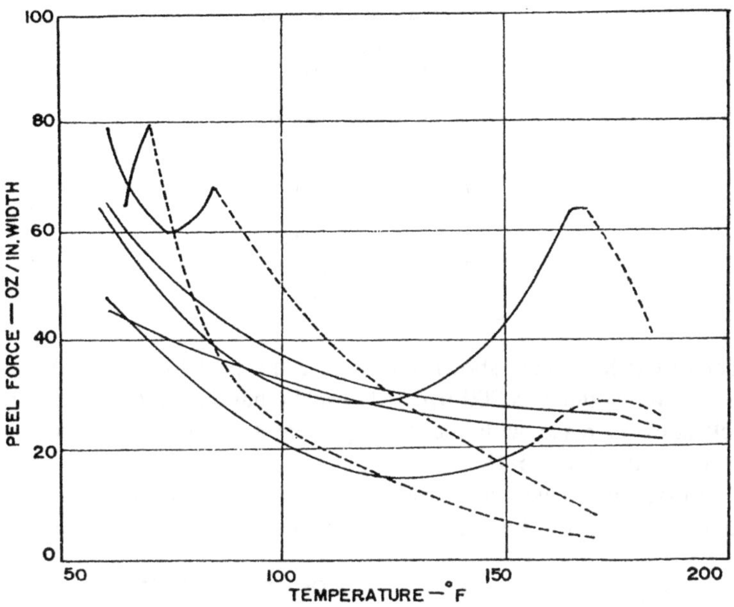

Fig. 5-16. Peel force of adhesives with various degrees of ionic binding. Sodium content: *curve top right,* 0.500%, 0.34%, 0.69%, 0.18%; *curve bottom right,* 0% (Satas and Mihalik[36]).

The peel force versus temperature are different for polymers with different degrees of cross-linking, or hydrogen and other secondary bonding. An increased degree of cross-linking increases the splitting temperature. Acrylic polymers containing carboxylic groups were treated with Na ions in order to affect ionic binding.[18] Figure 5-16 shows the peel force versus temperature curves for polymers with various degrees of ionic binding. The dotted line denotes cohesive failure.

Gerace[35] has presented T-peel data as isocleaves at varying peel rates and temperatures for an acrylic and a butyl rubber adhesive.

EFFECT OF ADHESIVE THICKNESS

It is known that peel adhesion increases with increasing adhesive thickness and levels off at some thickness, usually in the range of 50 μm (2 mil). Table 5-4 shows the peel adhesion data reported by Chan and Howard,[36] where the peel adhesion appears to level off at 30 μm (1.2 mil). Similarly Johnston[37] has reported a leveling of the peel force at about 50 μm (2 mil), as shown in Figure 5-17. The dependence of peel force on adhesive thickness has been discussed by many other researchers.[6,28,38–42]

Table 5-4. Variation of Peel Strength with Adhesive Thickness

THICKNESS (μm)	PEEL STRENGTH (N/m)
7.6	201–15
10.2	323–91
15.2	511–11
22.9	565–18
30.5	560–52

Other researchers have shown data where the peel force was affected by adhesive coating as thick as 200 μm.[43] Druschke has shown that peel adhesion and quick stick increase with the coating weight beyond a thickness of 25 μm,[6] as shown in Figure 5-18.

The stresses are transmitted through the adhesive to the backing. The adhesive is deformed and the force required to deform the adhesive contributes to

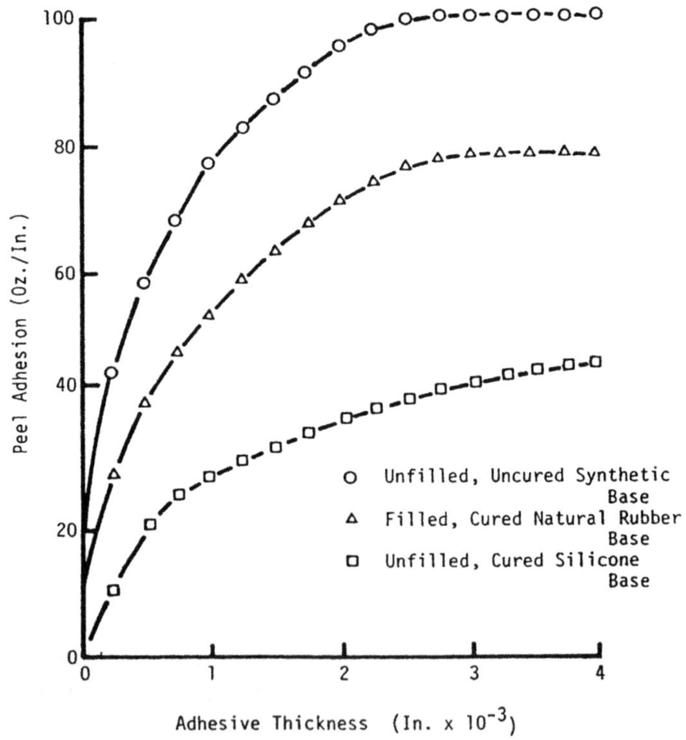

Fig. 5-17. Effect of adhesive thickness on peel force. Backing: 12 polyester film.

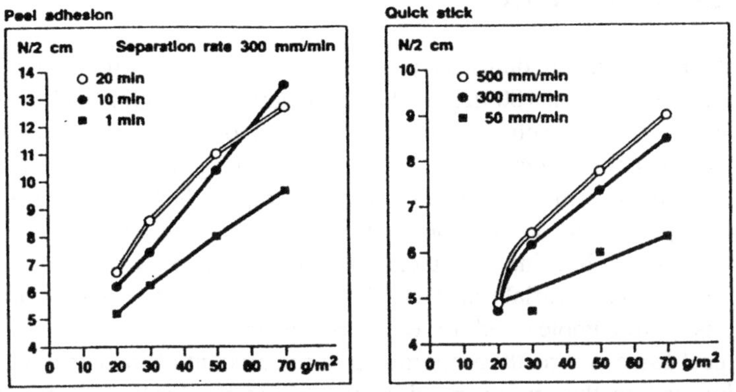

Fig. 5-18. Adhesion as a function of coating weight. (Courtesy BASF A.G.)

the measured peel force. The deformation is higher near the adhesive–adherend interface, and it becomes less severe at the points further away from the locus of failure. Eventually an adhesive thickness is reached where no further deformation takes place, and thus the contribution to the total peel force becomes negligible. Thus the leveling of the peel force will be reached at a lower adhesive thickness for firmer, higher modulus adhesives than for softer adhesives.

The adhesive thickness has a twofold effect on the peel force. Heavier adhesive coating increases the volume of adhesive under deformation, thereby increasing the peel force. It also increases the total thickness of the tape, and the effect is similar to that of increased backing thickness, which causes a lowering of the peel force, especially at 180-degree peel test conditions.

Increase of the adhesive thickness causes a shift of the cohesive–adhesive failure transition in the peel-force–peel-rate curve (Fig. 5-7) to higher peel rates.

A number of analytical expressions have been proposed showing the dependence of peel force on the adhesive thickness. Gardon[40] has shown that at high thickness values

$$P = \alpha t_a^{0.25} \tag{4}$$

where

P = peel force
t_a = adhesive thickness
α = parameter dependent on substrate thickness, adhesive, and substrate moduli

At low adhesive thickness peel force is proportional to $t_a^1 - t_a^{0.25}$. Similarly, Bikerman[41] has shown that $P \sim t_a^{0.25}$.

EFFECT OF THE BACKING

Backing thickness and modulus have an effect on the peel force that is more pronounced in tests run at 180-degree than 90-degree peel angle. Backing thickness determines the geometry in the failure area, has an effect on the amount of the adhesive under stress at any given time during the peeling, and also affects the moment arm (Fig. 5-19). The backing modulus has an effect on the energy expanded to deform the backing and, therefore, on the resistance to peel. The type of backing, whether film, paper, foil, or woven textile, also has an effect on these variables. Thus the combined effect of the backing on peel force is quite complex and not easily predictable.

The peel force is directly proportional to the amount of the adhesive under deformation and to the energy required to deform the backing. Both factors increase with increasing backing thickness. The peel force is inversely proportional to the acting moment arm, which increases with increasing backing thickness as shown in Figure 5-19. Since $P_1 l_1 = P_2 l_2$ and $l_1 < l_2$, P_1 must be larger than P_2. Because the increase of backing thickness affects the peel force in several opposing ways, it is expected that a plot of peel force versus backing (or total) thickness will exhibit a maximum at some thickness value. This indeed is the case as shown in Figure 5-20, which shows the data of 180-degree peel tests. The effect of the moment arm is less pronounced in the case of 90-degree peel angle tests as shown in Figure 5-21.[44] Curve A in Figure 5-19 shows a very clear maximum. The bending radius of polyester film is highly dependent on the film thickness and the effect of the moment arm is important. Polyurethane elastomer film (curve D) elongates when peeled, decreasing the effect of backing thickness. Similar results have been reported by Johnston[37] and are shown in Figure 5-22. The continuing increase of the peel force for dead soft aluminum foil tape is said to be due to the effect of the energy required to deform the backing.

The contribution to the work of detachment arising from the plastic yielding of the backing strip is discussed by Gent and Hamed.[45] Aubrey *et al.*[38] have reported peel force data for various backing thicknesses and for various peel rates. Polyester film bonded with poly(n-butyl acrylate) was used. At low peel rates where the failure was cohesive, the peel strength was unaffected by the

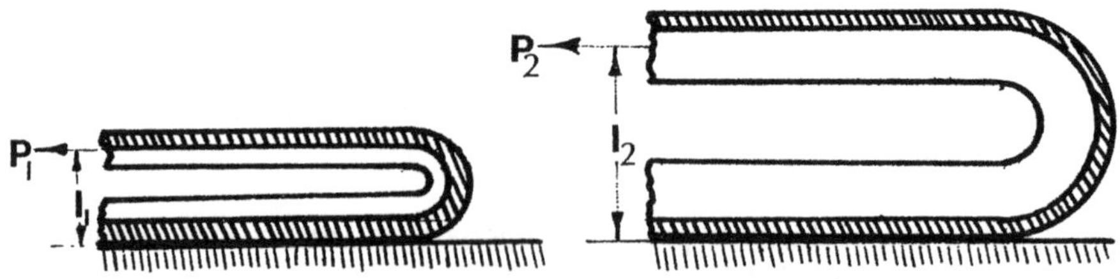

Fig. 5-19. The dependence of moment arm on backing thickness.

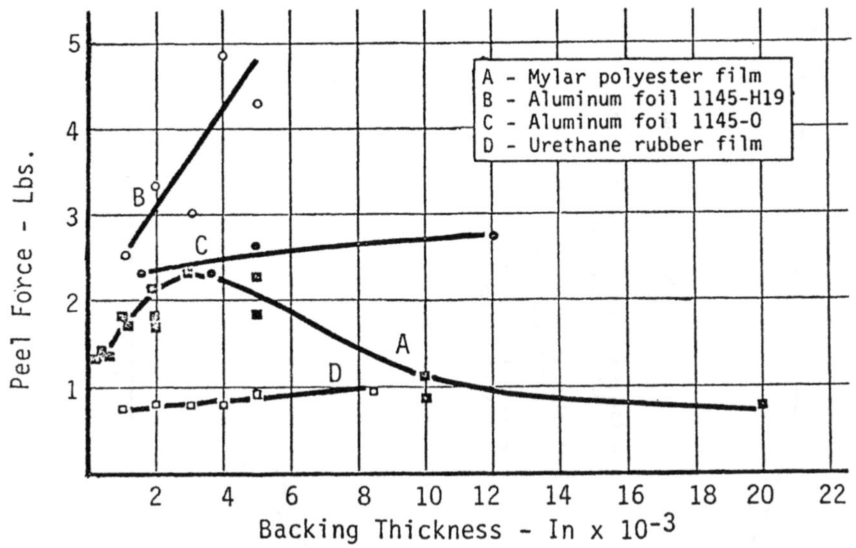

Fig. 5-20. The dependence of peel force on the backing thickness in a 180-degree peel test. (Reprinted from *Adhesive Age* **9**(8) (1966).)

backing thickness in the range of 0.015–0.052 mm. The peel strength decreased with higher backing thickness.

Duke[30] suggests that four different relations exist between the peel force and adherend thickness. The peel force may be proportional to T^0, $T^{2/3}$, $T^{3/2}$, and the relation may be controlled by the increasing moment arm. At low

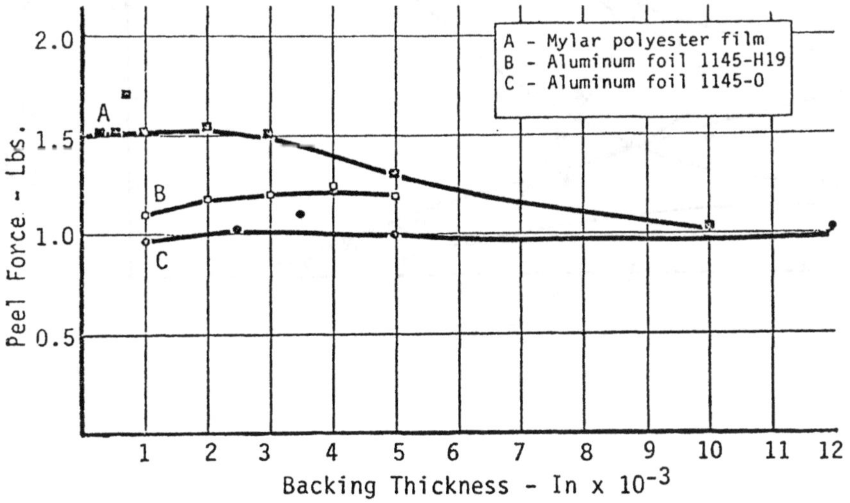

Fig. 5-21. The dependence of peel force on the backing thickness in a 90-degree peel test. (Reprinted from *Adhesive Age* **9**(8) (1966).)

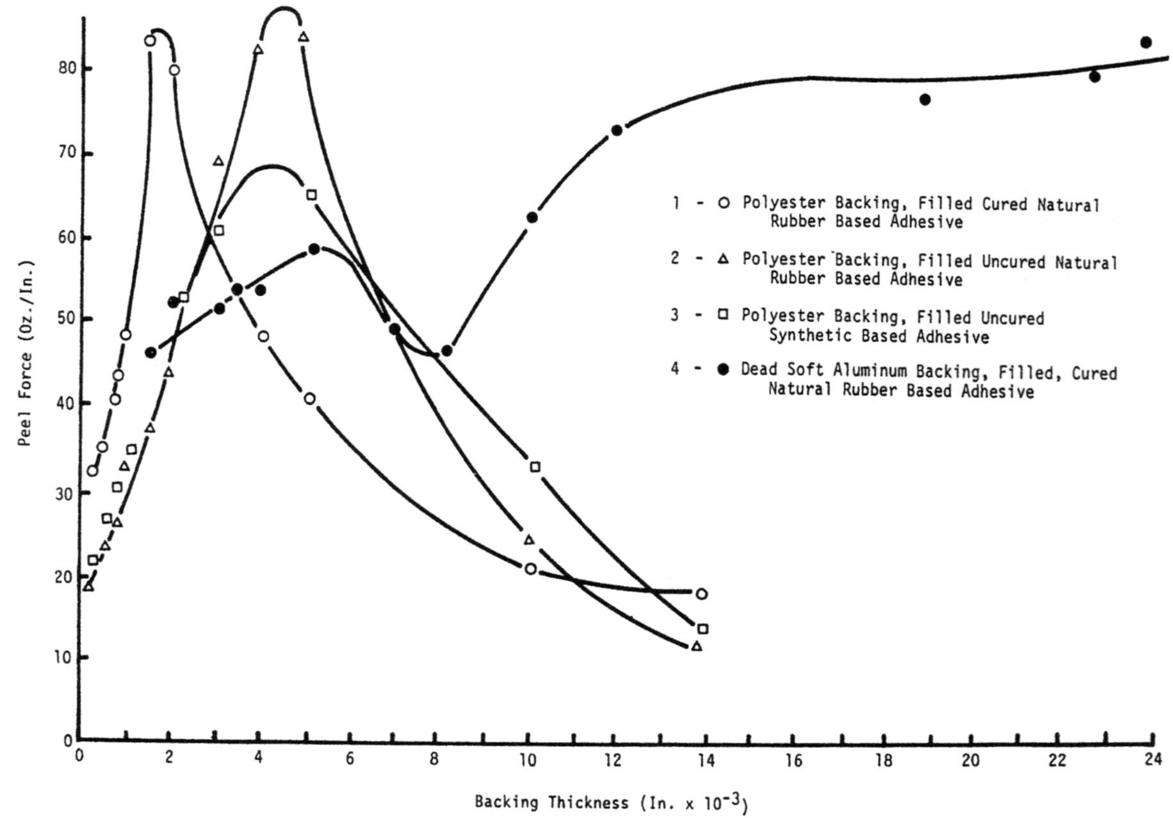

Fig. 5-22. Relationship between peel force and backing thickness for polyester and dead soft aluminum backing tapes. (Reprinted from *Adhesives Age* **11**(4) (1968).)

adherend thickness with pronounced adhesive legging during peel, the peel force is independent of backing thickness. In the region following that, at higher backing thickness, legging is decreasing and $P \sim T^{2/3}$, where P is the peel force and T is the thickness of the adherend. In the region of truly elastic failure, $P \sim T^{3/2}$. The increase of P with T is ultimately limited by either the inadequacy of adhesive bond or, more commonly, by the ability of joint configuration to impose increasing moment arms as peel or cleavage begins.

The surface of the backing material also has an effect on the peel behavior. It has been reported that roughening of the cellophane film surface in a cellophane film-acrylic binder system causes a change of the mode of failure from nonrandomly oscillating to steady peel.[28] A nonrandomly oscillating force is usually not observed when the adhesive is peeled off a rough, fibrous textile surface. Apparently, a smooth surface is needed to propagate the fracture. The backing thickness, while affecting the peel force level, did not appear to affect the transition point significantly.

EFFECT OF PEEL ANGLE

Peel tests are usually carried out at either 180 or 90 degrees. In actual use conditions, however, the tape may be subjected to a peel at various other angles. Kaelble has suggested that the peel force (P) should vary inversely as ($1 - \cos \omega$), where ω is the peel angle. Some of the experimental data are qualitatively in agreement with this relation, as shown in Figure 5-23.[46] Other data show

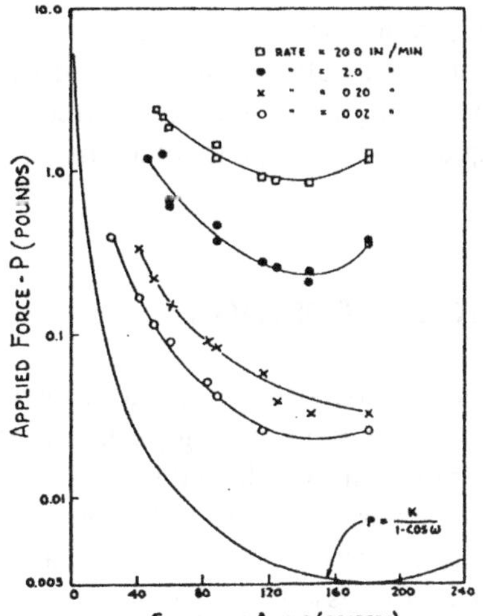

Fig. 5-23. Peel force vs. peel angle for polyester tape. (Reprinted from *Transactions of the Society of Rheology* **III**: 161–180 (1959).)

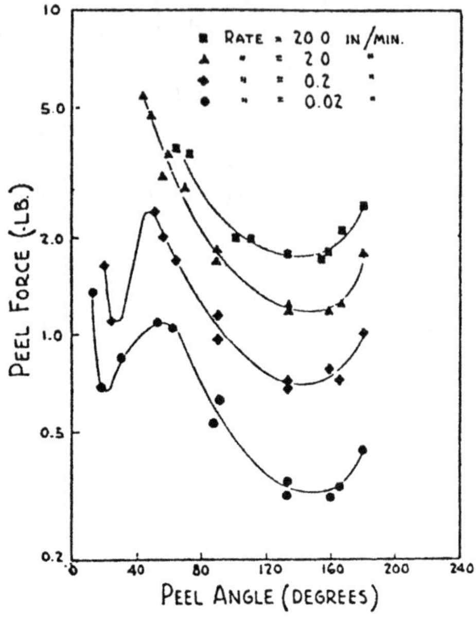

Fig. 5-24. Peel force as a function of peel angle for a glass cloth tape. (Reprinted from *Transactions of the Society of Rheology* **IV**: 45–73 (1960).)

discrepancies, such as maxima-minima observed at 30–40-degree peel, as shown in Figure 5-24.[47] These discrepancies are attributed to the transition of the failure mechanism from cleavage to boundary shear. All tapes displayed a minimum of peel force at angles of 120–140 degrees.

The effect of peel angle has also been discussed by Gardon.[48] The effect of peel angle, especially peeling at 90 and 180 degrees, on the deformation of the backing and consequently on the peel force, has been discussed by Gent and Hamed.[45]

EFFECT OF ADHESIVE COMPOSITION

Bulk adhesive properties, that is, storage and loss moduli, have an effect on peel adhesion, since most of the work during peeling is expended in deforming the adhesive. Increasing molecular weight and increasing degree of cross-linking are expected to cause a decrease in peel adhesion. The effect of molecular weight in case of acrylic adhesives is shown in Table 15-2. The effect of molecular weight was also discussed by Aubrey.[13,38] Figure 5-25 shows 90-degree peel data of poly(butyl acrylate) adhesive of varying molecular weight and degree of cross-linking.[13] Increase of molecular weight increases the modulus

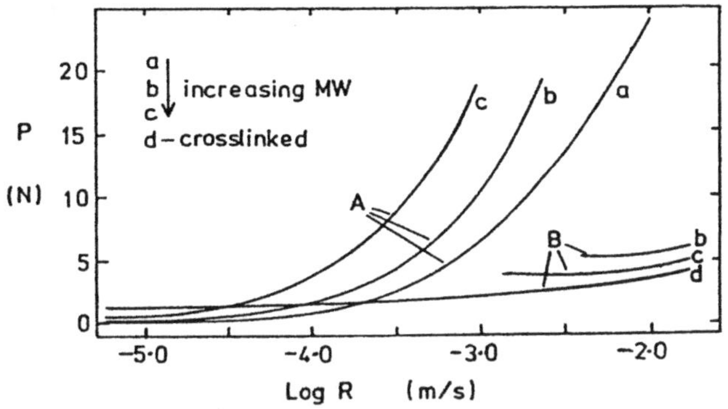

Fig. 5-25. Peel force vs. peel rate for a polybutyl acrylate. (A) Cohesive failure; (B) adhesive failure. (Reprinted from reference 12, p. 25).

of elasticity and decreases the polymer chain mobility. This causes an increase of peel force in case of cohesive failure, but a decrease in the region of adhesive failure. Harder, less extensible adhesive results in a smaller adhesive volume being deformed at any time during peeling and a lower peel force. The increase of molecular weight has a most noticeable effect on the splitting temperature: transition temperature from adhesive to cohesive failure as shown in Figure 15-3.

Figure 5-26[49] shows peel force data of two 0.1-mm-thick soft aluminum foil tapes. Tape A has a soft uncross-linked acrylic polymer adhesive, while tape B has an adhesive of similar composition, but cross-linked and considerably more elastic. It is obvious that these two curves represent different sections of the generalized peel-force–peel-rate curve shown in Figure 5–7. Curve A is the steady peel and curve B is the nonrandomly oscillating peel-force portion of the generalized curve. The solid line denotes steady peel force, and the discontinous line denotes oscillating peel force.

Figure 5-27[49] shows the unwind of polyethylene coated fabric tapes. Tape C has a soft natural rubber-resin adhesive; tape D, a cross-linked natural rubber-resin adhesive; tape E, an acrylic polymer adhesive. The data illustrate the shift of the transition point to the higher peel rates with decreasing elasticity of the adhesive.

The effect of hydrogen bonding on the peel force is similar to that of cross-linking and is discussed in Chapter 15 on acrylic polymers. It has been shown that the introduction of hydrogen bonding via acrylonitrile comonomer increases polymer stiffness. Figure 15-9 shows the effect of acrylonitrile on penetrometer compliance, illustrating its stiffening effect. The peel behavior of these polymers is shown in Figure 5–28.[49] The decrease of peel force and the shift of the transition point is affected by the increasing acrylonitrile content.

Mao and Reegan[50] have shown the dependence of peel force on the type

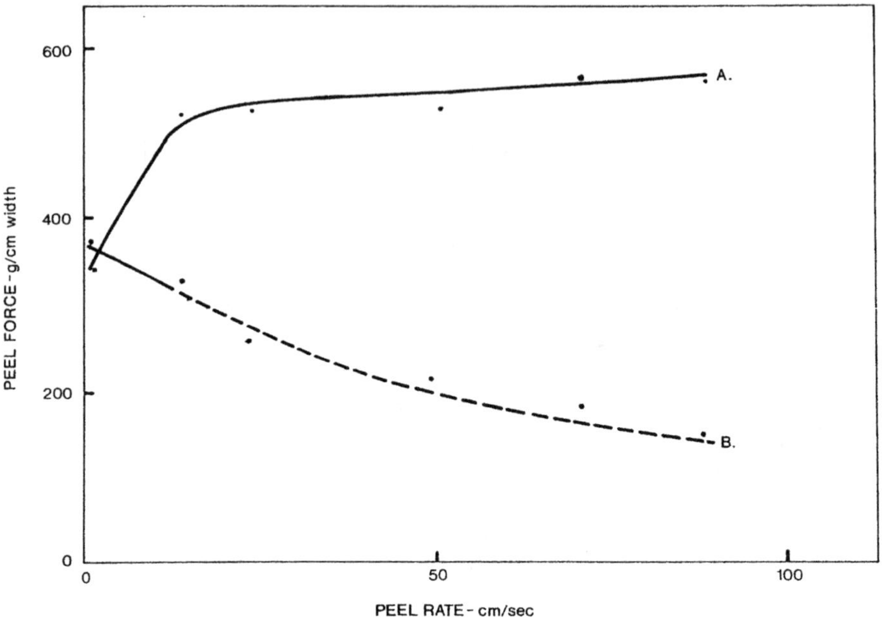

Fig. 5-26. Peel force aluminum foil tapes with two different adhesives.

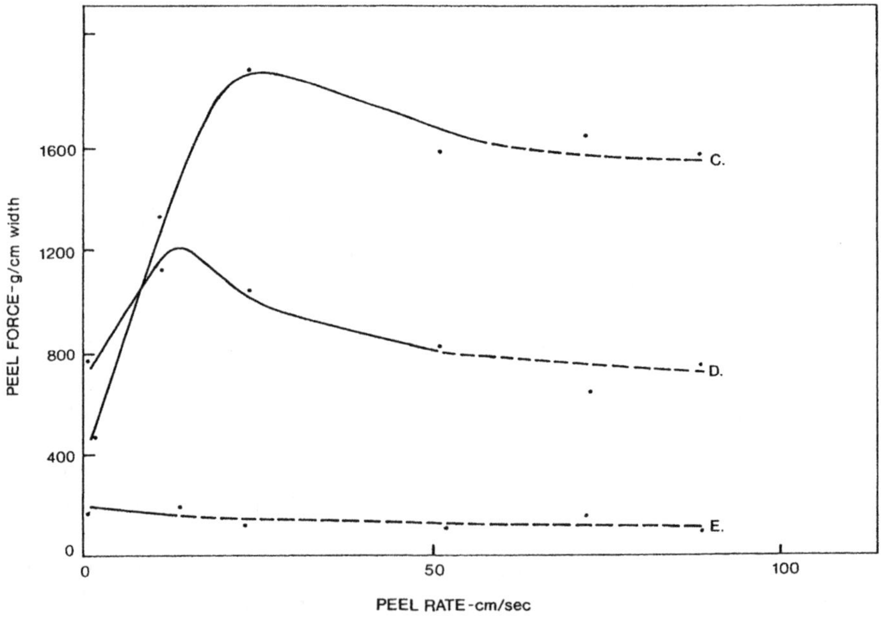

Fig. 5-27. Peel force of polyethylene coated fabric tapes with different adhesives.

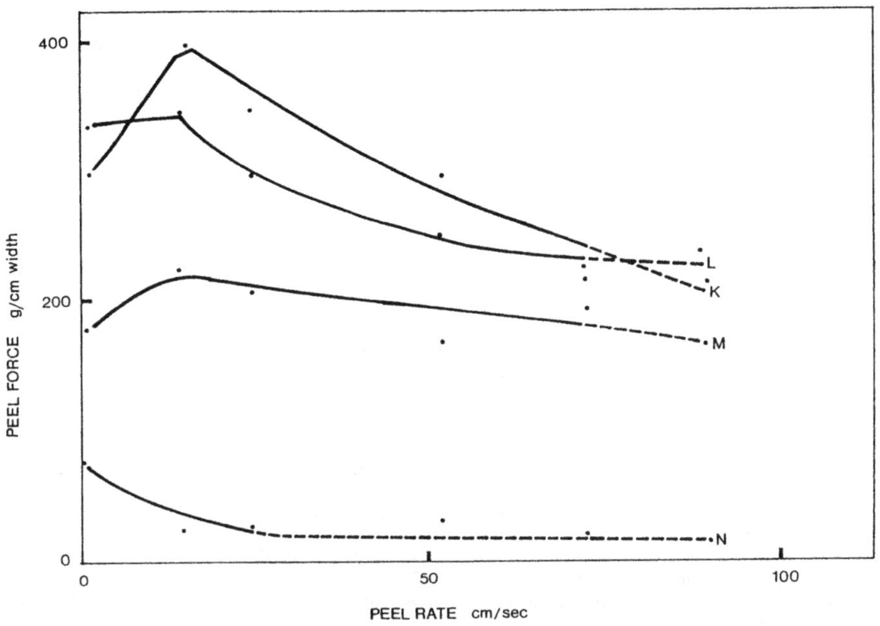

Fig. 5-28. Effect of acrylonitrile on peel force. Amount of acrylonitrile: (K) 0%; (L) 1%; (M) 2.5%; (N) 10%.

of comonomer used in acrylic polymers. Aubrey[13] has discussed the effect of an adhesive's glass transition temperature in acrylic-vinyl acetate copolymers of varying composition. An increase of T_g causes a lateral shift of the peel-force–peel-rate curve to the left. The comparison of peel-force–peel-rate curves for polybutyl acrylate and vinyl acetate indicates a shift over 12 decades.

Introduction of carboxylic groups into the adhesive polymer chain is known to increase the peel adhesion, and this is effectively used in acrylic and SBR pressure sensitive adhesives. Part of this effect may be attributed to the interfacial effects, that is, improved wetting, but the most important part is due to the effect of bulk adhesive properties. Aubrey and Ginosatis have investigated the effect of carboxylic groups on the peel force.[51] The authors made an attempt to distinguish between bulk and interfacial effects by preparing layered adhesives. A sample was prepared with carboxylated adhesive on the surface and a noncarboxylated one in the bulk; another set of samples was prepared in the opposite manner. The test results are shown in Figure 5-29. The differences between curves **a** and **b** and between curves **c** and **d** reflect the interfacial effects, while the difference between curves **a** and **c** and **b** and **d** reflect the bulk effects. The contributions to the peel force are not constant, but are rate dependent. The sum of the bulk and surface contributions of carboxylation gave curve **e**, which closely matches experimental curve **d**.

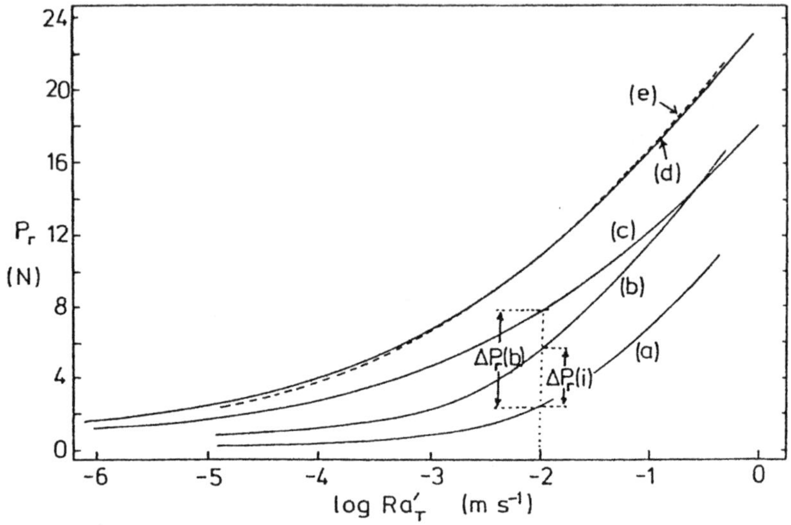

Fig. 5-29. Reduced peel-force–peel-rate curves. (a) All polybutyl acrylate adhesive. (b) Polybutyl acrylate with carboxylated PBA surface coating. (c) Carboxylated PBA adhesive with PBA surface coating. (d) All carboxylated PBA adhesive. (e) Hypothetical composite curve. (Reprinted from *J. Adhesion* **12:** 189–198 (1981).)

Aubrey and Sherriff[20] have reported the effect of compositional change for the case of natural rubber-tackifying resin adhesives. Figure 5-30 shows the shift to the lower peel rates of the slip-stick transition point with the increasing amount of tackifying resin. Such displacement reflects the increase of glass transition temperature with increasing resin concentration.

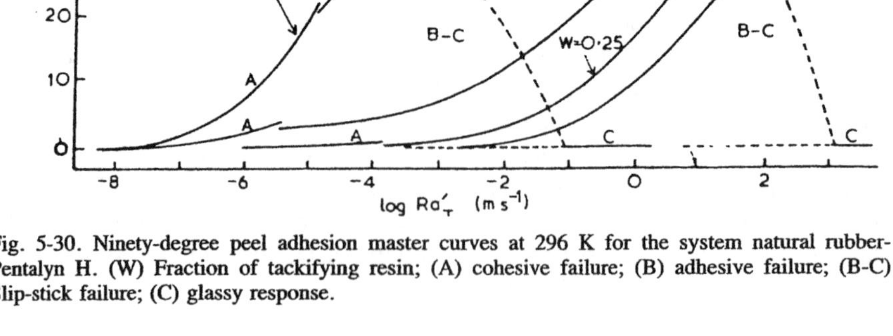

Fig. 5-30. Ninety-degree peel adhesion master curves at 296 K for the system natural rubber-Pentalyn H. (W) Fraction of tackifying resin; (A) cohesive failure; (B) adhesive failure; (B-C) Slip-stick failure; (C) glassy response.

EFFECT OF INTERFACE

Peel work consists of two main components: the work required to deform the adhesive and the backing, and the work required to debond the adhesive from an adherend. Usually the bulk effects dominate the pressure sensitive adhesive behavior, and the contribution of interfacial effects is of a smaller magnitude. This relationship, however, changes drastically if the adhesive does not wet the surface well. In such cases, the interfacial adhesion becomes dominant.

The interfacial processes may affect the bond in several ways:

1. Improved wettability may increase the contact area between the adhesive and the adherend, and consequently increase the measured peel force.
2. Additional bonds, that is, hydrogen or chemical bonds, may be established, supplementing the van der Waals forces.
3. The adhesive may or may not be compatible with the impurities on the adherend surface. A compatible adhesive dissolves the impurities, thus increasing the bond strength.

The time period required to establish the proper bond and to reach a constant level of peel adhesion depends upon the adherend surface and upon the flow characteristics of the adhesive. The bond is established faster to a smooth surface than to a rough one. Hendricks and Dahlquist[11] have shown that a time period of one to two days is sufficient to establish a good contact to most relatively smooth surfaces, but it might take much longer to achieve a constant peel level to a porous surface, such as wood. The bond might continue to increase over a long period of time, if other factors, besides wetting, become important in the formation of the bond.

Figure 5-31 shows the peel force as the function of peel rate of an aluminum foil tape to various release coatings.[49] All four tapes had the same adhesive of the same thickness. The difference in the observed peel force is due to the different release levels. It is of interest to observe that the differences in the release levels appear to decrease with increasing peel rate. Silicone release coated tape (curve R) shows a very easy release at low peel levels, but that advantage is lost as the peel rate increases. For a minimum of flagging when the tape is overwrapped on itself, a higher peel force at low peel rates is desirable. Therefore, such effective release coatings as silicone are rarely used for tape backings. For easy dispensing of tapes or labels at high peel rates, a low peel force at high peel rates is desirable.

Peel adhesion is usually measured by applying the tape to a standard steel surface. Adhesion to other surfaces might be important for many applications. Pressure sensitive labels are often tested against steel and against polyethylene surfaces. All adhesives show a lower peel adhesion to polyethylene than to

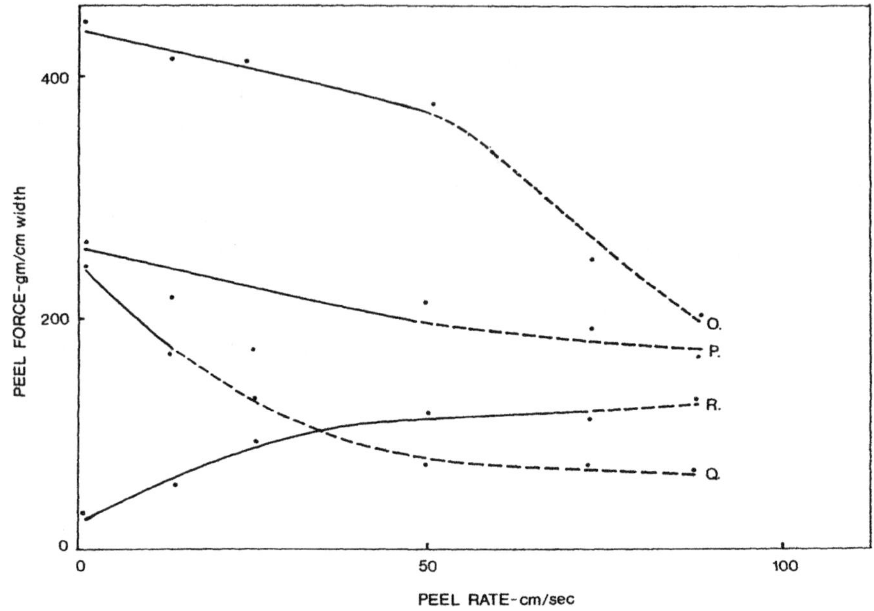

Fig. 5-31. Effect of release coatings on peel. (O) Acrylic release coating; (P) Quilon C release; (Q) polyethylene release coating; (R) silicone release coating.

steel, because a polyethylene surface is more difficult to wet. Polar acrylic adhesives may exhibit 50% lower peel adhesion to polyethylene than to steel or other higher critical surface tension surfaces. This difference can be decreased by the addition of tackifying resin, which improves wetting.

STRESS DISTRIBUTION

The distribution of stresses in the adhesive bond must be considered in order to gain a better understanding of the test and of the relation of peel force to the properties of pressure sensitive adhesives. It has been recognized by DeBruyne[52] and discussed by many other authors that compressional stresses are developed in the adhesive during peeling.

An apparatus for measuring the normal stresses during peeling of an adhesive bond has been described by Johnson and Kaelble[53] and later discussed in greater detail.[54] Figure 5-32 shows a schematic diagram of such an apparatus designed for mounting on an Instron Tester. The substrate to which the adhesive tapes is applied is divided into two parts. The rightmost section is rigidly supported and its position is insensitive to the normal forces of peel. The substrate to

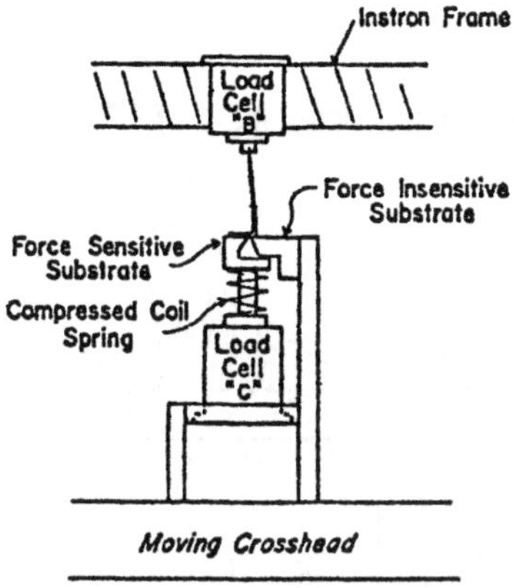

Fig. 5-32. Schematic of the bond stress analyzer. (Reprinted from *Transactions of the Society of Rheology* **9**(2) (1965).)

the left of the division is sensitive to normal forces and deflects slightly in response to the normal bond stresses in a peel test.

Figure 5-33 shows a schematic diagram of a profile of a pressure sensitive adhesive tape undergoing peel at 90 degrees. A typical stress distribution profile as recorded by the measuring instrument is shown. The stresses are highly localized at the boundary of the bond. Adjacent to the boundary is a region of tensional stress. A zone of compressional stress with a maximum compressive stress $\sigma_c = 5.7$ kg/cm^2 and a compressive zone length $\lambda/2 = 0.034$ cm is produced because of the leverage of the flexible member. A zone of low tensional stress appears next, and then the normal stress drops to zero. Kaelble has shown that the peel force was essentially constant throughout the tests, while internal tensile (σ_t) and compressive (σ_c) stresses and the compressive zone length ($\lambda/2$) varied considerably. The variability of the peel force in a 180-degree test was slightly higher than that in a 90-degree test, but σ_c, σ_t, and $\lambda/2$ showed less variation.

Kaelble and Reylek[55] describe the instrument modified to measure the normal stresses over three decades of the peeling rates. A schematic diagram of a 180-degree peel test and the distribution of normal stresses is shown in Figure 5-34. The 180-degree peel test stress diagram differs from the diagram of the 90-degree peel test by a more pronounced presence of two distinct tensile stress maxima. Niesiolowski and Aubrey[56] have investigated the stress distribution and they have not observed the presence of the secondary stress maximum (σ_{t2}). It is suggested that the difference is in the type of adhesive used. Kaelble used an adhesive based on natural rubber, which can crystallize on extension, whereas noncrystallizing polymer was used by Niesiolowski and Aubrey.

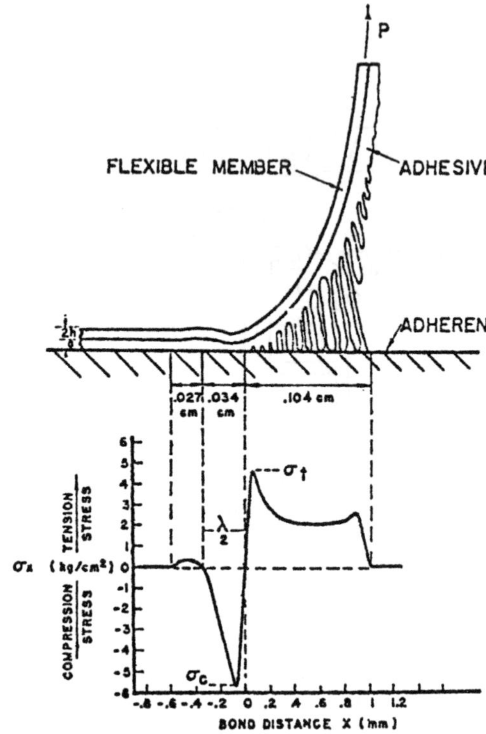

Fig. 5-33. Peel profile in a 90-degree peel test and the distribution of normal stresses. (Reprinted from *Transactions of the Society of Rheology* **9**(2) (1965).)

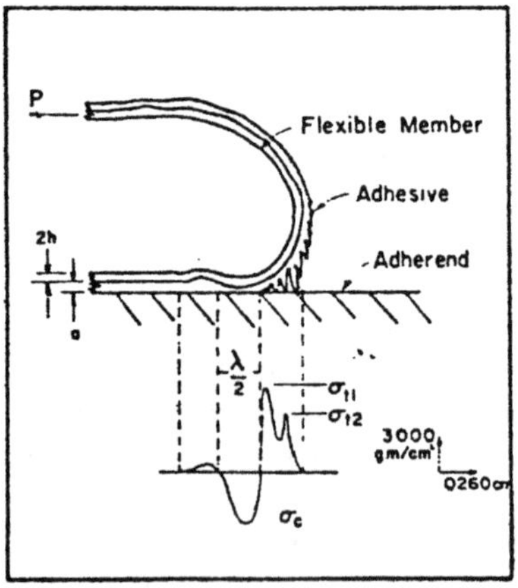

Fig. 5-34. Peel profile in a 180-degree test and the distribution of normal stresses. (Reprinted from *Journal of Adhesion* **1**, 124–135 (1969).)

ANALYSIS OF PEEL

The analysis of the behavior of adhesive joints during peeling and the derivation of mathematical expressions, which would allow the prediction of the strength of such joints, has been attempted by many authors. Unfortunately, none of the analyses is entirely satisfactory. Many simplifying assumptions have been made, and they do not always represent the real mechanism of failure.

One of the early analyses was by Spies,[57] where he calculated the internal bond stress distribution in peel for a peel angle of 90 degrees. A solution for the peel force was offered, assuming adhesive failure of a Hookean interlayer.

In the simplest theory, assuming that the adhesive and the backing are Hookean solids and that shear stresses in adhesive are negligible, Bikerman[41,58] shows that the peel force P is proportional to:

$$P \sim wt_a\sigma, \tag{5}$$

where

w = width of the tape
t_a = adhesive thickness
σ = tensile strength of the adhesive

Jouwersma[59] elaborated further on Bikerman's work.

Chang[60] discussed peel of pressure sensitive tapes using a mechanical model consisting of three elements. Peel of pressure sensitive adhesives has been reviewed by Dahlquist.[61] Dahlquist[62] has also discussed the estimation of the work of adhesion between a pressure sensitive adhesive and a substrate. Peel has also been reviewed by Gardon,[48] who has also derived equations showing the change of peel force with adhesive thickness. Non-pressure sensitive polymers were used for his study. Mathematical expressions describing the tensional and compressional forces during peel have been given.[40]

Kaelble[34,46] has been the most prolific author on the peel of pressure sensitive adhesives and his theory has gained most acceptance. Assuming a Hookean response of the adhesive, Kaelble has derived the following expression for 90-degree peel:

$$P = wt_a\, K^2\sigma_0^2/2E\ (1 - \cos\omega) \tag{6}$$

where

K = a dimensionless parameter dependent on cleavage stress concentration, moment arm, and peel angle
σ_0 = boundary stress for cleavage

For 180-degree peel, Equation (6) simplifies to:

$$P = wt_a\sigma_0^2/2E \qquad (7)$$

Peeling has been also reviewed by Duke.[30] He points out that the theories of peel adhesion (Spies, Jouwersma, Kaelble, Gardon) have assumed that the adhesive is elastic. The stresses applied to the adhesive bond during peeling are less than those required to cause an inelastic response. This assumption also implies that the adherend is stiff and does not undergo a plastic flexular yield. According to Duke, this latter assumption is incorrect and the elastic peel mechanism describes only a very limited number of cases. It is of practical interest mainly in the case of joints between thick, rigid adherends. An assumption of plastic behavior of adhesive and adherend better describes most bond failures by peeling.

REFERENCES

1. Shoraka, F. *Adhesion and Heat of Peeling of Pressure-Sensitive Tapes*. Ph.D. Thesis. State University of New York at Buffalo, May 1979.
2. Hamed, G. R. *J. Adhesion* **13:**101–105 (1981).
3. *Standard Methods of Testing Pressure-Sensitive Adhesive Coated Tapes Used for Electrical Insulation*. ANSI/ASTM D-1000-78.
4. *Tentative Method of Test for Peel Resistance of Adhesives (T-Peel Test)*. ASTM D1876-61T.
5. *Test Methods for Pressure Sensitive Tapes*. Pressure Sensitive Tape Council, Glenview, Ill., 1976.
6. Druschke, W. Adhesion and Tack of Pressure-Sensitive Adhesives. Presented at the Association des Fabricants Europeens de Rubans Auto-adhesifs *Meeting*, Edinburgh, Scotland. October 1–4, 1986.
7. Satas, D. *Adhesives Age* **13**(6): 38–40 (1970).
8. Diefenbach, W. T. TAPPI **45**(11): 840–842 (1962).
9. Šalkauskas, M. J., and Paulavičius, R. B. *Zavodskaya Laboratoriya* **(8)**:879–880 (1976).
10. Weidner, C. L. U.S. Patent 3,725,322 (1973) (assigned to Johnson & Johnson Co.).
11. Hendricks, J. O., and Dahlquist, C. A. Pressure-Sensitive Tapes. *Adhesion and Adhesives*, Vol. 2 (R. Houwink and G. Salomon, eds.) Elsevier, Amsterdam, 1967.
12. Mizumachi, H. *J. Appl. Polym. Sci.* **30:** 2675–2686 (1985).
13. Aubrey, D. W. Effect of Adhesive Composition on the Peeling Behaviour of Adhesive Tapes. *Adhesion 8* (K. W. Allen, ed.) Elsevier, London and New York, 1984.
14. McLaren, D. D., and Seiler, C. J. *J. Polym. Sci.* **4:** 63–73 (1949).
15. Huntsberger, J. R. *J. Polym. Sci.* **A1:** 2241–2250 (1963).
16. Bright, W. M. *Adhesion and Adhesives, Fundamentals and Practice*, 130–138. (F. Clark, J. E. Rutzler, and R. L. Savage, eds.) Soc. of Chem. Ind. New York: Wiley, New York, 1954.
17. Satas, D. Adhesive Tapes. *Polymer Conference Series*. Wayne State University, Detroit, Mich., 1967.

18. Satas, D., and Mihalik, R. J. *J. Appl. Polym. Sci.* **12:** 2371–2379 (1968).
19. Aubrey, D. W. Pressure-Sensitive Adhesives–Principles of Formulation. *Development in Adhesives-1.* (W. C. Wake, ed.). Applied Science, London, 1977.
20. Aubrey, D. W. Viscoelastic Basis of Peel Adhesion. *Adhesion-3.* (K. W. Allen, ed.) Applied Science, London, 1978.
21. Aubrey, D. W., and Sherriff, M. *J. Polym. Sci.: Polym. Chem. Ed.* **18:** 2597–2608 (1980).
22. Kaelble, D. H. *J. Adhesion* **1:** 102–123 (1969).
23. Gent, A. N., and Petrich, R. P. *Proc. Roy. Soc. A* **310:** 433–448 (1969).
24. Satas, D. Pressure-Sensitive Adhesives in Pharmaceutical Packaging, in *Proc. Third Wisconsin Update Conference, Packaging.* Univ. of Wisconsin Extension, Madison. Oct. 24–26, 1983.
25. Greensmith, H. W., and Thomas, A. G. *J. Polym. Sci.* **18:** 189 (1955).
26. Greensmith, H. W. *J. Polym. Sci.* **21:** 175 (1956).
27. Dannenberg, H. J. *J. Appl. Polym. Sci.* **5:** 125 (1961).
28. Gardon, J. L. *Appl. Polym. Sci.* **7:** 625–641 (1963).
29. Lannus, A., and Zdanowski, R. *Resin Rev.* **16**(4): 15 (1966).
30. Duke, A. J. *J. Appl. Polym. Sci.* **18:** 3019–3055 (1974).
31. Author's data.
32. Williams, M. L., Landell, R. F., and Ferry, J. D. *J. Am. Chem. Soc.* **77:** 3701 (1955).
33. Kaelble, D. H. *J. Coll. Sci.* **19:** 413 (1964).
34. Kaelble, D. H. *Physical Chemistry of Adhesion.* Wiley–Interscience, New York, 1971.
35. Gerace, M. *Adhesives Age,* **26** (9): 15–20 (1983).
36. Chan, H-K., and Howard, G. J. *J. Adhesion* **9:** 279–304 (1978).
37. Johnston, J. *Adhesives Age* **11** (4): 20–26 (1968).
38. Aubrey, D. W., Welding, G. N., and Wong, N. T. *J. Appl. Polym. Sci.* **13:** 2193–2207 (1969).
39. Fukuzawa, K. J. *Adhesion Soc. Japan* **5:** 294 (1969).
40. Gardon, J. L. *J. Appl. Polym. Sci.* **7:** 643–665 (1963).
41. Bikerman, J. J. *The Science of Adhesive Joints.* Second edition. Academic Press, New York, 1968.
42. Gent, A. N., and Hamed, G. R. *Polym. Eng. Sci.* **17** (7): 462–466 (1977).
43. Toyama, M., Ito, T., and Moriguchi, H. *J. Appl. Polym. Sci.* **14,** 2039 (1970).
44. Satas, D., and Egan, F. *Adhesives Age* **9** (8): 22–25 (1966).
45. Gent, A. N., and Hamed, G. R. *J. Appl. Polym. Sci.* **21:** 2817–2831 (1977).
46. Kaelble, D. H. *Trans. Soc. Rheol.* **III:** 161–180 (1959).
47. Kaelble, D. H. *Trans. Soc. Rheol.* **IV:** 45–73 (1960).
48. Gardon, J. L. Treatise on Adhesion, Vol. 1 (R. L. Patrick, ed.). Dekker, New York, 1967.
49. Author's data.
50. Mao, T. J., and Reagan, S. L. *Proceedings of the Symposium on Adhesion and Cohesion,* 209–217 (P. Weiss, ed.). Elsevier, Amsterdam, 1962.
51. Aubrey, D. W., and Ginosatis, S. *J. Adhesion* **12:** 189–198 (1981).
52. DeBruyne, N. A. *Nature* **180:** 262 (1957).
53. Johnson, E. G., and Kaelble, D. H. Apparatus for the Measurement of the Normal Stress Distribution in Adhesive Bonds During Peel. Presented at the Meeting of the American Physical Society, Cambridge, Mass. April 1959.
54. Kaelble, D. H. *Trans. Soc. Rheol.* **9** (2): 135–163 (1965).
55. Kaelble, D. H., and Reylek, R. S. *J. Adhesion* **1:** 124–135 (1969).
56. Niesiolowski, F., and Aubrey, D. W. *J. Adhesion* **13:** 87–98 (1981).
57. Spies, G. J. *Aircraft Eng.* **25** (289): 64 (1953).
58. Bikerman, J. J. *J. Appl. Phys.* **28:** 1484 (1957).

59. Jouwersma, C. *J. Polym. Sci.* **45** (145): 253–255 (1960).
60. Chang, F. S. C. *Trans. Soc. Rheol.* **IV:** 75–89 (1960).
61. Dahlquist, C. A. *Treatise on Adhesion,* Vol. 2. (R. L. Patrick, ed.). Dekker, New York, 1967.
62. Dahlquist, C. A. *ASTM Special Technical Publication No. 360.* 66th Annual Meeting Papers, Atlantic City, N.J. June 26, 1963.

6. Creep

Carl A. Dahlquist
St. Paul, Minnesota

If a substance exhibits the property we call pressure sensitive adhesiveness, that is, it adheres when brought into contact with a surface under light pressure, and has sufficient cohesiveness that it can be peeled away from the surface without leaving a residue, the substance must have rather specialized and characteristic rheological properties. It must have sufficient compliance to conform to surface rugosity, and it must undergo relaxation so that the stored energy due to elastic forces will be dissipated before they can overcome the forces of adhesion. Yet the pressure sensitive adhesive must exhibit an elastic cohesiveness and a resistance to flow under stress. It is this resistance to flow that is commonly associated with creep or resistance to creep, but in a broader sense the entire stress-strain behavior of a pressure sensitive adhesive can be treated as a creep phenomenon.

The creep of a soft polymer is conventionally regarded as the increase in strain with time after stress is suddenly applied and maintained constant. It can be shear deformation under constant shearing stress, compressive strain under constant compressive stress, or tensile strain under constant tensile stress. The creep response can be expressed either as the creep modulus versus time or as creep compliance versus time. Modulus is the ratio of stress to strain, and compliance is the inverse. The engineer usually has been trained to think of modulus as the property describing purely elastic stress-strain behavior. The creep of pressure sensitive adhesives almost invariably involves some viscous flow in addition to the elastic strain and it may be technically distasteful to the engineer to speak of a modulus that involves nonrecoverable strain. Though it may be technically no more acceptable, it seems less offensive to regard compliance as the ratio of total strain to stress, and, unless otherwise indicated, this interpretation of compliance will be used throughout this chapter. It is fully recognized that, depending on the time scale, creep compliance has all of the counterparts of modulus, namely storage, loss, elastic, and viscous components.

Although the term "creep" has a connotation of slow processes involving minutes, days, or months, one can, in principle, determine the short-term, or dynamic, stress-strain response of a viscoelastic solid as well as the long-term response by means of creep measurements and time-temperature superposition. Difficulties arise in the so-called "instantaneous" application of the load and in inertial effects associated with rapid response, so that it is more expedient to use oscillatory methods when the response time of interest is less than one second, but in principle any deformation that has an associated retardation time is a part of the creep spectrum. It is also possible, by well-known mathematical treatments, to obtain the elastic and viscous components of the creep spectrum. In the discussion to follow, attention will be focused mainly on the long-term creep response of pressure sensitive adhesives, but compliance and the retardation spectrum play such significant roles in the quick stick or tack of pressure sensitive adhesives that some attention must be given to the short-term response as well.

SHORT-TERM CREEP AND TACK

The creep compliance–time spectra of a gum rubber and a pressure sensitive adhesive are very similar, but there are subtle differences that distinguish the nontacky rubber from the tacky pressure sensitive adhesive. Used in this context, the word "tack" does not mean self-adhesion, as in the tire maker's building tack, but in the ability to stick under light pressure to foreign surfaces. Figure 6-1 shows the creep compliance versus time of a lightly milled Hevea rubber and of the pressure sensitive adhesive prepared by blending 59 parts by weight (pbw) of the milled rubber with 41 pbw of a polymerized β-pinene resin. The pressure sensitive adhesive is less compliant than the rubber when the stress interval is 0.01 sec or less, and increasingly more compliant at stress intervals exceeding 0.01 sec.

The shift to a less compliant response at time intervals of a very short duration, say 0.001 sec or less, can be accounted for quite well by the shift in the glass transition temperature that occurs when the resin is added. The glass transition temperature of the rubber is 200 K, and that of the adhesive 212 K. The time shift, log a_T, that would correspond to the T_g shift from 200 K to 212 K, can be estimated from the WLF equation,

$$\log a_T = \frac{-17.4\ (T - T_g)}{51.6 + T - T_g} \tag{1}$$

amounting to 0.53 unit of logarithmic time. If the compliance–time curve of the rubber is shifted toward longer time by this amount (see Fig. 6-2), it shifts

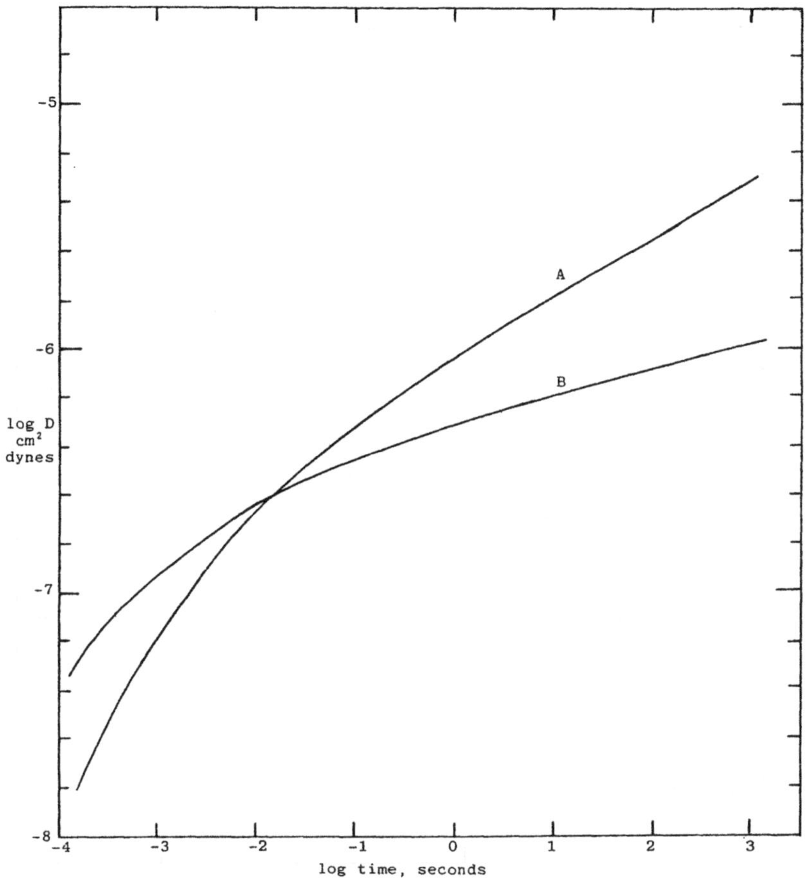

Fig. 6-1. Creep compliance in tension of a natural rubber-polyterpene pressure sensitive adhesive (A) and the natural rubber base of the pressure sensitive adhesive (B).

into near coincidence with the compliance–time curve of the pressure sensitive adhesive in the region where the stress duration is about 0.001 sec.

As the stress interval increases beyond 0.001 sec, the pressure sensitive adhesive shows increasingly greater compliance than the rubber base. Thus the tackifying resin acts as a stiffening agent when stress duration is in the millisecond range, and as a plasticizer when the stress interval is greater by about an order of magnitude.

It is of interest that the time scale in which the crossover in the compliance–time curves takes place is precisely the time scale of viscoelastic response in many of the conventional tests that measure the tack and peel adhesion of pressure sensitive adhesives and pressure sensitive adhesive tapes. In a simple probe tack test it is quite common to allow a contact time of about 1 sec, and the separation, which generates the force taken to be a measure of tack, may

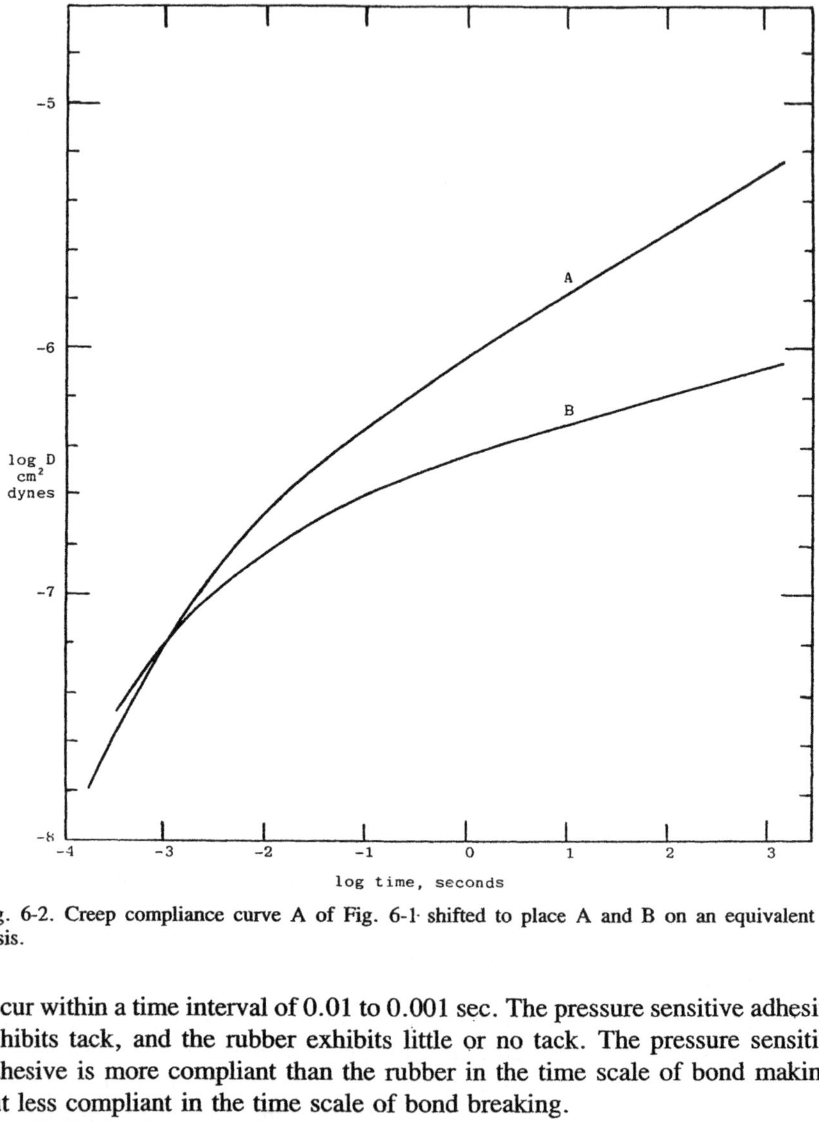

Fig. 6-2. Creep compliance curve A of Fig. 6-1 shifted to place A and B on an equivalent T_g basis.

occur within a time interval of 0.01 to 0.001 sec. The pressure sensitive adhesive exhibits tack, and the rubber exhibits little or no tack. The pressure sensitive adhesive is more compliant than the rubber in the time scale of bond making, but less compliant in the time scale of bond breaking.

Similarly, in peel adhesion tests run at very slow rates, say 1 cm/min or less, the adhesive may undergo very high deformation, and even cohesive failure, before releasing from the adherend, while at high rates, say 100 m/min, the peel process may become stick-slip and show all of the characteristics of brittle fracture.

The author has attempted, with some success, to evaluate the role of the modulus (reciprocal compliance) in tack in a more quantitative way.[1] An observation well supported by experiment is that the compression modulus in the time frame of the bonding process cannot exceed 10^7 dyne/cm^2, and preferably should approach 10^6 dyne/cm^2. When the compression modulus is greater than 10^7 dyne/cm^2, the degree of conformability to the surface contours of even so-called "smooth" surfaces is not sufficient to effect the desired adhesive bonding. Figure 6-3 illustrates this relationship between modulus and tack when both are measured over a range of temperature. A specific instrument is used for the tack measurements.[1,2]

High compliance alone, however, is not a sufficient criterion to ensure pressure sensitive tack. For example, one can break down natural rubber by milling so that it will have more than adequate compliance for bonding, yet it will display little or no tack at room temperature. If, however, it retains sufficient compliance when chilled, it may exhibit appreciable tack at a lower temperature. The author has attributed this development of tack at low temperature to the

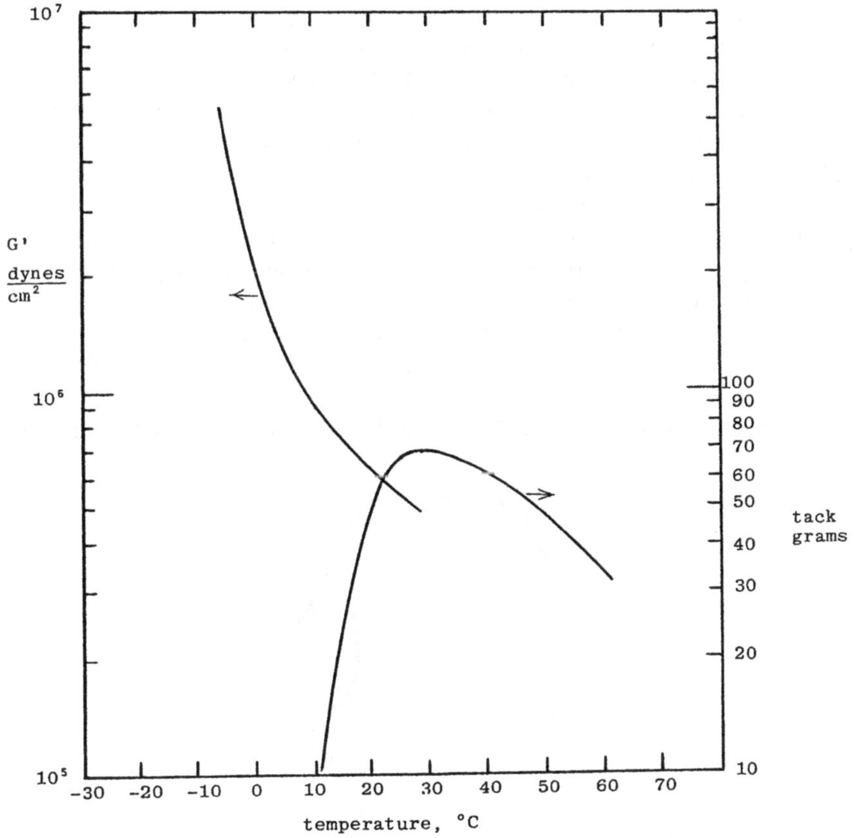

Fig. 6-3. Dynamic shear storage modulus, G', and tack variation with temperature.

increase in strength associated with lower temperature and high rate of extension in the bond breaking process.

Kraus *et al.*[3] have proposed that the loss modulus (or loss compliance) plays an important role in tack. In our laboratories Dr. Vasant Kolpe (unpublished work) has found that the temperature at which the loss compliance maximizes when the relaxation time is one second corresponds quite closely to the temperature of onset of tack when the contact interval is 1 sec. This implies that energy must be dissipated during the bonding process when a viscoelastic material exhibits tack.

LONG-TERM CREEP

Resistance to long-term creep is a highly desirable property in pressure sensitive adhesives. Generally the requirements are dictated by end use, where the stress conditions depend upon the application. For example, tape used for book mending or envelope sealing may never be subjected to severe stress, yet, even here, when a transparent tape is used to mend a torn page in a book, the pressure sensitive adhesive must resist the capillary forces, or else it will wick through the page and render it translucent. This process may take months to years. In other applications pressure sensitive adhesive tapes are required to support loads, sometimes for relatively short times at elevated temperatures, for example, masking tapes supporting protective paper aprons in automotive painting during baking cycles, and sometimes for very long periods of time at room temperature, as when objects are attached to walls with double-face tapes.

A dramatic example of undesirable creep is the "telescoping" of rolls of pressure sensitive tape (Fig. 6-4), where the compressive stresses in the rolls are not directed uniformly toward the core, but have unidirectional components

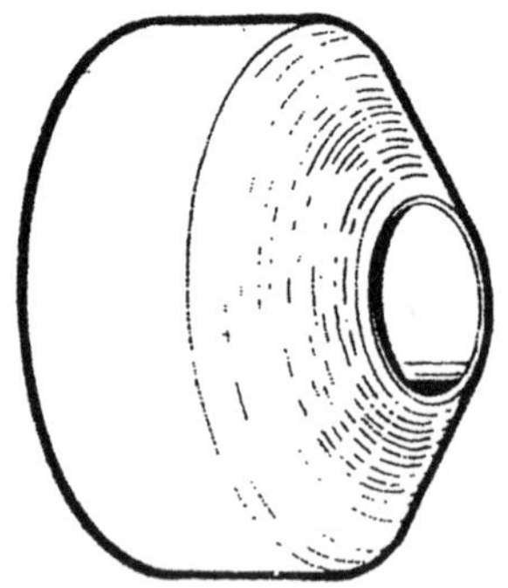

Fig. 6-4. Illustration of the telescoping of pressure sensitive adhesive tape.

of shear stress that cause each successive layer of tape to slide sideways on the one beneath. This problem was especially prevalent and annoying in the early days of the plasticized vinyl film tapes when plasticizer migration caused the pressure sensitive adhesives to become very soft and the techniques for winding tapes under controlled and uniform tension had not been perfected.

LONG-TERM CREEP AND STEADY-STATE VISCOSITY

When shearing stresses are small and the rate of creep is very slow, the creep rate depends primarily upon the steady-flow viscosity of the pressure sensitive adhesive. If shearing stresses are high and the rate of shear correspondingly high, the viscosity may vary as the shear rate changes (non-Newtonian viscosity).

The factors that determine the steady-flow viscosity are generally the same in pressure sensitive adhesives as in all soft polymers. The steady-flow viscosity is strongly dependent on the molecular weight of a polymer. Measurements on series of monodisperse homopolymers have shown that the steady-flow viscosity varies as the 3.4 power of the molecular weight. In our own laboratories, Mr. Guy Fortry (unpublished work) investigated the rheological properties of *cis*-polyisoprene prepared in fairly narrow molecular weight distributions by *n*-butyl lithium polymerization (Table 6-1). Although the polymers were not ideally monodisperse, the proportionality between the steady-flow viscosity (η_0) and the molecular weight raised to the 3.4 power is well supported by the grouping of data points about the line of slope 3.4 in the log η_0 vs. log M_w plot (see Fig. 6-5).

Polymers used in commercial pressure sensitive adhesives are rarely of narrow molecular weight distribution, and deviation from the $\eta_0 \propto M^{3.4}$ must be expected. Nevertheless, it is useful to know that the strong dependence of steady-flow viscosity on molecular weight exists, because a small increase in molecular weight can produce a significant improvement in capacity to support a load.

Table 6-1. Steady-State Viscosity vs. Molecular Weight of *cis*-Polyisoprene

M_w	M_w/M_n	η_0(poises)
125,000	1.23	1.2×10^6
350,000	1.27	4.0×10^7
590,000	1.28	2.0×10^8
830,000	2.50	4.0×10^8

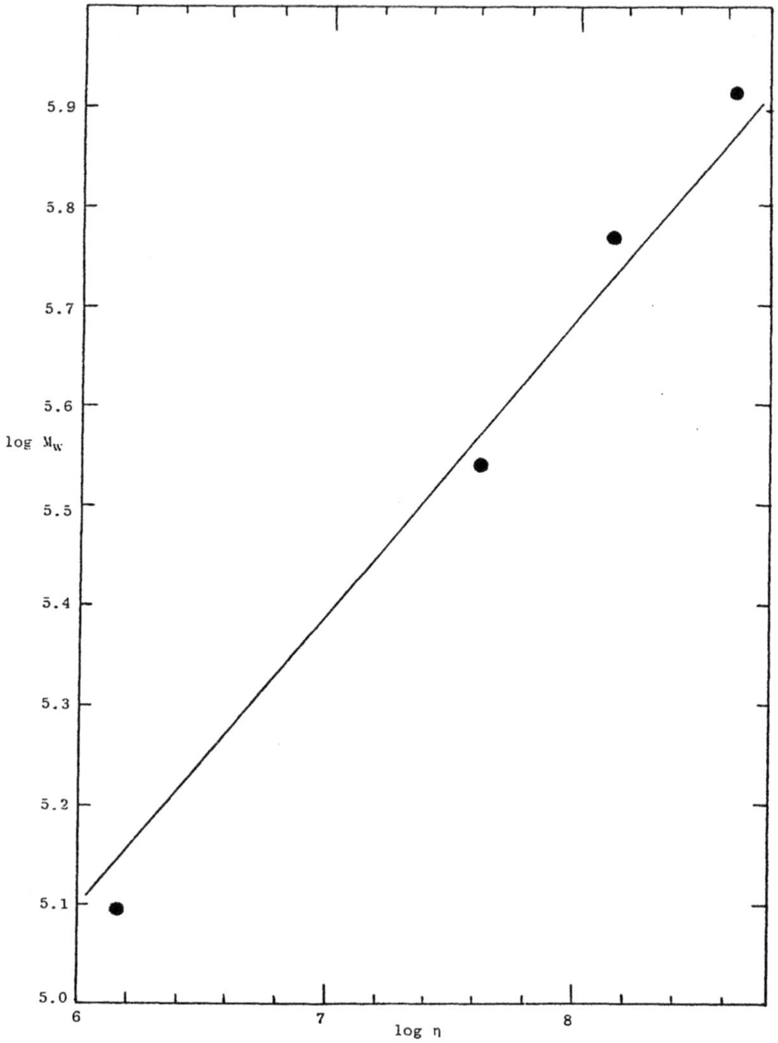

Fig. 6-5. Steady-flow viscosity (η_0) vs. molecular weight (M_w) in *cis*-polyisoprene.

For example, two laboratory batches of a tacky copolymer of *n*-butyl vinyl ether and *i*-butyl vinyl ether were prepared under similar conditions, but one batch had an intrinsic viscosity of 4.4 deciliters (dl)/g and the other 3.5 dl/g. The intrinsic-viscosity-weight–average-molecular-weight relationship was taken to be

$$[\eta] = aM^{2/3} \tag{2}$$

The M_w then stand in the ratio $[4.4/3.5]^{1.5}$, or 1.4, from which steady-flow viscosities would be expected to stand in the ratio $(1.4/1.0)^{3.4}$, or 3.1/1.0. The steady-flow viscosities were measured by creep and recovery and were found to be 2.1×10^9 poises and 5.9×10^8 poises, yielding a measured ratio of 3.6/1.0. In consideration of the uncertainties involved in the estimation of the molecular weights, and lacking knowledge of molecular weight distributions, the agreement between the measured and the estimated ratios of steady-flow viscosities is not bad. Since the ability of a pressure sensitive adhesive to sustain loads for long periods of time is directly proportional to its steady-flow viscosity, the significance of molecular weight is readily apparent.

A small fraction of high molecular weight material can make a substantial contribution to creep resistance. The first elastomer used as a base for pressure sensitive adhesives was natural rubber. High-quality natural gum has a bimodal distribution of molecular weights, and carries a fraction that is very high in molecular weight. For a practical reason, namely to produce solutions of suitable coating viscosities and adequate solids contents, the rubber was milled. It soon became apparent that the milling conditions had a great effect on the creep resistance, and therefore the shear holding ability of the pressure sensitive adhesives compounded from natural rubber. The high molecular weight fraction breaks down faster than lower molecular weight fractions, and therefore the steady-flow viscosity of natural rubber is rapidly reduced by milling. SBR gum, which does not have the bimodal distribution, breaks down much less rapidly than natural rubber gum, but it does not have the creep resistance provided by a high molecular weight fraction.

The contribution of high molecular weight to creep resistance is illustrated by an experiment in which small percentages of a synthetic pressure sensitive adhesive copolymer of high molecular weight were added to a lower molecular weight copolymer of similar composition. Again, these were copolymers of fairly high polydispersity, and the applicability of viscosity–molecular weight relationships that hold for monodisperse systems is open to criticism. Nevertheless, useful guidelines become apparent in analysis of the observed reults.

The weight average molecular weight of copolymer A was 7.2×10^6, that of polymer B 4.5×10^5. Polymer A was added to polymer B in the amounts of 0%, 4.14%, and 6.80% by weight. After these compositions were coated on polyester film to give a dry coating thicknes of 0.005 cm, overlap shear tests were made using polished steel as a test surface, an initial area of shear 1.27 cm × 1.27 cm, and a load of 1000 g. When times to failure in such a test are long enough so that the initial rate of shear is low and the slip is dominated by the steady-flow viscosity, the time to failure is given by

$$T = \frac{L^2 W \eta}{2tMg} \tag{3}$$

where

T = time to failure in seconds
L = overlap in cm
W = width in cm
η = viscosity in poises
t = thickness of adhesive interlayer
M = load in grams
g = gravitational constant (981 cm/sec^2)

Since the load and all of the geometrical parameters are fixed in each set of static shear tests, the time to failure becomes directly proportional to the viscosity of the adhesive.

The measured shear times, the steady-flow viscosities calculated from them, and the steady-flow viscosities that one would predict from the M_w increase resulting from addition of the high molecular weight copolymer are given in Table 6-2. The excellent correlation between the observed and calculated steady-state viscosities is possibly fortuitous, but it is obvious that small amounts of high molecular weight polymer can markedly increase the steady-flow viscosity and the creep resistance.

The test that was used in the experiment previously described is common in the pressure sensitive tape industry and is known as the ''Shear Adhesion'' (holding power) test PSTC-7[4] or ASTM D3654.[5] The equipment consists of a rack or jig that holds a panel 2 degrees from the vertical, so that the back of the panel forms an angle of 178 degrees from the extended piece of tape. The purpose of the small inclination from vertical is to assure that the tape will experience no peeling action, only shear.

The test panel is 2 × 5 × 1/16 in. (50.8 × 127 × 1.6 mm) stainless steel type 302 or 304 having a bright annealed finish. Its surface roughness height is 2.0 ± 1.0 μin. arithmetical average deviation from the mean. One or both of the panel ends is ground to form a 90-degree angle with the panel surface.

Table 6-2. Effect of High Molecular Weight Polymer Fraction on Shear Time and Steady-Flow Viscosity (η)

COMPOSITION OF BLEND (pbw)		OBSERVED (M_w)	CALCULATED (M_w)	OBSERVED SHEAR TIME (sec)	η POISES × 10^{16} OBSERVED	CALCULATED
A	B					
0	100	4.50×10^5	4.50×10^5	100	0.48	0.48
4.14	95.86	7.49×10^5	7.29×10^5	408	2.59	2.48
6.80	93.20	9.25×10^5	9.10×10^5	834	5.28	5.26

The tape specimen, approximately 6 in. in length, is removed from the roll of tape mounted on a freely rotating drum at a rate of 20 to 30 in./sec. At least two wraps are removed before the specimen is taken. The specimen is laid, with no pressure, exactly parallel to the length of the test surface and perpendicular to the ground edge such that about 2 in. of tape overlap the panel. A second panel is laid under the unattached portion of the tape so that the roller will not bump over the edge of the panel. The roller, 3.25 in. in diameter and 1.75 in. in width, covered with rubber one-quarter inch in thickness and of Shore A durometer hardness 75 to 85, and weighing 4½ lb, is mechanically driven over the tape at 12 in./min, making a pass in one direction and a return. (*Exception:* For foil tapes only, a 10-lb roller is used and the roller is passed over the sample five times in each long direction.)

Using a cutting jig or a hand-held razor blade, the rolled portion of the specimen is trimmed to 1 in. by 1 in. square on the test surface if the tape being tested is 1 in. wide. Half-inch-wide tapes are usually trimmed to ½ in. by ½ in. on the test surface.

The panel is placed in the shear adhesion rack with the extending tape hanging downward, and a weight is secured to the free end with an appropriate hook or clamp. The weight may typically be 1000 g when the test area is 1 in. by 1 in., or 500 g when the test area is ½ in. by ½ in. The load should be applied gently and without swinging.

One may allow the load to act for a specified time and measure the slippage, or one may measure the time for total separation to occur.

CREEP AND NON-NEWTONIAN VISCOSITY

When stresses are such that the rate of creep, and hence the rate of shear, exceed minimum values that are not easily defined, the viscosity becomes non-Newtonian. It becomes dependent on the rate of shear, and decreases as the rate of shear increases. Some years back W. A. Vievering and the author investigated the shear rate dependence of several pressure sensitive adhesives by subjecting them to various rates of shear in a shearing jig designed for an Instron tensile tester. At the time we were interested in the holding power of automotive masking tapes, and the measurements were made at 121°C.

The jig had a rectangular parallel plate configuration. Two layers of pressure sensitive adhesive were sandwiched between three plates, and the restraining force was measured as the middle plate was pulled out of the sandwich. An elastic strain followed by degeneration into viscous flow was observed. Typical load-displacement curves are shown in Figure 6-6. At the onset of viscous flow, the resisting force diminishes in direct proportion to the decrease in area under stress.

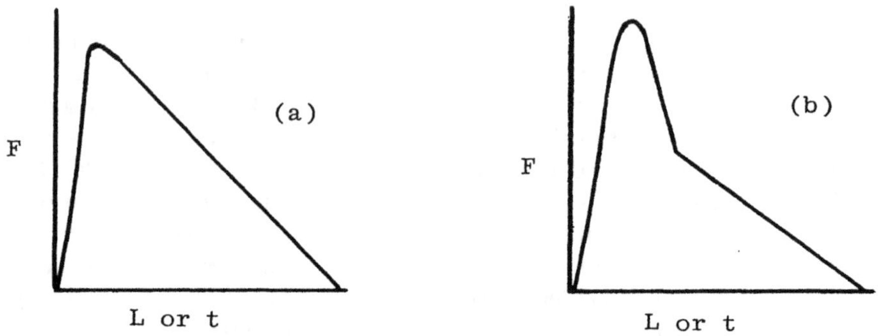

Fig. 6-6. Force vs. length of overlap (*L*) or time (*t*) in steady rate pull of a pressure sensitive adhesive overlap shear bond. (a) Elastic strain followed by viscous slip. (b) Elastic strain followed by shear breakdown, then viscous slip.

A log-log plot of viscosity versus rate of shear for one of the pressure sensitive adhesives is shown in Figure 6-7. Although a tendency toward steady-flow viscosity is indicated when the shear rate is less than 0.01 sec^{-1}, the viscosity is non-Newtonian over the entire range. Overlap shear bonds were prepared and the times to failure under various loads were measured. Times to failure were calculated using an equation that the author has derived for situations where the viscosity–shear-rate dependence can be expressed by a straight line on the log-log plot, such that

$$\log \eta = \log \eta_0 - a \log \gamma \tag{4}$$

where η_0 is the viscosity when $\gamma = 1\ sec^{-1}$, a is the slope, and γ is the rate of shear.[6]

In the shear tests the load was varied from 100 to 800 g per 1.27 cm. The initial overlap was 1.27 cm and the effective width of the double overlap specimen was 2.54 cm. The thickness of the pressure sensitive adhesive layer was 0.0127 cm. The times to failure (see Table 6-3) were such that the shear rates varied too much to fit a single straight line on the log-log plot. By fitting two lines to the data, as shown in Table 6-3, column 3, excellent agreement between calculated and observed times to failure were obtained. Newtonian flow would have given the results shown in column 4. Obviously the fitting of curves to viscosity versus shear rate data and the calculation of time to failure in shear when non-Newtonian flow prevails can be done much better with present-day computer technology.

Cognizance of the shear rate dependence of the steady-flow viscosity must be taken both in testing and in use of pressure sensitive adhesives. Thus one cannot assume that doubling the stress will halve the time to failure in a shear test, and when pressure sensitive adhesives are loaded in shear, overloading

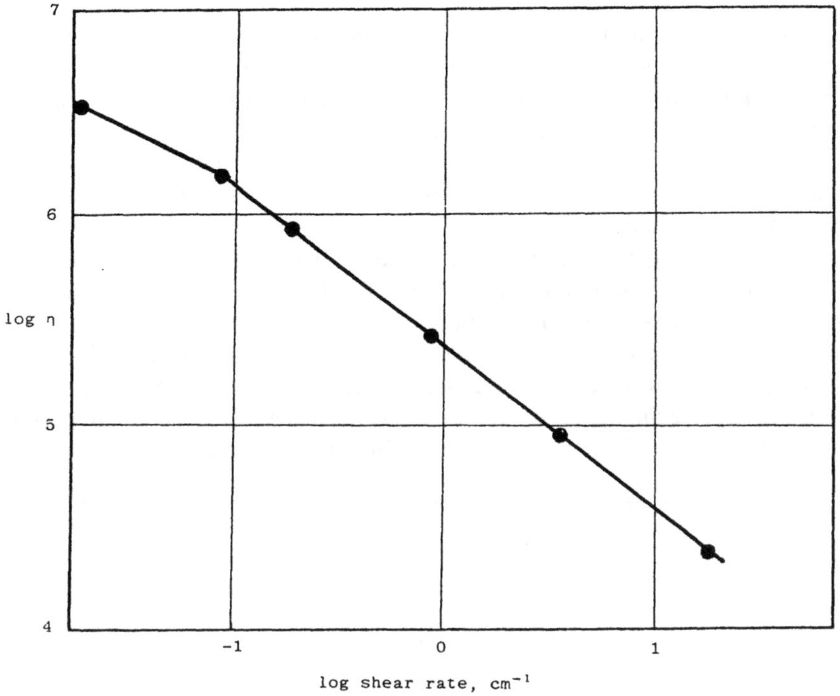

Fig. 6-7. Viscosity (η) vs. rate of shear of a rubber-resin pressure sensitive adhesive at 121°C.

Table 6-3. Time to Failure in Shear: Newtonian Flow (NF) and Non-Newtonian Flow (nNF)

LOAD (g)	MEASURED (min)	Calculated (nNF) (min)	CALCULATED (NF) (min)
100	128.5	127[a]	128.5[c]
200	27.3	31.8[a]	64.3[c]
400	2.9	2.96[b]	32.1[c]
800	0.25	0.185[b]	16.1[c]

[a] $\log \eta = 5.663 - 0.5 \log \gamma$.
[b] $\log \eta = 5.332 - 0.75 \log \gamma$.
[c] $\eta = 4.689 \times 10^6$ poises.

can reduce holding times far out of proportion to the performance anticipated from a high and constant steady-flow viscosity.

CHEMICAL CROSS-LINKING AND CREEP RESISTANCE

An obvious way to raise the steady-flow viscosity and increase the creep resistance of a pressure sensitive adhesive is to take advantage of cross-linking. If one can join each pair of molecules by a cross-link, thereby doubling the molecular weight, the steady-flow viscosity should increase by a factor of $2^{3.4}$, or 10.5. Cross-linking must be done judiciously, however, because overcure can have deleterious effects on the tack and adhesion.

Figure 6-8 illustrates the effect of cross-linking on the creep and recovery of a synthetic pressure sensitive adhesive polymer, which, in the un-cross-linked state, had a molecular weight (M_w) of about 500,000, a Gaussian distribution of molecular weights, and an M_w/M_n of about 10. The steady-flow viscosities before and after cross-linking indicate a 1.8-fold increase in molecular weight or about 0.9 cross-link per molecule.

The patent literature reveals that curing of pressure sensitive adhesives to improve creep resistance was practiced early in the development of pressure sensitive tape technology. It was found to be beneficial to cure rubber-resin pressure sensitive adhesives with oil soluble phenolic resins.[7] Peroxide curing

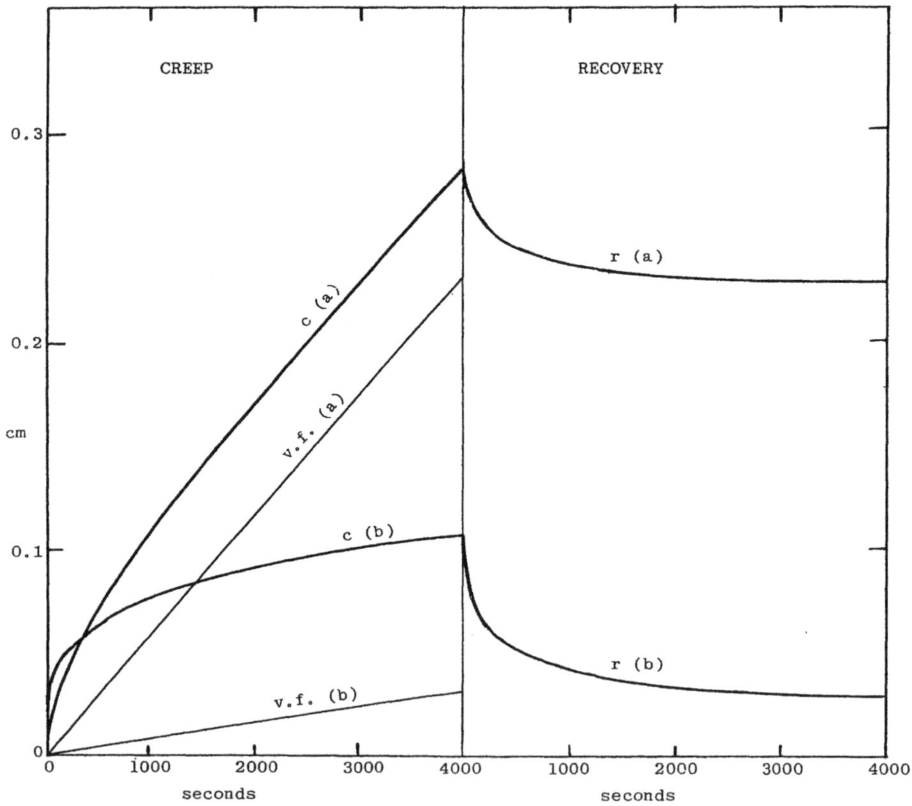

Fig. 6-8. Creep (c), recovery (r), and viscous flow (v.f.) of a pressure sensitive adhesive before (a) and after (b) cross-linking.

of synthetic polymer pressure sensitive adhesives has been practiced for a long time,[8] as has cross-linking by ultraviolet and high-energy radiation.

In the early days the control of the degree of cross-linking was very much an art. Modern methods for measurements of molecular weights, molecular weight distribution, and cross-link density have placed the technology on a much more scientific basis. Currently, when attention has been turned to solventless processing, which generally requires low molecular weight polymers for practical coating speeds, web curing is essential and the advances in the characterization of cross-linked systems have taken on added importance.

One might ask what level of steady-flow viscosity is a reasonable objective for adequate performance in a pressure sensitive adhesive. If we take as a requirement that an overlap shear test in which a dead load of 500 g is applied to a 1.27-cm $\times$ 1.27-cm $\times$ 0.0025-cm specimen should survive for at least 100 min, it can be calculated from Equation 4 that a steady-flow viscosity of 7.2×10^6 poises is required. This could be a typical objective for so-called general-purpose use. If, on the other hand, a double-face tape, having on each side a 0.005-cm layer of adhesive, is required to hold a 1.0-g load per 1.27 cm $\times$ 1.27 cm for a year with allowable slippage of only 0.127 cm (10%), the steady-flow viscosity must be 5.4×10^8 poises. This approximately 100-fold spread in steady-flow viscosity, and therefore in creep resistance, is quite typical of the spread found in commercial pressure sensitive adhesives, and even a 1000-fold spread is quite possible. In synthetic polymer pressure sensitive adhesives this would indicate a 4- to 8-fold spread in molecular weights.

If the molecular weight increase is to be realized simply by cross-linking a polymer of low molecular weight, say 2×10^5 to 5×10^5 daltons, the augmentation of molecular weight cannot be effected without some gel formation. Now, ideally, a gel should reach an equilibrium compliance and therefore exhibit infinite steady-flow viscosity, but due to imperfections in networks produced in commercial polymers of broad molecular weight distributions, pressure sensitive adhesives can have a substantial gel content and still exhibit nonrecoverable creep. For example, the cross-linked copolymer of Figure 6-8 had a gel content of about 50% as measured by 48-hr immersion in tetrahydrofuran.

PHYSICAL CROSS-LINKS

The concept of entanglement molecular weight has been introduced into the theory of long-term compliance and steady-flow viscosity of polymers. Long, linear, flexible polymers are believed to become entangled in interlinked loops that behave as pseudo cross-links, which, though they will eventually disentangle under stress, contribute long retardation times and high steady-flow viscosities. The contribution to high steady-flow viscosity and creep resistance will depend upon the number of entanglements per molecule. The lower the entanglement molecular weight, the higher the number of entanglements for a given molecular weight. Polymers with large side groups, such as the acrylic and vinyl ether polymers commonly used in pressure sensitive adhesives, have high entanglement molecular weight, in the range of 40,000 to 50,000. On the other hand, the entanglement molecular weight of common synthetic rubbers is of the order of 10,000 or less. Thus the number of entanglements per molecule of a typical synthetic polymer pressure sensitive adhesive of molecular weight one million is 20 to 25; that of natural rubber of the same molecular weight, about 100.

The addition of low molecular weight diluents reduces the number of entanglements per unit volume and increases long-term compliance and flow. Natural rubber, with its large number of entanglements per unit volume can tolerate a considerable dilution and yet retain adequate creep resistance.

Copolymerization of monomers having bulky aliphatic ether or ester groups with monomers that provide hydrogen bonding, ionic bonding, or acid–base exchange has an effect on compliance similar to that of increasing the prevalence of chain entanglement, thereby raising the holding power of pressure sensitive adhesives. The groups that enhance chain interaction generally raise the glass transition temperature. This will shift the whole mechanical spectrum such that the compliance is reduced at all time intervals of stress application. This can result in insufficient tack at room temperature, and shocky or stick-slip peel if the introduction of the strongly interacting groups is not done with impunity.

The introduction of the ABA type of block copolymers[9–11] confronted the pressure sensitive adhesive industry with technology for physical cross-linking and control of copolymer structure and molecular weight not previously available. The prospect of copolymers that could be melt processed with ease, yet retain the properties of cured rubbers over an acceptable range of use temperatures, was, to say the least, exciting to adhesive formulators.

Because the uniform size of the glassy polymer A blocks and the rubbery polymer B blocks, together with the essentially total incorporation of the A blocks into glassy domains, led to a near perfect network structure with a high level of physical cross-linking, tackification of the ABA elastomers was a difficult problem. This is obvious from the high level of plasticizing oils and soft tackifying resins used in the early formulations. While a high level of creep resistance is retained at room temperature, the migration of low molecular weight compounding agents may cause problems in long-term adhesion and in creep resistance at temperatures much lower than the T_g of the glassy blocks, this last problem resulting from plasticization of the glassy blocks by the tackifying agents.

A compromise is possible by formulating these pressure sensitive adhesives primarily on AB block copolymers, which associate through the A blocks to have an effective molecular weight at room temperature several times the molecular weight of the individual AB molecule. The steady-flow viscosity at room temperature then stands in relation to the melt viscosity as $n^{3.4}$, where n is the degree of association of the ABs. (Appropriate account must also be taken of viscosity dependence on temperature.) The AB block copolymers can be tackified with smaller amounts of tackifying agents than the ABAs, and if further augmentation of creep resistance is desired, this can be done by addition of ABA at low levels, or by low-level chemical cross-linking.[12]

DEGRADATION AND CREEP

Because of the high sensitivity of long-term compliance to molecular weight, compliance measurements are very useful in detecting polymer degradation in pressure sensitive adhesives. This came to the author's attention very forcefully when a problem arose due to iron salt contamination in a rosin tackifier used in a natural rubber-based adhesive. The iron catalyzed oxidation initially caused chain scission of the rubber, with a dramatic increase in long-term compliance. Short-term compliance was not greatly affected.

As time went on and oxidation continued, the pressure sensitive adhesive

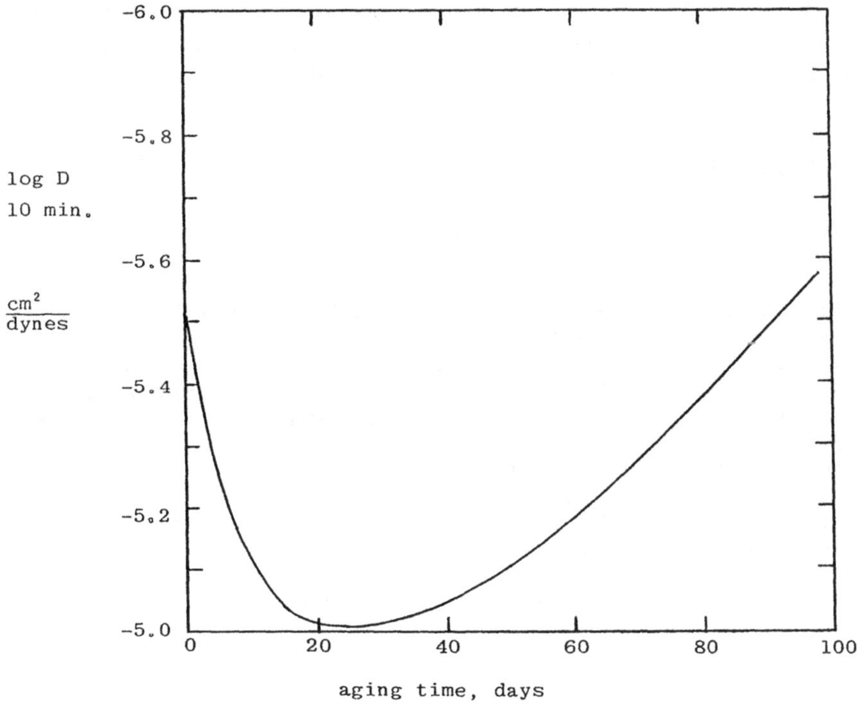

Fig. 6-9. Aging of an iron contaminated rubber-resin pressure sensitive adhesive. Change in 10-min tensile compliance with aging time.

turned leathery, and eventually resinified. Creep compliance then decreased accordingly, and both short-term and long-term compliances were affected (see Fig. 6-9). The degradation produced first a loss of holding power, and eventually a complete loss of tack, but the creep compliance measurements were more sensitive to the early changes than these functional tests.

CONCLUSION

The creep compliance of a pressure sensitive adhesive is significant in tack, adhesion, peel, and resistance to failure in loading. Both retarded elastic processes and steady-flow viscosity play roles, the former in quick stick, bonding, tack, and peel, the latter in long-term creep. In some circumstances the non-Newtonian viscosity of pressure sensitive adhesives must be taken into account.

The steady-flow viscosity is highly molecular weight dependent. Long-term creep can be greatly reduced by cross-linking, both chemical and physical. Long-term creep is also very sensitive to polymer degradation.

REFERENCES

1. Dahlquist, C. A. Tack. *Adhesion Fundamentals and Practice*. McLaren, London, 1966.
2. Hammond, F. H. Jr. Polyken Probe Tack Tester. Adhesion. *American Society for Testing and Metals Technical Publication No. 360,* 1964.
3. Kraus, G., Rollmann, K. W., and Gray, R. A. Tack and Viscoelasticity of Block Copolymer Based Adhesives. *J. Adhesion* **10**(3): 221 (1979).
4. *Test Methods for Pressure Sensitive Tapes*. Pressure Sensitive Tape Council. Glenview, Ill.
5. *Standard Test Method for Holding Power of Pressure Sensitive Tapes*. ASTM D3654-78.
6. Dahlquist, C. A. Pressure Sensitive Adhesives. *Treatise on Adhesion and Adhesives*. Dekker, New York, 1967.
7. Drew, R. G. U.S. Patent 2,410,053 (1945).
8. Ulrich, E. W. U.S. Patent 2,973,286 (1955).
9. Holden, G. U.S. Patent 3,231,635 (1966).
10. Zelinski, R. R. U.S. Patent 3,251,905 (1966).
11. Harlan, J. T. Jr. U.S. Patent 3,239,478 (1966).
12. Dahlquist, C. A., and Kolpe, V. V. U.S. Patent 3,787,531 (1974).

7. Bond Strength and Its Prognosis

Valentinas Rajeckas
Kaunas Polytechnic Institute
Lithuania, USSR

STRENGTH CONCEPT AND FAILURE MECHANISM

The resistance of a body against failure when it is subjected to the action of mechanical forces is referred to as the mechanical strength or simply as strength.

The kinetic concept of strength, which has lately been developed by the Zhurkov school,[1,2] considers failure as the final result of a gradual formation of cracks, their storage and growth. This process is due to thermal fluctuations, and the mechanical stress encourages and directs it. In light of this interpretation, the basic measure of strength is long-term strength of a stressed body, that is, the time from loading to failure. The fact that strength is often defined not by the time until failure, but by the force (or stress) at which the failure occurs, does not change the concept of strength radically, if the force (or stress) acting on the body is considered to be a function of time. The force during failure corresponds to the long-term strength that is achieved by a body subjected to a time-varying load.

It follows from the preceding considerations that a general discussion of strength is pointless if the failure conditions are not described, that is, temperature and the way of loading. The dependence of strength on these factors does not show up for polymers below the glass transition (T_g) temperature. If the test temperature is above T_g, where the effect of temperature and loading time (or rate) is important, the description of testing conditions is absolutely necessary. There are many experimental facts proving that the strength of a body is dependent not only on chemical bonds but also on the energy of intermolecular (van der Waals) bonds as well. The strength depends on the molecular and the supermolecular structure and on the bond energy in those microvolumes where the concentration of stresses is created, and the conditions are made for crack growth.[1–4] There are also other strength concepts worthy of mention.[5–9]

The destruction of polymers in the rubbery state at higher temperatures, when their deformation capability is increased, may occur due to cavitation and the formation of strings. The focuses of failure may become the points of chain fracture and separation of major structural elements. Eventually, when the flow temperature is approached, the deformation of polymers proceeds in accordance with the mechanics of plastic or quasi-plastic flow, that is, the elements of supermolecular structure are displaced irreversibly and the chains slide past one another. In such cases, the destruction of chemical bonds proceeds on a smaller scale than the process at low temperatures, and the strength may be determined by intermolecular bonds. It is known that rigid polymers subjected to a constant load show a long-term strength dependence on stress σ and temperature T. The temperature-time relationship of strength may be described by an exponential expression:

$$\tau = \tau_0 \exp(U_0 - \gamma\sigma/KT) \tag{1}$$

where

τ, U_0, and γ = certain parameters, discussed below
K = Boltzmann's constant, for which the universal gas constant R may be substituted

The empirical formula of this type, expressing a temperature-time relationship of strength when polymer destruction is treated as the kinetic process of crack growth, was suggested by W. Busse,[10] S. Zhurkov,[1] and by F. Bueche.[11]

If the failure occurs through the chemical bonds, then the parameters of this relationship in terms of Zhurkov's interpretation may acquire the following physical meaning: τ_0 is constant, equal to 10^{-12}–10^{-13} sec, and is numerically close to the oscillation period of an atom in a solid body; U_0 has the energy dimension kcal/mole, which readily correlates with activation energy of breaking chemical bonds (for polymers, it correlates with the activation energy of thermal decomposition); γ is the structure sensitive coefficient (kcal-mm^2/mole-kg) that depends on the dimensions and structure of the body. Thus γ may be dependent on the degree of orientation or other technological factors. It quantitatively expresses the nonuniform distribution of stress and its concentration in the area where the focus of failure occurs.

Underexponential numerator in the expression $U = U_0 - \gamma\sigma$ in Equation (1) may be treated as the energy of activation of the thermofluctuation process.

However, Equation (2) may be better suited for elastomers tested in the creep mode within a moderate load range:

$$\tau = C\sigma^{-b} \exp(U/RT) \tag{2}$$

where C, b, and U are certain parameters of the equation. The value of activation energy of the destruction process may be attributed to the parameter U. For nonfilled rubbers, the value is 12–15 kcal/mole, and it shows that the prevalent mechanism of failure is through the intermolecular bonds. For some types of filled rubbers, however, the relationship of Equation (1) holds better. C and b are the parameters, the values of which depend on the type and structure of the material.

The relationship of Equation (2) can be transformed into Equation (1), assuming that the parameter γ is dependent on the temperature, stresses, and on the ratio of parameters τ_0/C. Such a transformation may be useful for the prediction of strength.

Equations (1) and (2) are versatile, and can be applied for the materials at low or high deformations, respectively. The parameters of these relationships, such as γ, U_0 in Equation (1), and C, b, U in Equation (2), may be dependent on the variables (temperature and stress) and their physical meaning may be interpreted in different ways.

For very small stresses, the relationships in question have no practical value, because they cannot be tested experimentally due to very great long-term strength values (when $\sigma \rightarrow 0$, $\tau \rightarrow \infty$).

Practically, the determination of relations between long-term strength, temperature, and stress (or load) is of great importance in structural calculations as well as in strength prognosis. Sometimes the problem of the physical meaning of the parameters of long-term strength equations may not even arise.[1,12]

BOND STRENGTH

Adhesive bond failures are most commonly divided into adhesive (through the interphase surface between the adhesive and the adherend), cohesive (through the adhesive or adherend in the area of the bond), and mixed. According to Bikerman,[13] adhesive failure is impossible if the bonding process is carried out properly and a molecular contact between the adhesive and the adherend surfaces is achieved. Theoretically, this assumption can be justified, but in practical situations, full contact in all actual surface areas being adhered can hardly be achieved, because of the incomplete removal of the absorbed gas layer from the surface, especially from surface fractures, pores, impurities, and weakened surface layers (e.g., oxidized surface).

If the adhesive is used in a solid form, that is, the bonding is performed using pressure sensitive adhesive tape, then the contact is achieved due to the plastic properties of the adhesive layer, and apparently not over the whole surface of the adherend. Therefore, adhesive failure is possible and, sometimes as in the case of temporary adhesive bonds where great strength is not required, it may even be desirable.

For strength evaluation and prognosis, it is essential to state what kind of bonds are established between the adhesive and the adherend, because the total energy value of bond failure and the destruction kinetics of the bond are greatly dependent on it.

Numerous experimental data are available that prove the existence of various bonds between the adhesive and the adherend. Bonds of molecular interaction as well as those of a chemical nature are possible. For both types of bonds the mechanism of failure, in terms of kinetic conceptions, is thermofluctual and failure above brittleness temperature does not occur instantaneously, but during a time interval. These time intervals are considerable when glass transition temperature is approached, especially after a transition to the rubbery state.

For the prognosis of bond strength, it is important that the temperature-time relationship of polymers is qualitatively analogous, irrespective of the kinds of bonds responsible for failure, and that it can be described by Equations (1) and (2). The parameters of these formulas, however, may acquire different meanings in different cases.

Thus, Bikerman's view,[13,14] to treat the adhesive bond strength in the same way as the strength of homogeneous materials, is correct. The data[2,3] about the possibilities of polymer failure through molecular interaction show that there is no principal difference between the failure of materials or their adhesive bonds, and consequently, between their physical strength evaluations.

The numerical values of strength characteristics depend not only on energy required for bond failure, but also on that required for the deformation of the body under a load. Therefore, the mechanical deformation properties of the bond elements—adhesive and adherend—may have a great effect on the strength, especially for soft materials.

The stress concentration on the macro* level is also of great importance to a numerical evaluation of strength.

Generally, in terms of phenomenology, adhesive bond strength is dependent on the overall bond strength between adhesive and adherend, on the contact surface, and on the strength of the adhesive. Sometimes the failure may take place in the adherend. This occurs if both the adhesive bond and the adhesive itself are stronger than the adherend. It can also occur due to the concentration of stresses in the bond, which in turn depends on the bond construction and the mode of force application.

Bond strength between the adhesive and the adherend, as well as the strength of the adhesive in the bond, may depend on many other factors along with the

---

* The stress on the micro level is always distributed unevenly, not only in adhesive bonds but in homogeneous materials as well.

ones already discussed. In the case of pressure sensitive adhesives, the bonding technique is rather simple and its influence is minor. The significance of the relaxational properties of the adherend and the adhesive, as well as the influence of other parameters, especially temperature and deformation rate, may be considerable because pressure sensitive adhesives are rubbery with pronounced plastic properties.

STRESS CONCENTRATION IN ADHESIVE BONDS

Bond stresses may be divided into internal (occurring without the effect of outside forces) and external (occurring under the effect of outside forces).

If the adhesive layer in the bond does not harden after bond formation, but remains elastic and exhibits good relaxational qualities, the internal stresses do not greatly affect the strength. The stresses dependent on outside forces are not uniformly distributed, and thus their concentration occurs. In the bonds of soft materials, stress concentration cannot be avoided.

The quantitative index of stress concentration is of great importance in evaluating the bond strength. It is expressed by a concentration coefficient, which is defined as a ratio of maximum stress in its concentration zone to average stress that is equal to the force that is acting on the unit area of the bond.

It is possible to evaluate the stress concentration coefficient of the bonds of rigid materials for small elastic deformation within the limits of elasticity. For example, Volkersen's or Goland-Reissner's formulas may be used for this purpose in the case of lap joints. The latter is more accurate because it takes into account the bend of the bond in deformation.[15]

An experimental calculating method that enables one to determine the stress concentration coefficient is suggested for the bonds of soft materials for limited (not small) deformations, making use of the polar-optical method of tension measurements. The measurement method and mathematical apparatus of calculation are described.[16]

Here we shall mention some of the more interesting results achieved by testing the properties of stress distribution and its concentration in the bonds of soft materials. Stress tests were carried out for lap joints. Under the action of forces, the shear stresses at the adhesive and interphase surface are most hazardous for such bonds. Figure 7-1 shows the shear stress distribution of rigid bonds for small deformations within the elastic range (Fig. 7-1a) and of soft bonds for rubbery deformations (Fig. 7-1b). In the first case, the tension is distributed unevenly only in the length direction.

In the bond loaded by a single-axis load, a linear stress state is formed and stress concentration uniform over the width of the sample occurs at its

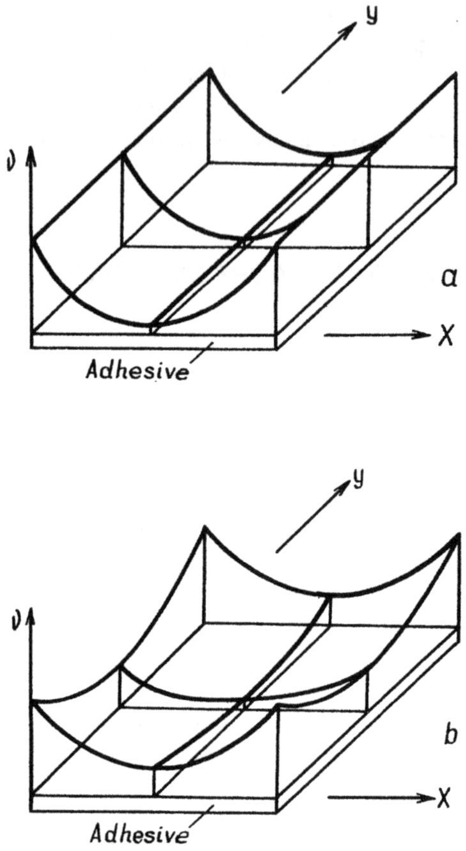

Fig. 7-1. Stress distribution in adhesive bonds: (a) for rigid solids; (b) for soft materials.

ends. In the second case, the stresses are distributed unevenly, not only in the length direction, but transversely as well. A flat stress state is formed, although the load is a single-axial one. The maximum stress concentration occurs at the corners of the bond.

Transverse deformation is the reason for this kind of tension distribution. Relative transverse deformations of soft materials may be of the same order of magnitude as the longitudinal ones. At the ends of a lap joint, where there is a transition from a double thickness of the adherend to a single thickness, causing a decrease of rigidity, additional stresses in the transverse direction occur. For joints of hard material, when the adherend deformations are small until failure, this effect does not show up. Thus, Figure 7-1 shows two possible stress distributions, depending upon the magnitude of deformations of the adherend and of the bond.

The nature of tension distribution for soft material bonds depends on their width. Reduction of the width causes the flat state of stress to approach the longitudinal one as shown in Figure 7-1a.

In double-axis loading of the bonds of soft materials, the nature of stress distribution remains the same as in the case of single-axis loading, except the numerical values of the concentration coefficient are greater.

The value of stress concentration in the case of bonds of rigid materials[15] depends on the mechanical deformation properties and the thickness of the adherend and the adhesive. In lap joints, it depends on the length of bond. Such relationships hold for bonds of soft materials as well, except that the stress value depends on the bond width, unlike in the case of rigid materials (Fig. 7–2). This relationship is due to considerable transverse deformations, which depend on the bond width, occurring along with longitudinal ones. In bonded soft materials, Poisson's coefficient decreases when the sample width increases.[16] Thus, the narrower the sample, the greater is the relative transverse deformation (compared with the longitudinal one), and the greater are the stresses occurring at the concentration points.

With the decrease of the sample width, the longitudinal stress state is approached. The stress concentration is greater for narrower samples than for wider ones under the same deformations and other conditions, but the stresses at the end of the bond will be distributed more uniformly in the width direction. The wider the bond of soft materials, the greater the possibility of failure at the corners of the bond.

The stresses in the bond increase from the time of force application until bond failure. In the case of bonds of soft materials, this process is followed by elastic deformation and the time interval until failure may be of the same order as the stress relaxation time characteristic for the particular adhesive and adherend. This may influence the stress concentration coefficient. Figure 7-3

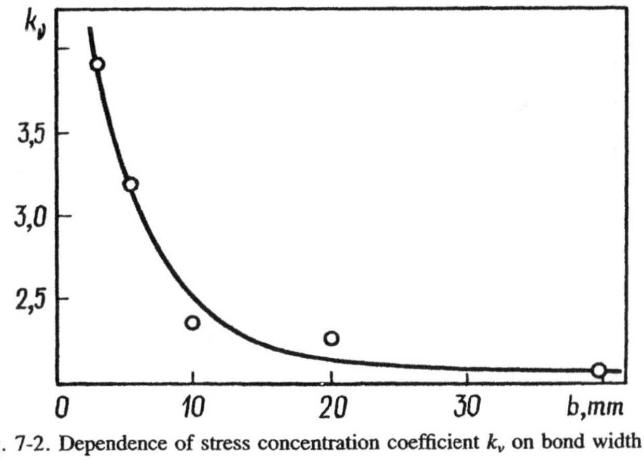

Fig. 7-2. Dependence of stress concentration coefficient k_ν on bond width (b).

shows the dependence of the stress concentration coefficient for lap joints on the load (Fig. 7-3a) and the deformation (Fig. 7-3b). The data on joints formed by bonding polyurethane rubber with different adhesives are shown in Table 7-1. The rubber was deformed only reversibly (the actual elastic deformation). The stresses were measured by the polar-optical method.[16] The samples were loaded by constant loads of various values until the time of failure. The deformations were measured in the area of stress concentration; in the adherend, at the end of the joint. Along with this, the relaxation time index was determined for the samples cut out of films cast on Teflon. The relaxation time index is defined as the time range in which the stresses decrease *e* times, *e* being the base of the natural logarithm.

It is apparent from Figure 7-3 and Table 7-1 that the stress concentration coefficient is decreased considerably, if adhesives with good relaxation capabili-

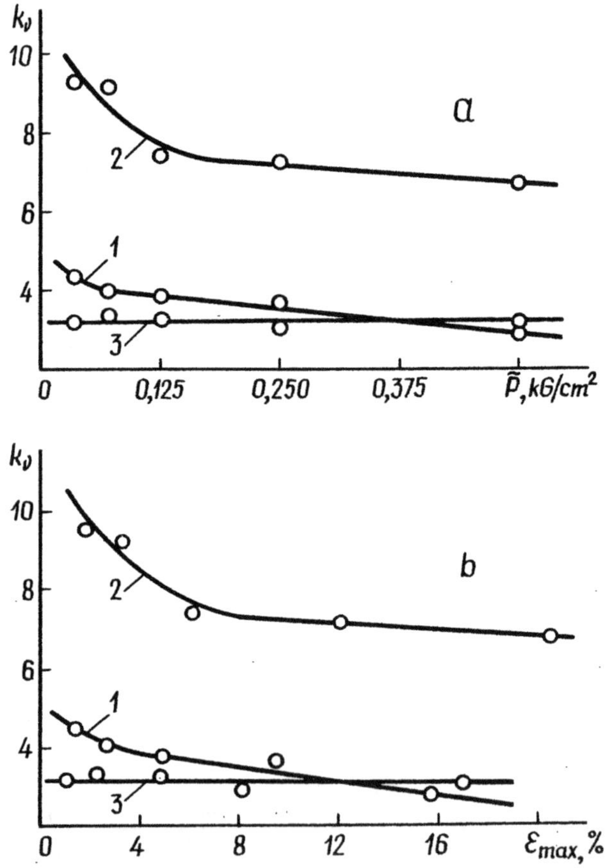

Fig. 7-3. Dependence of stress concentration coefficient on load (a) and deformation (b).

Table 7-1. Data on Polyurethane Rubber Joints Bonded with Various Adhesives

ADHESIVE	RELAXATION TIME INDEX (sec)	RELATIVE DECREASE OF STRESS CONCENTRATION INDEX Δk_v (%)
1. Nitrile rubber	2.5	34
2. Polychloroprene (100), butylphenolformaldehyde resin (40), zinc oxide (10, thiuram (5)	15.5	28
3. Silicone rubber adhesive	510	0

Note: $\Delta k_v = [(k_{v\bar{p}(\min)} - k_{v\bar{p}(\max)})/k_{v\bar{p}(\max)}] \cdot 100$, where $k_{v\bar{p}(\min)}$ and $k_{v\bar{p}(\max)}$ are the concentration coefficients for minimum and maximum loads used experimentally.

ties are used, such as adhesives 1 and 2. In this case, stress relaxation is caused by small creep deformations that appear long before the time of bond failure. In the case of the silicone rubber-based adhesive, characteristic in reversible deformations,the decrease of stress concentration coefficient was not detected.

The stress relaxation due to conformative rearrangement of the chains was of no essential value, but stress distribution due to rearrangement of structural elements (creep kinetics) produced some effect during the short time of the load duration. This does not necessarily mean that at longer time intervals the first relaxation mechanism cannot have any influence on the stress concentration coefficient.

The characteristic of stress distribution in lap joints is stress concentration at the ends of the bond, the dependence of concentration coefficient on the sample width, and instability of stress concentration in those cases when the adhesive exhibits good properties of creep and elastoplasticity. Generally, these properties appear in other types of bonds of soft materials as well. The instability of stress concentrations may also appear in bonds of rigid materials, depending on the relaxational properties of the adhesive.

The preceding facts should be taken into account when evaluating the bond strength. The stress concentration at the ends is significant for the evaluation of failure points in the structure. The effect of width shows that in comparing the relative strength of different bonds, their former width should be considered. Instability of stress concentration points to the possible relative strength increase when the adhesive has a high relaxational capacity. This often causes deviation from expected behavior.

THE EFFECT OF MECHANICAL PROPERTIES OF THE ADHESIVE AND THE SUBSTRATE

The fact that the cohesive strength of the adhesive influences the bond strength is supported by various references in the literature.[13,17,18] The cohesive strength of the adhesive may be varied in different ways: by increasing its molecular weight, increasing the number of functional groups, etc. This increases the bond strength until the cohesive strength becomes greater than the strength of the adhesive bond, or until the stresses within the oriented adhesive layer concentrate and achieve a critical value exceeding that capable of destroying the interaction bonds that are responsible for the adhesive strength in the nonoriented mass portion. In such cases, the direct link between bond strength and the cohesive strength cannot be expected.

Correlation of the mechanical deformation characteristics of free adhesive films and bond strength may be interesting.[16]

Various adhesives, those based on nitrile and polychloroprene rubbers and on chlorinated poly(vinyl chloride) resin, were studied. Seventy different bond variations were obtained, different in the composition of the adhesive or the adherend. The bond strength was experimentally determined by shear tests, evaluating the mode of the failure. The correlation between the bond strength and mechanical deformation characteristics of the adhesive films was analyzed.

Irrespective of the type of adhesive and adherend, the following relationships were observed.

1. In the case of cohesive failure, there exists a direct linear relationship between bond strength and the tensile strength of adhesive films, as well as their moduli of elasticity
2. In the case of interfacial failure (determined visually), there exists a direct linear relationship between bond strength and deforming capacity of the adhesive films, which is characterized by elongation at stresses of about 75% from the fracture stress. In both cases, the correlation coefficients are on the order of 0.85–0.95

Thus, in the first case, the bond strength can be increased by increasing the strength of the adhesive. In the second case, the bond strength may be increased by developing the deformation qualities of the adhesive. The latter may be understood if the formation of a contact between the adhesive and adherend is regarded as a kinetic process: better deformation qualities of the adhesive correspond to greater mobility of segment molecules. The increase of the adhesive bond strength may, in this case, result in the decrease of the cohesive strength of the adhesive and the required bond strength may not be achieved. The practical approach, for example, in developing adhesive formulations is the solution of an optimization problem. Here the application of the methods of mathematical experiment planning may be effective.

It is observed that there exists a close correlation link ($\Gamma \geq 0.9$) between the strength indices obtained by shear and peel tests, if the mode of the failure is the same in both cases. Such a link is absent if the nature of the failure is different. Thus it is possible to evaluate the strength of one case, if the data obtained in the other tests are known.

The bond strength is affected by mechanical deformation properties of the adherend. This fact was noticed long ago[15] and is explained as the effect of mechanical properties of the adherend on the value of the stress concentration coefficient. It decreases with the increase of adherend stiffness.

Numerous equations for peel strength evaluation[13,18–22] derived in accordance with some geometrical models suggest that the relationship of the form

$$P = aE_s^n \tag{3}$$

exists between peel strength P and adherend modulus E_s (or stiffness), where a is the coefficient depending on the mechanical properties and geometrical parameters of the adhesive and $n < 1$. In the equations suggested in References 13, 19, and 20, $n = 0.25$.

This relationship for lap joints was investigated experimentally by the author.[16] The stiffness of the adherend was varied in two ways: a thin layer of the adherent was backed up by another reinforcing layer, or the adherend was backed up by nitrile rubber strips of various degrees of vulcanization.

In the case of an adhesive of a very high relaxational capability when the bond failure is cohesive, the relationship between the adherend stiffness D* and bond strength P is of an extremal nature [Fig. 7-4(1)]. The strength maximum may occur at different values of D and is dependent on the time range until failure, which in its turn depends on the deformation capability of the adherend and adhesive and of the bond strength. This generally determines the actual values of the deformation and the time to failure. When the adhesive does not display the preceding qualities and when the failure is interfacial proceeding at high speed, the relationship $P(D)$ does not exhibit an extremal nature [Fig. 7-4(2)] and is readily approximated in semilogarithmic coordinates as a linear function. The part of the extremal relationship $P(D)$ up to D values close to the strength maximum is approximated rather well by this function.

The nature of an extremal relationship is relaxational. It is caused by the decrease of stress concentration under the load, before the beginning of the bond failure, as shown in Figure 7-3.

---

* Stiffness for tensile stress $D = E_s \cdot F$, where E_s is a relative modulus of adherend elasticity (in this case, at deformations $\varepsilon \sim 75\%$) and F is the cross-sectional area.

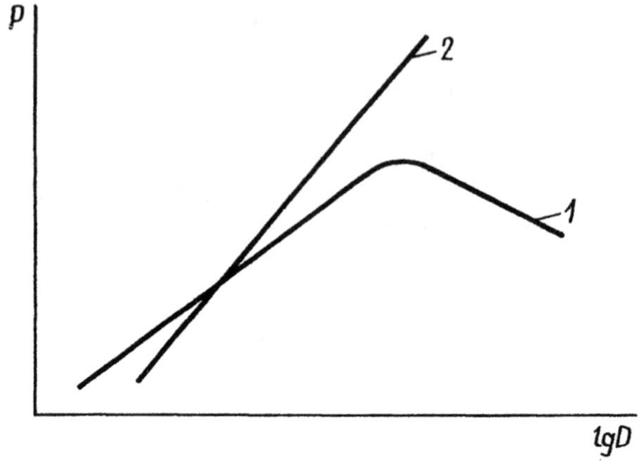

Fig. 7-4. Stress dependence on the stiffness of the adherend: (1) extremal; (2) nonextremal.

THE INFLUENCE OF SCALE FACTOR

The influence of scale factor on bond strength is very diverse. The dependence of strength on the adhesive thickness is most widely studied. The relationship depends on the test method.

For lap and butt joints, this dependence is hyperbolic [Fig. 7-5(1)] as shown by numerous experiments[13,16,23] to be:

$$P = ah^{-m} \tag{4}$$

where

P = strength
h = thickness of the adhesive layer
a, m = parameters

For most adhesives exhibiting elastoplastic properties, the value of m is 0.1–0.3.

In T-peel tests, the strength usually increases with increasing thickness of the adhesive layer up to a certain limit. This follows from the theoretical equations[13,19,20,22] for the evaluation of the peel strength. At higher adhesive thickness, this no longer holds and sometimes the relationship $P(h)$ is of an extremal character [Fig. 7-5(2)]. This could be understood in reference to the general laws of tear deformation kinetics. For low adhesive thicknesses, the deformation possibilities are limited and the process of failure proceeds in the form of small angle crack. The deformation in the adhesive increases with increasing thickness and the configuration of the crack changes: the angle in-

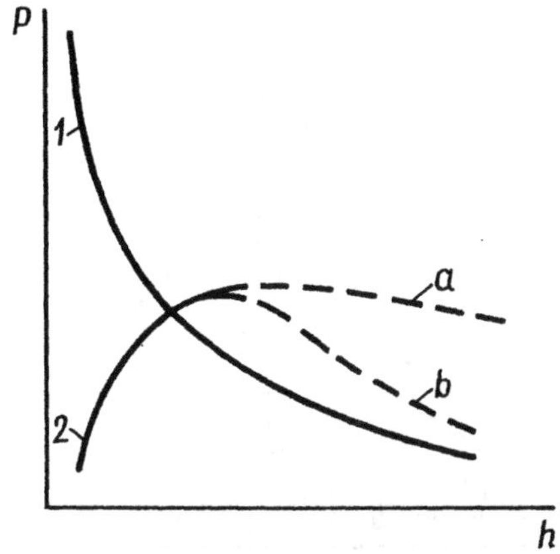

Fig. 7-5. Strength dependence on the thickness of the adhesive layer: (1) lap joints and butt joints; (2) T-peel joints.

creases, and at the same time, the stress concentration decreases corresponding to an increase in strength. After the adhesive layer thickness reaches a certain critical value, at which the local deformation in the crack area is sufficiently large, the effect of thickness may not be noticeable anymore [Fig. 7-5(2)a], or it may be similar as in the case of other tests [Fig. 7-5(2)b]. The problem is discussed in greater detail in reference 16.

The dependence of concentration coefficient on the bond width (Fig. 7-2) is also of some importance including its stability or instability under the load effect (Fig. 7-3), which is related to the relaxational capability of the adhesive.

In cases when the stress concentration coefficient does not decrease until the failure, relative strength usually increases with increasing bond width until a certain limit is reached. When the stress concentration coefficient decreases due to stress relaxation, the relationship $P(b)$ is nonmonotone in the limited range of bond width. Generally, the dependence of strength on the bond width is most noticeable in the samples of narrow width up to $b \sim 25$–30 mm.

THE INFLUENCE OF TEMPERATURE AND LOADING REGIME*

It has been noted[16,22,24–26] that the nature of bond strength dependence on temperature is extremal (Fig. 7-6). This character with one or more maximum values does not depend on the test method. It has been shown that the strength maximum corresponds to the transition of the adhesive from a solid to a highly elastic state.[25] This shows that the strength maximum is of a relaxational nature,

* The value of load and its variation in time is referred to as the loading regime.

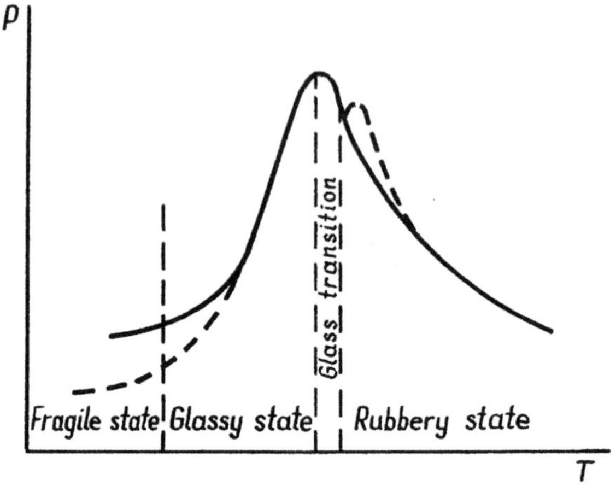

Fig. 7-6. Strength dependence on temperature.

and the greater the stress concentration, the more pronounced the maximum. Sometimes the change of the mode of failure is observed from an interfacial to a cohesive failure. In cases when the failure mode does not change when passing through a maximum, a second local maximum related to the transition of the mode is observed.[24] For adhesives of crystalline structure, local strength maximum is observed in the area of crystal flow.[25]

For the purposes of prognosis, the dependence of strength on temperature should be studied separately for the temperature range in which the adhesive is in a solid state (glassy or crystalline) and in a highly elastic state. Pressure sensitive adhesives at normal temperatures are in the state of high elasticity with some viscous properties.

The most abundant information about the dependence of bond strength on temperature and loading mode is provided by the study of temperature-time dependence. The experimental data of bonds of soft materials, as well as bonds of thin layers of any stiffness, are most conveniently obtained by testing lap joints in creep mode, that is, making long-term strength experiments at constant loads.

The dependence of long-term strength τ of bonds on temperature T and static load $\tilde{P}$ is described by the form of Equations (1) and (2), in which a load term is substituted for the stress:[16]

$$\tau = \tau_0^* \exp(U_0 - \gamma\tilde{P}/RT) \tag{5}$$

$$\tau = C\tilde{P}^{\,-b} \exp(U/RT) \tag{6}$$

The parameters U_0, γ, C, and U could be given the same meaning as in Equations (1) and (2). As to the physical meaning of the parameter τ_0^*, no uniform opinion has been achieved so far. Sometimes, when its values are 10^{-12}–10^{-13}, it can be treated in the same way as in Equation (1). Often, however, for bonds $\tau_0^* > 10^{-12}$, it is possible to assume that $\tau_0^* = \tau_0 \cdot C$, where $\tau_0 \sim 10^{-13}$, and the values of C may vary over a wide range and depend on the variables T or $\tilde{P}$.

Equation (6) can, of course, be transformed to the form (5) analogous to the way Equation (2) can be transformed to (1). This is handy in theoretically studying the temperature-time dependence.

At constant temperature, Equation (5) acquires the form

$$\tau = A \exp(-\alpha\tilde{P}) \tag{7}$$

and at constant load

$$\tau = A_1 \exp(U/RT) \tag{8}$$

Here, A and α are the parameters dependent on temperature; A_1 and U, on the load (U = constant is also possible).

Figure 7-7 schematically shows various possible and experimentally obtained

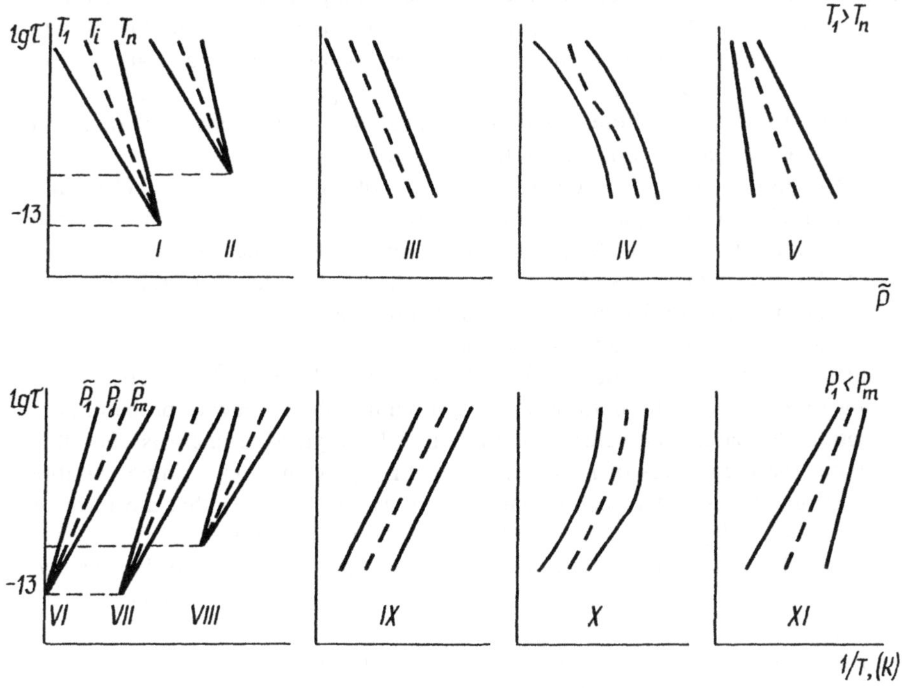

Fig. 7-7. Long-term strength (τ) dependences on static load ($\tilde{P}$) and temperature (T) in semilogarithmic coordinates.

relations between long-term bond strength and the load (at fixed temperatures, T_i = constant) or the temperature (at fixed loading values, $\tilde{P}_j$ = constant).

It is obvious that long-term strength dependence on load and temperature in semilogarithmic coordinates may have various shapes. They may run to a single-point pole, with the ordinate value of -13 (i.e., $\tau_0^* \geq 10^{-13}$) or greater, and the abscissa value log τ and $1/T$ equal to zero or greater; they may show parallel or converging linear relationships with decreasing load and increasing temperature; they may be nonlinear and have different directions with regard to each other; they may be lines running from each other as the load increases and the temperature decreases.

The possibilities of defining the dependences $\tau(\tilde{P}, T)$ by Equations (5) and (6) were studied theoretically.[16] Referring to the abundant experimental data, the parameters τ_0^*, U_0, and γ of Equation (5) are shown to be constant only in some cases. In elastomeric adhesives with plastic properties, some of these parameters are dependent on the variables.

In seeking qualitative or quantitative information on the long-term bond strength dependence on temperature and load, the first stage is an experiment. The values of long-term strength should be determined for 3 or 4 levels of temperature and 4 or 5 of constant load for each temperature value. The number of samples should be selected considering the high long-term strength test data dispersion. The variation coefficient, even for homogeneous materials, is 20–30%. It can be reduced to 10–12% only if the experiment is very thoroughly prepared, that is, the differences in mechanical deformation characteristics and thickness of the adhesive layer are kept as small as possible.

The next stage is to construct a graph of the relationships $\tau(\tilde{P})_{T_i}$ and $\tau(T)_{\tilde{P}_j}$, as shown in Figure 7-7. The relationship $\tau(\tilde{P}, T)$ is described by the form of Equation (6), if $\tau(\tilde{P})_{T_i}$ are not straight lines in the coordinates log τ and $\tilde{P}$, but rectify in the coordinates log τ and log $\tilde{P}$. The plot of the relationships $\tau(\tilde{P})_{T_i}$ and $\tau(T)_{\tilde{P}_j}$ is sufficient for qualitative prognosis.

The quantitative prognosis and the evaluation of strength characteristics require the mathematical description of experimental data $\tau(\tilde{P}, T)$. The values of log τ are distributed according to the normal distribution law of probabilities. Equation (5) parameter values are evaluated by regressive analysis. For this purpose, the parameters of long-term strength dependence on a single variable for various values of the second variable are evaluated. After that the form of the parameter dependence on another variable and its numerical values are determined and substituted into the equation of long-term strength dependence on the first variable. This gives the final form of the dependence on two variables.

It is methodically more correct to describe first the dependence on that variable that shows a simpler graphic form of the dependence with respect to its description by Equation (5). This can be verified visually or by linear approximation of the experimental data using the mean square method and evaluating approximation error. Usually the form of the dependences for the adhesive bonds is evaluated first according to Equation (7). Then the dependence of the parameters A and α on temperature is evaluated and finally, substituting $A(T)$ and $\alpha(T)$ into equation (7) gives Equation (5).

In describing mathematically the dependences $\tau(\bar{P}, T)$, whatever graphic form of the dependences $\log \tau(\bar{P})_{T_i}$ and $\log \tau(1/T)_{\bar{P}_j}$ may assume, the value of the parameter τ_0^* should be determined first. In the case of running these dependences into the pole, τ_0^* value could be evaluated from experimental data by approximation methods. The values of the parameters U_0 and γ could be then evaluated accurately for the parameters as constants and as functions of variables. If the determination of accurate numerical values τ_0^* from the experimental data is not possible, then the accurate values of other parameters also remain unknown. Then, it is possible to consider that the description of the dependence $\tau(\bar{P}, T)$ by Equation (5) is strictly empirical and the prediction possibilities of the equation qualitatively remain rather limited.

Another view of this question is also possible. Numerous experimental data show[1] that the pre-exponential multiplier τ_0 in Equation (1) is a constant of the order of 10^{-12}–10^{-13} only for rigid materials, when the failure occurs through chemical bonds at small deformations. For elastomers, it has been noticed that this parameter may be equal to 10^{-5}–10^{-6}, or even greater, depending on the material. Bartenev[4] explains this as a dependence of activation energy (parameter U_0) on temperature. In this case, as has already been mentioned, $\tau_0^* = \tau_0 \cdot C$. According to this interpretation in describing experimental data by Equation (5), it is possible for the pre-exponential multiplier to acquire the value of 10^{-13} and the multiplier C may be placed under the exponent. Such an approach in describing experimental data mathematically could be applied in the case where the pole does not exist.

Here are some characteristic examples. In Figures 7-8 and 7-9, the relationships $\tau(\bar{P}, T)$ of cellulose diacetate films for the lap joints is represented. In the first case (Fig. 7-8), the films were bonded by a two-component adhesive, which consisted of polyurethane elastomer cross-linked by isocyanates. This polymer does not exhibit plastic properties due to its three-dimensional structure. In the second case (Fig. 7-9), the adhesive was nitrile rubber exhibiting plasticity, which is very clearly expressed at higher temperatures. The thickness of the layer of the adhesive in both cases is about 150 microns. This failure is cohesive. As seen from Figure 7-8 in the first case, the pole exists at $\tau_0^* \sim 10^{-13}$. The relationship $\tau(\bar{P}, T)$ is described by Equation (5), the parameters of which are U_0 = constant and γ = constant. The numerical values are $\tau_0^* = 3.2 \times 10^{-13}$ sec, $U_0 = 28.6$ kcal/mole, $\gamma = 0.72$ kcal-mm^2/mole-kg.

The relationships obtained in the second case are described by Equation (6); here the relationships $\log \tau(\bar{P})_{T_i}$ are straight lines in the coordinates $\log \tau$ and $\log \bar{P}$. The parameter U in Equation (6) is, in this case, a constant (log

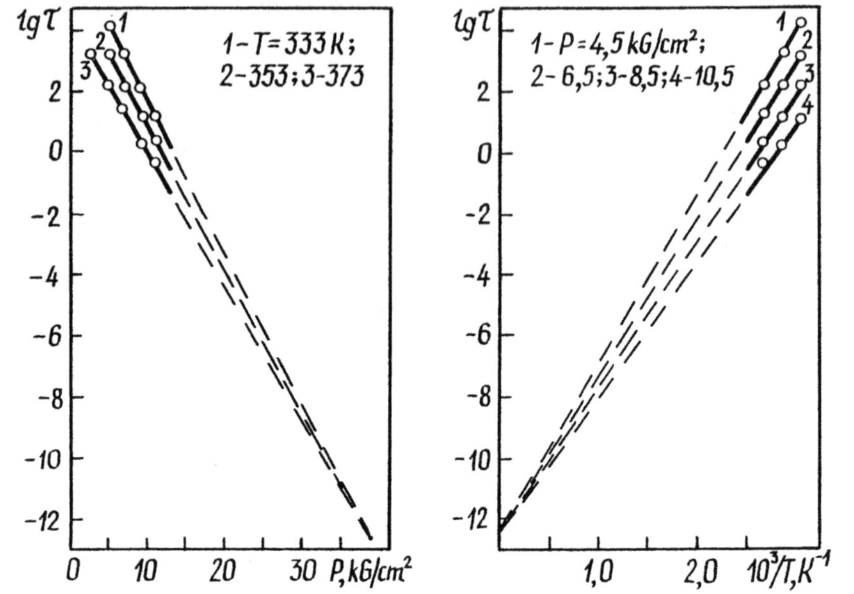

Fig. 7-8. Long-term strength (τ) dependences on static load ($\bar{P}$) and temperature (T) for cellulose diacetate film bonded with polyurethane adhesive.

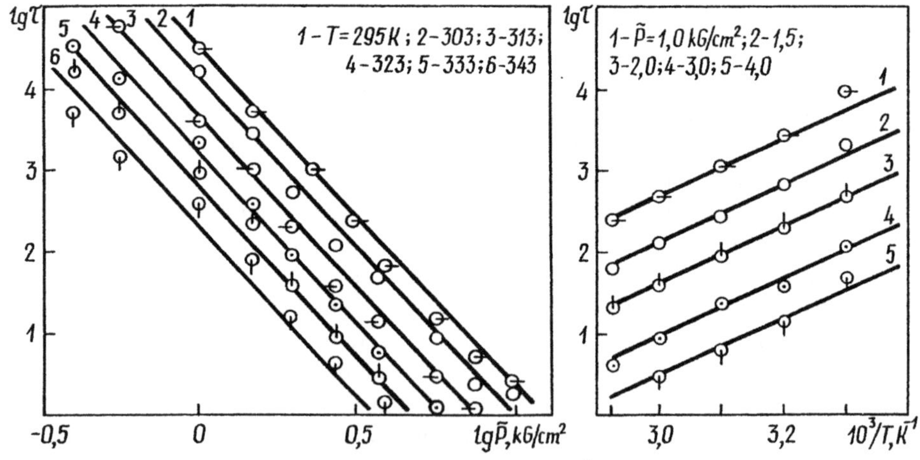

Fig. 7-9. Long-term strength (τ) dependences on static load ($\bar{P}$) and temperature (T) for cellulose diacetate film bonded with nitrile rubber adhesive.

$\tau(1/T)_{\bar{P}_j}$, the lines are parallel). This means that the activation energy of the failure does not depend on the load. This is generally characteristic of elastomers.[4]

After transforming Equation (6) to (5), assuming that $\tau_0^* = 10^{-13}$, we obtain that the parameters of (5) are U_0 = constant and $\gamma = \gamma(\bar{P}, T)$ of $\gamma = (a_1 \ln \bar{P} - a_2)T/\bar{P}$. The numerical values of these parameters for a given system are the following: $U_0 = 18$ kcal/mole, $\gamma = (0.031 \ln \bar{P} - 0.075)T/\bar{P}$ kcal-mm^2/mole-kg.

It is interesting to consider a case when the relationships log $\tau(\bar{P})_{T_i}$ and log $\tau(1/T)_{\bar{P}_j}$ are straight lines running away from each other with the increase of the load and decrease of temperature (Fig. 7-7). Relationships of this type are obtained experimentally for adhesive bonds where the failure was interfacial.[27] By changing the bonding technique, cohesive failure was obtained for the same system and the graphs of the relationship log $\tau(\bar{P})_{T_i}$ and log $\tau(1/T)_{\bar{P}_j}$ had another direction. They ran into the pole at $\tau_0^* \sim 10^{-10}$. This experimentally demonstrated fact can be interpreted in the following way: if the adhesive interaction is of adsorbing character and the bond failure occurs at the interfacial surface, the strength depends on intermolecular forces. As the temperature increases, the number of intermolecular bonds carrying the load decreases. The higher the temperature, the greater should be the intensity of long-term strength decrease with increasing load, because the stress per bond will increase more intensively, when the number of bonds on which the load is distributed is lowered. Similarly, the greater the load, the more significant will be the effect of temperature on long-term strength.

When the relationships log $\tau(1/T)_{\bar{P}_j}$ approach each other with decreasing temperature, the underexponential numerator in Equation (5), $U = U_0 - \gamma\bar{P}$, becomes a growing function. This may occur when $U_0 \neq U_0(\bar{P})$, if $\gamma = \gamma(\bar{P})$ and $\gamma(\bar{P})$ is a decreasing function, such that $|d\gamma(\bar{P})| > d\bar{P}$. If log $\tau(1/T)_{\bar{P}_j}$ straight lines approach each other with increasing temperature, then $U(\bar{P})$ becomes a decreasing function.

All common types of bond dependence on load and temperature as shown in Figure 7-7 will be generalized by selecting two extremal cases. One of them is when adhesive failure occurs only through chemical bonds. This corresponds to the cases of dependences log $\tau(\bar{P})_{T_i}$ and log $\tau(1/T)_{\bar{P}_j}$ as shown in Figures 7-7(I), (VI). The pole is present and for the relationship log $\tau(1/T)_{\bar{P}_j}$, it is on the ordinate axis, so that $\tau_0^* = \tau_0 \sim 10^{-12}$–$10^{-13}$. Another case concerns the failure occurring only through the intermolecular interaction, as shown in Figures 7-7(V), (XI). One would expect that the failure taking place through both chemical and intermolecular bonds will result in intermediate cases demonstrated by the displacement of the pole from the ordinate axis for the relationship log $\tau(1/T)_{\bar{P}_j}$, to the area of lower temperatures [(VII), (VIII)], the parallelism of lines log $\tau(\bar{P})_{T_i}$ and/or log $\tau(1/T)_{\bar{P}_i}$ [(III), (IX)], and the nonlinearity of these relationships [(IV), (X)]. This has been experimentally verified. It is characteristic of the bonds when the adhesive is in a rubbery state, but not purely elastic (i.e., the effect of plasticity comes up under the load and failure occurs according to the mechanical characteristic of elastomers when the processes of creep and cavitation develop). The stronger the viscoelastic properties of the adhesive, the more complicated is the form of the relationships log $\tau(\tilde{P})_{T_i}$ and, especially of log $\tau(1/T)_{\tilde{P}_j}$.

The practical significance of the preceding facts and the discussion of long-term strength dependence on load and temperature in reference 16 is summarized below.

If the graphic picture of relationships log $\tau(\tilde{P})_{T_i}$ is the one shown in Figure 7-7(I), and log $\tau(1/T)_{\tilde{P}_j}$ as in (VI), it means that failure occurs through the chemical bonds. Such bonds do not exhibit creep, at least in the range of loads and temperatures in which the experiment was carried out, and their failure (i.e., the growth of the main crack) occurs at relatively high speed. The magnitude of strength characteristics, is therefore, only slightly different at the beginning than at the end of failure. Cases II, VII, VIII are characteristic when intermolecular forces are also responsible for the strength. In this case, a lower thermal resistance and higher creep is expected. Cases III, IV, IX, X are characteristic of bonds where the adhesive, especially at higher temperatures, exhibits creep behavior and failure through intermolecular forces prevails. Such adhesives have a lower thermal resistance, may be sticky, and have a greater autoadhesive capability. In addition, cases IV, X, depending on the direction and slope of the curves, show that the structure of the adhesive is especially temperature-sensitive and, at certain temperature intervals, nonmonotone variations of the structure take place. These variations are related to the changes of deformation properties and failure mechanism because of recrystallization, crosslinking, orientation, etc. Cases V, XI are characteristic of bonds where the adhesion is a result of intermolecular forces and the failure is interfacial.

The temperature and load sensitivity of different bonds can be seen from the slope of the curves.

The long-term strength can be quantitatively evaluated from Equation (5) for various T and $\tilde{P}$ values.

Using the numerical values of the parameters of Equation (5), which were obtained experimentally at static conditions at constant loads and temperatures, quantitative information about the strength properties of bonds at variable loads and temperatures can be obtained.

As the data accumulated in the author's laboratory show, the value of errors resulting from long-term strength evaluation from Equation (5) is related to the extent of the complication (Fig. 7-7) of the relationships $\tau(\tilde{P})_{T_i}$ and $\tau(T)_{\tilde{P}_j}$, and thus is related to the form of the parameters of Equation (5), The error is least (10–15% or less) when these parameters are constant, and greatest

(15–20%) when at least some of the parameters are dependent on both variables. Using Equation (5) for the evaluation of other strength characteristics, as described below, the error may amount in some cases to 25–30%.

Thus, Equation (5) and occasionally (6) are sufficiently versatile and, in terms of information, sufficiently receptive. Certainly its derivation is more time-consuming than the process of standard dynamometric tests. But if the aim is to obtain detailed information on particular adhesives or adhesive bonds, and to accumulate a bank of information, then the tests described here are useful and effective.

THE EVALUATION OF LONG-TERM STRENGTH AT VARIABLE LOADS

The bonds in structures and products are often subjected not to constant but to time-varying loads, which may be increasing, decreasing, cyclically applied, or may vary with rest periods. If the failure is assumed to be the overall result of individual nonreversible local failures, then the principle of superposition of failures, or Bailey's criterion, can be applied for the determination of long-term strength. The mathematical expression of this principle at n discrete loads is

$$\sum_{i=1}^{n} \Delta t_i/\tau(\bar{P})_i = 1 \qquad (9)$$

and at steady load

$$\int_0^{t_p} dt/\tau(P\,(t)) = 1 \qquad (10)$$

where

t = time
t_p = time corresponding to the failure of the load
$\tau(\bar{P})$, $\tau(P(t))$ = the functional dependences of long-term strength on the load

Using Equations (9) and (10) to determine the long-term strength of the bonds, no limitations concerning the failure type need be applied a priori, whether the failure is cohesive or interfacial. It is known, however, that the principle of superposition does not always hold for the adhesive bonds[16] or for polymers in general.[1] Different structural changes may occur in the adhesive, when the time variable load is applied. A self-healing effect may be observed at a higher frequency of cyclic loads. Alternating loading periods with rest periods can cause self-healing of the defects,[25] resulting in strengthening if the adhesive has good relaxational properties.

In such cases, Equations (9) or (10) cannot be used directly, but the additional effects must be evaluated first. This can be accomplished by introducing additional terms into the equations. Thus, the possibility of evaluating long-term strength at varying loads and the validity of the superposition principle should be tested experimentally for every particular case. One of the ways to do this is the following. Long-term strength is determined experimentally at constant load $\tilde{P}_j$. Then the bonds are loaded by the same load $\tilde{P}_j$ and are kept for a shorter time than τ, that is, $\Delta t_1 < \tau$. Then, after the removal of the load, the rest period follows and the same load is applied again until failure (time Δt_2). If $\Delta t_1 + \Delta t_2 = \tau$ is satisfied statistically, the principle of superposition holds. If $\Delta t_1 + \Delta t_2 > \tau$, then the rest period strengthens the adhesive bonds. If $\Delta t_1 + \Delta t_2 < \tau$, the rest period weakens it, compared to the long-term strength that would be attained at continuous load application.

The problems associated with the application of Bailey's criterion may be of a double nature: a so-called direct one, that is, the evaluation of long-term strength at varying loads when the parameters of the equation at constant static load are known, and the inverse, that is, the derivation of the long-term strength equation at constant static loads from the short-term tests at varying loads (using a conventional test with a dynamometer).

Attention should be directed to some interesting theoretical aspects of this question. Let us assume that the superposition principle holds and at a constant static load the relationship $\tau(\tilde{P}, T)$ is described by Equation (5). Bailey's criterion then has the form

$$\int_0^{t_p} \frac{dt}{\tau_0 \exp(U_0 - \gamma P(t)/RT)} = 1 \tag{11}$$

Two characteristic aspects of long-term strength may be solved in accordance with the scheme of the direct problem:

1. Long-term strength may be evaluated from Equation (11) at varying load and at any constant temperature T_i, if the parameters τ_0^*, U_0, and γ of Equation (5) and the law $P(t)$ of the load variation with time are known.
2. Long-term strength may be evaluated in cases when the load variation $P_1(t)$ changes into $P_2(t)$.

Assume that load $P_1(t)$ was in effect during time t_1 (where $t_1 < t_p$), and after that, its effect changed to $P_2(t)$. Then the remaining time until the failure after t_1 will be $\Delta t = t_p - t_1$. It may be calculated using the equation

$$\int_0^{t_1} dt/\tau\ (T_iP_1(t)) + \int_{t_1}^{t_p} dt/\tau\ (T_iP_2(t)) = 1 \tag{12}$$

This method, of course, is also applicable when the law of force changes n times. Then the first member of Equation (12) becomes the sum of integrals.

In solving an opposite problem at constant temperature values (T_i = constant, $i = 1, 2, \ldots$), with the help of Bailey's criterion

$$\int_0^{t_p} dt/A\ \exp(-\ \alpha P(t)) = 1$$

the values of the parameters of Equation (7), A and α, are found. Then, assuming that $A = A(T)$ and $\alpha = \alpha(T)$, and having determined the form of these relationships, the values of the parameters U_0 and γ of Equation (5) may be found.

While determining A and α values at fixed temperatures, the relationship $P = P(t)$ of the load variation with respect to time is selected, and the values of time to failure $t_{P(e)}$ are experimentally determined. The load variation should be so selected that the integral of Bailey's criterion is of the simplest possible form. For this purpose, the linear function of the form $P = wt$ suits best, where w is the intensity of load variation.* To avoid great errors in determining the values of parameters A and α, some short-time experiments should be carried out at different values of w ($P = w_kt$, where $k = 1, 2, \ldots, j, \ldots, l, \ldots, n$). From the data obtained, C_n^2 (number of combinations of n in 2) equations of the following type are formed:

$$\left.\begin{aligned} \int_0^{t_{P,j}} dt/A\ \exp(-\ \alpha w_jt) = 1 \\ \int_0^{t_{P,l}} dt/A\ \exp(-\ \alpha w_lt) = 1 \end{aligned}\right\} \tag{13}$$

where $1 \neq j$. Parameters A and α are determined as the mean values of the solutions of separate sets of equations. Since the parameters α and $\log A$ are distributed normally (this follows from the normal distribution of the values of $\log \tau$ for bonds), an α value is calculated as an arithmetic mean and an A value as a geometric mean.

If the relationship $\tau(\bar{P}, T)$ is described by Equation (6) (at constant temperature, $\tau = BP^{-b}$) and the load varies in accordance with the linear relationship $P = wt$, then the set of equations in (13) acquires the following form:

* The tension machine (dynamometer) must work according to this deformation pattern, rather than according to the more conventional method of constant deformation rate, $V = \epsilon/t$.

$$\left.\begin{aligned}(w_j t_{Pj})^{b+1} &= Bw_j(b+1)\\ (w_l t_{Pl})^{b+1} &= Bw_l(b+1)\end{aligned}\right\} \tag{14}$$

Eliminating the value of B from Equation (14) we obtain:

$$b = \ln t_{Pl} - \ln t_{Pj}/\ln w_j t_{Pj} - \ln w_l t_{Pl} \tag{15}$$

Here the value of b is calculated as an arithmetic mean and B as a geometric mean.

The application of Bailey's criterion has been rather limited, even for the evaluation of long-term strength of polymer materials and especially for bond strength. The data accumulated by the author indicate that such a conservative attitude does not have a firm basis. A sufficient accuracy is obtained even for bonds based on viscoelastic adhesives, if the principle of superposition holds for the given load application.

Good agreement between experimental and calculated data is obtained when the load P varies continuously in time t, for example, it increases accordingly with the linear relationship $P = wt$. If we have long-term strength in the form of Equation (7), obtained at constant loads $\tilde{P}_j$, using Equation (10) (subintegral function $1/\tau(P(t))$ of form (7) where $P \sim wt$), the time to failure and the force can be calculated. Controlled experiments, in which the bond failure strength was determined dynamometrically, show that calculated results usually differ from the experimental ones by 1–1.5% in the range at low rate of load variation. The higher rates compare quantitatively to the ones that occur in conventional dynamometric tests at deformation speed of 1–2 mm/sec.

Equally accurate results can be obtained by applying a nonlinear relationship of load variation. In this case, a mathematical expression of dynamometric curve of force versus deformation must be obtained.

The practical value of such calculations is that if the long-term strength formula is known, dynamometric experiments are not necessary to predict the bond performance. Besides, using Equation (5) instead of (7), there are possibilities of evaluating the bond strength (break stress) at various temperatures, or at least in the temperature range in which the parameter values of Equation (5) were experimentally determined. Naturally, all this is also possible in cases where the relationship $\tau(\tilde{P}, T)$ is described by Equation (6).

The calculating possibilities of long-term strength at some other relationship of load variation with time are discussed below. The experimental results obtained by testing bonds of cellulose diacetate films bonded by nitrile rubber adhesive are used for illustration. A load variation program is shown in Figure 7-10 and the calculated results in Table 7-2 along with the values of the parameters of Equation (5). As can be seen from Table 7-2 for load variation programs I and II (Fig. 7-10), the calculated long-term strength values correspond sufficiently

Table 7-2. The Results of a Test of Bonded Cellulose Diacetate Film

THE EXPRESSION OF BAILEY'S CRITERION (I)	T(°C)	Δt_1 (sec)	W (kg/cm²sec)	Δt_n EVALUATED (sec)	Δt_n FROM THE EXPERIMENT (sec)	/δ/%
$2m\int_0^{\Delta t_1} dt/\tau(wt) + m\Delta t_1/\tau(\bar{P}_1) + \int_0^{\Delta t_n} dt/\tau(wt) = 1$	30	180	8×10^{-3}	385	386	0.3
	40	90		300	332	9.6

THE EXPRESSION OF BAILEY'S CRITERION (II)	T(°C)	$\bar{P}_1$ (kg/cm²)	$\tau(\bar{P}_1)$ (sec)	Δt_1 (sec)	$\bar{P}_2$ (kg/cm²)	$\tau(\bar{P}_2)$ (sec)	Δt_2 EVALUATED (sec)	Δt_2 FROM THE EXPERIMENT (sec)	/δ/%
	30	1.00	11.5×10^3	3.6×10^3	2.0	329	606	599	1.2
$\frac{\Delta t_1}{\tau(\bar{P}_1)} + \frac{\Delta t_2}{\tau(\bar{P}_2)} = 1$	40		2.4×10^3	0.9×10^3		165	105	106	0.95

THE EXPRESSION OF BAILEY'S CRITERION (III)	T(°C)	$\bar{P}_1$(kg/cm²)	$\tau(\bar{P}_1)$(sec) EVALUATED	Δt_1(sec)	$\Sigma_{i=1}^{n} \Delta t_1$(sec) FROM THE EXPERIMENT	/δ/%
	30	2.0	879	150	1256	42.9
$\sum_{i=1}^{w} \Delta t_i/\tau(\bar{P}_1) = 1$	40		165	30	470	184.9

Note: /δ/ is the relative error of experimental long-term strength values compared with the calculated ones.

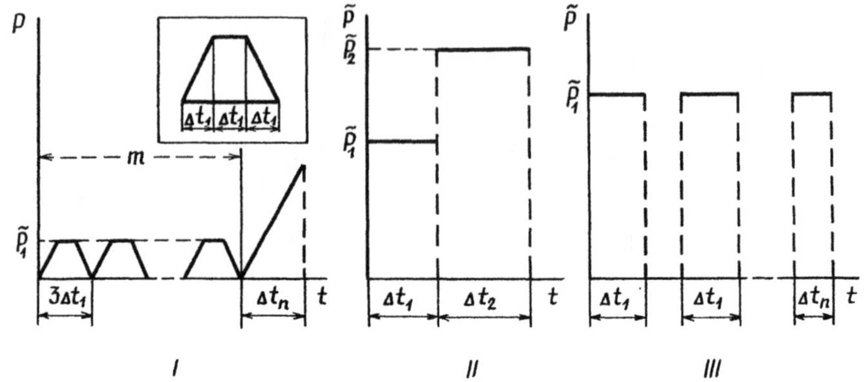

Fig. 7-10. Programs of load variation.

well to the ones obtained experimentally. At the same time, experimentally obtained values for load variation program III considerably exceed the calculated ones. These should have corresponded to the long-term strength value $\Sigma_{i=1}^{n} \Delta t_i$, obtained under the continuous effect of load $\tilde{P}_1$, if the superposition principle would have held. This indicates that the rest between loading cycles was significant and resulted in bond strengthening. Since the adhesive used was to some extent plastic, the strengthening. Since the adhesive used was to some extent plastic, the strengthening was caused by the self-healing of cracks. More examples are given,[16] showing that Bailey's criterion is applicable when the loading is continuous and its variation intensities are not great (such that internal heating does not result). It has been determined experimentally that when the rest period exists between load periods, long-term strength may either increase or decrease. For example, the decrease of long-term strength at normal temperatures is inherent for polychloroprene adhesives. So far the nature of this phenomenon is not sufficiently clear. With increasing temperature, this tendency becomes less obvious and eventually, even strengthening is possible.

Practical possibilities for solving the reverse problem will be illustrated by two examples. Figure 7-11 shows the relationship $\tau(\tilde{P})_{Ti}$ obtained experimentally at constant static loads (points), calculated by Equation (5) from the experimental data (continuous lines), and calculated from dynamometric tests, assuming a linear load variation (dotted lines). The fit is quite satisfactory for a complicated system, such as leather bonds. This shows that there are good possibilities of calculating parameter values of long-term strength Equation (5) from short-term dynamometric experiments. It should be noted that difficulties using this approach lie in the fact that parameters A and α of Equation (7) are sensitive with respect to the experimental long-term strength values (time to failure $t_{P(e)}$) and relatively small variations of these values may provide great variation of values A and α. The variation of experimental long-term strength values in

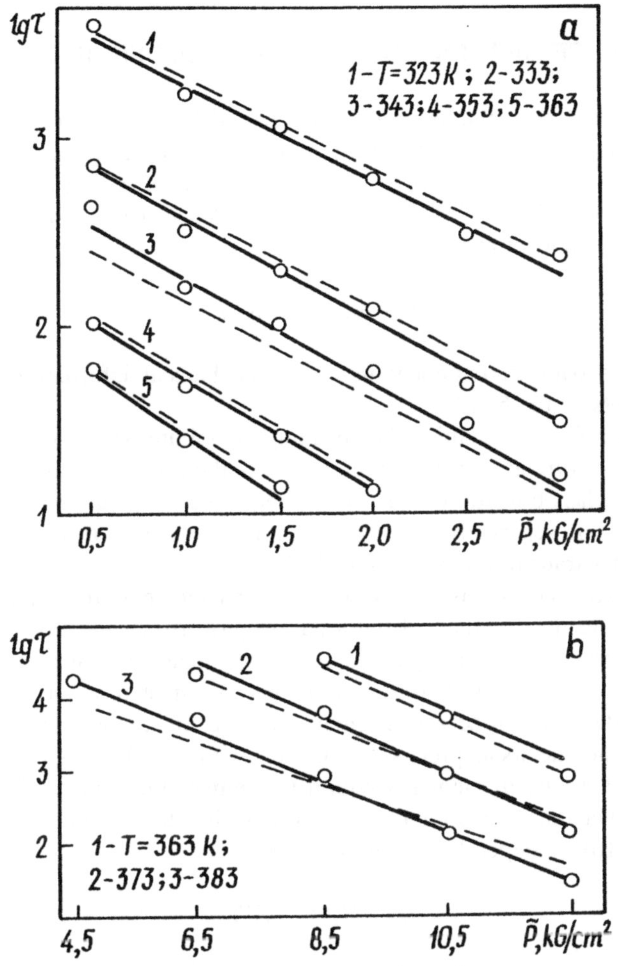

Fig. 7-11. Long-term strength (τ) dependence on static load ($\tilde{P}$). (a) Cellulose diacetate films bonded by polyurethane adhesive. (b) Chrome leather bonded by polychloroprene adhesive. The points are experimental results; continuous lines are calculated according to Equation (5); dotted lines are the result of a solution of the reverse problem.

dynamometric tests may amount to 10–15% and higher. The methods of the determination of parameters A and α suggested by the author and collaborators[16] decreases the inaccuracies, but on the whole, the methods of solving this problem are still to be developed. The reverse solution of the problem described here may be used as a fast method to approximate the long-term strength. The method is especially useful in comparative analysis of various bonds different from each other by the type of the adhesive and the adherends used.

THE INFLUENCE OF TEMPERATURE AND THE DEFORMATION RATE

In dynamometric tests the strength is expressed by force P or stress at the moment of sample failure which, in turn, depends on temperature and deformation rate. Such relationships have been widely discussed, including the behavior of rubbery adhesives.[13,16,20,21,23,25,29–33] The dependence on temperature $P(T)$ most often is of an exponential form[16,23]

$$P = A_0 \exp(U/RT) \tag{16}$$

where A_0, U = parameters depending on the mechanical deformation properties of the elements of the bond.

Parameter U may acquire the meaning of failure activation energy, but the quantitative correlation is not required within the wide range of temperatures, including the glass transition area. The dependence $P(T)$ may be expressed by this equation but separately for different physical states, that is, above and below the glass transition temperature.[16,23]

The deviations from this relationship in certain temperature areas are notable, especially where the mechanical deformation properties vary more intensively because of cross-linking, recrystallization, etc., than expressed by Equation (16). In such cases, the relationship $P(T)$ may be treated as an overall result of two simultaneous processes: normal decrease of strength with increasing temperature and the local process associated with the changes of mechanical properties of the system elements within the temperature range. Then, it may be assumed that the parameter U, determining the strength intensity variation in Equation (16), is dependent on temperature:

$$U = U_1 + U_2(T) \tag{17}$$

The positive sign corresponds to the relative strengthening of the bond; a negative sign would be used in case of bond strength decrease.[16]

U_1 expresses the change of strength intensity corresponding to the relationship $P(T)$ of the form of Equation (16) and $U_2(T)$ expresses the deviation from it. Then

$$P = A_0 \exp(U_1 + U_2\,(T)/RT) \tag{18}$$

$U_2(T)$ can be expressed by the equation

$$U_2(T) = B_u \exp(-\,C_u(1/T - 1/T_0))^2 \tag{19}$$

where B_u corresponds to the maximum deviation of U from U_1 (at $T = T_0$), and C_u characterizes the temperature range of the deviation of P values from the ones corresponding to Equation (16).

The dependence on the rate $P(v)$ is weak when the adhesive is in a glassy state. For adhesives in a rubbery state, within a wide range of rates comprising 6–7 decimal orders, it is represented by a curve of the S form. Therefore, the rate affects the strength values significantly only in a certain range of its mean values, about 0.015–15 mm/sec.

Within the range of rates, where the strength depends on the rate, relationship $P(v)$ is described by the function[16,23]

$$P = kv^m \tag{20}$$

where k and m are parameters depending on the properties of bond elements and test conditions. In certain cases, relationship $P(v)$ deviates from the form of Equation (20) even in a narrow range of deformation rates. It is suggested that such deviations are associated with the change of failure mechanism.[31]

Numerous data are available on the reverse dependences of failure rate on constant load value.[24,30,34] Their form is analogous to that of the relationship $P(v)$.

The dependence of strength on two variables—temperature and deformation rate $P(T,v)$—may be described by the equation

$$P = kv^m \exp(U/RT) \tag{21}$$

similarly, as for polymers in the rubbery state.[2] The application of this equation for different bonds of soft and laminated materials was studied by employing shear and peel tests.[16,23] The parameters M and U of Equation (21) are constants only if the relationships $\log P(1/T)_{vi}$ and $\log P(\log v)_{Tj}$ are linear and the straight lines are parallel. Practically, this is not always the case. Often the relationship $\log P(\log v)$ does not meet this requirement. Therefore, it is natural to suppose that the parameters of Equation (21) depend on the deformation rate. Such dependence can be approximated by the following functions:

$$U = U_1 v^{-b_1} \quad \text{and} \quad n = n_1 (1 - n_2 \log v) \tag{22}$$

For example, for the cellulose diacetate film bonded with nitrile rubber adhesive in the interval $T \in (30;60)$°C and $v \in (0.06;4)$mm/sec, the relationship $P(T,v)$ has the form:

$$P = 1.49 \times 10^{-5} \exp(7.6v^{-0.06}/RT)v^{0.85} \tag{23}$$

(the approximation error $\delta \sim 10\%$).

The dependence of parameters on variables (especially v) is stronger if the failure mechanism is more complicated. This is typical of the bonds of fibrous structures where the fibers reinforce the adhesive and must be broken.

In describing the experimental data of the relationship $P(T,v)$ by Equation (21), the error is usually less than 10–15%. On the whole, however, Equation (21) is not as suitable as (5) for the prognosis. The way to evaluate its parameters is the same as for Equation (5), but their physical meaning is not sufficiently defined.

The dependence $P(T,v)$ of the form of Equation (21) is very useful when the prognosis or calculations are based on the results of dynamometric tests. Sometimes, in solving problems of strength dependence, it is useful to hold one variable constant while the other is allowed to vary. Naturally, in describing strength dependences on a single variable T or v the approximation error is lower and usually does not exceed 10%.

HEAT RESISTANCE

Heat resistance of adhesive bonds is understood as the property of maintaining the bond at higher temperatures. The critical limit at which the bond loses this property is its flow temperature. At this temperature, failure of the bond may occur even without the effect of an outside force, for example, due to the weight of the adherends. The failure of adhesive bonds may also occur at temperatures below the flow, and the higher the temperature, the lower the creep resistance, because the molecular interaction weakens with increasing temperature. Therefore, heat resistance can be regarded as long-term strength dependency on load and temperature. This may be evaluated experimentally or by using Equations (5), (6), (16), and (21).

Comparing different adhesives or structural properties of the bonds, heat resistance may be evaluated in terms of the temperature at which the bonds lose their stability and the bond fails. If the heat resistance is determined at nonisothermal conditions, with temperature increasing in some pattern, the temperature at which the bond failure occurs may be assumed to be a limiting (external) characteristic of heat resistance at a given load and temperature variation. It is named the stability loss temperature T_{T_s}.

The way to determine this characteristic is not complicated. The joint is placed into a heat chamber; it is loaded by a constant or periodically applied load and the temperature is raised gradually in accordance with some given pattern (e.g., linear, exponential, parabolic, etc., that may be accomplished by a simple programmed control). As a critical temperature for the particular bond is reached, failure occurs. The temperature at the time of failure is recorded. For lap joints, it is the time when a nonreversible shift of the elements with respect to each other begins (nonreversibility may be checked by unloading). In the peel test, it is the beginning of peeling. Since this time may be difficult to fix, another characteristic may be used, that is, by fixing the time when the crack depth will be, say, 2 mm.

The dependence of temperature T_{T_s} on load $\tilde{P}$ and on the thickness of the adhesive h is of a hyperbolic form:

$$T_{T_s} = a_1\, \tilde{P}^{-k_1} \tag{24}$$

$$T_{T_s} = a_2\, h^{-k_2} \tag{25}$$

where a_1, k_1 and a_2, k_2 are the parameters of these relationships. The parameter a_1 corresponds to the value of T_{T_s} at $\tilde{P} = 1$ and may be regarded as a reference heat resistance characteristic of various bonds; k_1 shows the variation intensity of T_{T_s} as the load varies, thus characterizing the creep resistance of the adhesive. The relationship of the parameters a_2, k_2 to the adhesive thickness is similar.

It has been determined that for bonds with the adhesive in a rubbery state, the form of the relationship $T_{T_s}(\tilde{P})$ and $T_{T_s}(h)$ is not dependent on the chemical nature of the adhesive, type of bond, or on the method of temperature and force application. Thus, Equations (24) and (25) are quite versatile. It is interesting that the values of T_{T_s} are not dependent on the form of the temperature variation. If a temperature T_{T_s} is reached in time Δt, for example, linearly, the T_{T_s} value will remain the same after the change of the way the temperature increases (with an accuracy of 3–5°C). On the other hand, T_{T_s} values depend greatly on the rate of temperature change, indicating that this index represents the heat resistance evaluation characteristic. Equation (24) defines the extremal operational limits of the bond in terms of temperatures and load.

Table 7-3 shows parameter values of Equations (24) and (25) for some lap joints.

In principle, the evaluation of T_{T_s} values in terms of load is possible by applying Bailey's criterion

$$\int_0^{t_{T_s}} dt/\tau(\tilde{P};T(t)) = 1 \tag{26}$$

without performing the experiments described here, if the long-term strength form, for example, of Equation (5) is known.

Here τ is assumed to be the time during which temperature T_{T_s} is reached, that is, $\tau = t_{T_s}$, as the temperature varies with time in accordance with $T(t)$

Table 7-3. Parameter Values of Equations (24) and (25) for Some Lap Joints

	ELEMENTS OF THE BOND			DEPENDENCE $T_{T_s}(\tilde{P})$			DEPENDENCE $T_{T_s}(h)$			
			THICKNESS OF THE ADHESIVE	PARAMETERS OF EQUATION (24)		INTERVALS OF THE VALUES $\tilde{P}$(kg/cm²) AND T_T, (K)	LOAD	PARAMETERS OF EQUATION (25)		INTERVALS OF h(mm) VALUES
	ADHEREND	ADHESIVE	h(mm)	a_1	k_1		(kg/cm²)	a_2	k_2	
1	Cellulose diacetate film	Nitrile rubber	0.15	304	0.07	0.025–1.125 403–299	0.100	335	0.028	0.01–0.58
2	Cellulose diacetate film	Polycholoprene compound	0.05	322	0.07	0.025–0.875 417–331	0.125	317	0.020	0.03–0.58
3	Chrome Leather	Polyurethane adhesive with 5% of polyisocyanate	0.11	351	0.19	0.50–5.00 385–295	1.250	317	0.030	0.08–1.04

Note: Relative error of the experimental data approximation by Equations (24) and (25) did not exceed 1–1.5%, that is, the experimental values of the dependences $T_{T_s}(\tilde{P})$ and $T_{T_s}(h)$ in the logarithmic coordinates fitted almost ideally on the straight lines. The experiment was performed at linear temperature variation law at a rate of 2 K/min.

relationship. (The method of the solution of this problem is described in reference 35.) It can provide results with the error up to 2–3%.

THE APPLICATION OF ANALOGIES FOR STRENGTH PROGNOSIS

Different factors may have analogous effects on relaxational properties of polymers. For example, it is known that within certain intervals of temperature and deformation rate, there exists a similarity between the effect of temperature increase and rate decrease on the mechanical properties of polymers. This is one of the ways in which the temperature-time analogy reveals itself. The principle of this analogy is based on the experimental fact that the form of the dependence of the viscoelastic function of the polymer on the time logarithm log *t* does not change with temperature. The temperature effect is evaluated by a horizontal shift of the curves of the viscoelastic function along the log *t* axis by the value of log a_T, which is dependent on temperature. This shift results in one curve referred to as a master curve (Fig. 7-12).

The dependence of a_T on temperature is described by the WLF equation,[36] which has found a wide application:

$$\log a_T = -C_1^s(T - T_s)/C_2^s + T - T_s \tag{27}$$

Here a_T is the ratio of relaxation times at temperatures T and T_s; log a_T is referred to as a reduction factor; T_s is a standard (reduction) temperature; C_1^s and C_2^s are the coefficients. (Their physical meaning is explained in reference 36.)

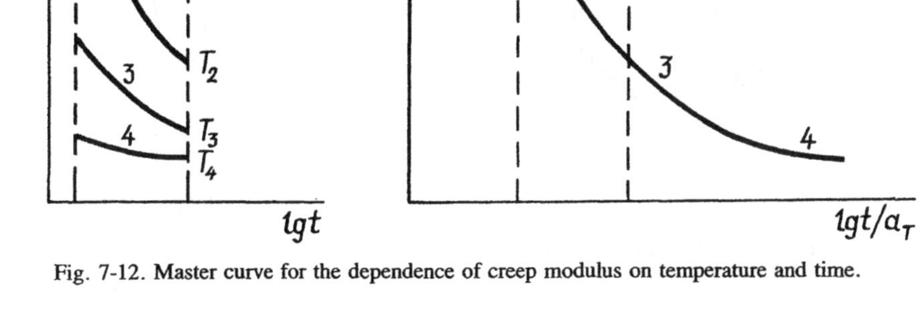

Fig. 7-12. Master curve for the dependence of creep modulus on temperature and time.

When the temperature-time analogy exists, the form of the WLF equation is not dependent on the selected value of T_s. Therefore, two ways of building the master curves are possible. The first is to assign to the coefficients C_1^s and C_2^s their universal values; $C_1^s = 8.86$ and $C_2^s = 101.6$. Then, most often $T_s = T_g + 50$ (T_g is the glass transition temperature). If $T_s \neq T_g + 50$, the first stage in building the master curve is to determine the reduction temperature T_s. The essence of the second way is that the T_s value is selected arbitrarily to be convenient in terms of building the master curve. Then, according to the empirical values of log a_T, optimal values of the coefficients C_1^s and C_2^s are to be determined.

The very idea of using analogies to obtain master curves proved very effective. It is used for the description of strength characteristics as well.[37] Theoretically, it is justified because the failure and deformation processes of polymers, especially above glass transition temperatures, are closely related, and are followed by conformative rearrangement of macromolecules.[36] If relaxation times change by a factor a_T, due to a temperature change from T_s to T, the amount of energy required for the failure must be accumulated during time τ/a_T (τ is long-term strength at temperature T_s for deformation rate va_T). Therefore, in describing the dependence of failure force (stress) on temperature and deformation rate, a generalized strength curve is built on the coordinates log PT_s/T and log va_T, and a long-term strength curve is built on coordinates log $\tilde{P}T_s/T$ and log τ/a_T (for polymers at constant stress log $\sigma T_s/T$ and log τ/d_T). The ratio T_s/T is referred to as a vertical shift factor. It evaluates the vertical shift of individual curves required for building a master curve. When the horizontal shift alone is sufficient, this factor can be neglected. Then master strength

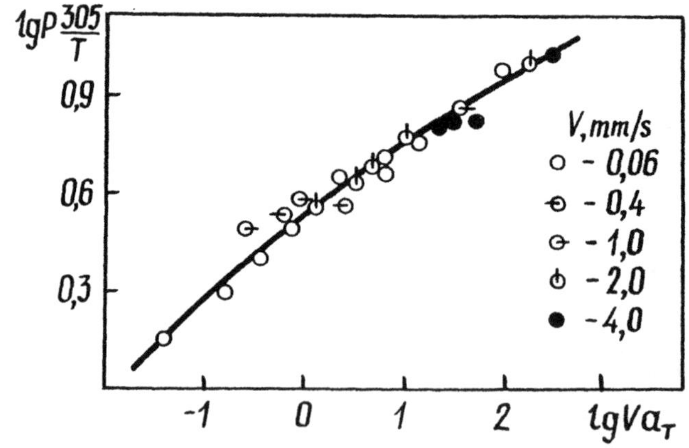

Fig. 7-13. The master curve of strength (P) dependence on temperature (T) and rate (v). Cellulose diacetate films bonded with nitrile rubber adhesive (Table 7-4).

curves are built on the coordinates log P and log va_T, and long-term strength curves in log $\bar{P}$ and log τ/a_T.

At present, a number of other analogies have been determined: vibro-time, concentration-time, stress-time.[38] The possibilities of the application of analogies for bonds have also been proved.[16,21,29,30,39,40] Temperature-time, concentration-time, temperature-concentration, and some other analogies were also determined.[16,41]

In all cases when one or another analogy exists, the reduction factor a_x is used for the construction of master curves. It is calculated from forms analogous to Equation (27) for which other variables, instead of temperature, exist that effect the relaxation (e.g., vibration frequency, concentration of plasticizer in the polymer, and stress intensity). The coefficients C_1^s and C_2^s also have other values. In such cases, the reduction factor a_x represents the ratio of relaxation time at different values of factors influencing relaxation, similar to the temperature-time analogy.

$$\log a_x = \alpha(x - x_s)/\beta + x - x_s \tag{28}$$

Equation (28) is the generalization of the WLF equation.

Figures 7-13, 7-14, and 7-15 illustrate examples of the temperature-time analogy obtained by testing cellulose diacetate film bonds. The data are represented in Table 7-4.

From the given illustrations, it is clear that experimental points lie sufficiently well on the master curve for strength characteristics obtained by dynamometric tests at different rates and temperatures (Figs. 7-13 and 7-14) and for long-

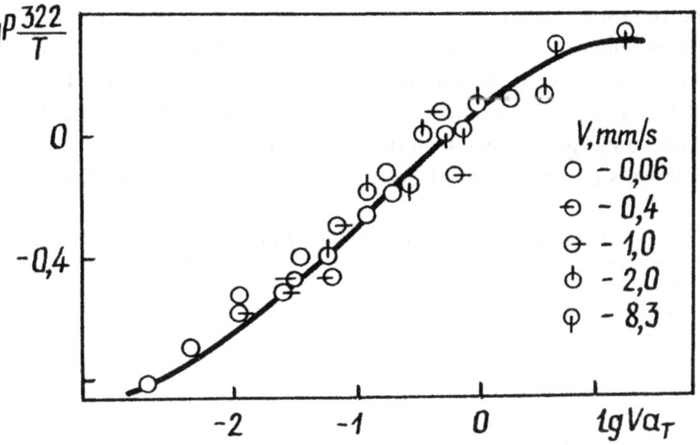

Fig. 7-14. The master curve of strength (P) dependence on temperature (T) and rate (v). Cellulose diacetate films bonded with polyurethane adhesive (Table 7-4).

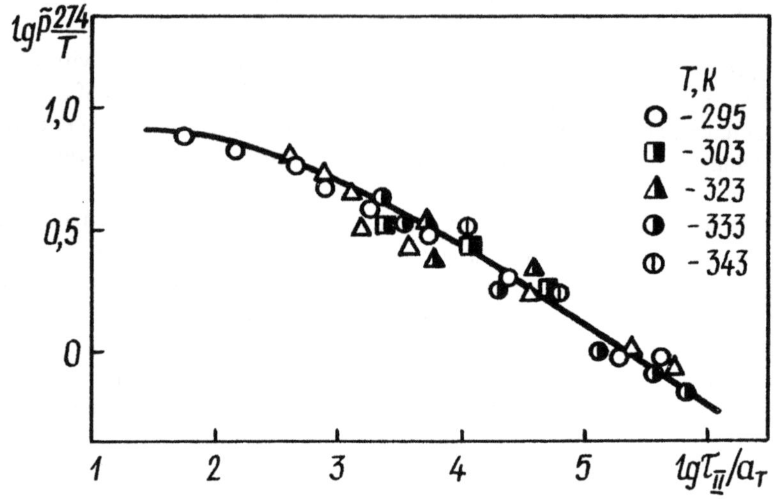

Fig. 7-15. The master curve of the long-term strength (τ) dependence on static load ($\tilde{P}$) and temperature (T). Cellulose diacetate films bonded with polyurethane adhesive (Table 7-4).

term characteristics at different static loads and temperatures (Fig. 7-15). In building master curves for the adhesive bonds mentioned here and for many others,[16,40,41] certain regularities concerning the reduction temperature values have been observed. For lap joints by dynamometric tests, $T_s - T_g = 50$; for dynamometric peel tests, as well as in the case of long-term strength tests at constant loads, $T_s - T_g < 50$.

This can be explained by the free volume concept. During the testing of lap joints, the thin adhesive layer enclosed by two adherend probes is deformed at conditions close to a simple shear. At such conditions, the free volume of the body does not change, the equation $T_s = T_g + 50$ is valid, and the coefficients C_1^s and C_2^s of the WFL equation have versatile values. In the peel test process, the adhesive undergoes greater changes and the free volume increases. In such cases, Poisson's coefficient of elastomeric materials depends upon deformation.

Table 7-4. Tests of Cellulose Diacetate Film Bonds

FIGURE NUMBER	ADHESIVE	TEST TYPE	T_g (K)	T_s (K)	$T_s - T_g$
7-13	Nitrile rubber	Shear	255	305	50
7-14	Polyurethane and polyisocyanate (7%)	Peel	282	322	40
7-15	Polyurethane and polyisocyanate (7%)	Shear at P = constant	255	274	19

The relationship between the Poisson's coefficient, volume of the body, and deformation is

$$\mu = \frac{1}{2}\left(1 - \frac{\delta}{\delta\epsilon} \ln V\right) \tag{29}$$

where V is volume and ϵ is relative deformation.

It has been determined[40] that for elastomers $\mu < 0.5$ at higher elongation, and that this value decreases as the elongation increases. Since stress deformation of the adhesive dominates in the peel tests, $\mu < 0.5$ is close to failure. Then from Equation (29) $\delta V/\delta\epsilon > 0$; that is, increase in volume occurs, and the portion of free volume in the deformed state exceeds that in the nondeformed one.

In the case of long-term tests under constant load, the portion of the free volume that develops microdefects[1] increases, and therefore the free volume increases. The inequality $T_s - T_g < 50$ is verified by another fact. In the case of adhesives that exhibit plastic behavior, or have a higher creep, the failure of bonds affected by an acting constant load occurs during a noticeable time interval rather than instantaneously. Thus, two long-term strength characteristics are determined experimentally: one corresponds to the initial moment of failure (I) and the other to the final bond destruction (II). In the construction of long-term generalized strength curves, the inequality $(T_s - T_g)_{\text{II}} < (T_s - T_g)_{\text{I}}$ is valid for both cases. This verifies the interpretation on the basis of the free volume concept.

The possibilities of applying generalized curves for prognosis and their reliability will be discussed below, along with some examples illustrating the existence of other analogies.

A generalized curve is drawn through the experimental points in reduced coordinates using the geometry approximation alone, that is, all experimental points (which are arithmetical means of experimental data) are located as close to the generalized curve as possible. The reliability index of the property evaluated from the curve may be a reliable interval obtained at a given probability level. The generalized curves, however, have a property that allows shifting curves of the main experiment (obtained at different values of a single variable for a fixed value of another variable) to extend the prognosis interval beyond that in which the experiment was carried out. That interval may be extended more if the standard values of the variable χ_s (for WFL equation T_s) is selected at the edge of the interval of the experimental values. Therefore, in evaluating reliable intervals, it is opportune to make use of the results of additional control–tests.

The control test with an applied analogy method is the most common criterion of the prognosis possibilities of generalized curves.[42] If the main experiment, from which the generalized curve is constructed, has been carried out correctly, the obtained values of the response characteristics are also the controlled characteristics. This is sufficient if there is no set aim to extend the prognosis interval. In the opposite case, the controlled test should be carried out only at such values of variables that correspond to the generalized curve, but are beyond the range of the main experiment. In this case, two approaches are possible.

The basis of the first approach is to carry out the controlled test at a value of one variable corresponding to the standard value of χ_s, and at a value of the other variable beyond the range of the main experiment (e.g., at a temperature T_s and deformation rates greater than the ones used in the original experiment). Then, with a set reliability, the reliable interval is determined for different values of the evaluated characteristic within the limit of the controlled test, and in this way the reliable area for all generalized curves is obtained. If the curve fits in this area, the prognosis possibilities of the generalized curve at a set reliability are sufficient.

In the second approach, the controlled experiment is carried out at both values of the variables beyond the limits of the main experiment. Then the generalized curve is shifted to such values of one variable ($\chi = \chi_i$) at which the control experiment is to be carried out. For this shift the value log a_x of the generalized curve is evaluated from Equation (28) with χ_i value substituted in the equation. This is, of course, just the extrapolation of the generalized curve to a set value of the variable χ_i. A control test is carried out at the value of one variable χ_i, and at different values of the other variable, at which the main experiment was carried out. If the shifted generalized curve fits in the reliable area obtained from the points of control test, prognosis possibilities of the generalized curve are sufficient (at a set probability). Evaluation methods for prognosis possibilities of generalized curves described here hold for any analogy.

For pressure sensitive adhesives, various temperature-concentration analogies are significant. For bonds with the same adhesive different in the degree of plasticization, the quantity of plasticizer in the adhesive c may be considered as a factor affecting relaxation. Then the reduction factor is log a_c, and Equation (28) has the form

$$\log a_c = - b_1(c - c_s)/b_2 + c - c_s \tag{30}$$

The master curve is constructed in the coordinates log $P\, c_s/c$ and log T/a_c or log P and log T/a_c. (This analogy is described in references 16 and 40.)

The other modification of the temperature-concentration analogy is useful for bonding with adhesives in solution form. After bonding, some solvent remains in the adhesive, which acts as a plasticizer. With time, the solvent gradually evaporates, often resulting in bond strengthening. The decrease of the quantity of solvent in terms of strength in such a case corresponds to the decrease of temperature. Therefore, from experiments conducted at room temperature, at various times after bonding, the bond strength at various temperatures may be predicted, or a reverse problem may be solved: from the experimental data at various temperatures, bond strength corresponding to various times after bonding may be predicted.

Such an analogy was obtained experimentally (Fig. 7-16) for polychloroprene adhesive-bonded textile.[41] The main experiment was carried out by peel tests conducted 1 to 25 days (7 tests) after bonding at 9 fixed temperatures from 20° to 80°C. Each experimental point is the arithmetic mean of 6 to 8 tests from 60 to 80 strength values with a variation coefficient up to 13%. A control experiment was carried out at a single c_s of residual solvent and at temperatures − 10°C, −5°C, 0°C, and 55°C. We can see that the points of the control experiment fit well on the master curve (error 4–9%) and that it is within a reliable interval. The interval of prognosis information in terms of temperature is extended by 10 degrees.

Another modification of temperature-concentration analogy (Fig. 7–17) is where the quantity of resin in a polychloroprene adhesive stands for the factor affecting relaxation. In the experiment, butylphenol formaldehyde resin was

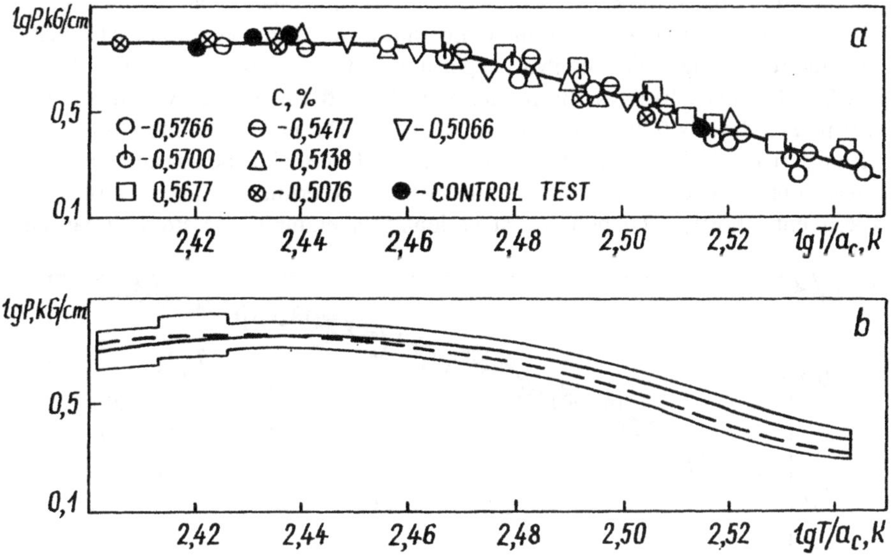

Fig. 7-16. Textile bonded with polychloroprene adhesive. (a) The master curve of strength (*P*) dependence on the remaining quantity of solvent in the adhesive (*c*) and the temperature (*T*). (b) The reliable area of the master curve. The continuous line represents the control test; the dotted line represents the master curve.

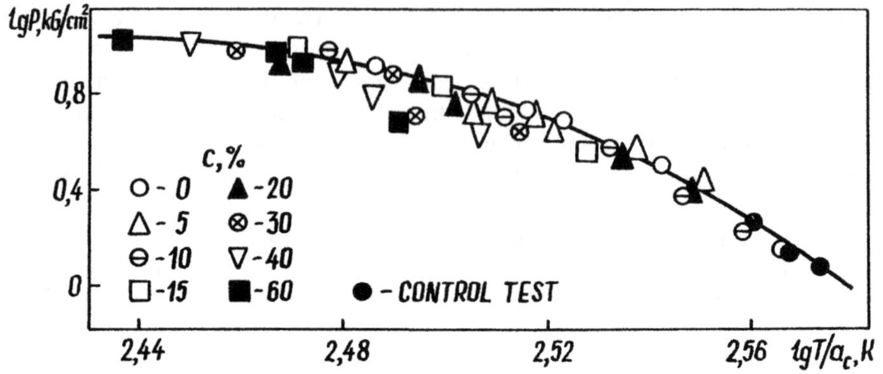

Fig. 7-17. The master curve of strength (*P*) dependence on the quantity of resin in the adhesive (*c*) and temperature (*T*). Textile bonded with polychloroprene adhesive.

used. Its quantity in the adhesive varied from 0% to 60%. The peel strength of textile strips was determined after 30 days in the temperature range of 20°–80°C. A control experiment was carried out at temperatures 90°C, 95°C, and 100°C. Its points fit well on the master curve (error 2–4%). The prognosis interval in terms of temperature was extended up to 100°C.

The last example (Fig. 7-18) is a temperature-time analogy obtained by making use of the same experimental data as for the analogy shown in Figure 7-16. The time after bonding and the temperature were taken as variables (factors affecting relaxation). The control experiment at T = 40°C was carried out after 25, 40, and 240 days. Its points fit well on the master curve (maximum error in 240 days is 6.7%, mean error is 2.5%). The prognosis interval here is extended from 7 minutes to 2.5 years in accordance with the master curve or at least until 240 days in terms of the control experiment. Thus, by making experiments shortly after bonding at various temperatures, the strength can be

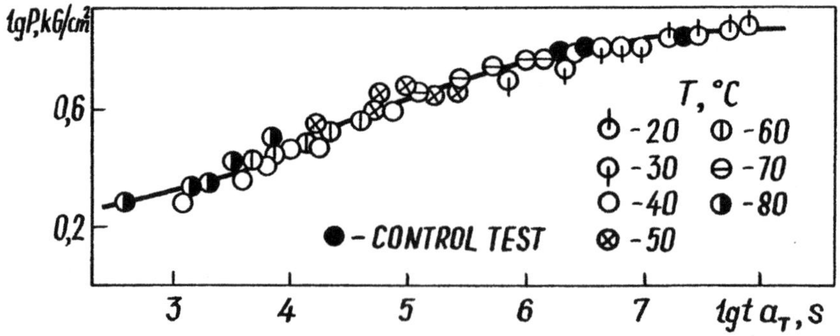

Fig. 7-18. The master curve of strength dependence of the time after bonding (*t*) and temperature. Textile bonded with polychloroprene adhesive.

predicted at any temperature and at any time after bonding within the limits of prognosis interval, or at least within the limits of the control experiment.

From this discussion, one can get a picture of its applicability for prognosis. In some cases, the physical meaning or phenomenon may remain unclear. The existence of analogies should always be formally checked. This is proved by the alignment of the dependence of the reduction factor on the variable in the coordinates $(x - x_s) - (x - x_s/\log a_x)$ and at the same time, by the existence of the form of the WLF equation (28) or in some cases, Equation (27).

Generally, in applying the analogy method, the first step is the experiment. Then, if the fact of existence of the analogy for the factors in question is not certain, the second step is to check the existence of the analogy. Then the master curve is constructed and the control test is carried out.

To make use of the master curves for prognosis one can rely on various known analogies or look for new ones. It depends on the character of the expected information of prognosis and on which investigations or experiments are more convenient to carry out. To obtain the information more rapidly from the master curves, special nomograms may be used.[16,40] To draw master curves, a special program may be set up and computer techniques may be applied.[16]

EVALUATION OF COMPLEX STRENGTH

Various methods have been used to evaluate bond strength characteristics. They are tests of strength at fixed conditions such as temperature, rate, environmental parameters, and quantity of aggressive agents. This way, some information is obtained that often may be sufficient.

In some cases, more extended information on strength is desirable. As new types of adhesives are being developed, more information on their application technology is sought, such as information about strength at various times after bonding, and heat-resistance and fatigue at cyclic or static loads. More extensive information requires more extensive experiments, which are not always effective enough in terms of information yield.

The way to increase experimental efficiency is to make use of the complex evaluation approach. There is no method yet developed to evaluate the bond strength completely. The possibilities of developing such a method are sufficiently real, as has been shown by the author's laboratory experiments. Besides, it should be stressed that there are strength evaluation methods that are quite informative and allow one to determine a series of strength values. They are the long-term strength equations described here, the use of stability-loss temperatures, Bailey's criterion, and the master curves.

In creating the complex strength evaluation methods, correlation links among the various experiment methods should be known. Then after certain formalization, the test complex can be expressed in the form of the informational matrix, which can be enriched with new results or new correlation links among the tests already known.

For the processing of the matrix, a special program for the computer should be developed. The computer informs us about the test combination that could supply the required information on the questions under consideration. With such information available, the laboratory could choose the desirable combination of tests or evaluation methods.

The same information may be obtained in a different way, if a sufficiently large quantity of correlation links among individual tests and methods is determined. The degree of reliability and the volume of tests may depend on the method chosen. Information of this kind can also be obtained from a computer.

REFERENCES

1. Regel, V. R., Slutsker, A. I., and Tomashevskiĭ. E. *Kineticheskaya priroda prochnosti tverdykh tel.* Moskva, "Nauka" (1974).
2. Gul, J. E. *Struktura i prochnost polimerov*. Moskva, Khimiya (1971).
3. Slonimskiĭ L., Askadskiĭ A. A., and Kazanceva, V. V. *Polymer Mechanics.* No. 5 (1977).
4. Bartenev, G. L., and Zuev, Yu. S. *Prochnost i razrushenie vysokoelasticheskikh materialov.* Moskva, Khimiya (1964).
5. *Fracture Processes in Polymeric Solids* (B. Rosen, ed.). Wiley, New York, 1964.
6. Andrews, E. H. *Fracture in Polymers.* Edinburgh-L (1968).
7. Bueche, F. *J. Appl. Phys.* **29:** 1231 (1958).
8. Tobolsky, H. Eyring *J. Chem. Phys.* **11:** 125 (1943).
9. Haplin, I. C. *J. Appl. Phys.* **35:** 3133 (1964).
10. Busse, W. F. *J. Appl. Phys.* **13**(11): 715 (1942).
11. Bueche, F. *Physical Properties of Polymers.* Wiley, New York, 1962.
12. Regel, V. *Polymer Mechanics.* No. 1 (1971).
13. Bikerman, J. J. *The Science of Adhesive Joints.* Academic Press, New York, 1968.
14. Bikerman, J. J. *Recent Advances in Adhesion* (Lee ed.). Gordon & Breach, New York, 1973.
15. De Bruine, N. A., and Houwink, R. (eds.). *Adhesion and Adhesives.* Elsevier, Amsterdam, 1951.
16. Rajeckas, V. *Mekhanicheskaya prochnost kleevykh soedineniĭ kozhevenno-obuvnykh materialov.* Moskva, Legkaya Industriya (1976).
17. Kaelble, D. H. *Physical Chemistry of Adhesion.* Wiley, New York, 1971.
18. Houwink, R., and Salomon, G. (eds.). *Adhesion and Adhesives.* Elsevier, Amsterdam, 1965.
19. Jouwersma, C. *J. Polym. Sci.* **45:** 253 (1960).
20. Gordon, J. I. *J. Appl. Polym. Sci.* **7:** 625 (1963).
21. Kaelble, D. H. *Adhesives Age.* **3**(5): 37 (1960).
22. Hata, T. *Zairyo* (Japan). **13:** 341 (1964).
23. *Voprosy prochnosti kleevyhk soedineniĭ.* sb. statei Izd. Kaunasskiĭ Politekhnicheskiĭ Institut (1971).
24. Fukuzava, K. *Setutiaku* (Japan). **11:** 250 (1967).
25. Malinskii, Yu. M., Prokopenko, V. V., and Kargin, V. A. *DAN SSSR.* **189:** 568 (1969); *VMS,* No. 10: 1832 (1964): *Polymer Mechanics.* Nos. **2, 3, 6** (1970).

26. Voyutskii, S. S. *Adhesives Age.* **5**(4): 30 (1962).
27. Tikhonov, A. T., Tager, A. A., and Stepanov, Ye. *Polymer Mechanics.* No. 4 (1968).
28. Bailey, J. *Glass Ind.* **20:** 21 (1939).
29. Kaelble, D. H. *J. Coll. Sci.* **19:** 413 (1964).
30. Hata, T., Gamo, M., Kozima, K., and Nakamura, T. *Chem. High. Polym.* (Japan). **22:** 160 (1965).
31. Aubrey, D. W., Wedling, G. N., and Wong, T. *J. Appl. Polym. Sci.* **13:** 2193 (1969).
32. Weidner, C. L. *Adhesives Age.* **6**(7): 30 (1963).
33. Voyutskii, S. S. *Autohesion and Adhesion of High Polymers.* Wiley, New York, 1963.
34. Imoto, T. *Sikizai Kyokaisi* (Japan). **40**(7): 304 (1967).
35. Rajeckas, V. *J. Adhesion.* **9:** 51 (1977).
36. Ferry, J. D. *Viscoelastic Properties of Polymers.* Wiley, New York, 1961.
37. Smith, T. L., and Stedry, P. J. *J. Appl. Phys.* **31:** 1892 (1960).
38. Urzhumtsev, Yu. *Polymer Mechanics.* No. 3 (1972).
39. Nonaka, J. *J. Soc. Mater. Sci.* (Japan). **16:** 460 (1967).
40. Rajeckas, V., and Pekarskas, V. *J. Appl. Polym. Sci.* **20:** 1941 (1976).
41. Zaletdinov, F. F., Pekarskas, V., and Rajeckas, V. *Izvestija vysshikh uchebnykh zavedeniĭ.* Tekhnologiya legkoĭ promyshlennosti. No. 5: 96 (1977).
42. Urzhumtsev, Yu. S., and Maksimov, R. D. *Prognostika deformativnosti polimernykh materialov.* Riga, Zinatne (1975).

8. Viscoelastic Properties of Pressure Sensitive Adhesives

Sung Gun Chu
Hercules Incorporated
Wilmington, Delaware

The performance of pressure sensitive adhesives is related to the viscoelastic response of the bulk adhesives, as well as to the surface energies of the adhesives and adherend. Bonding is a low rate process at low deformation that occurs when the pressure sensitive adhesive is brought into contact with a surface, while debonding in a tack or peel test is a high rate process, considering the thickness of the adhesive. The shear strength test is a creep test carried to high deformation. Thus, the viscoelastic behavior of an adhesive controls its response in adhesive testing and is an important factor in controlling its performance.

There have been a few earlier studies of the viscoelastic properties of rubber-resin pressure sensitive adhesive systems.[1–9] Sherriff and coworkers demonstrated the effect of adding low molecular weight resins to natural rubber.[1,2] Compositions were selected for study that exhibited pressure sensitive adhesive performance at an appropriate concentration of resin. Sherriff *et al.* found that addition of the resin to the rubber shifted the entry to the transition zone to a lower frequency, and also reduced the modulus in the rubbery plateau. Later, Aubrey and Sherriff examined the relationship between viscoelasticity and peel adhesion of rubber-resin blends.[3]

Kraus and coworkers reported on studies of resins blended with styrene-isoprene-styrene block copolymers.[4–7] These blends also exhibit pressure sensitive performance at the appropriate concentration. Kraus *et al.* showed that addition of a compatible resin increased the T_g of the rubbery midblock and decreased the plateau modulus.

A more systematic study of resin-rubber blends was reported by Class and Chu.[10–13] It revealed the relationship between the structure, concentration, and molecular weight of the resin and their effect on the viscoelastic properties of the rubber-resin composition. These authors also attempted to correlate the viscoelastic properties of the adhesives and its PSA performance.[13]

This information will be reviewed in this chapter, and the use of the dynamic mechanical properties for the development of pressure sensitive adhesives formulations will be discussed.

DYNAMIC MECHANICAL TESTING

All polymeric materials exhibit viscoelastic properties (Fig. 8-1). They are viscous as well as elastic. These elastic and viscous responses can be most easily identified by analysis of dynamic measurements. For example, measurements made on a sample in sinusoidal oscillation in shear between parallel plates give stress and strain curves as shown in Figure 8-2. The stress leads the strain by a phase angle delta (δ). The elastic response of the material is proportional to the strain, so it is in phase with the strain; while the viscous response is proportional to the rate of strain, it is also 90 degrees out of phase with the strain. The response can be resolved vectorially into the in-phase component, G', or storage modulus, and the out-of-phase, G'', or loss modulus. Tan δ, the ratio of G'' to G' is an indication of the ratio of energy lost to energy stored per cycle of determination.

In Figure 8-3 typical dynamic mechanical data of polymers as a function of temperature are shown. At a low temperature, in region 1, the polymer is in the glassy state. In this state, the polymer sample is brittle and stiff and the G' value is in the order of 10^{10} dyne/cm^2. As the temperature increases, the sample becomes flexible and changes into the rubbery state (region 3). In this state, the sample has a modulus on the order of 5×10^6 to 1.0×10^7 dyne/cm^2.

VISCOELASTIC PROPERTIES

Ideal Response to Deformation

Hooke – Stress is Proportional to Strain

Proportionality Constant is Modulus

Mechanical Model – Spring

Newton – Stress is Proportional to Rate of Strain

Proportionality Constant is Viscosity

Mechanical Model – Dashpot

Real Response of Polymeric Systems to Deformation

Elastomeric Solids Exhibit Relaxation

Polymeric Fluids Exhibit Die Swell

Mechanical Model – Spring and Dashpot Series (Maxwell) or Parallel (Voight)

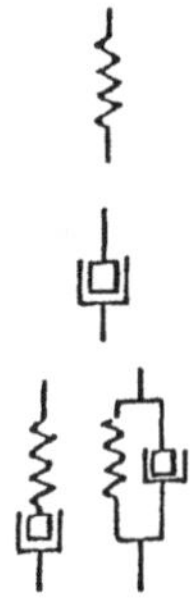

Fig. 8-1. Viscoelastic properties.

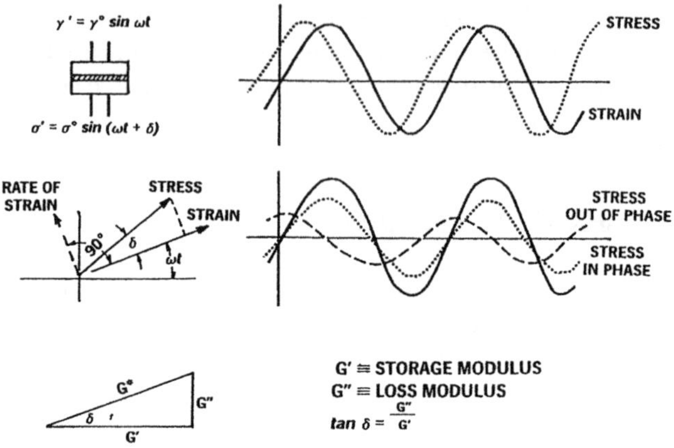

Fig. 8-2. Dynamic testing.

Depending on the molecular weight of the polymer sample, it has a different plateau modulus value that is a characteristic property of the polymer related to the molecular weight between entanglements. A high molecular weight sample would have a long rubbery plateau zone. As we further increase the temperature, the polymer sample starts to melt and enters into the terminal zone. In this state, the polymer sample has a higher viscous than elastic property. In general, the modulus value is below 5×10^5 dyne/cm^2. The tan δ curve, which is an indication of energy loss, has a peak near the glass transition temperature. It has a minimum value at the rubbery plateau modulus, and increases again at the melt region. Most polymers have tan δ values of an order of $0.1 \sim 10$. The aforementioned viscoelastic properties of the polymer sample are also changed by varying the testing frequency at a constant temperature.

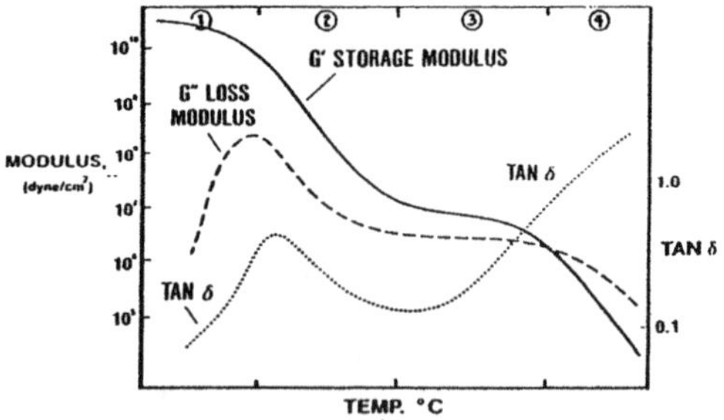

Fig. 8-3. Viscoelastic properties of polymers.

VISCOELASTIC PROPERTIES OF ELASTOMERS

The dynamic mechanical properties of various elastomers that are commonly used for pressure sensitive adhesive applications were examined first. The dynamic mechanical properties of these elastomers were characterized using the Rheometrics Dynamic Spectrometer (RDS). The measurement of dynamic mechanical properties is carried out on bulk or dried latex or solution samples. Caution should be taken during the drying step to avoid cross-linking or degradation.

Natural Rubber

In Figure 8-4 the viscoelastic properties of natural rubber latex (Hartex 103 from Firestone Synthetic Rubber and Latex Co.) and milled smoked sheet natural rubber are presented. The natural rubbers have tan δ peaks at −58°C and long plateau modulus values. Natural-rubber-latex-based pressure sensitive adhesives offer an advantage over solution system (milled smoked sheet) because of the molecular weight difference between the two systems. The high molecular weight portion of natural rubber is insoluble in solvent, and therefore cannot be used in solution adhesives. Natural rubber must be milled to a Mooney viscosity of 53 or below in order to obtain complete solubility. In this milling step, natural

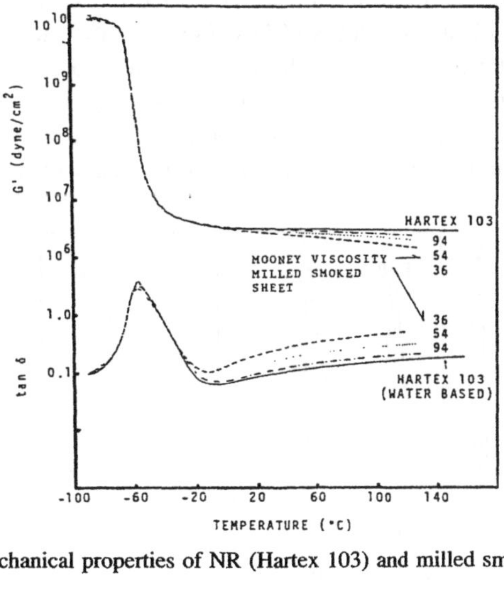

Fig. 8-4. Dynamic mechanical properties of NR (Hartex 103) and milled smoked NR with various Mooney viscosity.

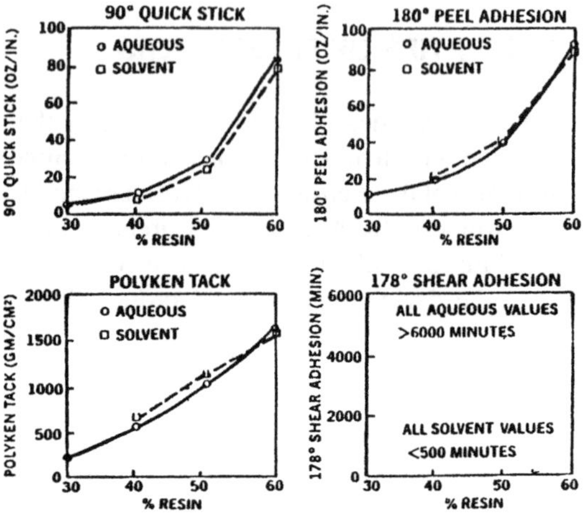

Fig. 8-5. Pressure sensitive adhesive performance of Piccolyte A85/natural rubber. Aqueous vs. solvent systems.

rubber degrades into a lower molecular weight polymer with narrower molecular weight distribution. Natural rubber latex systems retain the high molecular weight portion, which associates with cohesive strength. Therefore, latex-based systems can yield superior shear performance relative to un-cross-linked natural rubber solution systems[14] (Fig. 8-5).

The rheological data that characterizes these differences in properties are shown in Figure 8-4. The natural rubber from dried latex shows a longer plateau modulus and higher elastic modulus (G') value at 100°C than the milled natural rubber. The degradation does not change the tan δ peak temperature (glass transition temperature). It does reduce the modulus (G') at high temperatures (~100°C) as seen in Figure 8-4. This modulus (G') reduction relates to the lower shear performance of solvent-based systems, as seen in Figure 8-5.

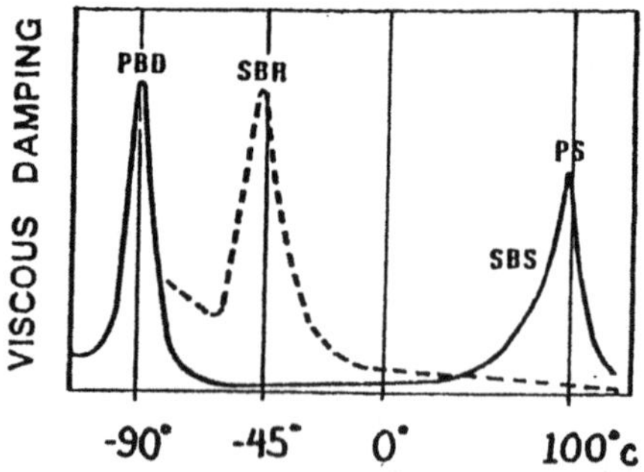

Fig. 8-6. Glass transition temperatures.

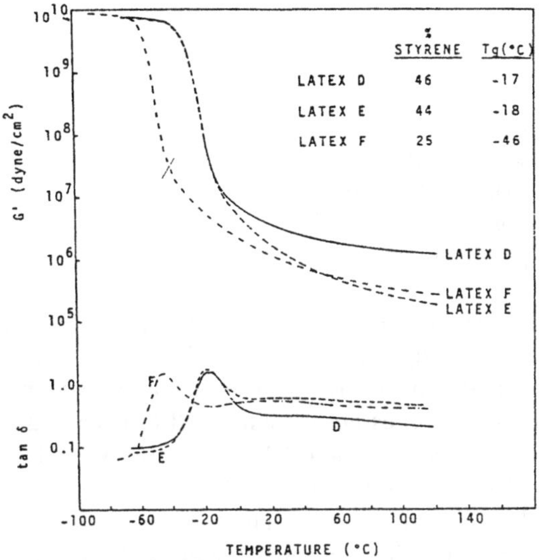

Fig. 8-7. Dynamic mechanical properties of SBR latex with different concentration of styrene.

The conventional method for characterizing natural rubber, Mooney viscosity, is not sensitive enough to be used to obtain the necessary information for natural rubber-based pressure sensitive adhesive work. The RDS generates more complete information than the Mooney viscometer.

Styrene-butadiene Rubber

In Figure 8-6, dynamic mechanical properties of a random copolymer, SBR 1011 (Ameripol 1011, BF Goodrich Rubber Co.), and SBS block copolymer (Kraton D1101, Shell Chemical Co.), are shown. SBR 1011 has only one tan δ peak temperature at −45°C due to its random structure. However, Kraton D1101 has two tan δ peak temperatures at −90°C and 100°C. The tan δ peak at −90°C relates to the polybutadiene domains (rubbery domains) and the tan δ peak at 100°C relates to the polystyrene domains (end blocks) in the block copolymer. In the case of SBR random copolymer, the tan δ peak temperature (glass transition temperature) changes depending on the styrene concentration in the rubber.

In Figure 8-7, the dynamic mechanical properties of three carboxylated SBR latexes are shown. These three latexes show different tan δ peak temperatures and elastic modulus values (G') at various temperatures. The styrene concentration of the latex can be calculated using the well-known Fox equation for random copolymers:[15]

$$\frac{1}{T_g} = \frac{W_1}{T_{g1}} + \frac{1 - W_1}{T_{g2}} \tag{1}$$

where

W_1 = styrene concentration
W_2 = butadiene concentration. $(1 - W_1)$
T_g = T_g of SBR random copolymer
T_{g1} = T_g of polystyrene (100°C = 373 K)
T_{g2} = T_g of polybutadiene (−90°C = 183 K)

SBR latex D has a tan δ peak temperature at −17°C (T_g) and high modulus values ($G' = 1.3 \times 10^6$ dyne/m^2 at 100°C). This latex contains 46% styrene according to the calculation using the Fox equation, which is confirmed by NMR analysis. This latex does not respond well to resin addition. Tack values are low, but shear adhesion is good. This is due to its high molecular weight and the high bound styrene concentration.[14]

Latex E is also high in styrene content—44% as indicated from its tan δ peak temperature (−18°C). Its molecular weight is lower than that of latex D, as indicated from their G' values at 100°C, although they have almost the same styrene content. Consequently, latex E generates better tack but lower shear compared to latex D when compounded with a tackifying resin such as Foral 85 (Hercules, Inc.)[14] (see Table 8-1).

Latex F has 25% styrene and a tan δ peak temperature of −46°C. The lower level of styrene (25%), lower glass transition temperature (−46°C), and balanced viscoelastic properties (G' values) of the latex generate good tack properties without sacrificing shear adhesion when formulated with a tackifying resin[14] (see Table 8-1).

Acrylic Latex

Acrylic latex G (Fig. 8-8) has different flow properties (G', T_g) and surface energy than SBR and NR. The glass transition temperature and modulus of the acrylic latex can be adjusted by polymerizing with different ratios of monomeric acrylates (Table 8-2). Therefore, acrylic latexes have been used with very little or no tackifying resin. However, it is possible that their performance can be optimized, and greater formulating latitude can be obtained by using a suitable tackifying resin, as seen in Table 8-3.[14]

Block Copolymers

Block copolymers with isoprene (S-I-S) and butadiene (S-B-S) midblocks (Kraton 1107 and 1101 by Shell Chemical Co.) and hydrogenated S-EB-S (Kraton G1650,

Table 8-1. Pressure Sensitive Adhesive Performance of Carboxylated SBR/Resin Blends

	1	2	3	4	5	6	7	8	9	10	11	12
Latex D	70	60	50	40	—	—	—	—	—	—	—	—
Latex E	—	—	—	—	70	60	50	40	—	—	—	—
Latex F	—	—	—	—	—	—	—	—	70	60	50	40
SE-10-55K	—	—	—	—	—	—	—	—	30	40	50	60
Foral 85-55K	30	40	50	60	30	40	50	60	—	—	—	—
90-degree quick stick (oz/in.)	16	11	9	8	30	42	45	3	22	26	41	49
180-degree peel adhesion (oz/in.)	45	48	55	72	56	75	78	30	37	54	93	98
Polyken tack (g/cm^2)	850	900	920	930	900	1005	1000	400	1030	1090	1200	2200
178-degree shear time (min)	>6000	>6000	>6000	>6000	4000	3500	2500	1500	>6000	>6000	>6000	2500
T_g (°C)	−5	+4	+14	+20	−5	+3	+13	+19	−22	−12	+2	+9
G' at 25 °C($\times 10^5$) (dyne/cm^2)	20	23	32	43	8	6	8	10	8	7	6.2	7.8
G' at 100 °C($\times 10^5$) (dyne/cm^2)	8	5.4	3	2	1.3	1.0	0.6	0.4	2	1.5	0.8	0.6

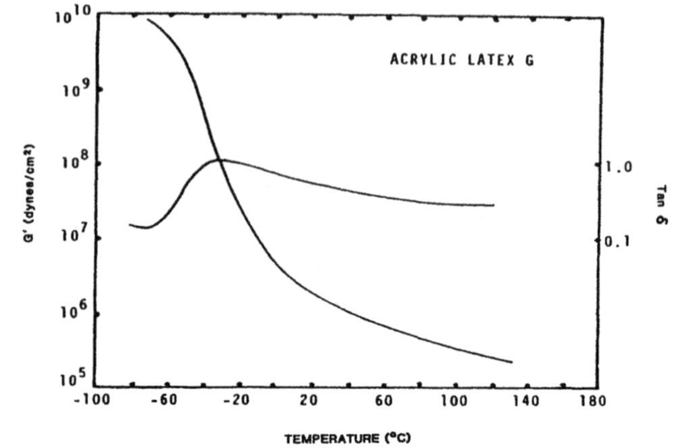

Fig. 8-8. Dynamic mechanical properties of acrylic latex G.

Table 8-2. Glass Transition Temperature of Acrylic Homopolymers (K)

ACRYLIC POLYMER	T_g(K)	METHACRYLIC POLYMER	T_g(K)
Acrylic Acid	379	Methacrylic Acid	501
Methyl Acrylate	283	Methyl Methacrylate	378 random
			311 isotactic
Ethyl Acrylete	249	Ethyl Methacrylate	338 random
			285 isotactic
Butyl Acrylate	219	Butyl Methacrylate	293 atactic
			249 isotactic
2-Ethylhexyl Acrylate	223		

Source: Polymer Handbook. Second edition (J. Brandroup, E. H. Immergut, eds.). Wiley, New York, 1975.

Table 8-3. Pressure Sensitive Adhesive Performance of Acrylic Latex G and Staybelite Ester 10 Blends

ACRYLIC LATEX G	100	30	40	50	60	70
STAYBELITE ESTER 10	0	70	60	50	40	30
90-degree quick stick (oz/in.)	8	25	18	9	4	0
180-degree peel adhesion (oz/in.)	20	42	45	50	25	18
Polyken tack (g/cm²)	500	760	820	830	1050	620
178-degree shear adhesion time (min)	>6000	>6000	>6000	>6000	>6000	>6000

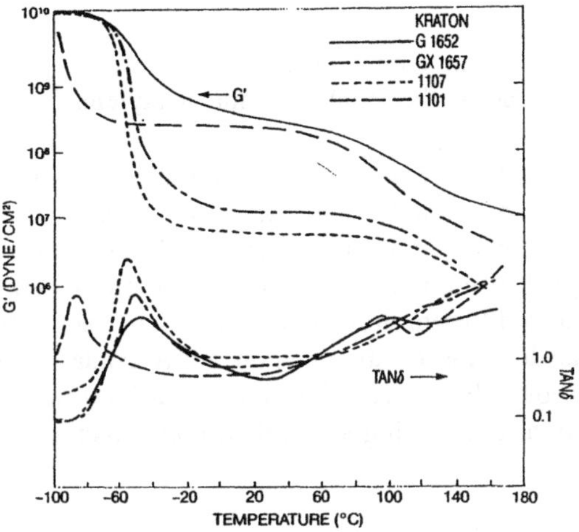

Fig. 8-9. Dynamic mechanical properties of Kraton samples.

G1652, and GX1657) are used for pressure sensitive adhesives compounded with tackifying resins. Since they are block copolymers, they have two glass transition temperatures (tan δ peak temperatures) as shown in Figure 8-9. The midblock or rubber phase has a glass transition near $-50°C$ and the end block or styrene phase has a glass transition temperature at 100°C. However, the T_g

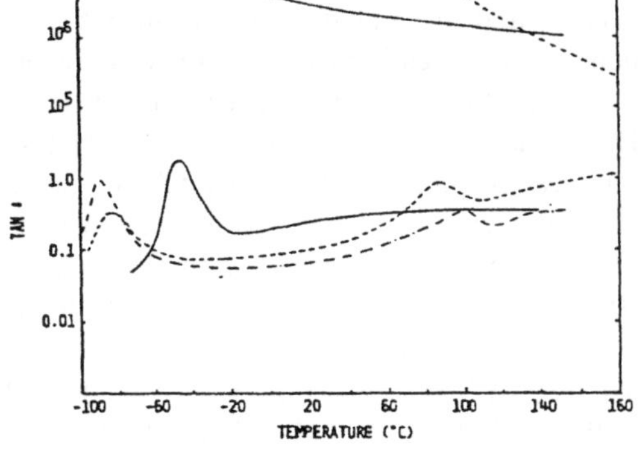

Fig. 8-10. Dynamic mechanical properties of styrene-butadiene rubbers.

of the styrene domain cannot be seen in the spectrum in the case of Kraton GX1657 and D1107 due to the small amount of styrene (~14%) in the polymers.

Stereon 840A (Firestone Synthetic Rubber and Latex Co.) is another block copolymer made of styrene and butadiene. It contains 43% styrene and has the highest amount of styrene among all styrene block copolymers. Consequently, it has the highest plateau modulus value among the block copolymers (Fig. 8-10). However, it has a different domain structure than Shell's SBS block copolymer. The styrene domains of the Stereon 840A contain small amounts of butadiene and consequently, it has a low glass transition temperature of 95°C. The rubbery domain also has a higher T_g than polybutadiene due to its styrene contamination.

VISCOELASTIC PROPERTIES OF TACKIFYING RESINS

A resin is described as a tackifier if, by adding it to an elastomer, the resulting composition has the properties of a pressure sensitive adhesive. In general, they can be divided into four different families in terms of chemistry: hydrocarbon resins (C_5, C_9), terpenes, rosin esters, and pure monomer resins. In Tables 8-4 and 8-5, the chemical identity and physical properties of some commercial tackifying resins are shown. They are low molecular weight oligomers of cyclic monomers and have weight average molecular weights (M_w) in the range of 300 to 2000. These materials are usually brittle glasses at room temperature and have glass transition temperatures from about −20°C to +70°C.

In addition, new resins are produced by hydrogenating hydrocarbon (C_9 or dicyclopentadiene) and pure monomer resins. They are water-white and have excellent ultraviolet (UV) and thermal stability (Table 8-5).

The measurement of viscoelastic properties of these resins is very difficult due to their brittleness. However, their viscoelastic properties were successfully obtained on the RDS by using specially designed parallel plates. In Figure 8-11 the dynamic mechanical properties of Piccolastic A75, which is a low molecular weight polystyrene, are compared with various molecular weight molding grade polystyrene resins. The low molecular weight tackifying resin does not have rubbery and transition zones. It has only glassy and melting zones. It also does not have the tan δ peak temperature that is shown by elastomers and high molecular weight polystyrene resins. Figure 8-12 gives the dynamic mechanical properties of aromatic resins with high softening points.

Table 8-4. Identification of Hercules Resins

	M_w	M_w/M_n	T_g °C	*Softening Point (°C)*
Rosin based				
Glycerol ester of highly stabilized rosin (Foral 85)	550	1.2	32	85
Pentaerythritol ester of highly stabilized rosin (Foral 105)	650	1.3	48	105
Pentaerythritol ester of stabilized rosin (Pentalyn H)	650	1.3	48	105
Terpene				
Poly α-pinene (Piccolyte A 115)	650	1.4	68	115
Poly β-pinene (Piccolyte S 115)	1600	2.6	62	115
Pure monomer				
Poly (vinyl cyclohexane) (Regalrez 1094)	650	1.4	38	80
Poly *t*-butyl styrene	850	1.3	59	105
C_5 stream				
Polymerized petroleum stream, low M_w (Piccotac 95)	1350	1.8	45	95
Polymerized petroleum stream, high M_w (Piccopale 100)	2850	2.8	42	100

Table 8-5. Typical Properties of Hercules Hydrogenated Resin

REGALREZ RESIN		1018	1033	1065	1078	1094	1126	3102	6108
Properties									
Ring and ball softening point	°C	18	32	65	76	94	126	102	108
Glass transition temperature	°C	−22	−9	16	26	38	63	44	49
Specific gravity		0.942	0.943	0.948	0.948	0.971	0.973	1.167	1.003
Melt viscosity									
@ 100 cps		90	105	160	165	190	209	196	200
@ 1000 cps		55	75	120	135	151	182	164	168
@ 10000 cps		35	55	100	120	126	159	139	143
Cloud points in:	°C								
Methylcyclohexane/aniline 1:2		63	71	80	84	83	93	23	54
Odorless mineral spirits		<−40	<−40	<−40	<−40	<−40	<−40	−5	<−40
Compatibility with:									
Styrene		VSC	VSC	1	1	1	1	C	PC
Isoprene		C	C	C	C	C	C	PC	C
Butadiene		C	C	C	PC	1	1	C	PC
Ethylene-butylene		C	C	C	C	C	C	C	PC
High VAC-EVA		PC	PC	PC	1	1	1	C	C
Low VAC-EVA		C	C	C	1	1	1	C	C

I—Incompatible;
VSC—Very slightly compatible;
PC—Partially compatible;
C—Compatible.

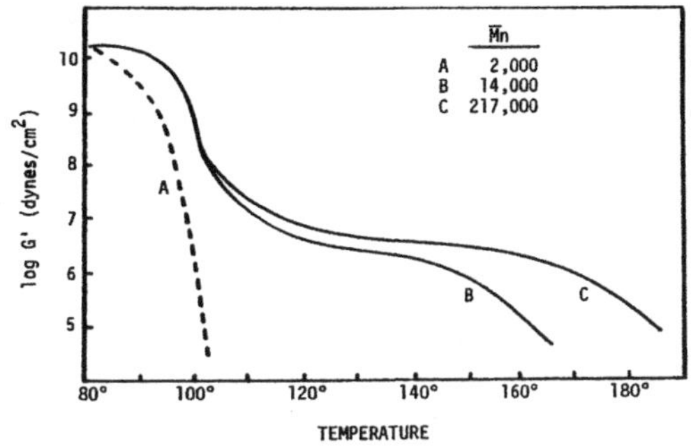

Fig. 8-11. Dynamic mechanical properties of polystyrene with various molecular weights.

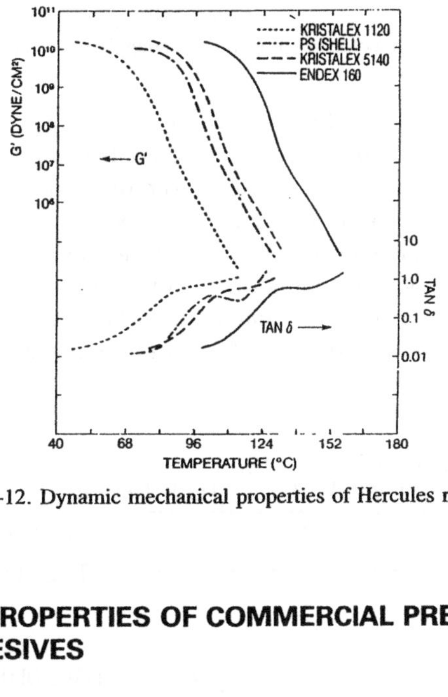

Fig. 8-12. Dynamic mechanical properties of Hercules resins.

VISCOELASTIC PROPERTIES OF COMMERCIAL PRESSURE SENSITIVE ADHESIVES

In Figure 8-13, dynamic mechanical properties of a commercial freezer tape and a strapping tape adhesive are plotted. These pressure sensitive adhesives both have $G' = 10^6$ dyne/cm^2 at room temperature (25°C), but the freezer tape adhesive has a lower tan δ peak temperature (-20°C) with lower plateau modulus values than those of the strapping tape adhesive. This is because the freezer tape requires good low-temperature performance, but the strapping tape needs good high-temperature performance. However, both tapes need good pressure sensitive performance at room temperature.

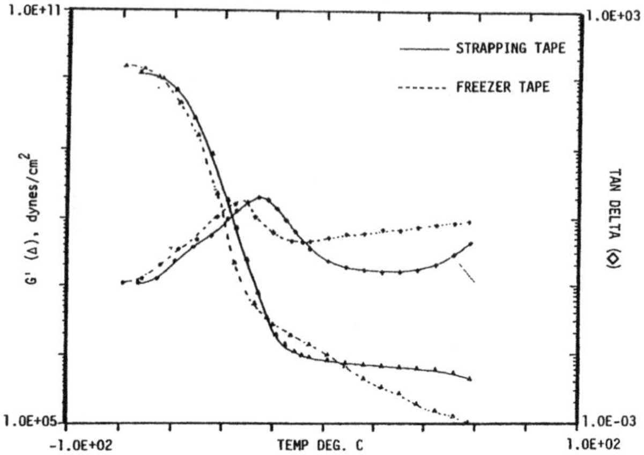

Fig. 8-13. Dynamic mechanical properties of strapping tape and freezer tape.

In 1966, C. Dahlquist[16] defined the requirement for a good pressure sensitive adhesive as an adhesive with a 1-sec creep compliance greater than 1×10^{-6} cm^2/dyne (Fig. 8-14). Recently, the viscoelastic properties of many commercial tapes and label adhesives were characterized, and it was found that the glass transition temperature (tan δ peak temperature) and modulus (G') at the application temperature are the most important requirements for pressure sensitive adhesive performance (Figs. 8-15 and 8-16). The requirements for tape and label adhesives are somewhat different for various applications, but, in general,

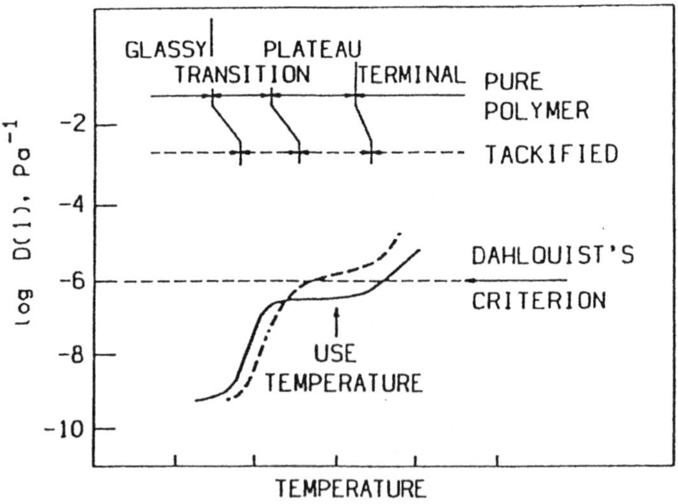

Fig. 8-14. Dahlquist's criterion of good pressure sensitive adhesives. (Data from Kraus.[6])

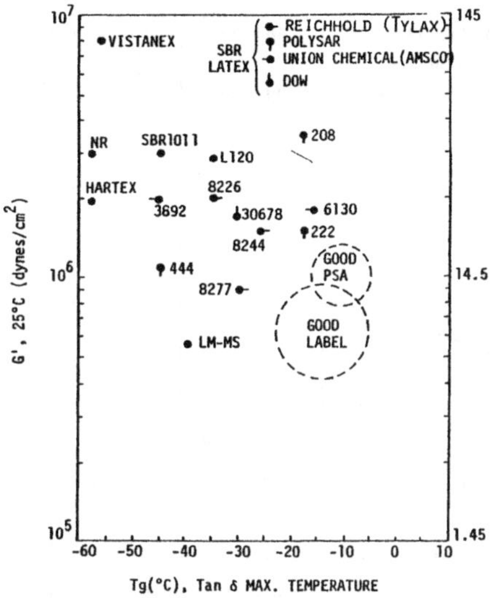

Fig. 8-15. Modulus and glass transition temperatures of label and pressure sensitive adhesive tape.

pressure sensitive tape adhesives require a higher modulus value at room temperature than labels. Pressure sensitive tape adhesives require G' at room temperatures of $5 \times 10^5 \sim 2 \times 10^6$ dyne/cm^2 and T_g about $-15°C$ to $10°C$, while label adhesives require a low room temperature modulus $G' \sim 2 \times 10^5$ to $\sim 8 \times 10^5$ dyne/cm^2 values (Fig. 8-15). The glass transition temperature requirement

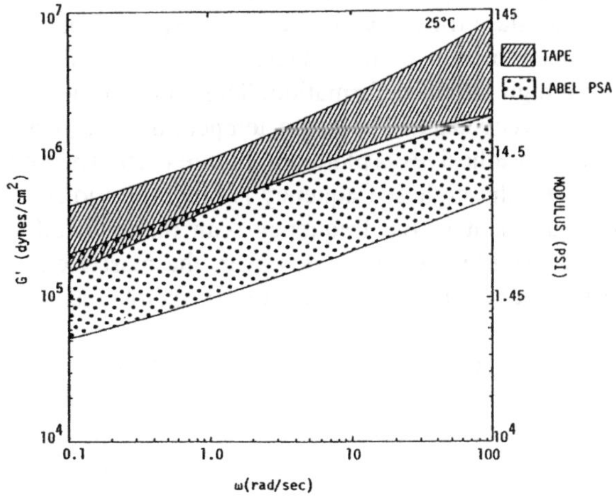

Fig. 8-16. Room temperature modulus values of label and pressure sensitive adhesive tape at various frequencies.

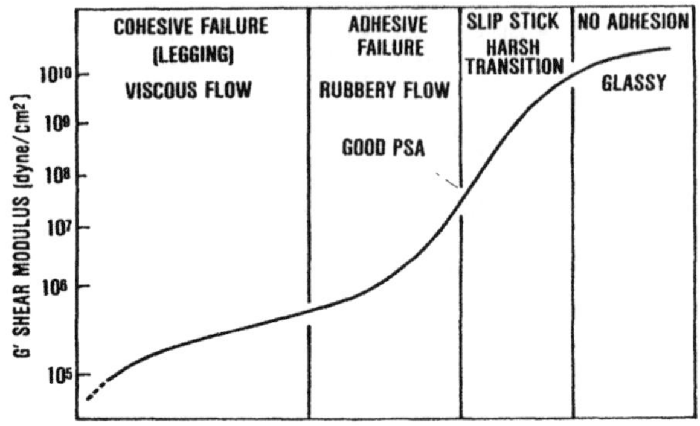

Fig. 8-17. Master curve from various frequency sweeps. Data of pressure sensitive adhesives.

is different for cold temperatures, permanent, and removable applications. Cold temperature label adhesives need low glass transition temperatures ($-30°C <$ Tg $< -10°C$). Permanent label adhesives need a higher adhesive strength than removable label adhesives, but removable label adhesives require adhesive failure without leaving any adhesive trace on the substrate. The requirement of G' values for having good pressure sensitivity is related to the diffusional property of the bulk adhesives onto substrates within a short period of time. In other words, it is related to the time-dependent wetting properties of the bulk adhesive. Therefore, it is also influenced by the surface energy of the adhesive and adherend. However, the performance of typical pressure sensitive adhesives strongly depends on the viscoelastic properties rather than the surface energy phenomenon of the adhesives. The glass transition temperature requirement is mainly related to the debonding phenomenon of the adhesives. It effects the peel adhesion, which is related to a high rate of deformation. This can be more easily explained from the frequency sweep data. The room temperature frequency sweep data of the pressure sensitive adhesives is plotted in Figure 8-16. The lowest frequency is $\omega = 0.1$ rad/sec and the highest frequency is only 100 rad/sec. The practical limitation of the equipment in the frequency scan makes it difficult to obtain the full spectrum at room temperature. However, the master curve can be generated using the time-temperature superposition principle, as shown in Figure 8-17.

The modulus value at low frequency is related to the wetting and creep behavior of the adhesives. The frequency range is not quite quantified since it is dependent on the creep testing conditions, but, in the past, Dahlquist used 1-sec creep compliance ($J(t)$) to correlate its wetting behavior. The creep compliance value can be approximated for the modulus G' value using the following equation.

$$J(t) \cong \frac{1}{G'(\omega)} \times \frac{1}{\sqrt{1 + \tan^2 \delta(\omega)}} \qquad (2)$$

Therefore, the adhesives with lower creep compliance or high G' value along with low tan δ value will have better cohesive strength (creep resistance). If their values are too high, they do not wet the substrate within a short period of time or require high pressure to achieve bonding.

The modulus G' at high frequency (~100 rad/sec) may be related to the peel or quick stick properties of the pressure sensitive adhesives (Fig. 8-18). An important problem in relating peel adhesion to dynamic shear measurements is the correlation of the linear debonding rate with an appropriate frequency for the dynamic measurement. Wong[17] has related the pulling rate R, in a peel test, to the angular frequency ω of the cleavage strain wave immediately preceding the peel line, via the expression $R = \lambda\omega/2\pi$, where R is the peeling rate and λ is the wavelength of the peeling wave. Wong measured λ at $R = 1.67 \times 10^{-4}$ m/sec and found $\lambda = 6.0 \times 10^{-4}$ m/sec. While Wong demonstrated that λ is essentially independent of R over a one hundredfold range of R, this independence will only be valid while the adhesive stiffness is negligible compared with that of the backing. In the PSTC test, $R = 5.08 \times 10^{-3}$ m/sec, assuming $\lambda = 6 \times 10^{-4}$ m, the angular frequency ω should be $\omega = 53$ rad/sec. Since the peel value correlates with the debonding behavior as well as the bonding (wetting) behavior of the adhesives, the modulus value G' at only $\omega = 53$ rad/sec does not correlate with peel values measured using the PSTC method. It is believed that the modulus values at low frequency, $\omega < 10^{-2}$ rad/sec, and at high frequency, $\omega > 10^{2}$ rad/sec, are both related to the pressure sensitive adhesive performance. A rough, quantitative relationship was obtained

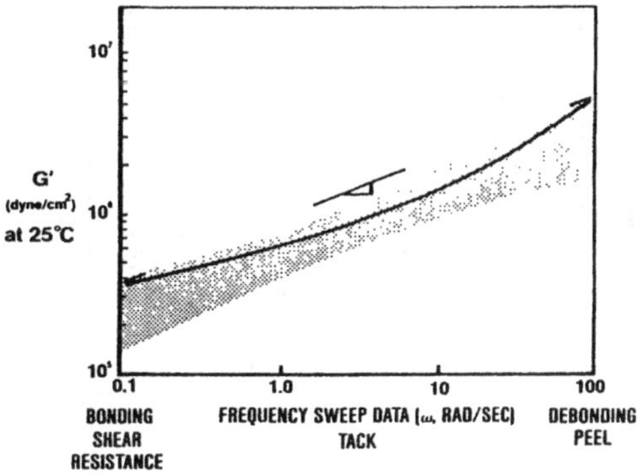

Fig. 8-18. Room temperature modulus requirement at various frequencies. Shaded area is the window that has good pressure sensitive adhesive properties. The slope ($G'_w = 100/G_{IS} = 0.1$) is an important factor.

by plotting G' at $\omega = 0.1$ rad/sec versus $G'(\omega = 100)/G'(\omega = 0.1)$ rad/sec (Fig. 8-31). In order to cover broader viscoelastic characteristics of the adhesives, G' at $\omega = 0.01$ rad/sec and $\omega = 1000$ rad/sec are the preferred values to be used. However, due to the practical limitation of the measurement from the RDS, G' values at $\omega = 0.1$ rad/sec and $\omega = 100$ rad/sec were used. According to Figure 8–31, the pressure sensitive adhesives with 2×10^5 dyne/cm^2 $< G'(\omega = 0.1) < 4 \times 10^5$ dyne/cm^2 and $5 < G'(\omega = 100/G'\omega = 0.1) < 300$ have good peel, tack, and shear properties. A detailed discussion of frequency scan data of pressure sensitive adhesives is given in the next section.

ELASTOMER–RESIN BLENDS

Natural Rubber and SBR

The compatibility of natural rubber with various resins is explained in papers published by J. B. Class and S. G. Chu.[10–13] The compatibility of resin and rubber can be determined by measuring viscoelastic properties of the blend. Compatibility is identified (in the G' versus temperature plot) by a pronounced shift of the tan δ peak temperature (T_g), associated with a decrease in the storage modulus in the plateau. An incompatible system is indicated by a minimal shift of the tan δ peak along with an increase in the storage modulus in the plateau (Fig. 8-19). A second peak in the tan δ curve may be apparent in an incompatible system. Compatibility of rubber-resin systems depend on the struc-

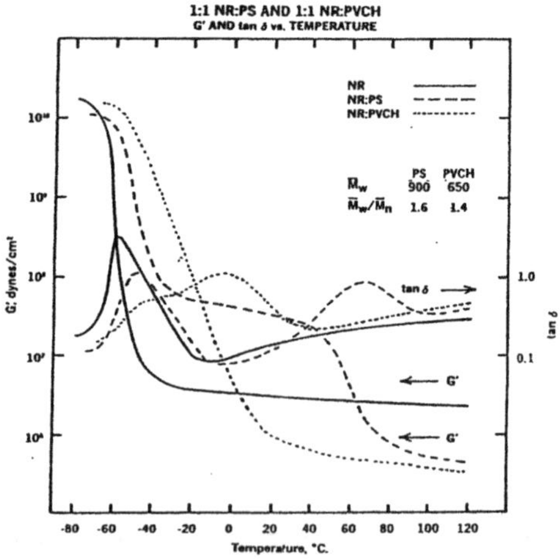

Fig. 8-19. Dynamic mechanical properties of natural rubber and its blend (1/1) with polystyrene (incompatible system) or with polyvinyl cyclohexane (compatible system).

ture (polarizability), molecular weight, and concentration of the resin in the blends. The compatible systems exhibit pressure sensitive adhesive performance at some ratio of rubber to resin. The incompatible systems, on the other hand, are not pressure sensitive. The addition of resin to the rubber is required to reduce the plateau modulus so as to form a bond under a light deforming stress, such as when a pressure sensitive substrate is pressed onto a surface. Therefore, the concentration of the resin is very important in adjusting the viscoelastic properties of a rubber-resin blend to achieve pressure sensitive adhesive performance. Figures 8-20 and 8-21 show the effect of concentration on the viscoelastic properties of rubber-resin blends. On these plots, blends of SBR and the glycerine ester of hydrogenated rosin are shown. As we move from the front of this plot, where we have the SBR control, to the rear, the resin concentration increases and that of SBR decreases. The tan δ peak temperature increases from −46°C for the SBR control to 50°C for the 80% resin. The plot shows a progressive increase in tan δ peak with increasing concentration of resin. The degree of compatibility of resin to rubber is a critical factor for pressure sensitive adhesive properties. The rosin esters (hydrogenated) are more compatible with natural rubber than the C_5 and terpene resins. Consequently, the adhesives compounded with rosin esters have better tack properties but poorer shear strength than their counterparts from C_5 and terpene resins. The degree of compatibility of resin to rubber can be determined from a plot of G'

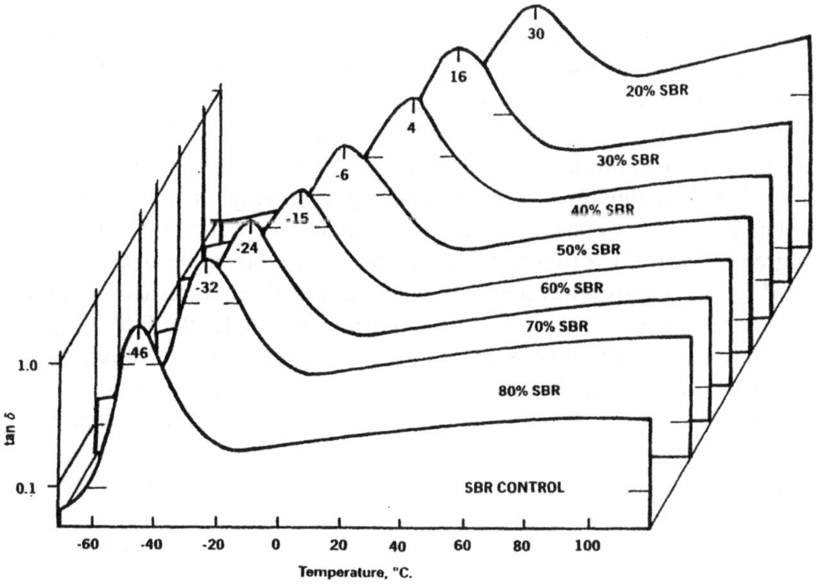

Fig. 8-20. SBR: Glycerine ester of hydrogenated rosin. Effect of concentration on tan δ.

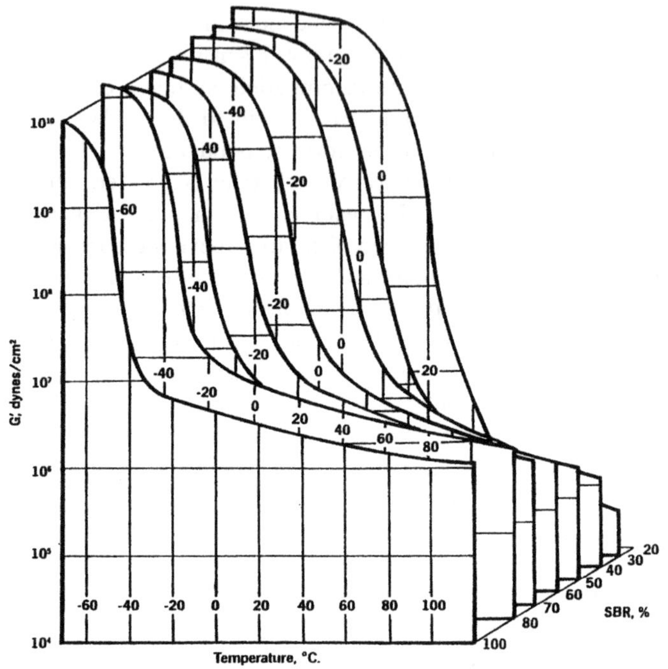

Fig. 8-21. SBR: Glycerine ester of hydrogenated rosin. Effect of concentration on G'.

at 25°C versus resin concentration. A tackifying resin, which has a higher T_g than natural rubber, will increase the glass transition temperature of the rubber. However, it decreases the room temperature modulus value at low concentration and then increases the modulus at high concentration (Fig. 8-22). The modulus versus resin concentration curve does depend on the structure of tackifying resin used; but, a modulus increase will begin, usually, in the range of 40–50%. The minimum values in G' (in G' versus resin concentration plot) reflect the degree of resin-rubber compatibility. The more compatible resins will have lower G' values. Figures 8-22 and 8-23 illustrate the G' versus resin concentration for the various resins. In the case of natural rubber blends, the following order of resin-rubber compatibility is observed: rosin ester > C_5 > α pinene > β pinene > aromatic resin. For SBR: rosin ester > α pinene > β pinene > C_5 resin. This order reflects the resin polarizability, molecular weight, and molecular weight distribution. Rosin esters are the most polar resins and have the lowest molecular weight among all resins. Consequently, they are the most soluble in NR and SBR rubbers. α-pinene resins have a narrower molecular weight distribution than β-pinene, and therefore they are more soluble in the rubbers. The aromatic resin is not soluble in natural rubber and the C_5 resin is not compatible with SBR.

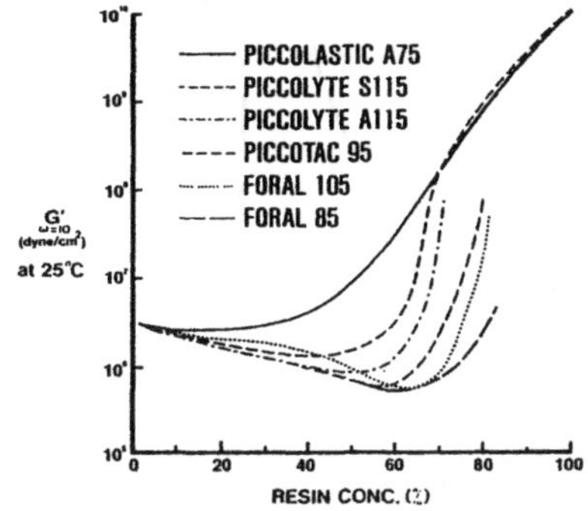

Fig. 8-22. Room temperature modulus values of NR/resin blends.

The maximum tack properties of pressure sensitive adhesives appear to occur when the room temperature modulus falls within the range of 5×10^5 to 1×10^6 dyne/cm^2 and the glass transition temperature is within the range of -10°C to 10°C. Figure 8-24 is a plot of adhesive properties versus resin concentration for a natural rubber /β-pinene (Piccolyte S115) blend. As mentioned previously, the adhesives show optimum tack and peel properties at the resin concentration that has a minimum G' (25°C) value (about 40–60% loading).

Figure 8-25 is a plot of G' at 25°C versus the glass transition temperature of the various SBR latex blends and is another way of showing that the modulus initially decreases as resin is added, reaches a minimum, and then increases at higher resin loadings. In a similar way, adhesive tack also increases with increasing resin concentration and then decreases at high concentration, as can be seen in Table 8-1.

It is more convenient to correlate adhesive performance to viscoelastic properties determined at the adhesive test temperature. The bond is formed and destroyed during testing using conditions that differ in stress level, deformation rate, and extent of deformation at room temperature. Therefore, the measure-

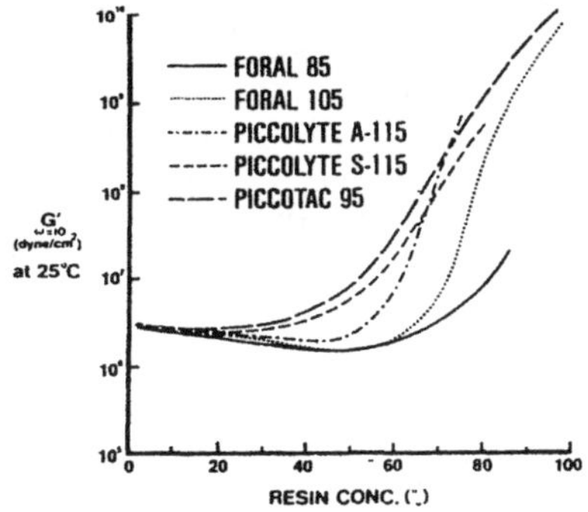

Fig. 8-23. Room temperature modulus values of SBR/resin blends.

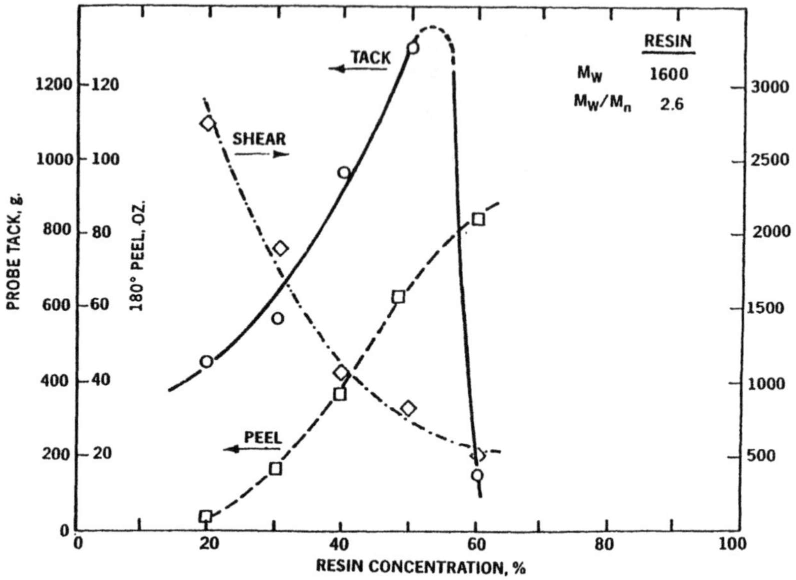

Fig. 8-24. Natural rubber: Poly β-pinene. Pressure sensitive adhesive performance vs. concentration.

ment of viscoelastic properties at room temperature may be more pertinent. Viscoelastic properties at constant temperature are determined by dynamic mechanical measurement over a range of frequencies, for example from 0.1 to 100 rad/sec. The frequency scan curves are approximately a reciprocal arrangement of the temperature scan curves, that is, the left-hand part of the curve for

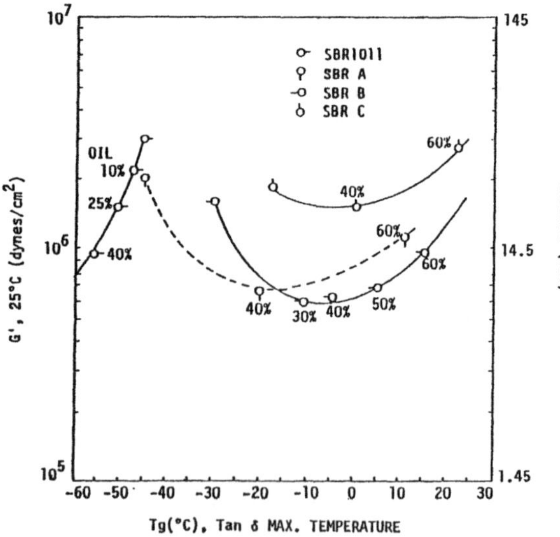

Fig. 8-25. G' and T_g, values of SBR latexes and Foral 85 WK or oil blends.

frequency is lower than the right, while the right-hand curve for temperature is lower than the left. The frequency scan is essentially an expanded view of the reciprocal of the temperature scan, allowing the data within that area to be seen a little more clearly and reliably.

Three different resin or oil rubber blends were examined by frequency scan at room temperature. Figure 8-26 shows the G' versus frequency at various concentrations of compatible aliphatic oil to natural rubber. As oil is added, the modulus curve as a function of frequency is depressed. This is to be expected as the oil softens the composition by increasing the spacing between entanglements, which in turn, causes the reduction in modulus. Oil does not change the tan δ peak (T_g) temperature of natural rubber: it only reduces the room temperature modulus value of natural rubber. Consequently, it does not function as an effective tackifying resin.

Figure 8-27 shows modulus values of the various blends of natural rubber with an incompatible polystyrene resin. The polystyrene resin (Piccolastic A75) has $M_w = 900$ and T_g of 30°C. In this case, the unmodified natural rubber is placed at the bottom of the plot. As the resin concentration is increased, the modulus increases across the entire frequency range. This behavior is what would be expected from a reinforcing filler. At low levels, there is probably partial compatibility, since the addition of 20% or 30% of the resin does not increase the modulus as much as would be expected, if the resin were completely incompatible.

Blends of natural rubber with a compatible resin give a completely different

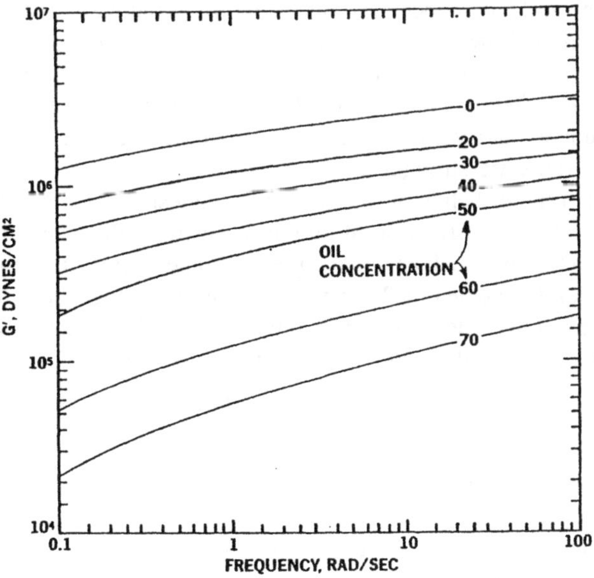

Fig. 8-26. Natural rubber: Aliphatic oil. G' vs. frequency at 25°C.

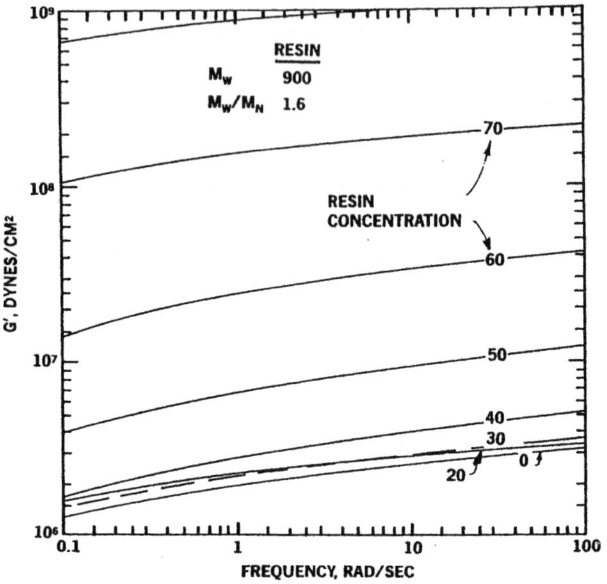

Fig. 8-27. Natural rubber: Polystyrene resin. G' vs. frequency at 25°C.

picture. Figure 8-28 is a plot of G' versus frequency at various concentrations of a compatible resin β-pinene resin. As the concentration of the β-pinene resin increases, the modulus of the blend is depressed up to about 40% concentration. But at 50% and 60% concentrations the modulus increases at higher frequencies, and at 70% it shows a large increase over the modulus for the unmodified natural rubber. At certain resin concentrations, the modulus of the composition is depressed at low frequencies, that is, the compositions are softer than unmodified natural rubber, and yet the modulus is increased at high frequencies, that is, the compositions are stronger.

This is the type of behavior that has been suggested for a pressure sensitive adhesive. The rate of deformation of pressure sensitive adhesive during bonding is considered to be low, such as when a piece of tape is pressed onto a substrate, but when the bond is peeled or destroyed, the rate is high because the thickness of the adhesive is low. Therefore, the modulus at low frequencies or low testing rates should be depressed, so that the composition is soft enough to flow and wet the substrate in the short time during bonding, yet at high testing rates the modulus should be elevated so that the material appears to be significantly stronger than natural rubber itself. This is what is found with a blend of about 40 parts of natural rubber and 60 parts of β-pinene resins. At this resin concentration, good PSA properties have been observed (see Fig. 8-24). This interpretation is consistent with the result of temperature scan data (see Fig. 8-20; G' at 25°C versus resin concentration).

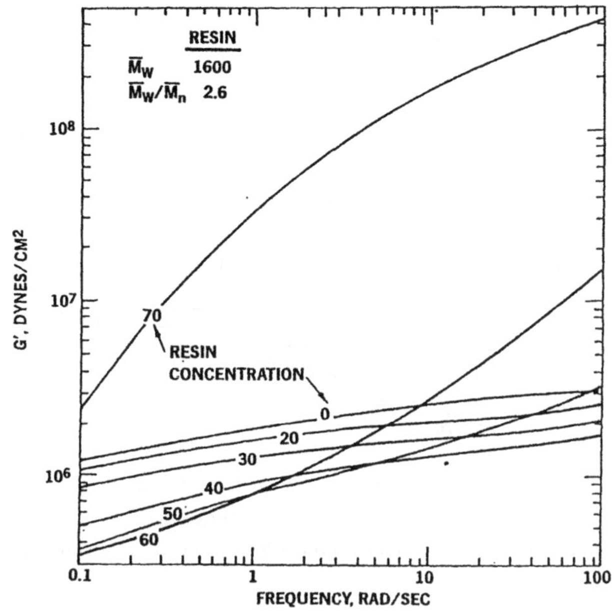

Fig. 8-28. Natural rubber: Poly β-pinene resin. G' vs. frequency at 25°C.

This performance is not unique for β-pinene resins, but is found for all compatible systems. Figure 8-29 shows another example of the effect of a compatible low molecular weight C_5 resin known to give good pressure sensitive adhesive properties with natural rubber. The modulus is depressed with an increasing concentration of resin across the entire frequency range and then, at an optimum range of resin, the modulus increases at high frequencies while it is still depressed at low frequencies. These are the compositions that would show a good pressure sensitive performance.

The curves for rubber blended with compatible resins suggest that it should be possible to identify some parameters that are related to performance, such as a low G' at low rates (essentially an intercept value on the left-hand side of the plot), along with a relatively high slope. If the modulus is below some value at 0.1 rad/sec and above some value at 100 rad/sec, or if the slope is great enough (the shaded area in Fig. 8-18), then the adhesive should have pressure sensitive properties. In an attempt to examine this, a plot of G' at 0.1 rad/sec versus the ratio of G' at 100 rad/sec to G' value at 0.1 rad/sec was constructed for the natural rubber resin blends (Fig. 8-30). On the left side is a grouping of points that consists of natural rubber with low levels of various resins. The modulus is high and the slope is low. As resins are added to the system, the modulus is depressed at the frequency of 0.1 rad/sec and

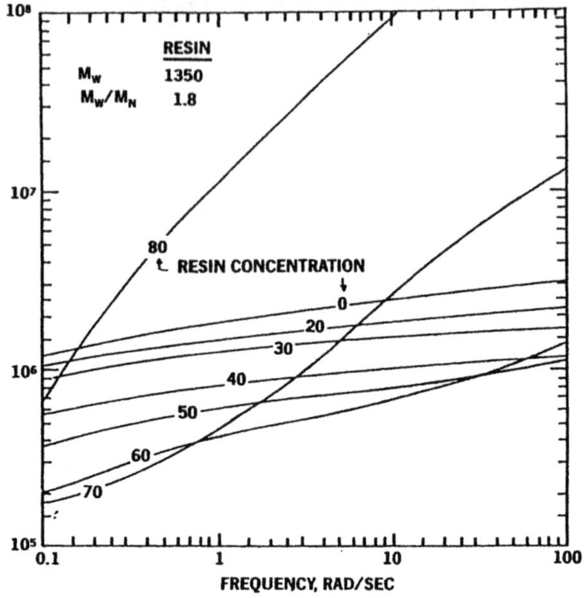

Fig. 8-29. Natural rubber: Low MS C_5 resin. G' vs. frequency at 25°C.

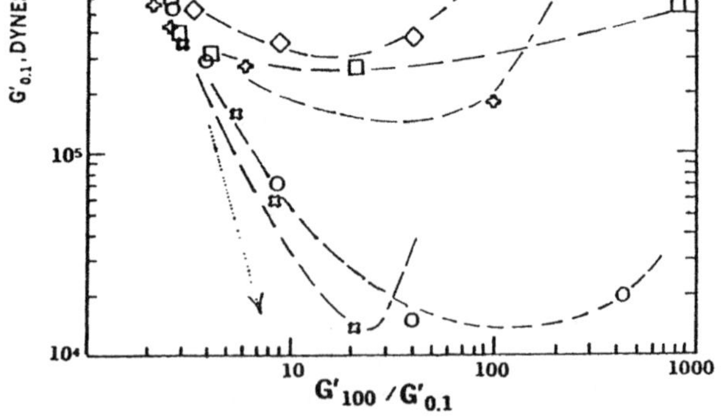

Fig. 8-30. Natural rubber: Resin blends.

the slope increases. In the systems containing compatible resins, the 0.1 rad/sec modulus reaches a minimum and then begins to rise, while the slope continues to rise. Additional data at high resin concentrations, not shown on this plot, indicate that these curves tend toward a circle. That is, as the modulus increases at 0.1 rad/sec, the ratio of G' at 100 to G' at 0.1 begins to drop so that the curve begins to bend back upon itself. This is the characteristic for all compatible systems that were examined and also for those that have pressure sensitive properties. The performance of the various incompatible systems is bounded on the upper left, shown by the dash-dot line in Figure 8-30. In this case, the modulus at 0.1 rad/sec increases, but the slope does not change, as was seen earlier. The boundary on the lower left, shown by the dotted line, represents a compatible oil. The modulus G' at 0.1 rad/sec is depressed and there is a slight increase in slope as observed on the curves for natural rubber-oil blend. The pressure sensitive adhesive properties of natural rubber with various resins were characterized (Figs. 8-31 through 8-33). Rubber blends with intercept $2 \times 10^5 < G'\omega = 0.1 < 4 \times 10^5$ dyne/cm^2, and slope $5 < G'\omega = 100/G'\omega = 0.1 < 300$, have good peel, tack, and shear properties.

Block Copolymer

Figure 8-34 shows the viscoelastic properties of Kraton block copolymer blends with low molecular modifying resins. The rubber-resin compatibility data can be obtained from these curves. As mentioned in the previous section, a block copolymer has two glass transition temperatures. A T_g at 90°C relates to the

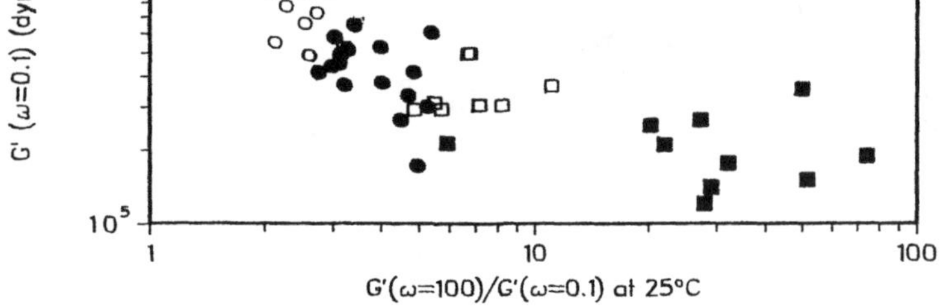

Fig. 8-31. 180° peel data vs. RDS data (NR/resin system).

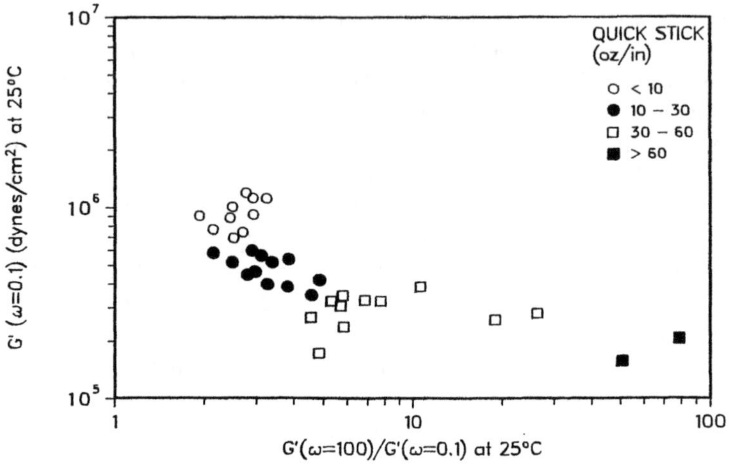

Fig. 8-32. Quick stick data vs. RDS data (NR/resin system).

polybutadiene domains (rubbery domains) and a T_g at 100°C relates to the polystyrene domains (endblock) in the block copolymer. Consequently, the Kraton block copolymer-resin blends also have two glass transition temperatures. The two T_g of block copolymer vary depending on the structure and molecular weight of the modifying resin.

The curves in Figure 8-35 show the tan δ and G' values of the Kraton block copolymer as a function of temperature. In general, three different phenomena can be observed from blends of block copolymer with various resins.

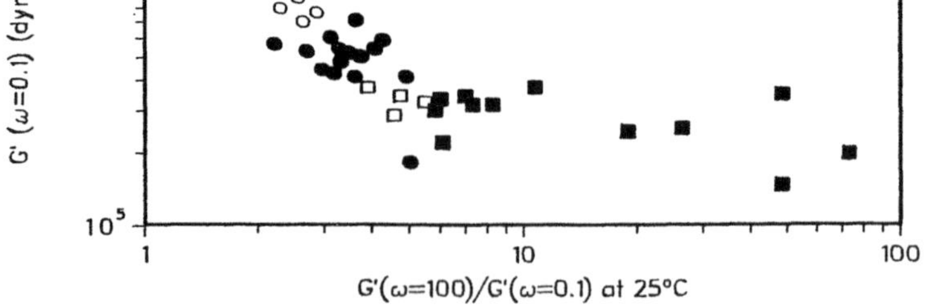

Fig. 8-33. Polyken tack data vs. RDS data (NR/resin system).

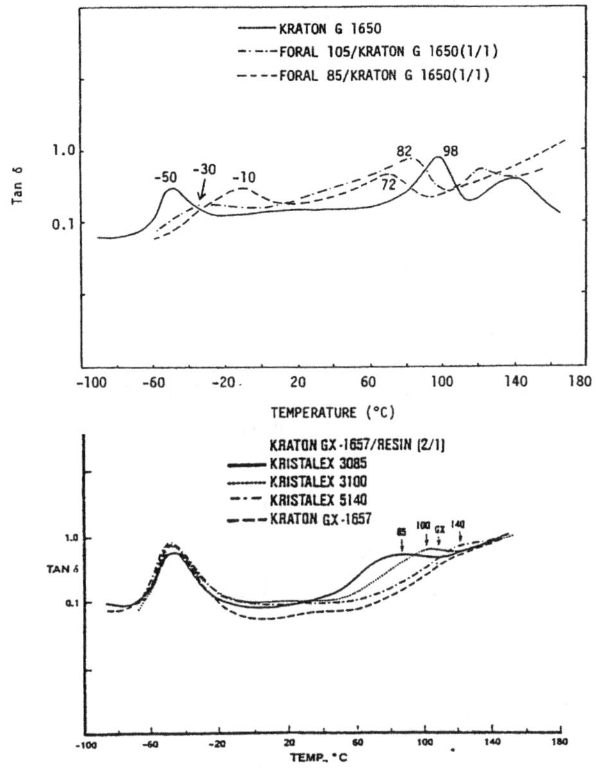

Fig. 8-34. Dynamic mechanical properties of rubber-resin blends.

Resin Group 1 (midblock compatible resin). These resins increase the midblock T_g of block copolymer without changing the T_g of the styrene domains, as they are compatible only with the midblock. These resins also simultaneously decrease the plateau modulus values. These are tackifying resins or midblock compatible resins because they reduce the room temperature modulus and, consequently, generate tack. Most aliphatic and cycloaliphatic resins belong to this group.

Resin Group 2 (endblock compatible resin). These resins increase the endblock T_g of the block copolymer without changing the T_g of the midblock domain. These resins also simultaneously increase (slightly) the modulus values of the block copolymer up to about 150°C (depending on the T_g of the resin). They are endblock reinforcing resins, and they increase high temperature performance when blended with block copolymer along with midblock resins. Most of these products are high softening point (>140°C) aromatic resins.

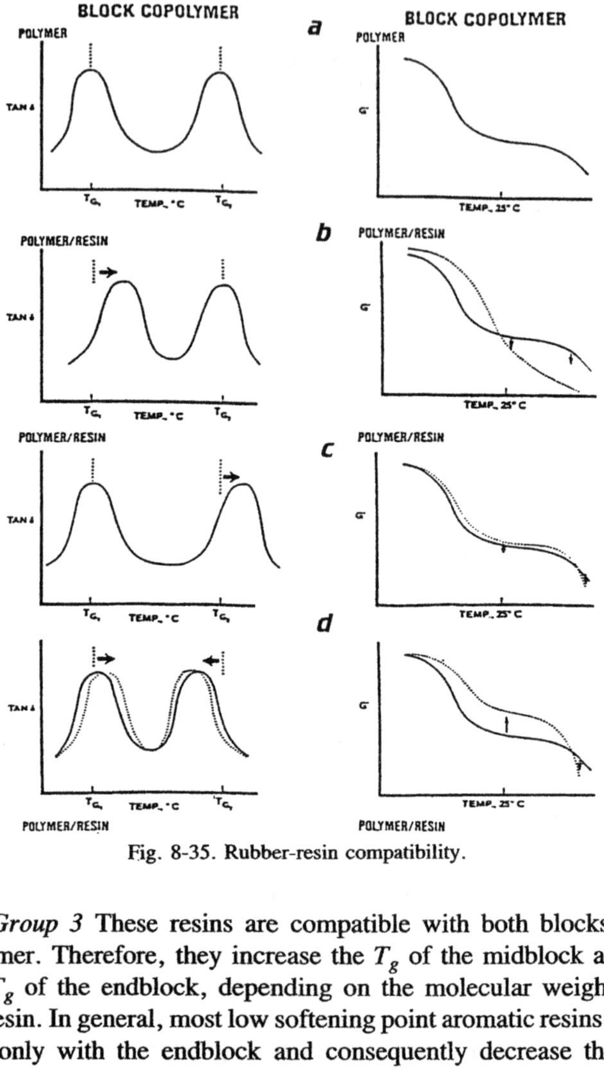

Fig. 8-35. Rubber-resin compatibility.

Resin Group 3 These resins are compatible with both blocks of the copolymer. Therefore, they increase the T_g of the midblock and decrease the T_g of the endblock, depending on the molecular weight and T_g of the resin. In general, most low softening point aromatic resins are compatible only with the endblock and consequently decrease the T_g of the end block. However, low molecular weight fractions of these resins may also be compatible with the copolymer midblock. Thus, the midblock T_g is also raised slightly. Rosin esters are typical of resins that are compatible with both the midblock and endblock.

Table 8-6 summarizes the experimental data on blends of Kraton G1650 (SEBS) and various Hercules resins. Regalrez resins are cycloaliphatic resin with various molecular weights (Table 8-5). The last three numbers in the resin name indicate the softening point and the first number indicates the degree of

Table 8-6. Glass Transition Temperatures of Kraton/Resin Blends

KRATON G 1650			MIDBLOCK T_g (°C)	END BLOCK T_g (°C)
			−50	100
Midblock	Regalrez	1018	−23	84
Rubber/resin (10/10)		1033	−20	85
		1065	−8	95
		1078	2	100
		1094	8	100
		1126	28	100
		3102	−45	85
		6108	−10	85
	Regalite	7070	−9	80
		1100	−10	100
	Foral	85	−10	70
	Foral	105	−30	83
	Piccotac	95 BHT	10	100
	Piccopale	100	−3	100
	Piccolyte	A 115	20	100
	Piccolyte	S 115	20	100
End block	Kristalex	3085	−50	72
Rubber/resin (70/30)		3100	−50	83
		1120	−50	102
		5140	−50	108
	Piccotex	LC	−45	76
		100	−50	79
		120	−50	88
	ENDEX	155	−50	110
		160	−50	113
	Shellflex	371	−50	82

hydrogenation. Piccolastic, Kristalex, and Piccotex are aromatic resins prepared from pure monomers, such as styrene, vinyl toluene, and α-methylstyrene. Regalite resins are hydrogenated C_9 resins. These data were obtained from a temperature sweep using the Rheometrics Dynamic Spectrometer ($\omega = 10$ rad/sec). The data can be used for the other Kraton samples (SEBS), such as GX1657, G1652, and G1726X.

The data indicate that the rubber-resin compatibility depends on the resin polarizability and molecular weight. Cycloaliphatic and aliphatic resins are compatible with isoprene and ethylene-butylene midblocks, and aromatic resins are compatible with end blocks of the copolymer. However, if the molecular weight of the resin is low, the preceding rule is not valid. The low molecular weight aliphatic resins (Piccolastic A5 and A25) and cycloaliphatic resins (Regalrez 1018 and 1033) are soluble in both blocks of the copolymer. Regalrez 6108, which is a 60% hydrogenated cycloaliphatic resin with a 108°C softening point, is also compatible with both blocks of the copolymer.

APPLICATIONS OF RHEOLOGICAL ANALYSIS

A major advantage of rheological analysis in formulating adhesives is that a reasonable starting point formulation can be rapidly determined and laboratory mixing and testing can be minimized. Selected examples of the use of this rheological approach to solve various problems follow.

Formulation Work for Kraton D1107 and GX 1657

The plateau modulus values ($G_{e_n}^0$) and midblock T_g values for various Kraton samples are plotted in Figure 8-36. This plot also contains a window (requirement in T_g and G' plateau) for good pressure sensitive adhesive properties. This window was obtained after the viscoelastic properties of many commercial pressure sensitive adhesives were characterized. According to the plots, the rubber with higher plateau modulus $G_{e_n}^0$ and T_g values is much more difficult to tackify than the rubber with lower $G_{e_n}^0$ and T_g, since its $G_{e_n}^0$ and T_g values are far from the required window. Using these viscoelastic data and the Equations (3) and (4) the resin concentration and the T_g of the tackifying resin, which are necessary to produce a good pressure sensitive adhesive with these block copolymers, can be calculated. The tackifying resin, which has a higher T_g than the elastomer, will increase the glass transition and decrease the plateau modulus value ($G_{e_n}^0$) of the elastomer. Consequently, an elastomer blend with a compatible tackifying resin will have a low modulus with a high T_g that can generate good pressure sensitive adhesive properties. This increase of the T_g

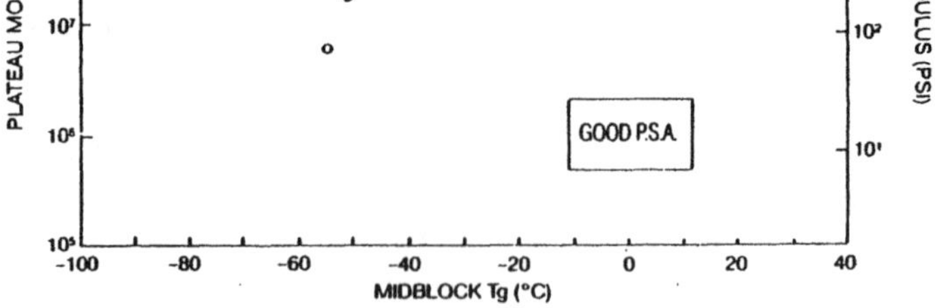

Fig. 8-36. Plateau modulus and glass transition temperature of Kraton samples.

value and decrease of the plateau modulus value can be predicted using Equations (3) and (4)[12]

$$\frac{W_1}{T_{g1}} + \frac{1 - W_1}{T_{g2}} = \frac{1}{T_g} \tag{3}$$

$$G_{e_n} = \nu_2^{2\cdot 3}\, G_{e_n}^0 \tag{4}$$

where W_1 and ν_2 are the weight fraction of the resin and the volume fraction of the elastomer, respectively; T_g, T_{g1}, and T_{g2} refer to the glass transition temperature for the blend and pure components in degrees Kelvin. G_{e_n} and $G_{e_n}^0$ are the plateau modulus values of the blend and the elastomer, respectively. The T_g and G_{e_n} values just given can be obtained from the dynamic mechanical spectra (Figs. 8-20 and 8-21). The glass transition temperature was defined as the tan δ maximum temperature and the plateau modulus value, G_{e_n}, is identified as the modulus value G' in the plateau when tan δ is a minimum.[12]

Figure 8-37 is a schematic illustration of how two different Kraton samples can be tackified. According to this plot, Kraton D1107 is easier to tackify than Kraton GX 1657 since its G_{e_n} and T_g values are nearer to the window than those of Kraton GX 1657. Using these viscoelastic data and Equations (3) and (4), the resin concentration and the T_g of the tackifying resin, which is necessary to produce a good pressure sensitive adhesive from the Kraton samples, were calculated. Based on these calculations, Kraton D1107 needs only one tackifying resin, Piccotac 95 Special hydrocarbon resin (about 50% by weight), in order to have good pressure sensitive adhesive properties. However, Kraton GX 1657 needs two tackifying resins. It needs about 65% resin

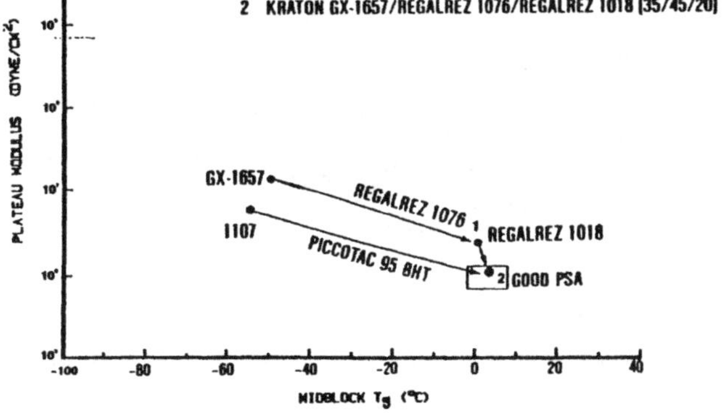

Fig. 8-37. Kraton G pressure sensitive adhesive formulation work.

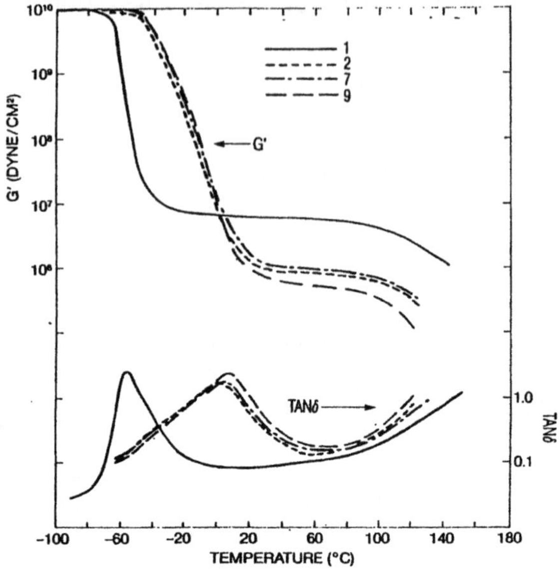

Fig. 8-38. Dynamic mechanical properties of Kraton 1107 and its blends with Hercules resins.

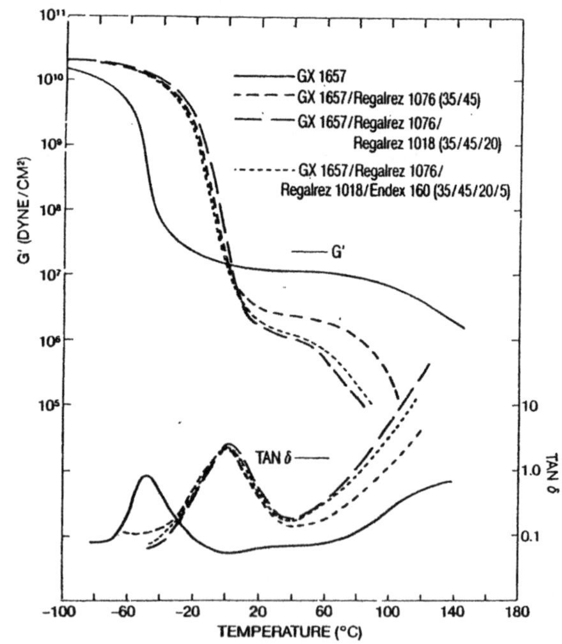

Fig. 8-39. Dynamic mechanical properties of Kraton GX 1657 and its blends with Hercules resins.

Table 8-7. Adhesive Properties of Kraton 1107 and Hercules Resin Blends

		FORMULATION			
		1	2	7	9 (CONTROL)
Kraton 1107		100	100	100	100
Piccotac 95 Special		—	100	100	140
Napthenic Oil		—	—	—	30
Regalrez 1018		—	—	10	—
ENDEX 160		—	—	45	—
Coumarone Indene		—	—	—	45
SAFT (°C)		128	108	121	108
Shear adhesion time (min)		>8300	>8300	>8300	>8300
Viscosity, η^*	100°C	3×10^4	5.5×10^3	8.7×10^3	3.7×10^3
(poise)	163°C	1.2×10^4	1800	1800	850
$\omega = 100$ rad/sec	177°C	9000	1100	800	280
Peel (oz/in.)		—	88	93	78
Quick stick (oz/in.)		—	62	63	47
Polyken tack (g/cm²)		—	>1700	>1700	>1700
T_g(°C)[a]		−50	5	4	5

[a] By RDS.

by weight, since it has a higher $G^0_{e_n}$ than that of Kraton D1107. Regalrez 1076 can reduce the plateau modulus and increase the T_g value of Kraton GX 1657, while Regalrez 1018 resin can further reduce the plateau modulus value without changing the T_g of the Kraton GX 1657 and Regalrez 1076 blend. The dynamic mechanical properties of the preceding blends are shown in Figures 8–38 and 8-39. Their pressure sensitive adhesive properties are shown in Tables 8-7, 8-8, and 8-9. They have excellent tack and peel properties along with the good UV stability.

Improving the High-temperature Performance and Creep Resistance of Kraton-based Pressure Sensitive Adhesives

The blend of Kraton D1107 and Piccotac 95 Special has good peel and tack properties and has a shear adhesion failure temperature of 108°C (Table 8-7). However, for certain applications, a 108°C shear adhesion failure temperature value is too low. Therefore, it is necessary to add an endblock reinforcing resin to improve the high-temperature performance. The performance of the Kraton endblock reinforcing resin, ENDEX 160 was verified by the following measurements.[18]

1. Dynamic mechanical properties and creep properties
2. Pressure sensitive adhesive properties measured by PSTC methods
3. Morphological picture obtained by transmission electron microscopy (TEM)

The dynamic mechanical properties of Kraton D1107, its blend with Piccotac 95 Special (50/50), and blends with ENDEX 160 are shown in Figure 8-38. In Table 8-7, adhesive properties of several formulations of pressure sensitive adhesives based on Kraton D1107 are summarized. The formulation including ENDEX 160 (formulation 7) has a higher modulus and lower tan δ at high temperature (~100°C) than the formulation without ENDEX 160 (formulation 2, Fig. 8-38). The creep data (Fig. 8-40) also indicate that the adhesive sample with ENDEX 160, formulation 7 (Table 8-7), has less creep compliance than the sample without ENDEX 160.

Formulation 7 (Table 8-7), which contains a blend of Kraton D1107, Piccotac 95 Special, Regalrez 1018, and ENDEX 160 (10/10/1/4.5), has a much higher shear adhesion failure temperature value with excellent tack and peel values compared to that of formulation 2, which does not contain ENDEX 160. Its performance is also better than that of the control sample (formulation 9), which contains a coumarone-indene resin. It is also interesting to note that the melt viscosities at high frequencies (at 177°C and 100 rad/sec) of formulation 7 are lower than those of formulation 2, which does not contain the modifying resin (Fig. 8-41). The formulation containing ENDEX 160 has a lower melt viscosity at processing temperatures (177°C) but higher shear holding power (SAFT) than the control formulation without ENDEX 160.

Table 8-8. Adhesive Properties of Kraton 1107 and Hercules Resin Blends

		FORMULATION			
		3	4	5	6
Kraton 1107		100	100	100	100
Piccotac 95 Special		140	140	120	110
Naphthenic Oil		30	30	15	10
ENDEX 160		—	45	45	45
SAFT (°C)		92	101	112	113
Shear adhesion time (min)		>8300	>8300	>8300	>8300
Viscosity, η^*	100°C	3.8×10^3	4.5×10^3	5.4×10^3	7.8×10^3
(poise)	163°C	420	600	1300	1400
ω = 100 rad/sec	177°C	270	270	560	780
Peel (oz/in.)		54	87	91	99
Quick stick (oz/in.)		42	57	67	70
Polyken tack (g/cm^2)		>1700	>1700	>1700	>1700
T_g(°C)[a]		6	8	7	7

[a] By RDS.

Table 8-9. Pressure Sensitive Adhesive Properties of Kraton G and Hercules Resin Blends

		FORMULATION		
		1	2	3 (CONTROL)
Kraton GX 1657		35	35	35
Regalrez 1076		45	45	45
Regalrez 1018		20	20	20
ENDEX 160		—	5	—
Coumarone Indene		—	—	5
SAFT (°C)		87	90	93
Polyken tack (g/cm^2)		>1700	>1700	>1700
180-degree peel (oz/in.)		83.7	79.5	80.7
Quick stick (oz/in.)		49.6	48.5	44
Rolling ball tack (cm)		8.2	10.1	19.8
Shear adhesion time (min)		>6000	>6000	>6000
Viscosity, η^*				
(poise)	163°C	610	610	640
ω = 100 rad/sec	177°C	350	370	390
Color		WW	WW	Yellow

Several formulations listed in Table 8-8 contain naphthenic oil. All the formulations are clear and have good peel and tack properties along with high shear adhesion failure temperature values. These data show that the beneficial effects of the oil (increased peel and quick stick) can be obtained, while the undesired effect (loss of shear strength) can be offset by the use of the reinforcing resin. Therefore, depending on the requirements of cost, processing conditions, and desired pressure sensitive adhesive properties, Kraton D1107 can be formu-

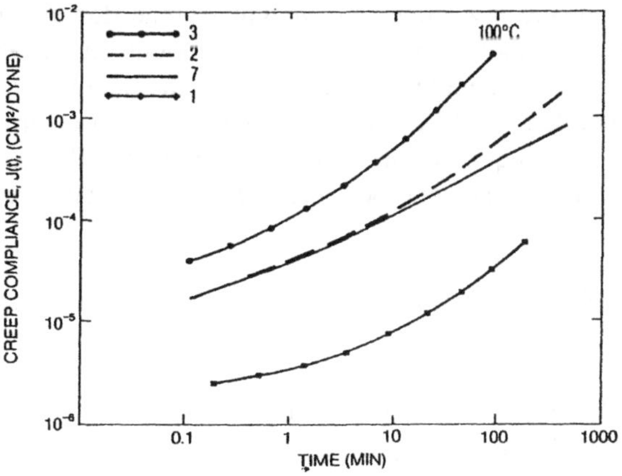

Fig. 8-40. Creep compliance of Kraton 1107 and its blends.

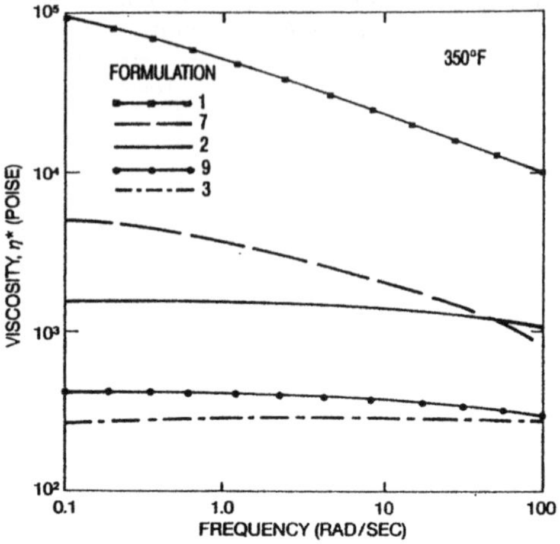

Fig. 8-41. Complex viscosities of Kraton 1107 and its blends with Hercules resin.

lated with ENDEX 160 for hot melt pressure sensitive adhesive applications. In general, this resin can be incorporated in amounts up to about three times the styrene content of Kraton D1107 without diminishing the tack and peel properties of the adhesives.

In Table 8-9 the pressure sensitive adhesive properties of the formulations based on Kraton GX 1657 and hydrogenated midblock resins, Regalrez, are compared with those of similar formulations containing ENDEX 160. Their dynamic mechanical properties are plotted in Figure 8-39 as a function of temperature. The formulation containing the reinforcing resin, ENDEX, has slightly better high-temperature performance than the formulation without ENDEX, as indicated by their SAFT, G', and tan δ values at high temperatures. This is due to the fact that the low molecular weight styrene domains (M_w = 5000) of Kraton GX 1657 cannot accommodate large amounts of ENDEX resin. The improvement of high-temperature performance of the adhesive depends on the amount of incorporated ENDEX resin. In general, we can incorporate ENDEX resin in amounts about equal to the styrene content of the Kraton G elastomers without deteriorating the tack and peel properties of the adhesives.

Morphological Study of Block Copolymer Based Pressure Sensitive Adhesives

In this section, the performance of the block copolymer endblock reinforcing resin, ENDEX 160, by morphological evidence is discussed.

Figure 8-42 shows the morphology that was obtained from Kraton D1107

Kraton 1107

Kraton 1107/Piccotac 95 SPL (10/10)

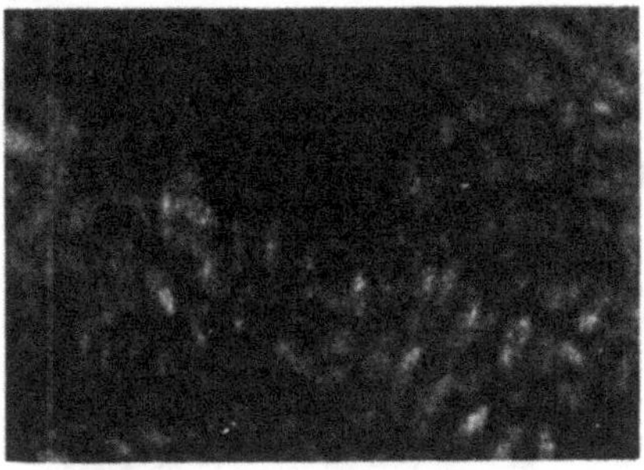

Kraton 1107/Piccotac 95 SPL/Endex 160 (10/10/4.5)

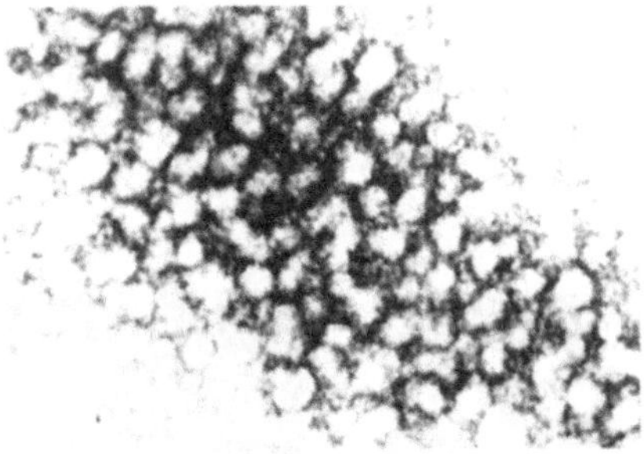

Kraton 1107/Endex 160 (10/4.5)

Fig. 8-42. TEM pictures of Kraton 1107 and its resin blends.

and its blends with tackifying resins and/or ENDEX 160. The transmission electron micrographs were taken after staining the samples with OsO_4 for 36 hr. The samples were microtomed (500 Å) at cryogenic temperatures. The magnification factor is 150,000. It has a spaghetti-and-meatball morphology with a continuous rubber phase and a dispersed polystyrene phase. The discontinuous phase is a polystyrene phase (endblock), and the electron dense continuous phase is an isoprene rubber phase (midblock). These dispersed domains are quite small, less than 200 Å in size.

The dispersed domains in the blend of the tackifying resin Piccotac 95 BHT and Kraton D1107 appear to be spherical and less densely packed due to the dilution effect. Swelling of the elastomer phase by Piccotac 95 BHT gives no change in the dispersed domain diameter, except to make them clear spherical shapes, since the resin is dispersed in the continuous rubber phase. It also does not affect the clarity. The dispersed polystyrene phase in the blend of ENDEX 160 and Kraton D1107 appears to be more elliptical (bean shape), while others appear to be more spherical. The size of dispersed domains are homogeneous (~250 Å). This indicates that all of the ENDEX 160 resin went into the polystyrene domains of the Kraton without agglomerating by itself. This is further evidence of the good compatibility of our ENDEX 160 with

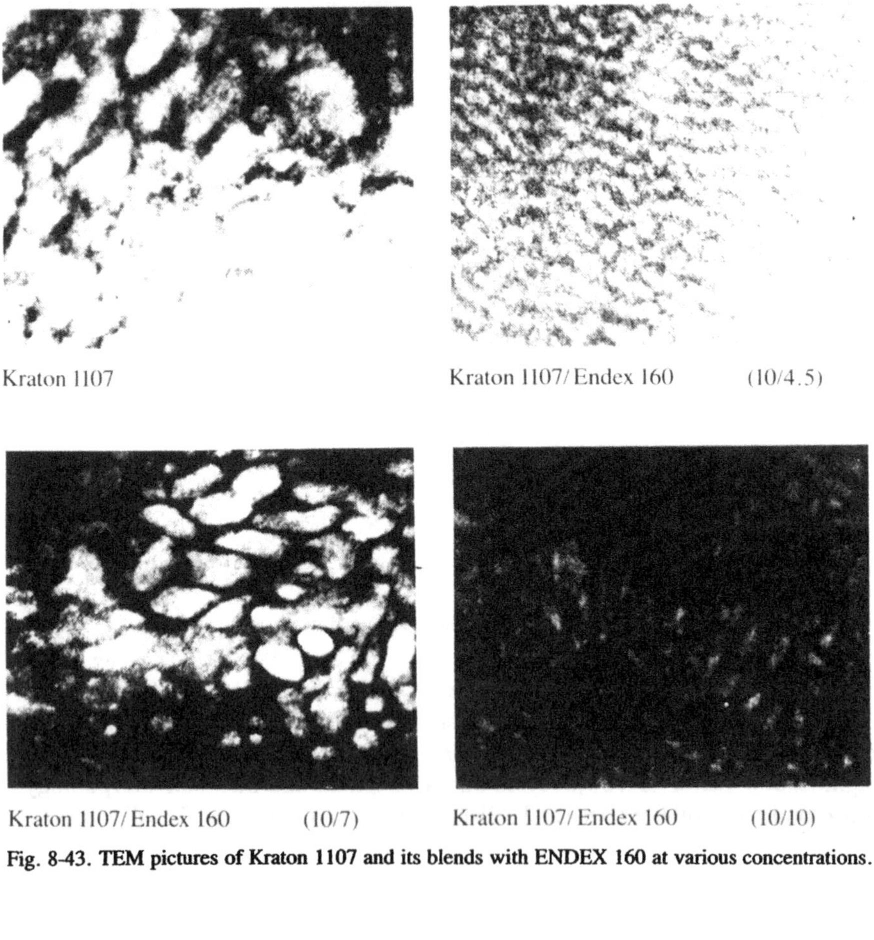

Kraton 1107 — Kraton 1107/Endex 160 (10/4.5)

Kraton 1107/Endex 160 (10/7) — Kraton 1107/Endex 160 (10/10)

Fig. 8-43. TEM pictures of Kraton 1107 and its blends with ENDEX 160 at various concentrations.

the polystyrene domains of the copolymer sample. The final PSA formulation consisting of Kraton D1107, Piccotac 95 BHT, and ENDEX 160 appears to have the largest domains (less than 300 Å). However, it is still clear (no light scattering), since the size of the dispersed particle is much less than the wavelength of visible light (about 5000 Å).

Figure 8-43 shows the morphological change of Kraton D1107 blends at various loadings of ENDEX 160. By blending 4.5 g of ENDEX 160 with 10 g of Kraton D1107, the size of the dispersed styrene phase increases and the shape changes from a spherical type to a jellybean type, while the size of the dispersed phase is homogeneous. All of the endblock reinforcing resin, ENDEX 160, enters into the polystyrene domains. So by blending with ENDEX 160, a bigger dispersed domain size is obtained and, consequently, a higher cohesive strength at room temperature is achieved. Also it will have better high-temperature performance due to the higher glass transition temperature of the dispersed domains, as shown in Figures 8-44 and 8-45. The T_g of the styrene domains in Figure 8-44 was measured using the RDS from the Kraton D1101 and ENDEX 160 blends. If more ENDEX 160 (7 g to 10 g of Kraton D1107) is added, the domain size increases. However, a small domain size may also be due to the overloading of copolymer with ENDEX 160. Such samples are very cloudy. ENDEX 160 starts to form its own domain. From this ENDEX 160 level (7 g to 10 g) the copolymer is overloaded. If more ENDEX 160 is added, it will

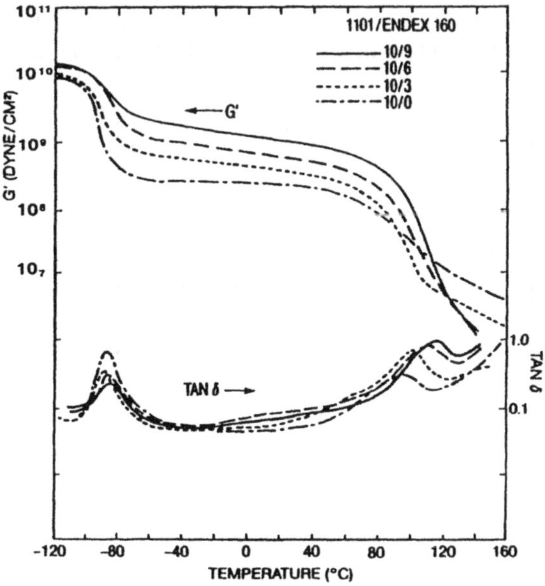

Fig. 8-44. Dynamic mechanical properties of Kraton 1101 and its blends with ENDEX 160.

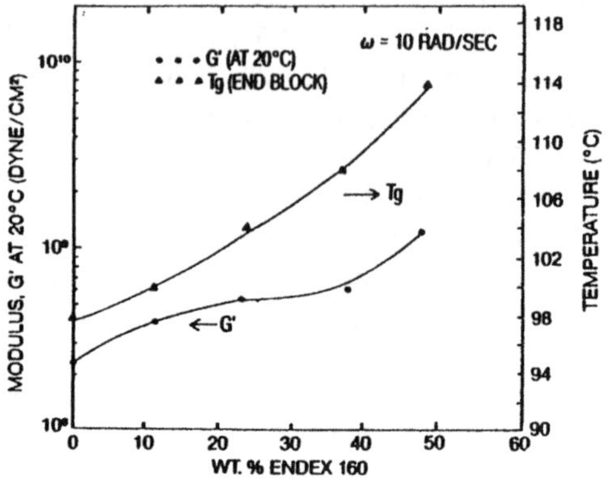

Fig. 8-45. Physical properties of Kraton 1101 blends with ENDEX 160.

undergo a phase inversion where the dispersed phase will be rubber and the continuous phase will be polystyrene. The ideal composition is 4.5 g of ENDEX 160 and 10 g of Kraton D1107 (about three times the styrene content of Kraton D1107).

Formulation Work for SBR Based PSA

In case the plateau modulus value of the elastomer cannot be determined from the dynamic measurement, a modulus value at application temperature and the

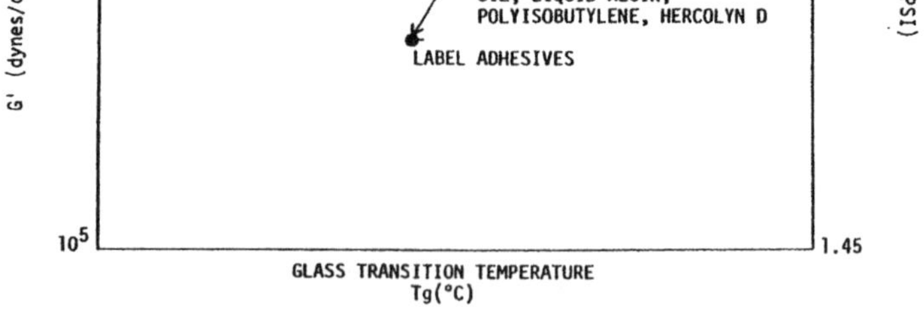

Fig. 8-46. G′ vs. glass transition temperature as affected by plasticizers and tackifiers.

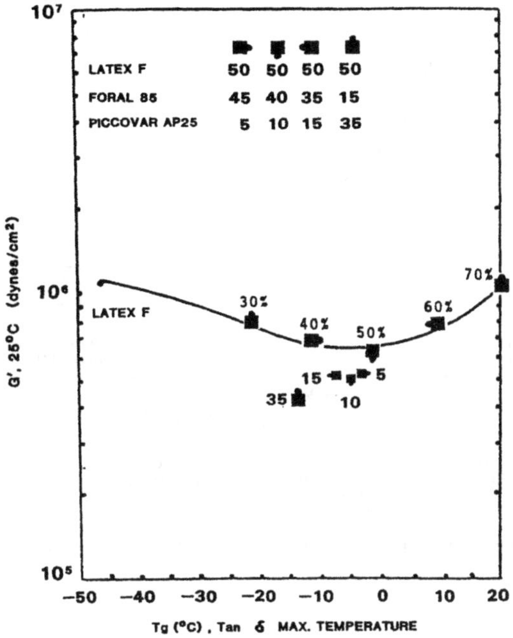

Fig. 8-47. Latex F/Foral 85 55WK blends.

tan δ maximum temperature can be used. This approach may not be perfect, but it is a very convenient method to obtain starting-point formulations. Figure 8-46 is a schematic of how the addition of various components changes the modulus and glass transition temperature. Figure 8-47 shows, more specifically, the effect of adding varying amounts of Piccovar AP25 to an SBR/Foral 85 blend. Replacing the Foral 85 with 5, 10, 15, and 35 parts of Piccovar AP25 causes a decrease in the room temperature modulus and also results in a lowering of the glass transition temperature.

Figure 8-48 shows several windows of performance that have been defined and that can be used as guidelines for making good pressure sensitive adhesives. The pressure sensitive performance of adhesives can be related to the bulk viscoelastic properties. By focusing on certain temperature and modulus ranges, various needs, such as cold temperature adhesion, room temperature application, shear resistance, and melt processing requirements, can be addressed. It is obvious from this figure, that the modulus response over a wide temperature range must be considered in order to develop a satisfactory adhesive system.

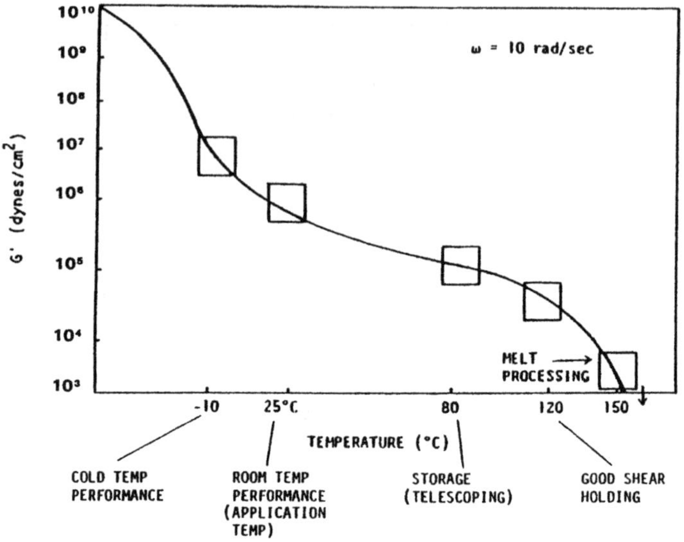

Fig. 8-48. Modulus (G') requirements (windows) for good pressure sensitive adhesive tapes.

ACKNOWLEDGMENTS

The author wishes to thank Hercules Incorporated for permission to publish this paper. Special thanks are due to Mr. Brian Hodgson for measuring RDS data. Technical discussion with Drs. J. B. Class, K. Foley, and F. Nelson were invaluable in both the initiation and completion of this work.

REFERENCES

1. Sherriff, M., Knibbs, R. W., and Langley, P. G. *J. Appl. Polym. Sci.* **17:** 3423 (1973).
2. Aubrey D. W., and Sherriff, M. *J. Polym. Sci., Polym. Chem. Ed.* **16:** 2631 (1978).
3. Aubrey D. W., and Sherriff, M. *J. Polym. Sci., Polym. Chem. Ed.* **18:** 2597 (1980).
4. Kraus, G., Jones, F. B., Marrs, O. L., and Rollmann, K. W. *J. Adhesion* **8:** 235 (1977).
5. Kraus G., and Rollmann, K. W. *J. Appl. Polym. Sci.* **21:** 3311 (1977).
6. Kraus, G., Rollmann, K. W., and Gray, R. A. *J. Adhesion* **10:** 221 (1979).
7. Kraus G., and Hashimoto, T. *J. Appl. Polym. Sci.* **27:** 1745 (1982).
8. Bates, R. *J. Appl. Polym. Sci.* **16:** 2631 (1978).
9. Coply, B. C. *Rubber Chem. Technol.* **55:** 416 (1982).
10. Class J. B., and Chu, S. G. *J. Appl. Polym. Sci.* **30:** 805 (1985).
11. Class J. B., and Chu, S. G. *J. Appl. Polym. Sci.* **30:** 815 (1985).
12. Class J. B., and Chu, S. G. *J. Appl. Polym. Sci.* **30:** 825 (1985).
13. Class J. B., and Chu, S. G. Presented at the Adhesion Society Meeting, February 1984.
14. Zawilinski, A. *Proceedings of PSTC Meeting,* 1984.

15. Fox, T. G. *Bull. Am. Phys. Soc.* (Ser 2). **1:** 123 (1956).
16. Dahlquist, C. A. *Proc. Nottingham Conf. on Adhesion,* in *Adhesion: Fundamentals and Practice.* MacLaren, London, 1966.
17. Wong, T. K. M. Ph.D. Thesis. Council for National Academic Awards. London, 1969.
18. Chu, S. G. *Proc. Adhesives and Sealant Council Meeting* (1984).

9. Analytical Techniques for Identification and Characterization

David N. Kendall
Dave Kendall Associates
Hilton Head, South Carolina

Most important to the identification of pressure sensitive adhesives is the skill of the analytical scientist. This factor is often overlooked because the trend in analytical instrumentation in recent years has been increasing computer control and automation, including automation of sample preparation. The goal of instrument designers and manufacturers appears to be a completely integrated analysis system, capable of processing a raw sample all the way to the finished report. While an admirable goal, even when these systems are at hand, they have little value without a well-trained and well-experienced analytical scientist behind them. One could argue that computerization and automation have in some instances gone too far. Just as in any other human activity, the pendulum is apt to swing too far until a sustainable balanced beat is achieved.

Several analytical techniques can be helpful in the identification of pressure sensitive adhesives. Infrared absorption spectroscopy (IR)[1,2] is probably the most useful analytical technique for the identification of pressure sensitive adhesives. It can help identify all organic materials and most inorganics, except for those few having monatomic cations and anions.

Ultraviolet absorption spectroscopy (UV)[3] is useful for helping to identify unsaturated (e.g., C=C) substances, such as determining whether a rosin component of an adhesive has been partially or totally hydrogenated. Ultraviolet spectroscopy has greater sensitivity than infrared for those materials to which it is applicable. Therefore it can be useful in quantitative work at very low concentration levels. Quantization at a component level of 0.005% by weight is not unusual, while a level of 0.1% by weight is often encountered as a limit for practical quantitative IR methods, without resorting to specialized IR techniques.

Laser Raman spectroscopy can be useful in certain solvent analyses and for the analysis of aqueous adhesive systems.

The power of chemical analysis, often currently referred to as "wet chemical analysis," should not be overlooked in adhesive identification. Spot tests, as found in Feigl and Fieser, can give fast identification of critical functional groupings or elements present in an adhesive. Judicious use of sulfide formation at controlled pH conditions serves, for example, to identify Zn where the presence or absence of ZnO is sought as a filler and/or vulcanizing agent in a neoprene-type pressure sensitive adhesive.

Ashing an adhesive to burn off the organics leaves the inorganics in a more easily identified state by IR or chemical tests. Usually only a limited number of possibilities exist, so a tedious number of chemical tests is not required for identification.

Simple flame tests, using a Bunsen or Meker burner, can quickly establish the presence of sodium, potassium, calcium, lithium, and occasionally copper (especially when moistened with hydrochloric acid) when any are possibly present in an adhesive.

Differential thermal analysis (DTA)[4] examination can be useful in the identification of polymers present in a pressure sensitive adhesive. Once the polymers have been separated, each can sometimes be identified by their glass transition temperatures (T_g). Moreover, the heats of transition or fusion determinable by DTA can measure the degree of crystallinity of a polymer, which can be used to identify polymers. Standard known polymers must already have been subjected to the technique, so that this information is available, just as a catalog of known spectra is normally necessary for a spectral identification.

Light microscopy can be useful in comparing the surface characteristics of pressure sensitive adhesives. Where the surface microstructure is important to an adhesive's functioning, a study of that surface by scanning electron microscopy (SEM) techniques will reveal minute structure and surface differences.

In order to characterize pressure sensitive adhesives, the concern is to a large extent with polymers and the determinants of polymer properties. Because the film-forming elastomeric material of a pressure sensitive adhesive composition is generally a polymer, as is also the resin or other material used to impart the desired degree of tack, wetting power, and specific adhesion, polymer properties are critical.

Polymer properties are determined by molecular weight and its distribution, degree of crystallinity, chain branching, side-chain substitution, and cross-linking. The ability to crystallize, the flexibility of chain bonds and the nature, number, and spacing of polar groups come into consideration. For crystalline polymers, the type of crystal structure, degree of crystallinity, size and number of spherulites, and orientation are important to polymer properties. Tackifiers, plasticizers, fillers, and other modifiers come under this category. The most widely applicable technique for identifying and analyzing pressure sensitive adhesive formulations is IR.

INFRARED METHODS

The atoms of any molecule not at absolute zero are continuously undergoing oscillations of various kinds. Fortunately, these molecular motions are of the same order of magnitude [10^{13}–10^{14} cycles per second (cps)] as those of IR radiations. When the frequency of molecular motion is the same as that of IR radiation impinging on that molecule and when there is a change in dipole moment during the vibration, the molecule absorbs by resonance all, or a portion of, the radiation incident upon it. A permanent dipole in a molecule is not necessary for IR absorption. Carbon tetrachloride (CCl_4) for example, absorbs IR because there are vibrations that the C—Cl groupings undergo, when they are not in phase, during which a change in dipole moment occurs.

While the IR region extends from about 0.75 to 1000 μm (13,333 to 10 cm^{-1}), the most useful range for chemical analysis is from 2 to 25 μm (5000 to 400 cm^{-1}). Approximately 95% of all such work is carried out in the 2–16 μm (5000 to 625 cm^{-1}) region.

All organic chemicals and most inorganics can be handled by IR. Inorganics having monatomic cations and monatomic anions (NaCl, KBr, and CsBr, used as windows or prisms for dispersing IR) do not yield characteristic absorption spectra, except for their lattice vibrations. Neither do most metals. Diatomic molecules such as H_2, O_2, N_2, Cl_2, cannot be identified or determined by IR. They do not give useful IR absorption spectra, since no change in dipole moment occurs during their lone stretching vibrations that are electrically symmetrical.

No two substances that absorb IR radiation absorb it at the same frequencies to the same extent. Therefore, an IR absorption spectrum is a "fingerprint" of a substance as unique as a fingerprint of a human being and as useful in differentiating one molecule from another as human prints are in differentiating one individual from another. Moreover, the spectrum of a mixture, excepting certain hydrogen bonding situations or chemical reactions taking place, is simply the sum or superposition of the spectra of the individual components comprising the mixture. IR is the most widely used technique for identifying unknown substances and mixtures.

Infrared spectrophotometers, the instruments used for chemical analysis by IR, comprise three essentials: the source of IR radiation, a dispersing means to separate the radiation by frequency (wavelength), and a detector that changes the received radiation into an electrical signal. Other components are necessary, such as slits, collimators, mirrors, amplifiers, and recording mechanisms.

Sources can be a Nichrome helix, a Nernst glower (a mixture of rare earth oxides), or a Globar (a silicon carbide rod that requires water or air cooling). Dispersing elements are commonly gratings now having replaced prisms used in the earliest IR spectrophotometers. Filter-grating types of grating spectrometers can cover a range up to about 333 μm (30 cm^{-1}). Detectors are radiation thermocouples, bolometers, pneumatic detectors, or triglycine sulfate (TGS).

One of the more recently developed instruments is the Fourier transform infrared spectrometer (FT-IR). This technique uses an interferometer of the Michelson type[5] to produce an interferogram of the sample. A fast digital computer quickly transforms this to the familiar transmission or reflection-type spectrum. The FT-IR began to be a practical proposition when Fellgett[6] pointed out the advantages of its multiplex operation. Today modern FT-IR instruments incorporating sophisticated computers offer high resolution (0.08 cm^{-1} is not uncommon), many rapid scans to average out noise, and quick subtraction of the components of a mixture.

MULTIPLE INTERNAL REFLECTION

When a pressure sensitive adhesive sample has the adhesive surface exposed, or when it can be exposed by simply removing a release layer, the multiple internal reflection (MIR) technique can be used. The MIR technique was formerly known as attenuated total reflectance (ATR). It is also known as internal reflection spectroscopy (IRS). The tacky pressure sensitive adhesive surface is placed against the MIR crystal surface on one or both sides of the MIR crystal, the angle of incidence of the incoming IR beam adjusted to yield maximum energy at 5 μm (2000 cm^{-1}), and the adhesive's spectrum scanned from 2.5 to 50 μm (4000–200 cm^{-1}). Often, a scan from 2.5 to 15 μm (667 cm^{-1}) is sufficient. Thallous bromide-iodide (KRS-5) is the most useful crystal with AgCl a distant second. Germanium and Irtran crystals also find use.

The MIR crystal is normally incorporated in a small accessory that slides into the sample holder and sampling compartment of commercial IR spectrometers, and that can be manually adjusted via knobs to vary the angle of incidence of the incoming IR beam. One can also be placed in the reference beam of double-beam instruments, if desired, to produce a greater spread between absorption band maxima and minima. A common type of IR accessory provides about 25 internal reflections. The different MIR crystals of differing refractive indices allow varying angles of incident IR radiation to be used, from about 20 to 55 degrees. Satisfactory MIR spectra are produced by judicious selection of the appropriate ratio between the refractive index of the crystal and that of the sample, of the angle of incidence that the radiation makes with the crystal–sample interface, and the contact between the sample and crystal. Good contact is vital. Unless it can be made, a useful MIR spectrum is not attainable. This limits the technique to relatively smooth surfaced samples.

While liquids can be scanned by smearing them onto an MIR crystal, normally no advantage is gained over sample transmission techniques. The MIR technique is more tedious. Moreover, the MIR spectrum is very similar but not identical to the transmission spectrum. The MIR technique should be viewed as a specialized one of limited application.

Illustrative of a specialized application is the use of MIR to determine if the plasticizer in a pressure sensitive adhesive may have migrated from the backing layer of the substrate through the surface of the adhesive layer. One wishes to examine the pressure sensitive adhesive layer as is. Comparing the MIR spectrum of this surface to that of a freshly prepared specimen as a control could show whether or not plasticizer migration had taken place. Poor adhesion of a pressure sensitive tape could suggest that the examination just described should be made, particularly if it is known that the backing layer did contain a plasticizer. In such circumstances, a barrier layer should be present to prevent migration, but barrier layers can be misapplied.

Occasionally it may be easier to place the MIR crystal in contact with a very small liquid film sample, which will transfer to the crystal and can thus be scanned. This situation could arise where one wished to avoid extracting the liquid film with solvent by eliminating the solvent and then scanning in the usual transmission mode. For example, one might wish to avoid injuring or contaminating the substrate beneath the liquid film.

Since the impinging IR radiation beam penetrates the sample's surfaces up against the MIR crystal roughly to the depth of the wavelength of the IR beam, MIR is essentially a technique for examining surfaces more deeply than the scanning electron microscope (SEM) or Auger spectroscopy.

ANALYSIS OF BACKING MATERIAL

When identification or other analysis of the backing material is sought, the adhesive is first removed by scraping off with a knife wet with an appropriate solvent, one that will loosen the adhesive. Acetone, methyl ethyl ketone (MEK), carbon tetrachloride, or toluene may be appropriate.

If the backing is paper, an analysis of coating, saturants, fillers, etc., may be desired. A coating can be examined by MIR spectroscopy if the coating is smooth enough. If not, the coating will be extracted with a solvent. One normally tries solvents with low boiling points initially, since they are easier to remove from the extracted coating. Toluene, MEK, CCl_4, or orthodichlorobenzene (ODCB) may be required to remove the coating. After concentrating, the extract is cast onto NaCl plates, the solvent removed by heating, and the room temperature coating film on the NaCl plate scanned in the IR, usually from 2.5 to 16 μm.

Water-soluble additives in the paper can be extracted with boiling water and after concentration cast onto AgCl or Irtran plates, the water removed by heating in an oven, and the coating scanned in the IR.

Inorganic fillers in the paper backing can be identified by carefully ashing a portion of the sample over Bunsen or Meker flames or in a muffle furnace, then scanning the ash, not only from 2.5 to 16 μm but out to 25 or even 50 μm, since useful identifying inorganic absorptions are often found in the region from 15 to 50 μm (200 cm^{-1}).

Once the coatings, saturants, and fillers have been identified, one may wish to check to determine if the backing is really paper. Mulls can be made either from the backing without the pressure sensitive adhesive or from such a backing after removing the coatings. To about 15 mg of the paper, a drop of mineral oil, such as Nujol, is added to the sample on a glass plate and thoroughly ground and dispersed using a glass paint muller. The objective is to obtain as fine a particle size as possible to minimize scattering of the IR beam impinging on the sample. Scattering is radiation that impinges on the sample, but rather than being transmitted through it, is scattered aside without getting into the monochromator of the IR instrument. Scattering is effectively eliminated if all mulled particles are below 2 μm in size. This is not always possible to obtain.

It may be necessary to add more Nujol or occasionally more sample during the mulling process. A satisfactory mull will usually be translucent to visible light, after being placed between two NaCl plates and squeezed out. The adequacy of the mull can be judged by noting the slope of the spectrum between 2 and 4 μm. The worse the deviation from linearity, the greater the scattering and the poorer the mull and the recorded spectrum. Some slight upward slope from 2 to 4 μm can be tolerated. A comparison of the spectra against the spectra found in reliable literature will enable one to judge how much scattering can be tolerated without loss of accuracy. Many infrared spectroscopists now use KBr routinely, since they wish to scan out to 400 cm^{-1} (25 μm).

After the mineral oil mull, a mull may be made using perfluorokerosene or hexachlorobutadiene (HCBD) and scanned over only those regions where mineral oil absorbs, namely, 2–4 μm, 6.5–7.8 μm, and 13.5–14.3 μm. Thereupon one has the complete 2.5–15-μm spectrum of the solid sample free of mulling agent absorptions.

If the pressure sensitive adhesive backing should be a fabric, one can identify the fibers by scanning Nujol and HCBD mulls. (Admittedly, the mulling process may be a bit difficult, but mulling such fabrics keeps the hand and arm in shape!) Nylon-type backings can be identified by scanning IR spectra as evaporated films on KBr plates, cast from solutions in cresols or formic acid.

A detailed examination of the fabric may be desirable. For silk and synthetic filaments, the term denier is employed. The denier is the weight in grams of 9000 m of filament or yarn. International scientific and standards organizations use the tex system. The tex in grams/kilometer is the internationally preferred unit to describe linear density. The heavier the filament or yarn, the greater is the denier or tex number. A vibrascope can be used to determine quickly and accurately the denier of individual fibers 2–4 cm in length according to the equation

$$m = T/4l^2f^2 \times 10^5 \tag{1}$$

where

m = linear density (g/cm)
T = fiber tension (dyne)
l = effective fiber length (cm)
f = fundamental resonant frequency (cps)

For natural fibers such as wool, cotton, and flax, the denier is not commonly measured, since staple length and diameter are usually considered sufficient to define fineness. In cotton the longer the staple length, the better the quality. Staple length can be measured using a Fibrograph.[7] The diameter can be measured with the light microscope. The quality or grade of wool is determined on a statistical basis of distribution of fiber diameter. Both the mean value and distribution around the mean are measured.

The fineness of cotton can be determined using the Micronaire, which is manufactured by a Bendix Corporation subsidiary. It measures, under precise conditions, the airflow resistance of a plug of cotton of specified dimensions. Other things being equal, the airflow is a function of the fineness of the fiber. Also, the density of fibers can be determined by precise microscopic measurement of length, cross-sectional area, and weight.

IR microspectroscopy is increasingly being used to identify microgram to nanogram and picogram samples. This has come about with the advent of FT-IR systems that use available energy much more efficiently than do dispersive systems. Raman microspectroscopy, which complements the findings from IR, is now useful, made possible by the advent of the laser source. Both the IR and Raman microspectroscopies are molecular spectroscopy through a microscope.[8] For the identification of a single fiber of a backing material of a pressure sensitive adhesive, say, all that is available in a forensic situation, these microspectroscopies could provide the answer. In the Raman area, fluorescence is still a formidable problem. For IR microspectroscopy, the most common problem is excessive sample path length. Use of the diamond cell to thin the sample can alleviate this problem.

Nonwoven fabrics are fibrous webs bonded by the addition of chemicals or fusion of some of the constituent fibers. Web bonding comprises production of a network structure in the web, using adhesive and/or entanglements to bond the constituent fibers. Wet or dry bonding systems can be used. Since polymeric emulsions featuring acrylics, butadiene copolymers with styrene or carboxylated styrene, acrylonitrile-butadienes, acrylates and their copolymers, vinyl acetate-based polymers, or vinyl chloride polymers may be present, the choice of solvent to extract the binder is wide. Typically, MEK, toluene, and ODCB should be tried. Because some binders are self-cross-linking or formed from a reactive additive, they may be insoluble in most solvents. ODCB, however, will normally dissolve enough of the low molecular weight ends of an insoluble binder to allow it to be identified by an appropriate IR scan. Here, microsize plates and ordinate scale expansion may be necessary to obtain IR spectra adequate for identification of the binder.

It is sometimes preferable to scan the nonwoven backing in IR without the pressure sensitive adhesive prior to solvent extraction and after solvent extraction, to learn whether or not the binder has been practically completely removed. This can be done as mineral oil and HCBD mulls. The IR spectra of the fabric itself, with binder, needs to be subtracted out where appropriate.

Foams can be urethanes or thermoplastics foamed by incorporating a blowing agent that decomposes to a gas at elevated temperatures or by incorporating an inert gas. Rubber lattices are foamed by mechanical whipping. Some double-face tapes have a backbone of foam, since foams provide a cushion between two surfaces and fill the gaps between two irregular surfaces. For identification of foams by IR, proceed as for organic polymers or films, as previously described; scan as evaporated films from solution or as mineral oil and HCBD mulls, if the foam is solvent-insoluble.

IDENTIFICATION OF COMPONENT PARTS

A pressure sensitive adhesive is usually composed of an elastomer, a tackifying resin, and an oil. More than one resin may be present to provide properties not available from a single resin. Fillers are often added to raise the viscosity, add color, increase the specific gravity, or reduce the cost. Antioxidants stabilize adhesives against heat and light degradation, as well as against oxidation.

Adhesives for electrical applications or high temperature use sometimes incorporate heat-curing, oil-soluble phenolic resin. To promote reaction between the phenolic resin and the rubber component, a catalyst such as an alkaline filler or zinc resinate is used.

Elastomer Identification

Natural or reclaimed rubber, SBR (styrene-butadiene rubber), polyisobutylene or butyl rubber, higher polyvinyl alkylethers, Buna *N* (butadiene-acrylonitrile copolymer), higher polyacrylate esters, styrene-butadiene-styrene (S-B-S), and styrene-isoprene-styrene (S-I-S) block copolymers are typical of the rubbery elastomers used in pressure sensitive adhesives.

To determine whether an elastomer is or is not a thermoplastic rubber of the block copolymer type, a glass transition temperature T_g measurement will suffice. The thermoplastic rubber will have two T_g's, while a random SBR copolymer will have only one. An S-B-S block copolymer with the same styrene butadiene ratio as a random SBR copolymer will show two T_g's.

An IR identification of the elastomer in a pressure sensitive adhesive can be made sometimes on the spectrum of the adhesive scanned as is, and squeezed out between KBr plates. Elimination of fillers, when present, may be helpful. The elastomer can be extracted from the fillers using an appropriate solvent, such as toluene or ODCB. A concentrated solution of adhesive without fillers is cast onto KBr plates, the solvent evaporated off by heating in an oven, and the evaporated film scanned in the IR. A substantial catalog of IR spectra of known elastomers for pressure sensitive adhesives is a must.

If both the fillers and tackifying resin can be removed to leave essentially only the elastomer, then identification can be made by scanning the pyrolyzate of the elastomer and comparing the resulting spectrum against a catalog of pyrolyzate spectra of known elastomers. Pyrolysis equipment is available from Foxboro Analytical, Harrick Scientific Company, and Harshaw Chemical. These accessories enable control of time and temperature at which pyrolysis is carried out and the volume of the pyrolyzing chamber. When precisely reproducible conditions are unnecessary, one can prepare a satisfactory pyrolyzate using a borosilicate test tube of appropriate sample size.

A gram or less of the sample as is, or cut into pieces to conform to the test tube, is placed in the bottom of the tube and a Bunsen flame applied to the tube held with its open end slightly higher than the action end. Should some moisture be produced by heating the sample initially, it is well to evaporate this out of the tube before continuing to heat and collect the liquid pyrolyzate produced in the cooler open end of the test tube. This pyrolyzate material is placed between appropriate salt plates, such as NaCl, KBr, and scanned in the IR. Comparison against known pyrolyzate spectra serves to identify the elastomer.

If one has known pyrolyzate spectra of elastomer and tackifying resin, then identification can proceed from pyrolysis of the unknown combination of elastomer and tackifier.

While Pyrolysis/IR is very useful in identifying the polymers and tackifiers present in pressure sensitive adhesives, another multitechnique method can be helpful. This is Pyrolysis/GC/MS, pyrolysis/gas chromatography/mass spectroscopy. Pyrolysis of polymeric materials breaks them down into smaller molecules, often monomers. These have more IR absorption bands than the polymers that begat them, and thus are sometimes easier to identify. This is true especially when one is confronted by a mixture of polymers such as co- or terpolymers in a complex pressure sensitive adhesive formulation. To simplify the identification process, a pyrolysis effluent can be inserted onto a GC column along with a carrier gas. By means of temperature programming, individual components present in the sample are swept from the column by the carrier gas into the inlet of the mass spectrometer. The time of arrival of species and the mass spectral intensities are recorded. If the gas chromatographic column has been properly selected, it will often succeed in a time separation of species contained in a sample. The intensities of the mass spectrum will be those associated with single components observed one after the other, free from peak overlaps. Therefore, by using this combined technique of Pyrolysis/GC/MS, complex pressure sensitive formulations can be identified with respect to their polymeric components. Removal of nonpolymeric components prior to the use of this technique is a simplifying advantage.

Pyrolysis/GC can be a powerful technique for analysis of polymers used in pressure sensitive adhesives. Studies of pyrolysis mechanisms and degradations can throw light on copolymer composition, cross-linking, etc. Flynn and Florin[9] have reviewed degradation and pyrolysis mechanisms of vinyl-type polymers, heterochain polymers, copolymers, and blends analyzed by Pyrolysis/GC.

Evolved gas-IR spectrometry (EGA-IR), a technique new to the published literature in the last ten years, has been successful in characterizing polymers of interest to the pressure sensitive adhesive field. The author used this technique in his laboratory some 30 years ago to identify pyrolytic degradation products of commercial polymer formulations to learn what, if any, potentially hazardous substances were evolved. The results went into confidential reports. Davidson and Mathys[10] used EGA-IR, for example, to show that samples of PMMA (polymethylmethacrylate) cross-linked with different agents could be distinguished from each other. Both static and dynamic systems were examined and evolution profiles of selected gaseous components were used to further characterize the materials. They noted that EGA-IR is useful for any process in which gases are evolved. It is not limited to pyrolysis or combustion.

An identification of whether or not an acrylonitrile-butadiene (AB) copolymer is present in an elastomer component simply involves scanning the pressure sensitive adhesive between 4 and 5 μm. The presence of an absorption at about 2250 cm^{-1} indicates a nitrile, and an acrylonitrile copolymer should come under consideration. The lack of a band at 2250 cm^{-1} indicates no nitrile, and thus no possibility of an AB copolymer.

IR spectra of the following elastomers are presented as thin films evaporated onto KBr plates from appropriate solvents: Natural crepe rubber (Fig. 9-1); polyisobutylene (Fig. 9-2); Kraton 1102 [styrene-butadiene-styrene (S-B-S) block copolymer, Shell Chemical Co.] (Fig. 9-3); Kraton 1107 [styrene-isoprene-styrene (S-I-S) block copolymer, Shell Chemical Co.] (Fig. 9-4); and acrylic copolymer elastomer (Fig. 9-5).

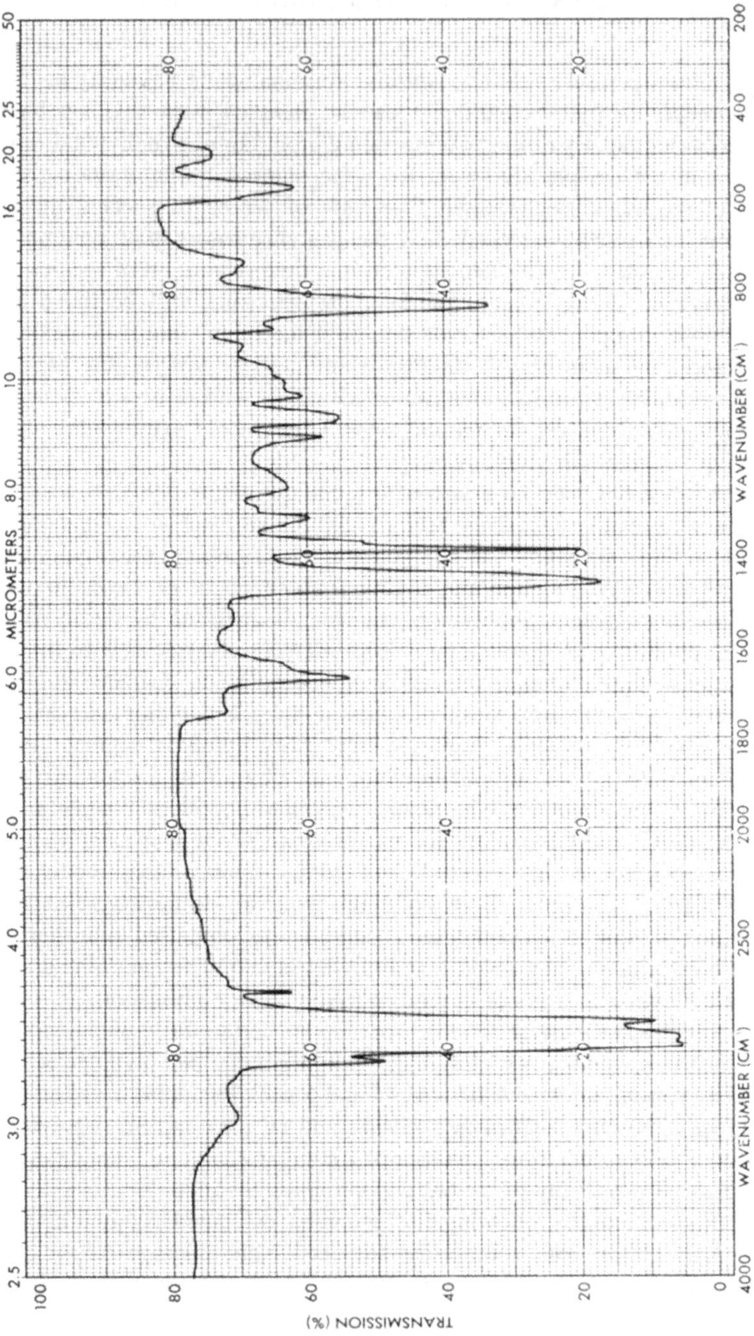

Fig. 9-1. Natural crepe rubber thin film.

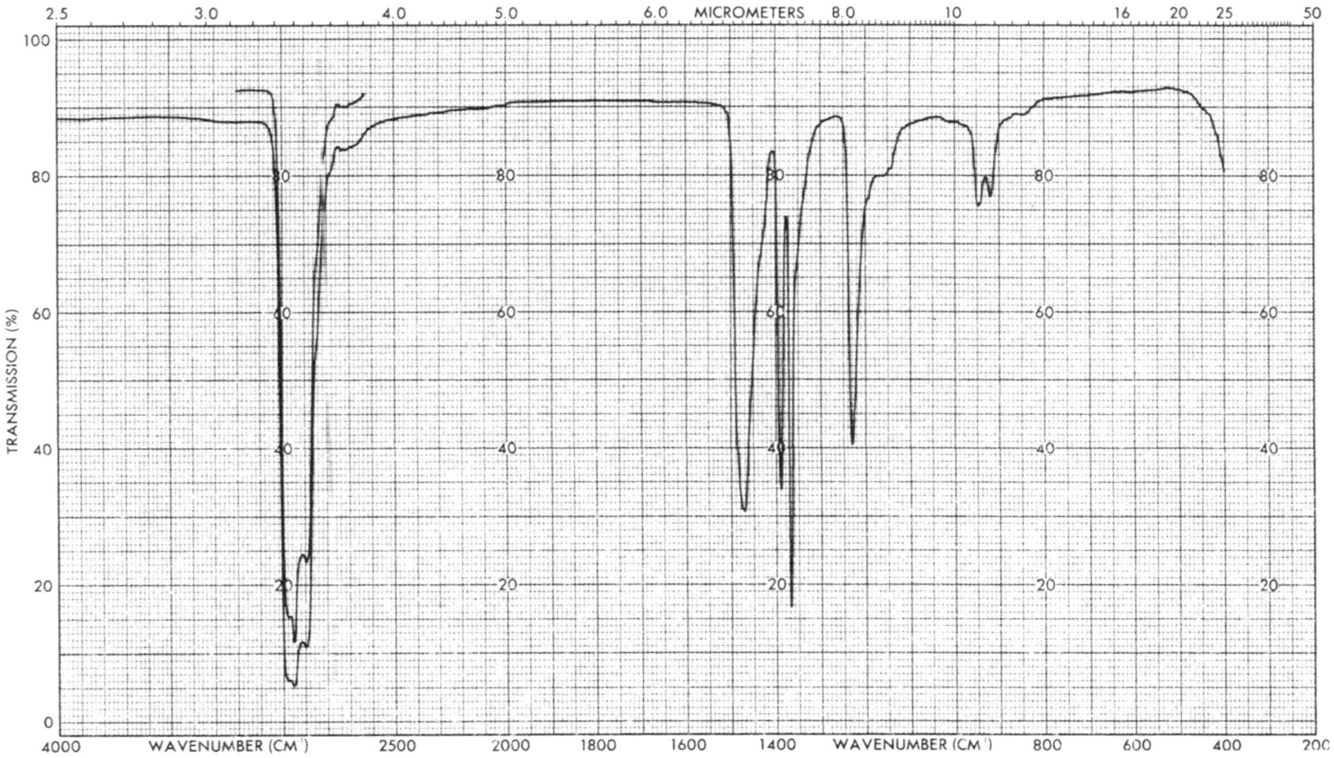

Fig. 9-2. Polyisobutylene thin film.

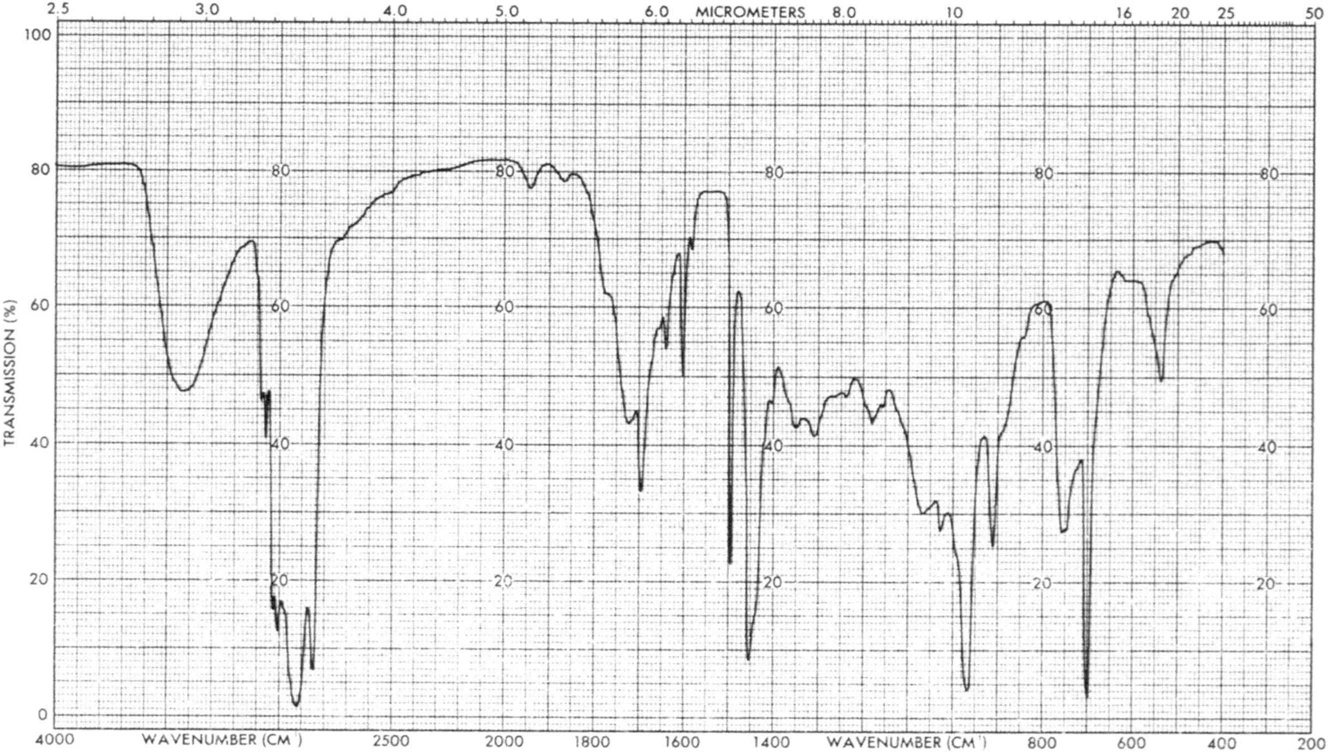

Fig. 9-3. S-B-S block copolymer (Kraton 1102 by Shell Chemical Co.).

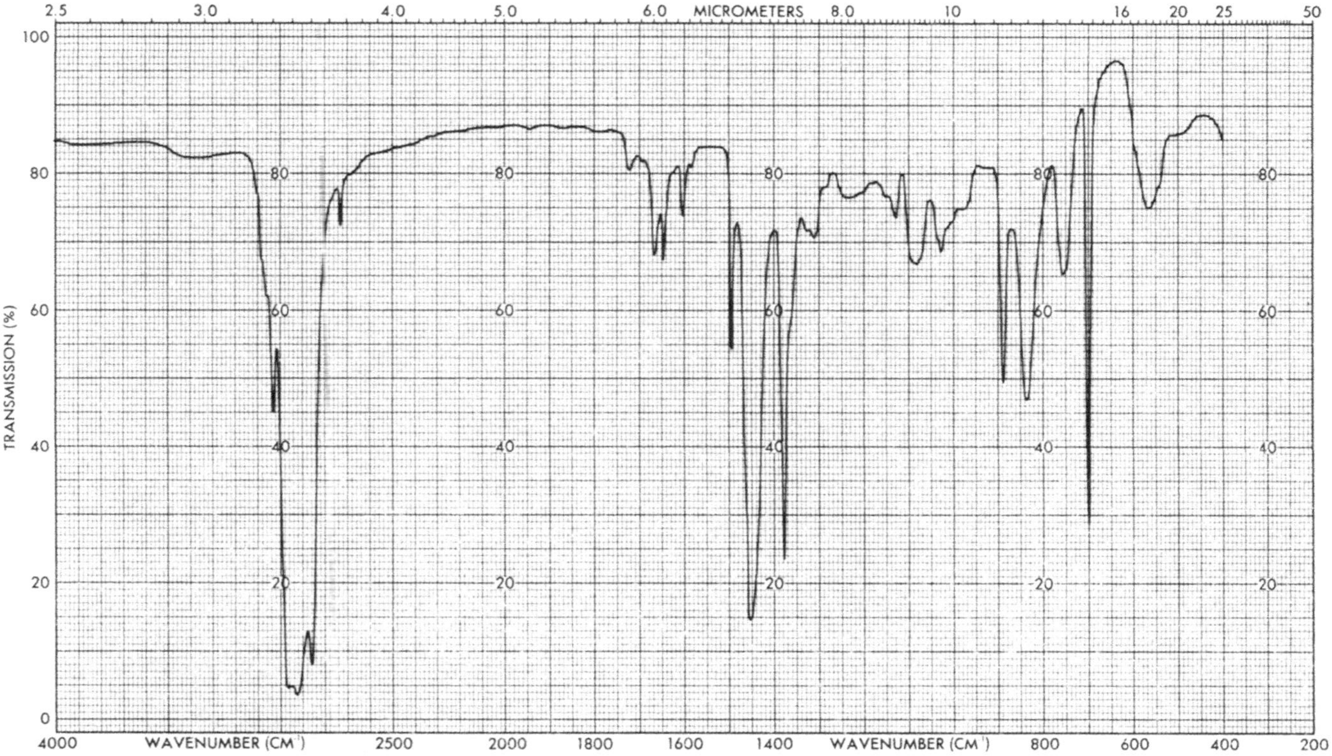

Fig. 9-4. S-I-S block copolymer (Kraton 1107 by Shell Chemical Co.).

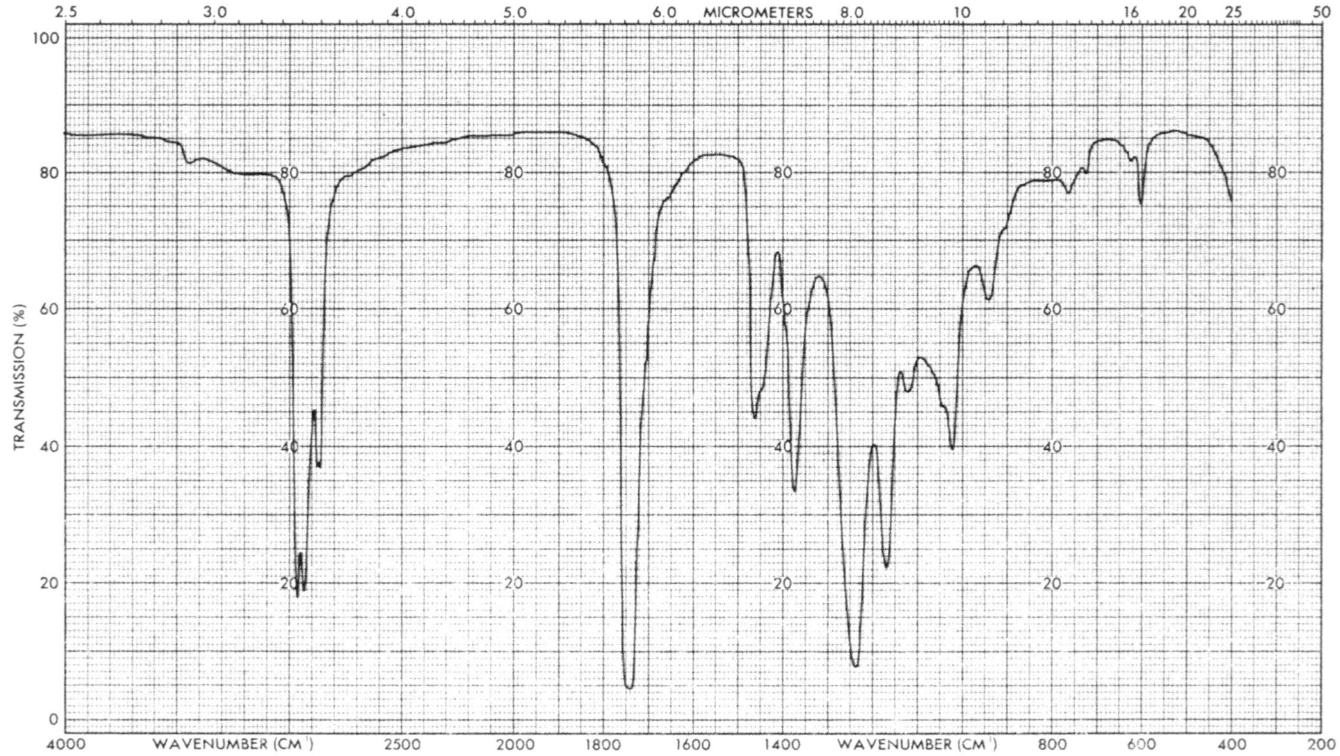

Fig. 9-5. Acrylic copolymer.

Tackifying Resin Identification

Commonly used tackifying resins or tackifiers in pressure sensitive adhesives are gum rosin, rosin esters, and other rosin derivatives, polyterpene resins, coumarone-indene resins, oil-soluble phenolic resins, and petroleum hydrocarbon resins. While these can sometimes be identified by IR scanning of the as-is adhesive and comparing against known tackifier spectra, such is not always possible. Critical portions of the tackifier spectra may be swallowed up by absorptions of the elastomer. Therefore, separating the elastomer and tackifier may be necessary. A styrene-containing elastomer can often be separated from the tackifier and oil of a pressure sensitive adhesive by extraction with an alcohol such as *n*-butanol. The tackifier and oil will be soluble in the *n*-butanol, while the elastomer will be practically insoluble. An IR scan of the spectrum of tackifier and oil, as evaporated films from *n*-butanol on KBr plates, will lead to identification of the tackifying resin, since the oil will normally contribute few interfering absorptions. As a bonus to this separation, IR scanning of the preceding elastomer, which has been separated, may lead to more certain identification of the elastomer.

The separation of the three main pressure sensitive adhesive ingredients may also be done by liquid chromatography.[11]

Pyrolyzates of separated tackifiers can be made, scanned in the IR, and identified by comparison against pyrolyzate spectra of knowns produced under the same conditions. Keep in mind that more than one tackifier may be present in an adhesive formulation. A mixture of resins may provide a balance of properties not obtainable with a single resin.

IR spectra of several tackifying resins are given as thin films evaporated onto KBr plates from appropriate solvents, or scanned as is in the case of the liquid coumarone-indene resin: Dymerex (dimerized rosin, Shell Chemical Co.) (Fig. 9-6); Foral AX (a rosin tackifier, Hercules, Inc.) (Fig. 9-7); Foral 85 (rosin ester, Hercules Inc.) (Fig. 9-8); Pentalyn H (rosin ester, Hercules Inc.) (Fig. 9-9); and coumarone-indene resin (Fig. 9-10).

Plasticizer Identification

The plasticizers used in pressure sensitive adhesives can be mineral oils, liquid polybutenes, liquid polyacrylates, certain phthalates, and lanolin, as typical examples. Some adhesives contain no plasticizers.

Perhaps mineral oils are the most commonly used. These often can be identified in the combination of tackifying resin and plasticizer (oil) that is separated from the elastomer using *n*-butanol as previously described. If separation of tackifier and plasticizer is necessary for identification of the latter by

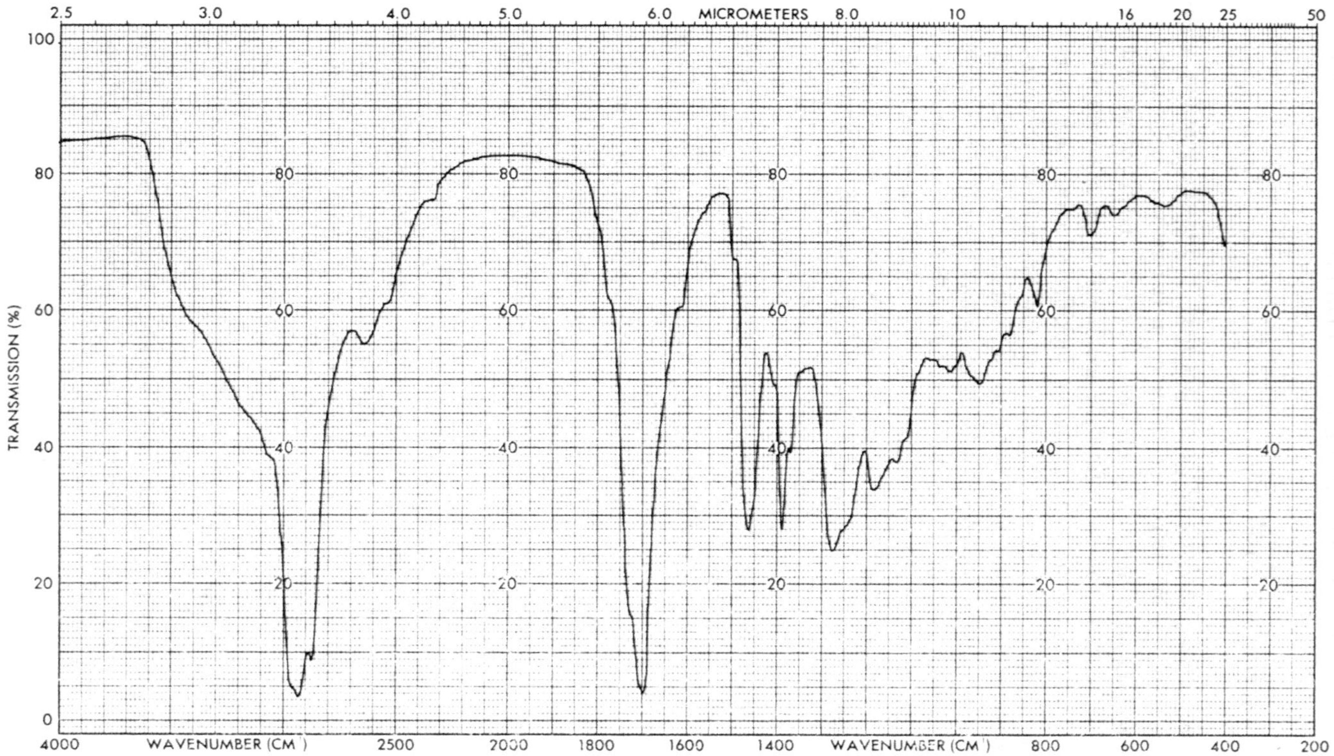

Fig. 9-6. Dimerized rosin (Dymerex by Shell Chemical Co.).

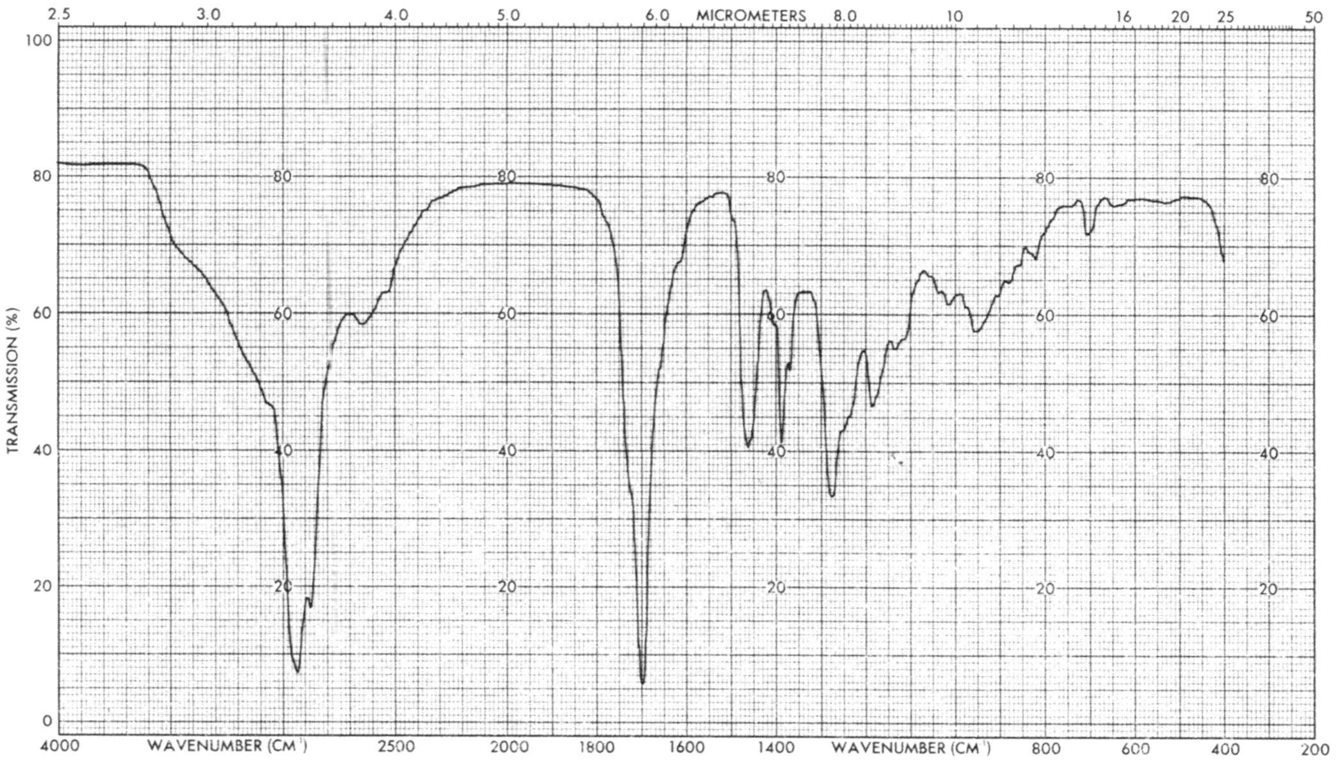

Fig. 9-7. Rosin tackifier (Foral AX by Hercules, Inc.).

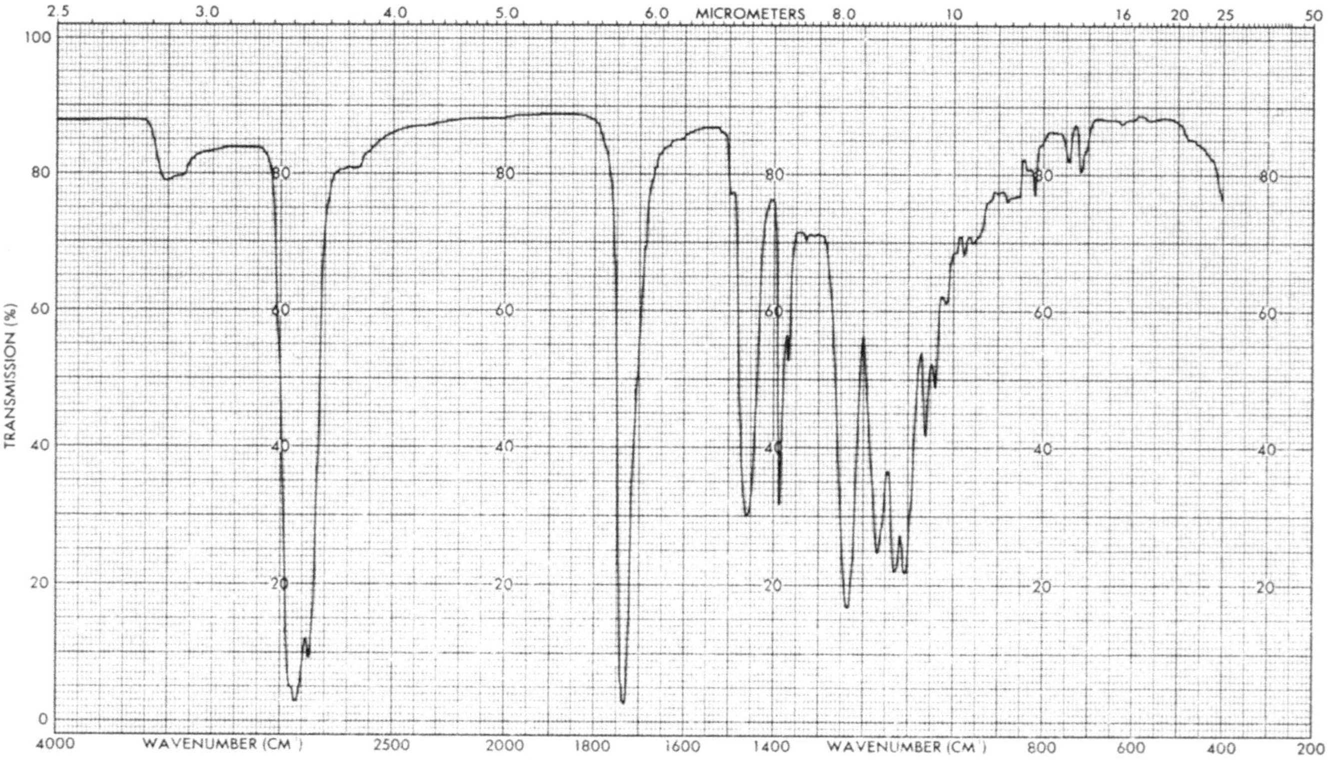

Fig. 9-8. Rosin ester (Foral 85 by Hercules, Inc.).

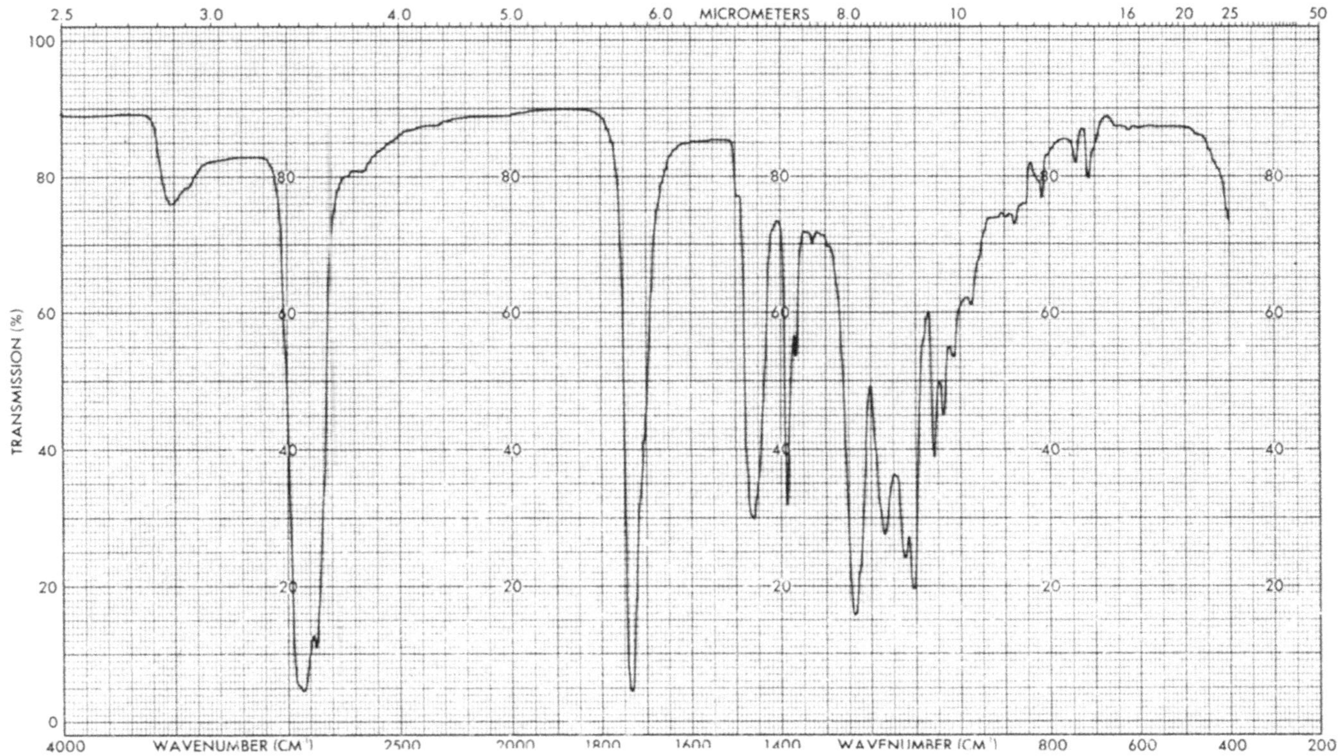

Fig. 9-9. Rosin ester (Pentalyn H by Hercules, Inc.).

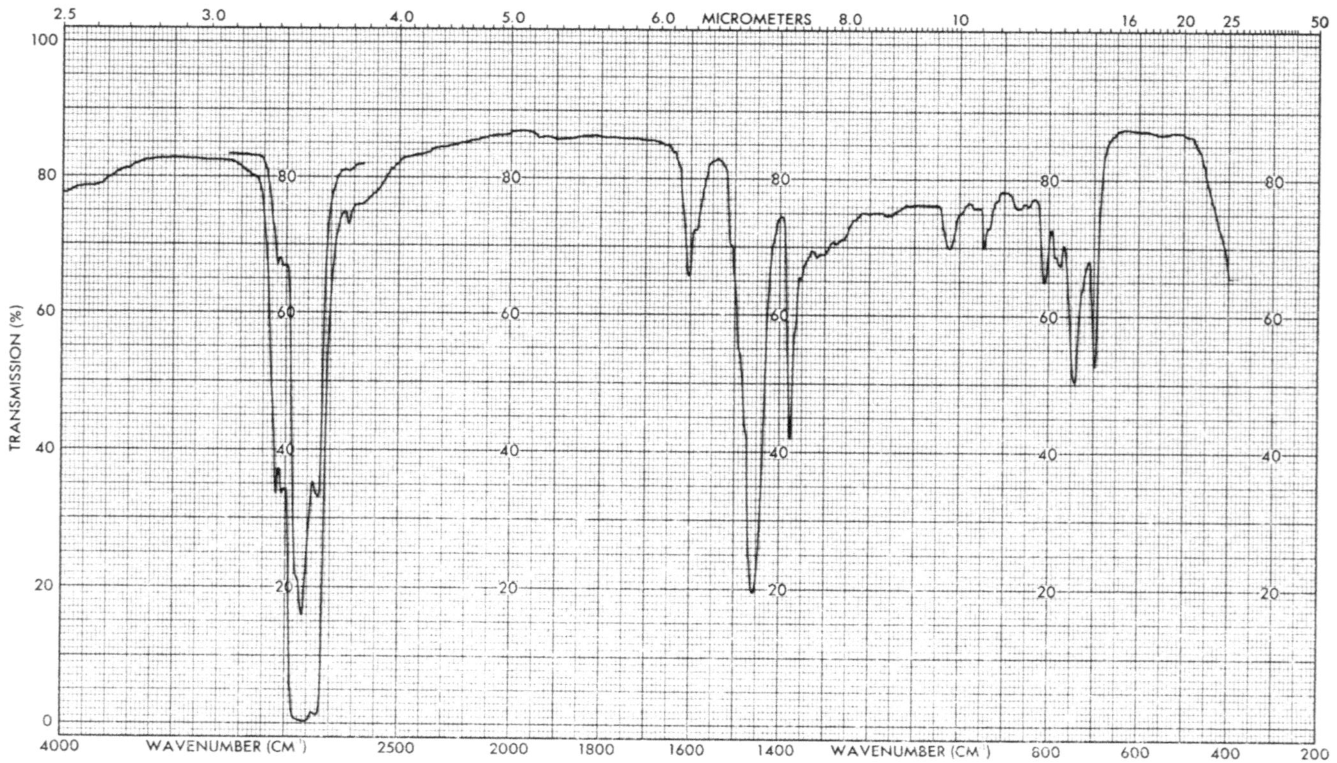

Fig. 9-10. Coumarone-indene resin.

IR, this can be done by liquid chromatography. Sometimes column chromatography will accomplish the separation. Once the plasticizer is alone, infrared identification by comparison of its spectrum against the spectrum of knowns can proceed. This can be a painstaking process, since a number of plasticizers, particularly of the petroleum oil type, have very similar IR spectra. When aromatic C=C unsaturation is present, a fine distinction between otherwise very similar oils can be made by UV analysis.

IR spectra of two plasticizers [mineral oil and liquid polybutenes (principally polyisobutyl ones)] used in pressure sensitive adhesives are shown in Figures 9-11 and 9-12, respectively.

Filler Identification

Fillers in pressure sensitive adhesives may be $CaCO_3$, TiO_2, ZnO, clays, pigments, and aluminum hydrates. Ashing the total composition may be the easiest way to bring about filler identification. The elastomer, tackifier, and plasticizer usually are organic and burn off, leaving the inorganics, the fillers. The latter can then be scanned from 2 to 50 μm in the IR as mineral oil and HCBD mulls between CsI plates. For inorganics in particular and more specifically for oxides, the region between 16 and 50 μm is very useful for identification purposes.

Quantitative ashes where calcium carbonate is involved must take into account the breakdown of $CaCO_3$ to CaO and CO_2 and the taking up of CO_2 from the air to form some $CaCO_3$ again after cooling begins. Changes can also take place in certain clays, and degrees of hydration should be kept in mind as one makes comparison of IR spectra among those of ashes against known inorganics.

Fillers can often be left in the solvent-insolubles from the separation of elastomer, tackifier, and oil. Infrared scanning of these, after heating off the solvent, should then lead to identification of the fillers.

ICP/MS, inductively coupled plasma/mass spectrometry is a technique of recent origin.[12] The first commercial instrument was introduced in 1983. A laboratory fortunate enough to have this instrument can use it to identify about 54 elements. The presence of metallic elements in a filler or antioxidant of a pressure sensitive adhesive, for example, might be sought. ICP/MS gives the sensitivity and selectivity of MS to the elemental analysis of solutions. ICP provides the ion source. In ICP 54 elements are expected to be ionized with an efficiency of 90% or more. Detection limits of 0.03–0.3 μg/l are realized.

IR spectra of two fillers, titanium dioxide and calcium carbonate, are presented in Figures 9-13 and 9-14, respectively. Each was scanned as a mineral oil mull.

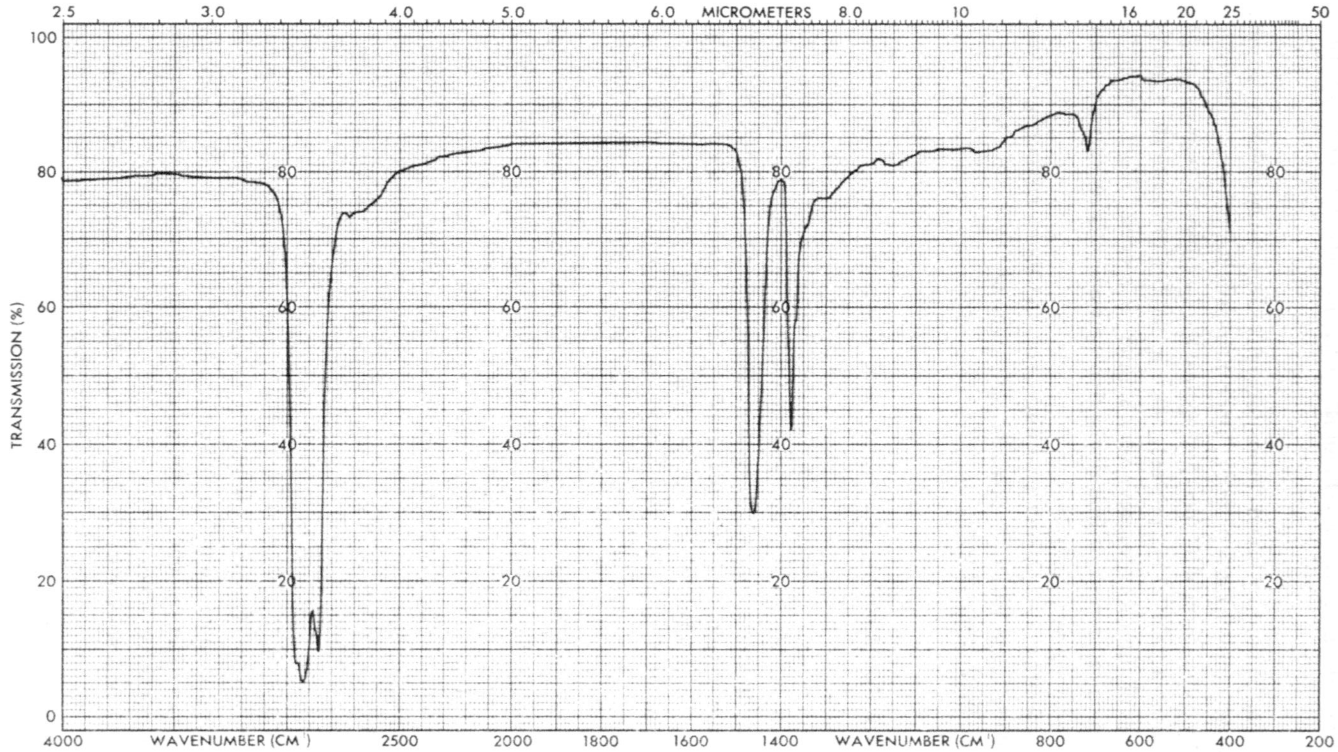

Fig. 9-11. Mineral oil.

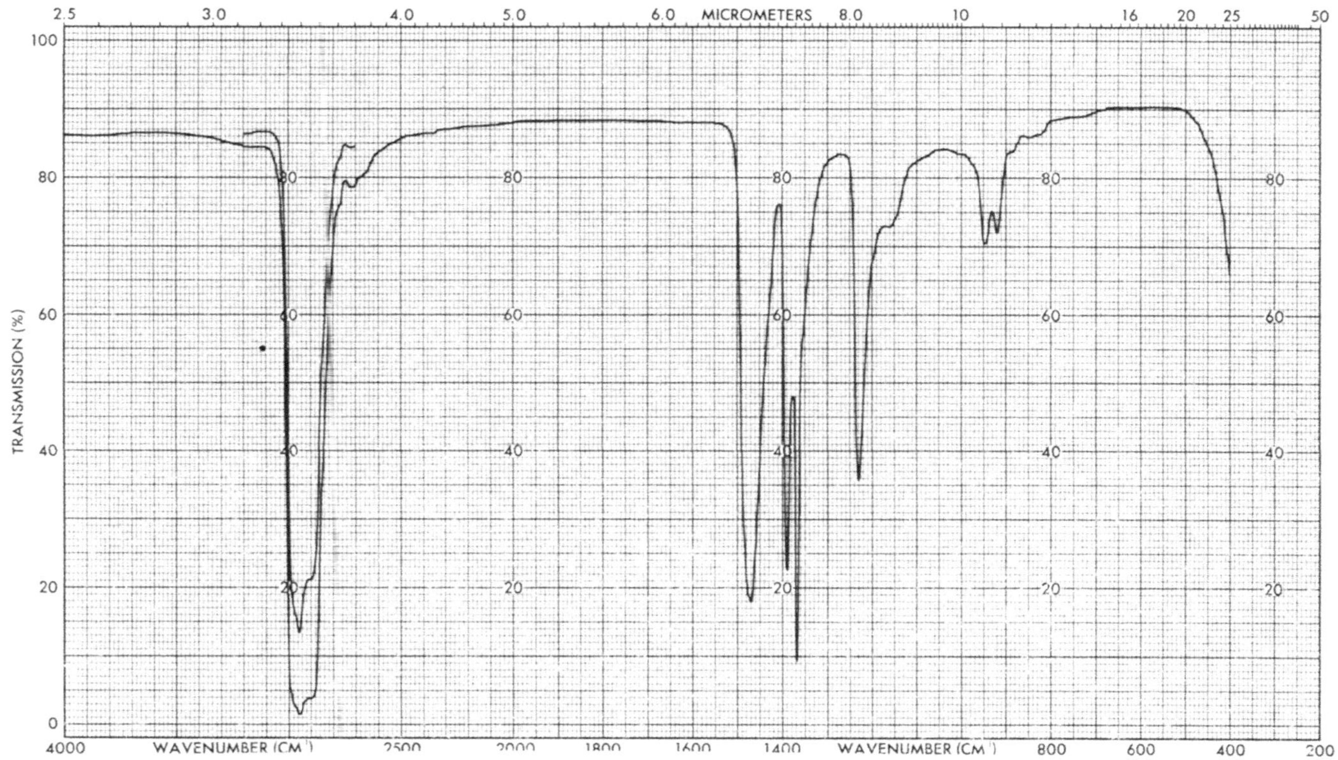

Fig. 9-12. Liquid polybutenes (principally polyisobutylenes).

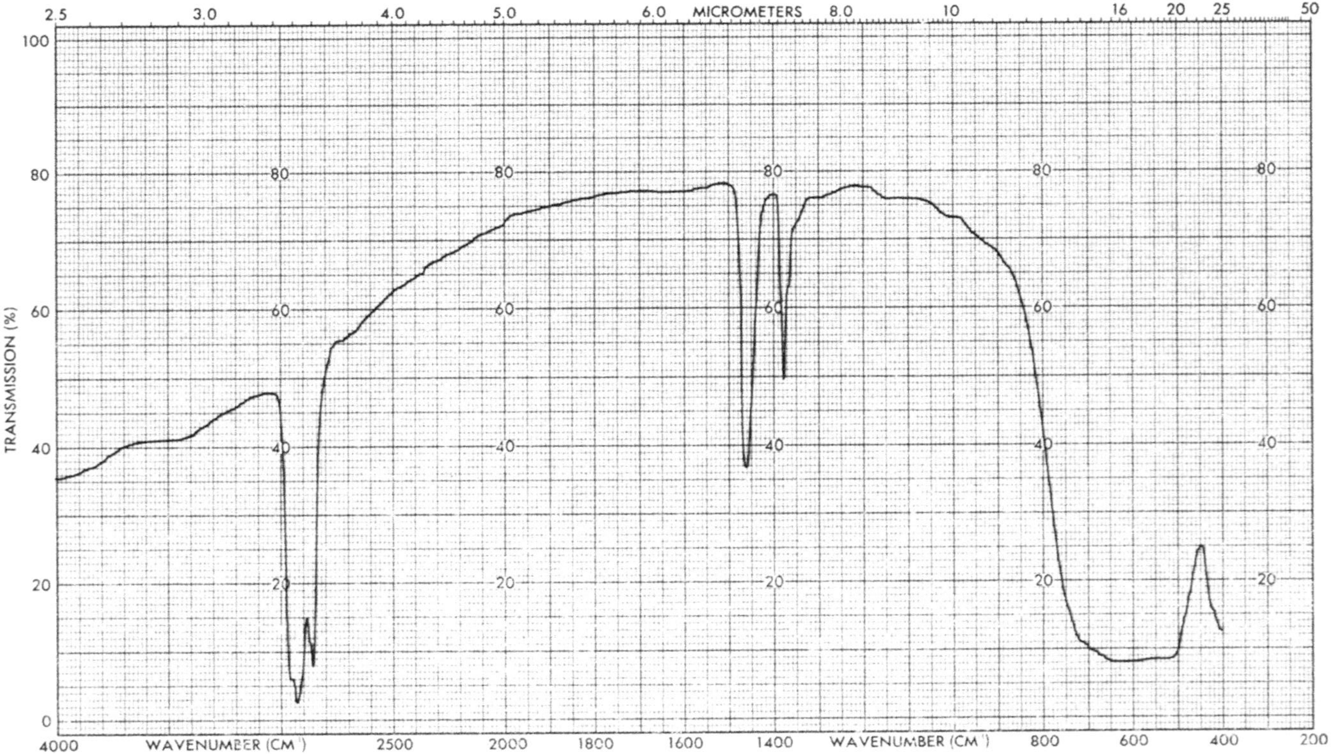

Fig. 9-13. Titanium oxide.

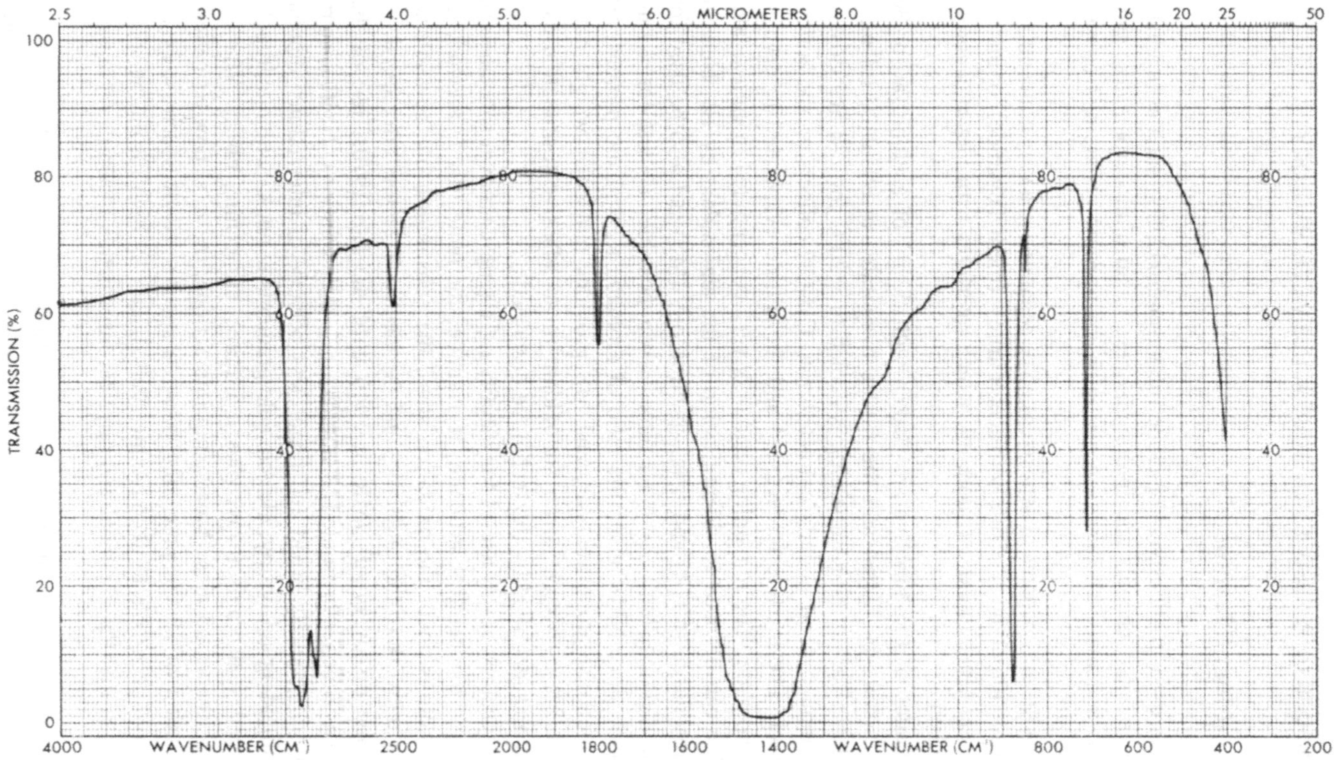

Fig. 9-14. Calcium carbonate.

Antioxidant Identification

Antioxidants used in adhesives[13] include rubber antioxidants, such as aromatic amines, substituted phenols and hydroquinones, metal dithiocarbamates, such as zinc salt, and metal chelating agents, such as *n*-propyl gallate and copper-8-quinolinolate.

Most of the antioxidants in adhesives can be separated in a very concentrated form by extraction of the adhesive with acetonitrile[14] (methyl cyanide), or methanol. The latter tends to extract more oil, which interferes with identification of the antioxidant.

Ten grams of the adhesive are shaken with 50 ml of acetonitrile for 30 min. The mixture is filtered and the filtrate evaporated to about 20 ml on a 60°C hotplate. The solution remaining is cooled to −20°C for 2–3 hr. Most of the small amount of oil initially extracted with CH_3CN collects on the bottom of the vessel. The mixture is decanted and the filtrate evaporated to dryness on a 60°C hotplate. The remaining residue is very concentrated in antioxidant. This residue can be used for identification of the antioxidant by IR. Nuclear magnetic resonance (NMR) and mass spectroscopy (MS) can also be used for antioxidant identification. Using all three, when possible, is best for a sure identification, since NMR gives the ratio of aromatic to aliphatic protons, IR supplies functional group analysis, and MS provides molecular weight data. As low as 0.2% by weight of an antioxidant in an adhesive formulation can be detected and identified by the technique described.

NMR spectroscopy is useful for identifying a pure unknown substance as well as finding its chemical structure. It examines the molecular framework. It answers questions about the oxidation state, the substitution on the nucleus, the bonding network of the molecule, and the electronic factors affecting it.

From the discovery of NMR in 1946 until the late 1970s, use of the technique was practically limited to liquids, including solutions. NMR spectra of solids were generally broad and featureless because of interactions: dipolar coupling, shielding anisotropy, electric field gradients in nuclei with quadrupole moments, and scalar anisotropy, plus a lifetime broadening effect. When a nucleus experiences all these broadening interactions, the total powder pattern is a convolution of all the preceding possible spectra. NMR spectroscopists, however, have been able to deconvolute such spectra by high-resolution solid-state NMR, so that the chemical and physical information available from each interaction is obtained.[15]

One of the recent focuses of high-resolution solid-state NMR has been ^{13}C, since carbon constitutes the bulk of organic substances. Experimentally ^{13}C is favored because of its low natural abundance (1.11%). Therefore homonuclear dipole–dipole interactions become negligible. Strong dipolar coupling exists with ^{1}H, which has a large magnetic moment. Strong RF irradiation at the proton frequency can eliminate this. Multiple pulse techniques can eliminate homonuclear coupling.

Multiple-pulse line narrowing and heteronuclear decoupling do not remove broadening due to chemical shift anisotropy. Rapid spinning of a polycrystalline sample around an axis adequately tilted with respect to the external field effectively removes chemical shielding anisotropy. The resonance pattern of each nucleus collapses into a single line in spectra thus obtained. The magic angle is 54.74 degrees.

Since interactions between the various spins are much stronger in the solid state than in the liquid phase, much larger RF fields must be used. The gating of the RF pulses must be accurate on a submicrosecond scale and under computer control. Rotation rates for the magic angle spinning (MAS) of 4–5 kHz are routinely attainable.

The MAS technique bears by far the largest potential for the analytical chemist. For NMR studies of solid polymers and semicrystalline ones, including polymer structure and dynamics and involving ^{13}C techniques, several articles may be consulted.[16–20] Useful information unattainable by other techniques can be uncovered to the advantage of the pressure sensitive adhesive analyst and formulator.

Scanning of the antioxidant already separated from the other adhesive components in the UV[21] from 200 to 400 μm and comparison against UV spectra of knowns can also aid in identifying antioxidants.

IR spectra of two antioxidants, AgeRite Resin D (R. T. Vanderbilt) and butylated hydroxy toluene (BHT) are given in Figures 9-15 and 9-16, respectively.

CHARACTERIZATION OF PRESSURE SENSITIVE ADHESIVES

Once the identification of the components of a pressure sensitive adhesive formulation has been completed, there may be a need to characterize further the total formulation or the polymers comprising the elastomer or tackifier. While IR can tell that the elastomer present is an SBR, it cannot take molecular weight into account. An SBR with a molecular weight of 1000 and one of 2000 would yield essentially identical IR spectra in the 2 to 50 μm region. IR sees nearest neighbors for the most part.

Molecular Weight Determination

Molecular weight determination of any polymers present in an adhesive formulation such as the elastomer or the tackifier can be helpful in improving adhesive properties. Knowing the molecular weight and solubility parameter δ for a

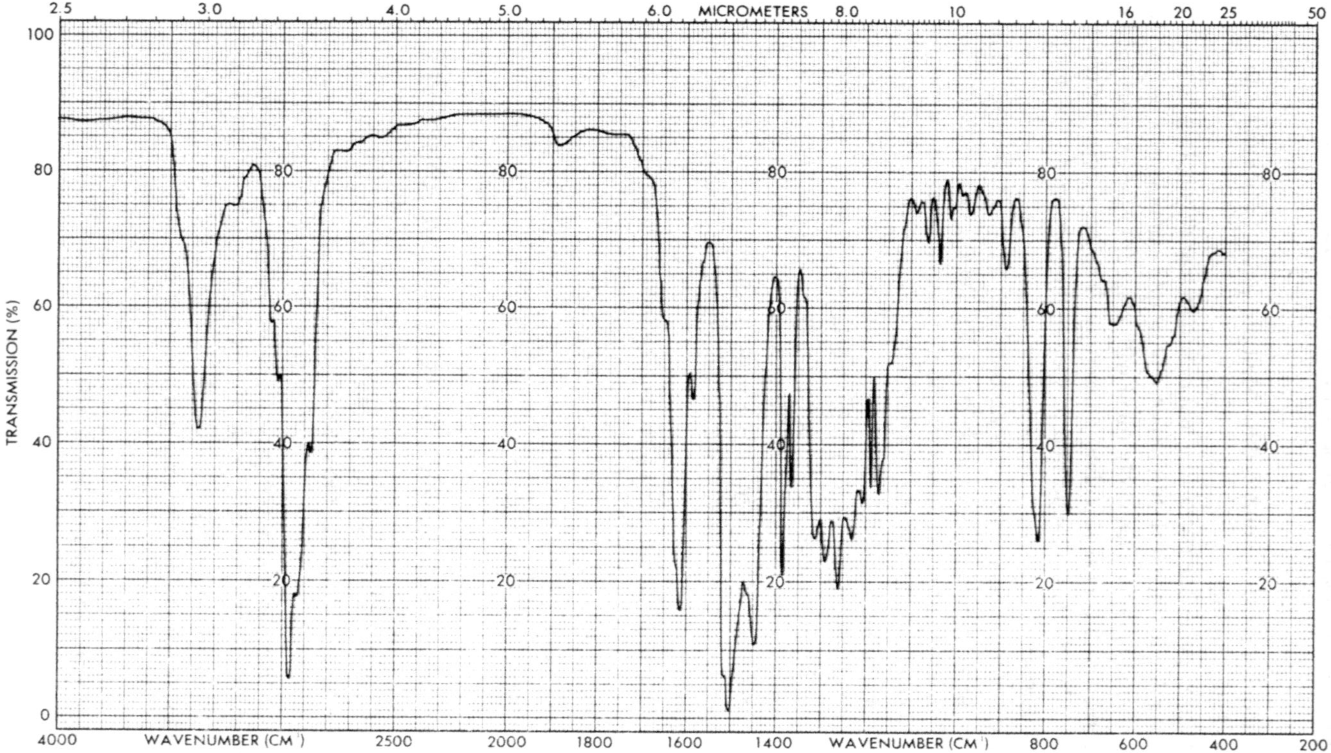

Fig. 9-15. Antioxidant (AgeRite Resin D by R. T. Vanderbilt).

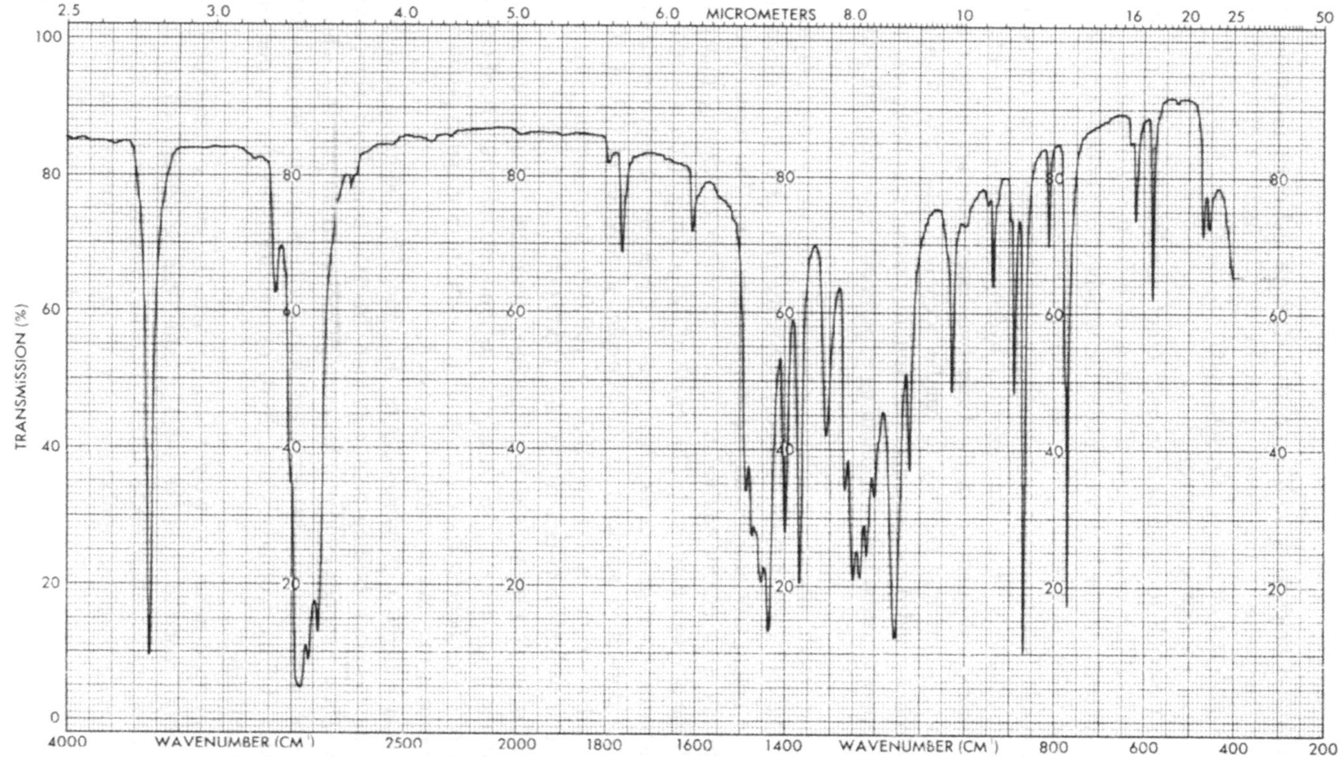

Fig. 9-16. Antioxidant (butylated hydroxy toluene).

series of tackifiers helps to optimize cohesive strength and other properties of the adhesive. The value of the solubility parameter is a function of molecular weight, because δ is a measure of the total intermolecular bond energy holding the molecules of a substance together.[13]

For two materials to be soluble in each other, the free energy of mixing must be negative. This occurs when the values of δ for the two materials are close. As a general rule, the higher the molecular weight of one or both materials, the closer the values of δ must be for miscibility to occur.

For nonpolar substances, δ is a relatively simple number, but for polar materials and situations where intermolecular hydrogen bonding is possible, δ is a three-dimensional quantity. In the latter two cases, data are meager and therefore there is currently no way to handle accurately situations involving polar or hydrogen-bonded components. The solubility parameter, however, may serve on an approximation basis for planning experiments and interpreting findings.

Molecular weight affects the solubility parameter and many other polymer properties. The molecular weight of a polymer is the product of the molecular weight of the repeat unit and the degree of polymerization. The repeat unit is usually equivalent or nearly equivalent to the monomer or starting material from which the polymer is formed. The length of the polymer chain is specified by the number of repeat units in the chain, that is, the degree of polymerization (DP). Taking poly(ethyl acrylate) as an example, a polymer of DP of 1000 has a molecular weight of $100 \times 1000 = 100{,}000$. Most useful polymers have molecular weights between 10,000 and 1,000,000,

Control of molecular weight is of decided commercial importance, for example, in the polymerization of butadienes to synthetic rubbers. Often chain transfer agents[22] that help to depress molecular weight are essential in producing a synthetic rubber that can be readily processed.

Molecular weights of polymers can be determined by end-group analysis, vapor-pressure lowering, boiling-point elevation, freezing-point depression, and osmometry. These methods yield number average molecular weight ($\bar{M}_n$), where the number of molecules in a known mass of material is counted to yield the average value. Of the techniques described, osmometry is by far the one of greater applicability.

Weight average molecular weight ($\bar{M}_w$), where the mass of molecules in a known mass of material gives the average value, can be determined by light scattering, ultracentrifugation, and solution viscosity. Strictly speaking, solution viscosity determinations are applicable only to linear polymers and yield a viscosity average molecular weight ($\bar{M}_v$) that is closer to $\bar{M}_w$ than to $\bar{M}_n$.

An osmometer holds a solution and the solvent in compartments separated by a semipermeable membrane such as cross-linked polyvinylalcohol or gel cellophane. Polymers with molecular weights between 15,000 and 1,000,000 can be accommodated by osmometry.[23] The method is limited at the low molecular weight end by diffusion of the solute through the membrane, and at high molecular weight by the small magnitude of the osmotic pressure. The latter is developed when solvent molecules pass readily through the membrane, but solute molecules cannot.

Determination of the melt viscosity of a polymer, separated from the adhesive formulation, gives a strong indication of weight average molecular weight and cross-linking as well as of chain branching.

Molecular Weight Distribution Determination

Gel permeation chromatography (GPC) provides a convenient method for determining molecular weight distribution of a polymer up to molecular weight of about 20,000,000. The stationary phase in GPC is either a cross-linked polymer that swells when placed in contact with a solvent or a porous polymer of rigid structure. The pore size is critical, since molecules above a certain size will be unable to enter the pores and the pores will be only partly accessible to all but the smallest molecules.

The mobile phase solvent is selected for its ability to dissolve the sample and for low viscosity at operating temperature. Flowing through the column, the solvent takes the larger molecules through unhindered. Smaller molecules travel slower, the rate depending upon their ability to penetrate the gel. This GPC technique separates materials of similar chemical composition in order of their molecular weight, with the higher molecular weight components eluting first. Differential refractometer and UV detectors can be used.

To study how copolymer properties are affected by composition, a new technique is available. It could be applied, for example, to S-I-S or S-B-S block copolymers or acrylic copolymers of interest to the pressure sensitive adhesive field. Liquid adsorption chromatography (LAC) and size exclusion chromatography (SEC) are involved.

Most copolymers have a chemical composition distribution (CCD) and a molecular weight distribution (MWD). Accurate information on MWD cannot be obtained by SEC alone. Separation by SEC is achieved according to the sizes of molecules in solution and the molecular weights of the copolymers are not proportional to molecular size unless the composition fluctuation is negligible and the chemical structure is the same across the whole range of molecular weights. One procedure is to fractionate according to chemical composition by LAC. Then the MWD of the LAC fractions is analyzed by SEC. Mori *et al.*[24] separated styrene-methylmethacrylate (S-MMa) random copolymers using this procedure. The mobile phase was a mixture of 1,2-dichloroethane (DCE) and chloroform ($CHCl_3$) varying from 100:0 to 0:100. When DCE was used as the mobile phase, the copolymers adsorbed on the external surface of the silica gel (the adsorbent). With increasing $CHCl_3$ content in the mobile phase, the copolymers desorbed as a fraction of their composition. The early-eluted fractions in LAC had lower molecular weight averages and higher styrene content than those for the unfractionated copolymer. The late-eluted fraction had the opposite values. High pressure liquid chromatography (HPLC) instrumentation suitably modified was used for both the LAC and the SEC.

Chain Branching Determination

Chain branching in a polymer of interest to pressure sensitive adhesives can sometimes be detected by IR spectroscopy. For long chain branches, an observable effect will be seen on the solution viscosity of the polymer. Comparison of the viscosity of a branched polymer against that of a linear polymer of the same molecular weight can be used to study chain branching. Short chain branches, such as ethyl or butyl groups, usually have little effect on solution viscosity.

Solubility measurements can be used to determine the controlled chain branching or cross-linking introduced into styrene copolymers by addition of small amounts of, say, divinylbenzene during styrene polymerization. About 0.0025% divinylbenzene yields soluble branched polymers. Approximately 0.1% divinylbenzene results in an insoluble, nonthermoplastic polymer that swells greatly in solvents. When 5–10% divinylbenzene is used, the product is a hard, brittle, cross-linked material.

Degree of Crystallinity

The degree of crystallinity of a polymer can be measured by specific volume determination, X-ray diffraction, and IR spectroscopy. One can determine the crystalline specific volume V_c from X-ray unit cell dimensions, the amorphous specific volume V_a from extrapolation of the specific volumes of polymer melts, and the specific volume of the sample concerned V from curves relating specific volume as a function of time at temperatures below the crystalline melting point. Then the crystallinity (S_c) of the sample is given by

$$S_c = (V_a - V)/(V_a - V_c) \tag{2}$$

This equation assumes additivity of the specific volumes and a sample free of voids.

The X-ray technique permits calculation of the relative amounts of crystalline and amorphous material in a specimen, if it is possible to resolve the contributions of the two types of structure to the X-ray diffraction pattern. If the intensity of X-rays scattered by a sample is plotted versus the diffraction angle, the amorphous peak is broad while the crystalline peaks are narrower and usually more intense. Areas under peaks or peak heights can be used to estimate the degree of crystallinity.

The IR method can be used to determine the degree of crystallinity. In polyethylene, for example, an absorption at 7.67 μm can be identified with a methylene wagging motion in the amorphous regions.[25] In nylon 610,[26] the band at 10.6 μm was found to have absorbance varying linearly with specific volume. For polypropylene,[27] it is possible to use the ratio of the 995 to 974 cm^{-1} bands as a measure of the crystalline-amorphous content. The mostly crystalline isotactic polypropylene has strong bands of nearly equal intensity at 995 and 974 cm^{-1}. The mostly amorphous atactic polypropylene has a strong, slightly broader band at 974 cm^{-1}, but practically none at 995 cm^{-1}. By measuring the absorbance ratios of these bands from suitably prepared standards and plotting a working curve, the amorphous-crystalline content can be measured with reasonable accuracy.

The absorbance ratio of 720/730 cm^{-1} doublet can be used for IR crystallinity determinations in polyethylene,[28] but films taken from melted samples should be used. This is because orientation in molded films causes variations in the absorbance ratio of the 720/730 cm^{-1} doublet.

Side-chain Substitution and Cross-linking

Side-chain substitution in polymers can be evaluated by comparing the effects on T_m (crystalline melting point) or T_g (glass transition temperature) of substituting nonpolar groups for the hydrogens of a polymer chain. Cross-linking can be ascertained by the effect on T_g and the decreased solubility as compared to the uncross-linked polymer.

Substitution of nonpolar groups for hydrogens of a polymer chain can lead to a reduction in T_m (or T_g) or even a complete loss of crystallinity. Measurement of T_m can sometimes distinguish between a side-chain substituted polymer and an unsubstituted one. For polyethylene, the crystalline melting point is lowered about 25°C going from the linear to the branched material. An increase in the length of the side-chain of a polymer often results in a looser crystal structure with an increasingly lower T_m. But an increase in the bulkiness of the side chain increases T_m, because rotation of the side chain is hindered in the liquid state.

Measurement of the glass transition temperature can throw light on side-chain substitution. In poly(*n*-alkylmethacrylates),[29] increasing chain length from 1 to 12 carbons reduced T_g from 105 to −65°C.

Cross-linking of polymers can be assessed at least qualitatively by solubility determinations. An uncross-linked polymer can be completely soluble in a given solvent, say, toluene, while its cross-linked counterpart may show varying degrees of insolubility from complete to partially soluble.

If cross-linking in a polymer system has been brought about by radiation, ions or radicals will normally be present, which can be detected by conductivity, chemical tests, or by electron paramagnetic resonance (EPR) spectroscopy. IR is almost always productive in disclosing the effects of cross-linking. The formation of trans-vinylene unsaturation often accompanies radiation cross-linking. This can be followed by observation of the characteristic IR spectra of trans-unsaturated functional groups.

Irradiation of pressure sensitive components, such as polystyrene, polyacrylates, and natural and synthetic rubbers, leads to cross-linking as the principal reaction. Gel formation accompanies cross-linking and ultimately leads to insolubilization of the entire sample.

Elastomers can be cross-linked by a great variety of substances, including many rubber accelerators. The latter can sometimes be detected by means of extraction with a 1:1 mixture of alcohol and chloroform, provided a residual cross-linker remains in the composition.

Cross-linking in phenolic resins such as those used in pressure sensitive adhesives for high-temperature use or electrical applications can be followed and studied by IR spectroscopy.

Orientation in Polymers

Orientation in polymers of pressure sensitive adhesives can be determined by X-ray diffraction, birefringence changes, and IR dichroism studies. Orientation can occur in noncrystalline polymers. This can be measured by the force with which a specimen tends to retract on heating above T_g as well as by birefringence.

X-ray diffraction patterns of oriented versus unoriented crystalline polymers can follow the alignment of the polymer chains in the direction of the applied stress that produced the orientation. Measurements of tensile strength and stiffness of polymers with increasing orientation will show the increase in tensile strength and stiffness that normally occur.

Effect of Additives

The effect of additives on the polymer properties of the polymers in an adhesive, such as plasticizers, can be measured by the changes in T_g as well as by measurement of stiffness, hardness, and brittleness.

Plasticizers can often be extracted from a pressure sensitive adhesive formulation with a selected alcohol or other solvent in which the elastomer and tackifier are insoluble or substantially so.

Measurement of the melt viscosity of a plasticized polymer, determination of the dilution ratio, readings of the viscosity of dilute solutions of the polymer in the plasticizer, and measurement of electrical or mechanical properties of the plasticized polymer can all help in selection of the optimum plasticizer for a pressure sensitive adhesive.

Plasticizers used in pressure sensitive adhesives are most commonly oils such as mineral oil, liquid polybutenes and polyacrylates, and lanolin. Characterization of these is often made difficult by the small percentages used in formulations and by the necessity to separate them from other components present, often not a simple task. Once separated, however, they can be characterized by IR spectroscopy, especially from spectra scanned on thick films of these oils.

The effect of fillers on pressure sensitive adhesives can be measured by tensile strength, elongation at break and 400% modulus (tensile stress at 400% elongation, a measure of stiffness). Inert fillers such as clay and calcium carbonate have little effect on physical properties, although they make an elastomer mixture easier to handle. Reinforcing fillers do have an effect on physical properties, such as increasing tensile strength and stiffness.

Fillers in adhesive formulations can be separated for property measurements by ashing off the other components that are normally organic. Identification can be made by IR scans from 2.5 to 50 μm.

The use of antioxidants to stabilize an adhesive against oxidation and heat and light degradation is common. Normally, they are the type used in rubber formulations. An antioxidant is itself readily oxidized or it combines with the oxidizing polymer to form a stable product. Rubber literature[30] on antioxidants should be consulted. Simulated stepped-up conditions to produce oxidation can be used to quickly evaluate antioxidants. Actual trial of antioxidants is best.

The use of one's own judgment and common sense is often the best and simplest approach in selecting techniques for the identification or characterization of the components in a pressure-sensitive adhesive.

Figures 9-17 and 9-18 illustrate the use of IR spectroscopy for identifying and duplicating commercial adhesives. In Figure 9-17, the IR spectrum of a commercial pressure sensitive adhesive shows absorptions (bands) at 1600, 1580, and 1493 cm^{-1}, indicating a benzene derivative. Bands at 696 and 756 cm^{-1} suggest the aromatic is monosubstituted, as do the four bands between about 1740 and about 1940 cm^{-1}. The sharp shoulder absorptions above 3000 cm^{-1} indicate aromatic C—H. Bands at 964 and 900 cm^{-1}, together with the monosubstituted benzene derivative already indicated, suggest a butadiene-styrene copolymer. Comparison against spectra of such knowns confirms that this elastomeric copolymer is present.

The weak band at about 1658 cm^{-1} could arise from the C=C of an

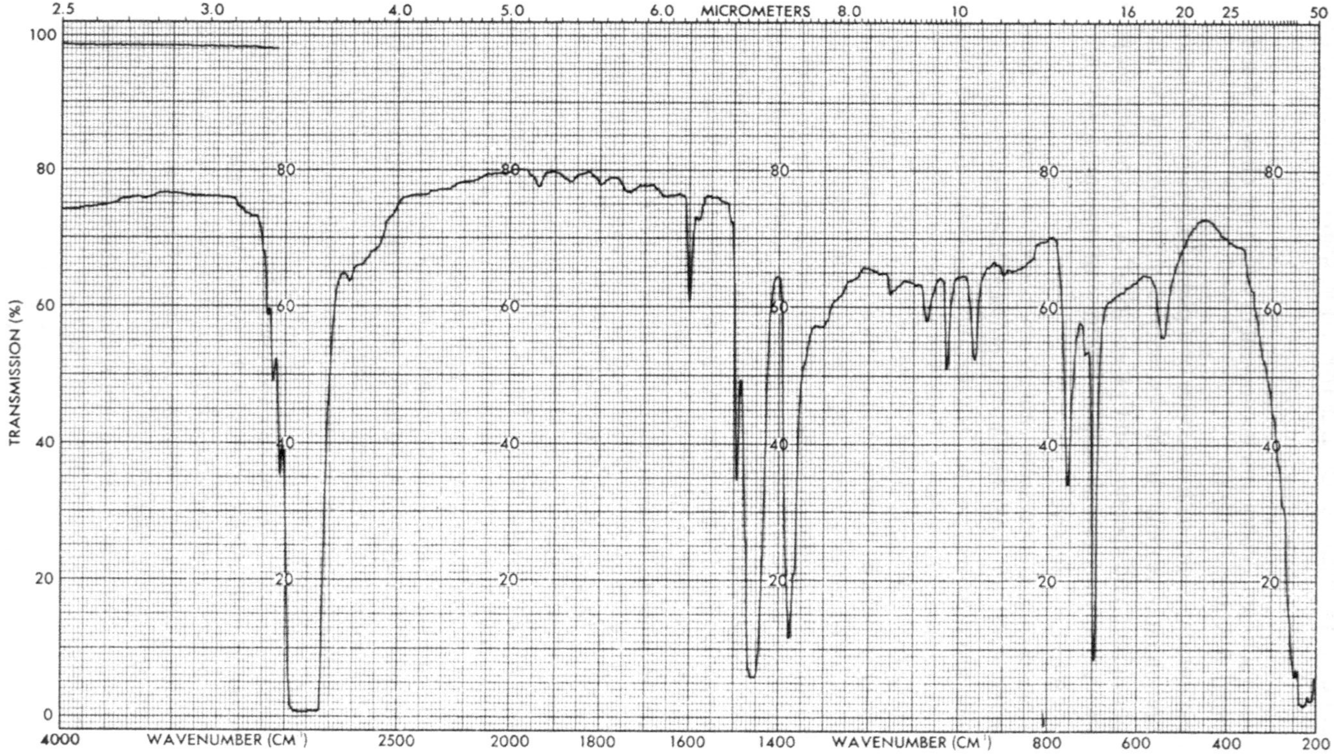

Fig. 9-17. Commercial pressure sensitive adhesive.

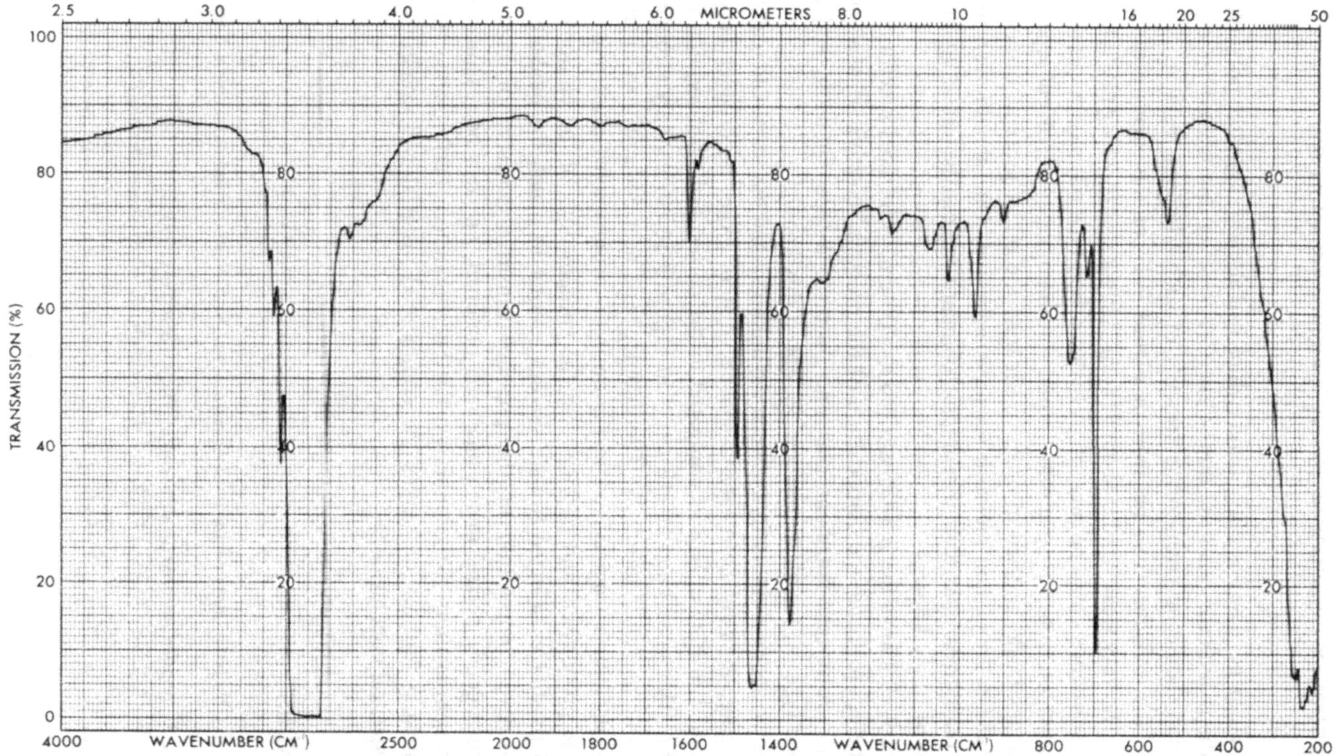

Fig. 9-18. First attempt to match commercial pressure sensitive adhesive.

unsaturated terpene. The 964 cm^{-1} band (which had already suggested polybutadiene) perhaps indicates also the trans-CH═CH— of an unsaturated terpene. Consideration should be given to the presence of such a terpene resin.

Very little evidence can be seen in Figure 9-17 of the identity of the plasticizer. The very weak band at about 718 cm^{-1} suggests that a long methylene chain-type substance could be present. A hydrocarbon-type oil may be present, but one cannot tell from this spectrum. Liquid chromatography or solvent separation techniques could be tried in order to separate the oil, if present, from the elastomer and resin of this pressure sensitive adhesive. One should scan the separated plasticizer, and then meaningful comparisons can be made against the spectra of likely knowns.

Figure 9-18 shows the 2.5–25-μm (4000–400-cm^{-1}) infrared spectrum of the first attempt to match the composition of the pressure sensitive adhesive in Figure 9-17. A styrene-butadiene elastomer, a terpene type for resin, and a hydrocarbon type oil for plasticizer are present in this known.

While the spectral similarity between the unknown and the attempted match is encouraging, it is by no means a spectral match. Note the broader band near 753 cm^{-1} in Figure 9-18 as compared to the corresponding absorption in Figure 9-17. The weak band at 718 cm^{-1} in Figure 9-18 is relatively stronger than its counterpart in Figure 9-17. The doublet band near 540 cm^{-1} in Figure 9-18 contrasts with a singlet band in Figure 9-17. Adjustments must now be made in the proportions and perhaps also the compositions of the elastomer, resin, and plasticizer selected for the second attempt to match the qualitative and quantitative composition of the pressure sensitive adhesive of Figure 9-17. Once a good spectral match has been obtained, one can be reasonably certain the composition of the unknown has been matched. Performance tests of the matching formulation versus the commercial material will confirm or deny this spectral conclusion.

REFERENCES

1. Kendall, D. N. (ed.). *Applied Infrared Spectroscopy*. Van Nostrand Reinhold, New York, 1966.
2. Smith, A. L. *Applied Infrared Spectroscopy*. Wiley, New York, 1979.
3. Jaffe, H. H., and Orchin, M. *Theory and Applications of Ultraviolet Spectroscopy*. Wiley, New York, 1962.
4. Wendlandt, W. W. *Thermal Methods of Analysis*. Second edition. Wiley–Interscience, New York, 1974.
5. Michelson, A. A. *Light Waves and Their Uses*. Univ. of Chicago Press, Chicago, 1902.
6. Fellgett, P. B. *Ohio State Symposium on Molecular Spectroscopy*, 1952.
7. American Society for Testing and Materials. *Book of ASTM Standards*, Parts 24–25. Philadelphia, Pa., 1979.
8. Katon, J. E., Pacey, G. E., and O'Keefe, J. F. *Anal. Chem.* **58:** 465A (1986).

9. Flynn, J. H., Florin, R. E. *Chromatogr. Sci.* **29:** 149 (1985).
10. Davidson, R. G., and Mathys, G. I. *Anal. Chem.* **58:** 837 (1986).
11. Snyder, L. R., and Kirkland, J. J. *Introduction of Modern Liquid Chromatography.* Wiley, New York, 1974.
12. Houk, R. S. *Anal. Chem.* **58:** 97A (1986).
13. Skeist, I. (ed.). *Handbook of Adhesives.* Second edition. Van Nostrand Reinhold, New York, 1977.
14. Carlson, D. W., Hayes, M. W., Ransaw, H. C., McFadden, R. S., and Altenau, A. G. *Anal Chem.* **43:** 1874 (1971).
15. Reinhold, M. *Am. Lab.* **14**(2): 164 (1982).
16. Dickert, F. L., and Hellman, S. W. *JEOL News* [Ser.] *Anal. Instrum.* **57:** 18A (1982).
17. Gard, J. K., and Ackerman, J. J. H. *J. Magn. Reson.* **51:** 124 (1983).
18. Jutila, M. *Acta Chem. Scand., Ser. B* **B35:** 503 (1981).
19. Pearce, E. M., Wright, C. E., and Bordoloi, B. K. *J. Educ. Modules Mater. Sci. Eng.* **3:** 753 (1981).
20. Wengel, T. J., Ashley, M. E., and Sievers, R. E. *Anal. Chem.* **54:** 615 (1982).
21. Brock, M. J., and Louth, G. D. *Anal. Chem.* **27:** 1575 (1955).
22. Billmeyer, R. W. Jr. *Textbook of Polymer Science.* Wiley–Interscience, New York, 1962.
23. Kirk-Othmer (eds.). *Encyclopedia of Chemical Technology,* Vol. 16. Second edition. Wiley–Interscience, New York, 1968.
24. Mori, S., Uno, Y., and Suzuki, M. *Anal. Chem.* **58:** 303 (1986).
25. Miller, R. G. J., and Willis, H. A. *J. Polymer Sci.* **19:** 485 (1956).
26. Sandeman, I., and Keller, A. *J. Polymer Sci.* **19:** 401 (1956).
27. Luongo, J. P. *Anal. Chem.* **33:** 1919 (1961).
28. Luongo, J. P. *Polym. Lett.* **2:** 75 (1964).
29. Rogers, S. S., and Mandelkern, L. *J. Phys. Chem.* **61:** 985 (1957).
30. Morton, M. (ed.) *Rubber Technology.* Second edition. Van Nostrand Reinhold, New York, 1973.

10. Aging

Keiji Fukuzawa
Consulting Chemical Engineer
Tokyo, Japan

Many different elastomers are used for pressure sensitive adhesives: natural rubber, SBR, polyisobutylene, butyul rubber, block copolymers, silicones, urethanes, and acrylics. Natural rubber and acrylic resins are the most important, with block copolymers growing with the increasing use of hot melt adhesives.

Unsaturated elastomers such as natural rubber, SBR, and block copolymers (S-I-S and S-B-S) have double bonds in the molecule. Therefore, these polymers are inferior in aging properties to the saturated polymers. Since the pressure sensitive application of these elastomers is usually of a short duration, it is more important to determine how well these adhesives retain their pressure sensitive properties on aging in-roll, rather than how they age after application. Pressure sensitive adhesives based on acrylic resins and saturated rubbers such as polyisobutylene, butyl rubber, and others are employed for long-term uses. In this case, the properties after application also become important.

The rubber-type pressure sensitive adhesives are composed of rubber, tackifying resin, softener, and antioxidant. It is important to select stable components and a suitable antioxidant in order to maximize the retention of adhesive properties on aging.

EFFECT OF TACKIFYING RESINS

Aging properties are heavily influenced by the tackifying resin used. Rosin has been widely used in pressure sensitive adhesives and can produce adhesives with good properties, especially at low temperatures. However, rosin, a natural resin, tends to oxidize easily because of its conjugated double bonds. The stability of rosin has been improved by disproportionation and by hydrogenation of the conjugated double bonds. Wetzel and Alexander[1] investigated the effect

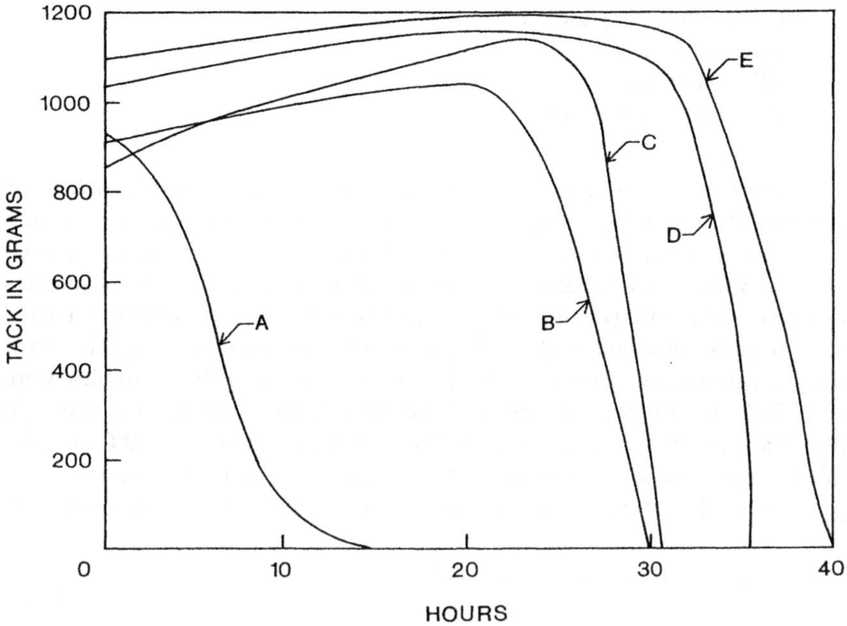

Fig. 10-1. The tack of natural rubber/tackifying resin blends (3/2) as a function of exposure at 93°C. According to Wetzel and Alexander, *Adhesive Age* 7(1): 28 (1964): (A) Pentalyn H; (B) Resin H64-pentalyn H; (C) Staybelite Ester 10; (D) Cellolyn 21; (E) ester of perhydrogenated rosin.

of various resins on the aging properties of adhesives as shown in Figure 10-1. Higuchi, Watanabe, and Toyama[2] have studied similar relations by the oxygen absorption method shown in Figure 10-2. Figures 10-3 and 10-4 show data by Fukuzawa and Matsukawa[3] that compare various tackifying resins, that is, rosin, terpene, and C_5 petroleum resins and their behavior on aging. The peel adhesion of natural rubber-resin-type adhesives has a tendency to increase in the region of adhesive failure and decrease in the region of cohesive failure, resulting from the softening of the pressure sensitive adhesive on aging. This is, of course, related to the decrease of the cohesive strength of natural rubber on aging. Petroleum resins show better aging properties than natural resins.

EFFECT OF ANTIOXIDANTS

Antioxidants provide an effective way to protect the unvulcanized rubber used in pressure sensitive adhesive formulations from aging. Various chemicals have been found to be useful as antioxidants:

1. Secondary amines, diamines, and their derivatives
2. Compounds of quinoline
3. Dithiocarbamates
4. Alkyl phenols
5. Phospshoric acid esters

Amine-type antioxidants are rarely used for pressure sensitive adhesives because of their staining. The most important antioxidants used for pressure sensitive adhesives are listed in Table 10-1. The effectiveness of various antioxidants was evaluated by Fukuzawa, Nemoto, and Kosaka.[4] The adhesive samples consisted of natural rubber (50 parts by weight), glycerine ester of hydrogenated rosin (50 parts), and antioxidant (2 parts). The adhesives were applied over fabric by calendering. The mass thickness was 0.07 mm. The samples were aged at 70°C for various time periods and the cohesion strength was tested in a T-peel test as shown in Figure 10-5. The test temperature was 30°C, sample width 10 mm, jaw separation rate 20 mm/min. The test results are shown in Figure 10-6. The effectiveness of antioxidants is demonstrated by comparing

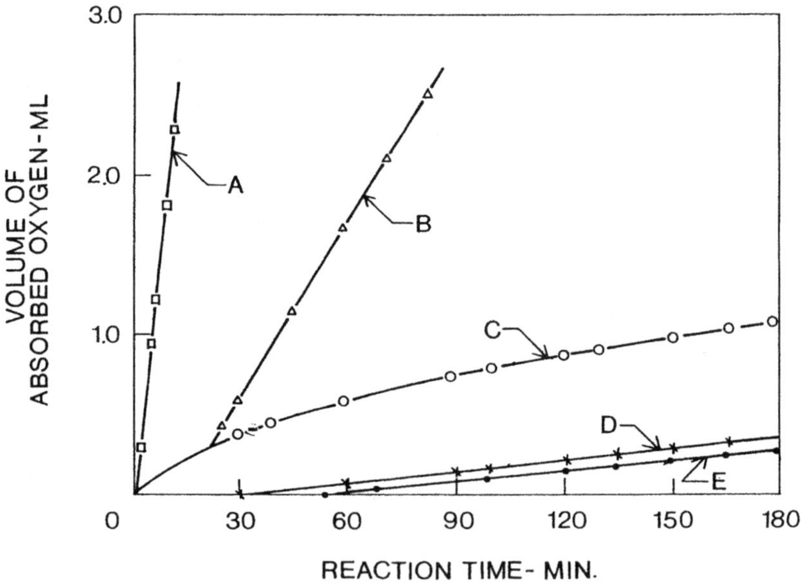

Fig. 10-2. Oxygen absorption curves of natural resins. Higuchi, Watanabe, and Toyama, Symposium on Adhesion and Adhesives (Japan), p. 17 (1965): (A) rosin; (B) glycerol ester of rosin; (C) pentaerythritol ester of partially hydrogenated rosin; (D) glycerol ester of partially hydrogenated rosin; (E) hydrogenated rosin. The sample consisted of 10-g resin solution in 10-ml *o*-dichlorobenzene. Reaction temperature 70°C, apparent oxygen pressure 760 mmHg.

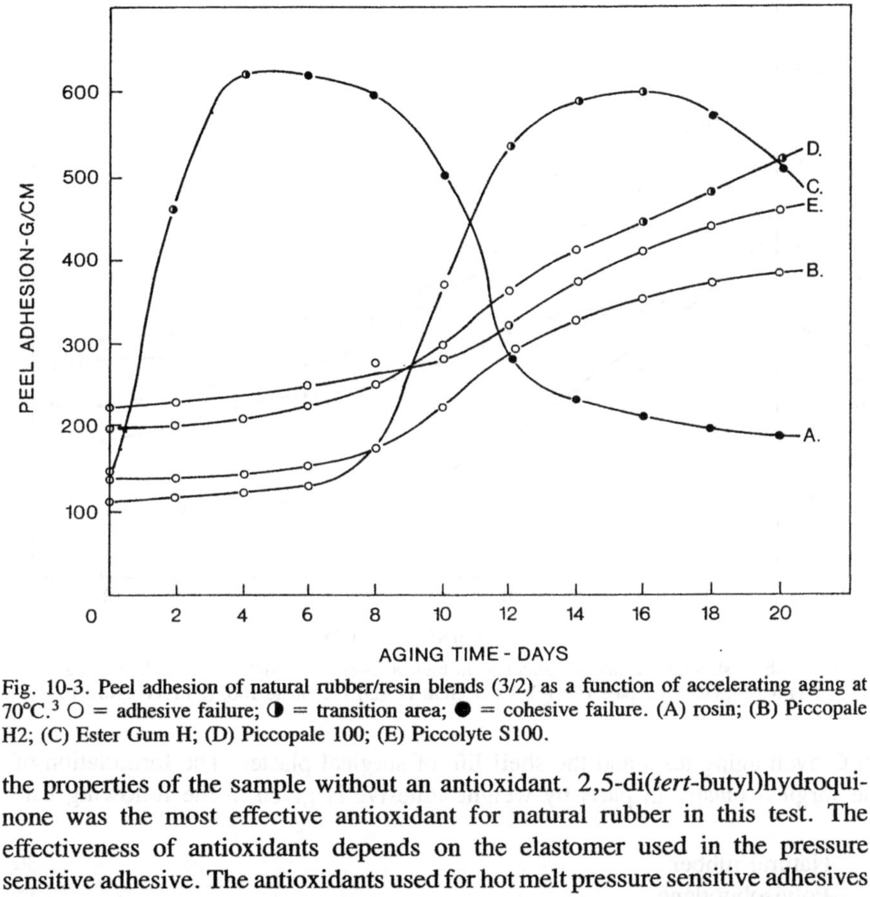

Fig. 10-3. Peel adhesion of natural rubber/resin blends (3/2) as a function of accelerating aging at 70°C.[3] ○ = adhesive failure; ◑ = transition area; ● = cohesive failure. (A) rosin; (B) Piccopale H2; (C) Ester Gum H; (D) Piccopale 100; (E) Piccolyte S100.

the properties of the sample without an antioxidant. 2,5-di(*tert*-butyl)hydroquinone was the most effective antioxidant for natural rubber in this test. The effectiveness of antioxidants depends on the elastomer used in the pressure sensitive adhesive. The antioxidants used for hot melt pressure sensitive adhesives are listed in Table 10-1, and zinc dibutyl dithiocarbamate and polyalkyl phosphite have been found especially useful.[5]

THE CORRELATION BETWEEN ACCELERATED AGING TESTS AND SHELF LIFE

A correlation between the accelerated aging tests and the actual shelf life of the tape is difficult to draw. Youman and Massen[6] reported the results of their investigation of the correlation between accelerated aging tests and the shelf life of vulcanized rubber using physical property data collected over a period of 22 years. A correlation between the accelerated aging at 70°C and the shelf life is shown. Fukuzawa and Matsukawa[3] have studied the correlation between

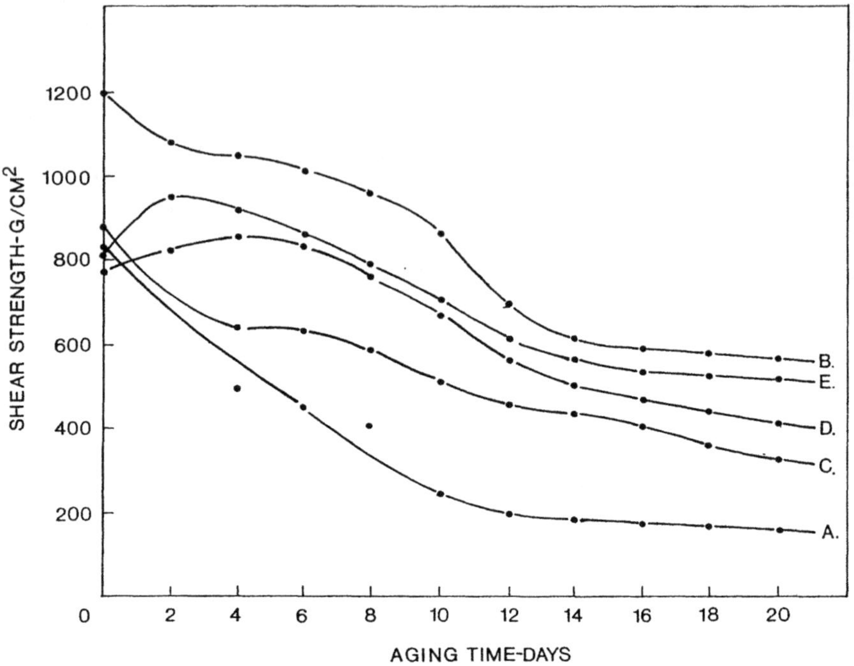

Fig. 10-4. Shear strength of natural rubber-resin blends (3/2) as a function of accelerated aging at 70°C.[3] (A) Rosin; (B) Piccopale H2; (C) Ester Gum H; (D) Piccopale 100; (E) Piccolyte S100.

70°C oven aging tests and the shelf life of surgical plaster. The formulation of the surgical plaster in parts by weight adhesive is given in the following list.

Natural rubber	30
Polyisobutylene	15
Glycerine ester of hydrogenated rosin	23
Zinc oxide	10
l-menthol	10
Methylis salicylas	5
Camphor	2
Antioxidants (2 types)	4

The 0.15-mm-thick adhesive mass was applied by calendering. The adhesive surface was covered with a cellophane film liner and cut into 42 × 65-mm sheets. The samples were packed in moistureproof cellophane and kept on the shelf for 40 months. Part of the material was aged in a 70°C oven. An adhesive-to-adhesive lap joint 25 × 25 mm was made, pressed for 20 sec with a 200-g weight, and the shear strength was measured as shown in Figure 10-7. The

Table 10-1. Various Antioxidants Used for Pressure Sensitive Adhesives

CHEMICAL NAME	TRADE NAME	M.P. (°C)	FDA STATUS	USES
2,5-di(*tert*-amylhydroquinone	DAH,Santovar A	>170	Approved	PSA[a]
2,5-di(*tert*-butyl)hydroquinone	DBH,NS-7	>200	—	PSA
2,6-di(*tert*-butyl)-4-methylphenol	BHT,Ionol	> 69	Approved	PSA
Butyl hydroanisole	BHA	57–65	Approved	PSA
Styrenated phenol	S,SP, WS	Liquid	Approved	PSA
Hindered phenol	AgeRite,Geltrol	217–225	—	HMPSA[b]
2,2′-methylenebis(4-methyl-6-*tert*-butylphenol)	NS-6,W-600,MBP LAO-5,Antioxidant 235	>120	Approved	PSA
2,2′-methylenebis(4-ethyl-6-*tert*-butylphenol)	Plastonox 425 NS-5,W-500,EBP Endox 22	>119	Approved	PSA
4,4′-butylidenebis(3-methyl-6-*tert*-butylphenol)	W-300,BBM,BB Santowhite powder	208–210	Approved	PSA
1,1-bis(4-hydroxyphenyl)-*cyclo*-hexane	Antigen W	>175	—	HMPSA
4,4′-thiobis(6-*tert*-butyl-3-methyl-phenol	Yoshinox SR	>150	Approved	HMPSA
4,4′-thiobis(6-*tert*-butyl-*m*-cresol)	Santowhite crystals	>150	Approved	HMPSA
Dilauryl thiodipropionate	DLTDP,LTDP, DLTP,TPL	> 38	Approved	HMPSA
Tetrakis[methylene-3-(3′,5′-di-*tert*-butyl-4′-hydroxyphenyl)-propionate] methane	Irganox 1010	110–125	Approved	HMPSA
Tris(mixed mono- and dinonylphenyl) phosphite	Polygard, Naugard P,TNP-S	Liquid	Approved	HMPSA
Zinc di-*n*-butyl dithiocarbamate	ZDBC,ZBUD,B8	95–104	Approved	HMPSA

[a] PSA: pressure sensitive adhesive.
[b] HMPSA: hot melt pressure sensitive adhesive.

tests were conducted at 30° ± 1°C and at a jaw separation rate of 20 mm/min. The failure was cohesive.

The test results are shown in Figures 10-8 and 10-9. The cohesive strength has a tendency to decrease on aging. The cohesive strength was 1.6 kg after 25 months in the shelf-aged samples, and the same value was reached after 15 days of accelerated aging. This would indicate that 15 days of accelerated aging is equal to two years of shelf aging for this product under the described conditions.

Tests of pressure sensitive adhesive tapes were performed after exposure for 24 months in the Weathering Test Center of Choshi Exposure Laboratory (Japan) by request of the Japanese Ministry of International Trade and Industry.[7] Cellophane, kraft paper, cloth tapes, and acrylic adhesive labels were selected

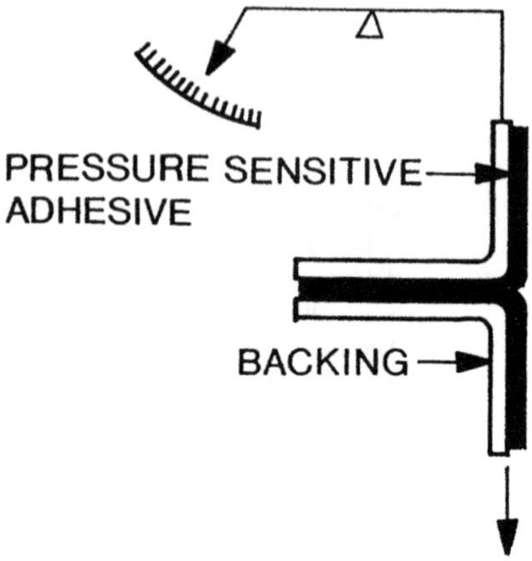

Fig. 10-5. T-peel test for measuring the cohesive strength.

for testing. The exposure shelf was placed outdoors in a horizontal position. The shelf was covered with glass and a light nontransmitting cover. The conditions were considered to be more severe than aging indoors. The accelerated aging tests were conducted at 65°C and 80% relative humidity for comparison purposes.

The aging conditions might have been too severe for cellophane tape generally used indoors. Telescoping was observed after 10 days in unpackaged rolls. Packaging is important to prevent moisture penetration. The tape retained its adhesive properties for nine months under no-light exposure conditions.

Kraft paper tape retained adhesive properties in roll form for two years. The aging properties of this tape applied over a stainless-steel plate were also investigated. This adhesive is based on natural rubber, is affected by light, and softens on aging. An increase of peel adhesion is observed initially until the failure changes from an adhesive to a cohesive one. After that, a decrease of peel adhesion is observed.

The cloth tape in a roll form showed little change after 24 months. Samples attached to a stainless-steel panel were also aged. In this case, the adhesive was greatly affected by light. This adhesive was based on reclaimed natural rubber and it appears that natural rubber adhesives are more sensitive to light than to heat. The pressure sensitive tapes, however, are supplied in rolls and exposure to light is rarely a factor.

Polyvinyl chloride film labels were constructed from plasticized polyvinyl chloride, acrylic adhesive, and release paper. One of the samples was exposed directly and the other was under glass. The release paper could not be removed after 12 months in the case of direct exposure and after 24 months in the case of exposure under glass. Samples adhering to a glass plate were evaluated. The changes in peel adhesion were small, even after long exposure, as compared to rubber-based adhesives. The peel adhesion changed greatly on accelerated

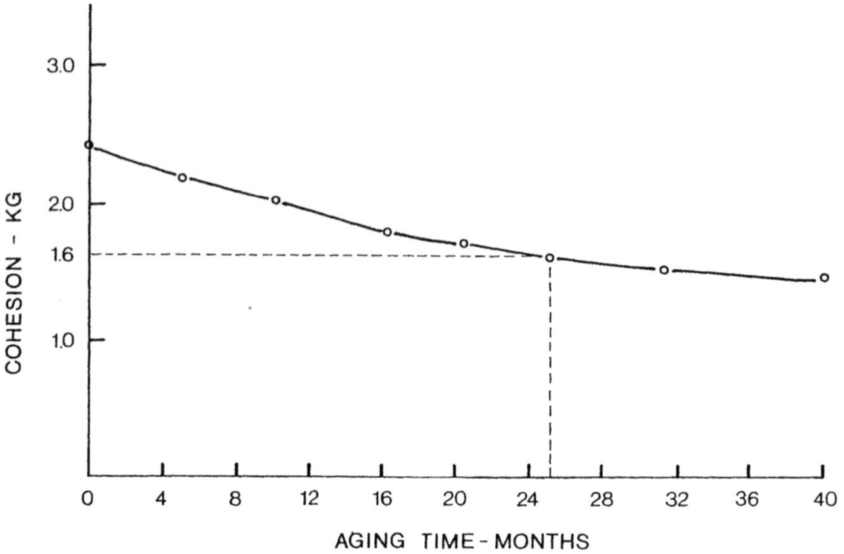

Fig. 10-8. The dependence of cohesion on aging time at room temperature.

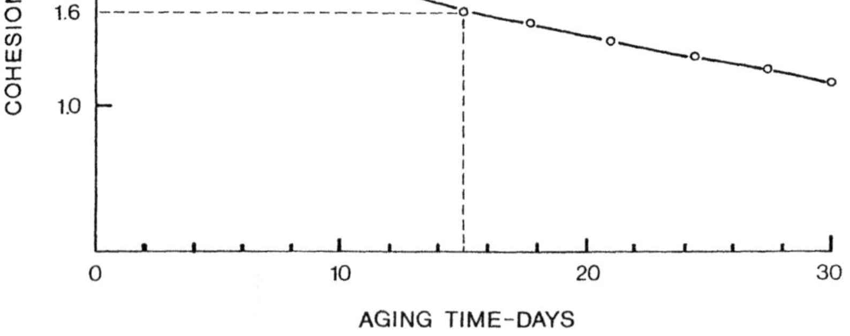

Fig. 10-9. The dependence of cohesion on aging time at 70°C.

merized to a liquid at the elevated temperature and high humidity, and it could not be tested at 16-week or longer exposures. Samples of two-component silicone and polyurethane adhesives were removed from testing after 16 weeks. The saturated release papers deteriorated and became brittle after 28 weeks of exposure at high humidity and elevated temperature. The test data are shown in Figures 10-10, 10-11, and 10-12.

Egan[14] has reported on the aging of acrylic pressure sensitive adhesives for electrical tapes. The testing is complicated by the fact that the adhesive tapes are two-part systems, consisting of backing and adhesive, and both parts must function properly if the entire tape is to function. Therefore, aging studies of such tapes included both backing and adhesive. Sometimes there is an interaction between the adhesive and the backing, and a difference in adhesive aging properties is observed, depending on the backing used. For example, silicone pressure sensitive adhesive ages faster on a woven glass fiber backing than on a Teflon film backing. If a metallic mandrel is wrapped with each tape and

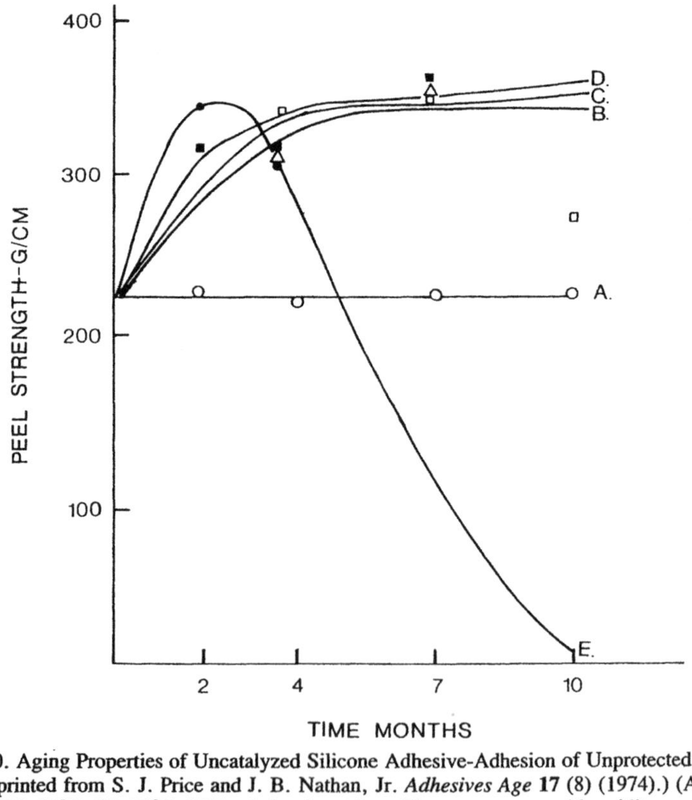

Fig. 10-10. Aging Properties of Uncatalyzed Silicone Adhesive-Adhesion of Unprotected Adhesive Film. (Reprinted from S. J. Price and J. B. Nathan, Jr. *Adhesives Age* **17** (8) (1974).) (A) −54°C; (B) 23°C; (C) 74°C; (D) 23°C, 95% relative humidity; (E) temperature and humidity cycling.

Fig. 10-11. Aging
sive Film. (Repri
−54°C; (B) 23°C
cycling.

placed in an ov
becomes embri
The Teflon film
after two years
to the greater a
A 1.25-cm
a 0.3-cm overv
of 900-g/cm w
periods of time
temperature sec
D1304 (Therma
if the material
then a typical
25°C above th
180°C and 205°

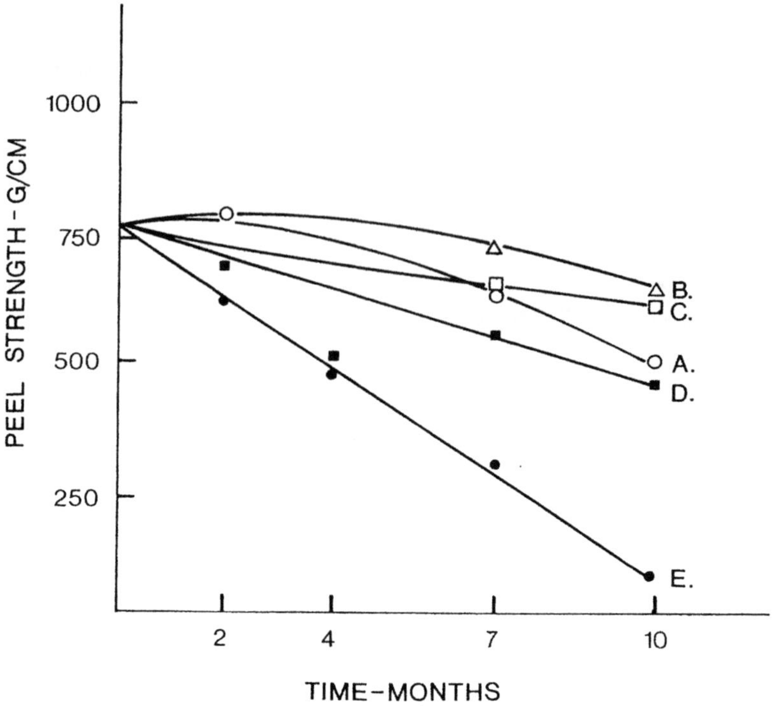

Fig. 10-12. Aging Properties of Peroxide Cured Silicone Adhesive-Adhesion of Unprotected Adhesive Film. (Reprinted from S. J. Price and J. B. Nathan, Jr. *Adhesives Age* **17** (8) (1974).) (A) −54°C; (B) 23°C; (C) 74°C; (D) 23°C, 95% relative humidity; (E) temperature and humidity cycling.

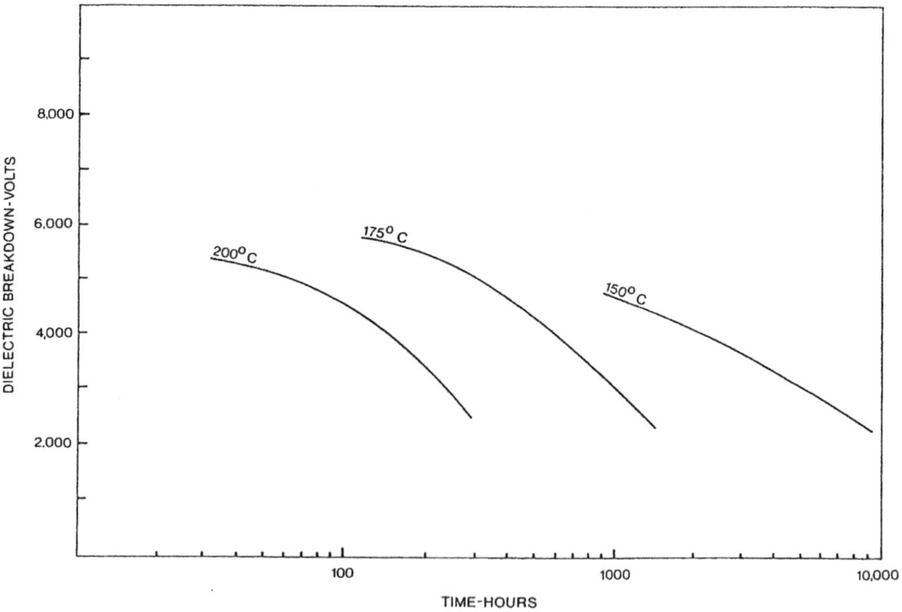

Fig. 10-13. Dielectric Breakdown Voltage vs. Aging Time at Various Temperatures. (Reprinted from Egan.[14])

tion rate is controlled by a single chemical reaction, the plot will approximate a straight line, as shown in Figure 10-14.

The usual practice is to extrapolate the curve to some agreed time figure, typically 40,000 or 50,000 hr, and report the corresponding temperature. In Figure 10-15 of a polyester film tape with an acrylic adhesive, the extrapolated 40,000-hr temperature is 123°C. The neighboring curve shows the data obtained with rubber adhesive applied on the same polyester film backing. The extrapolated 40,000-hr temperature for this tape is 118°C or 5°C lower than for acrylic adhesive.

This method appears to offer a means of evaluating the thermal aging properties quantitatively, not only of the adhesive alone, but of the complete tape.

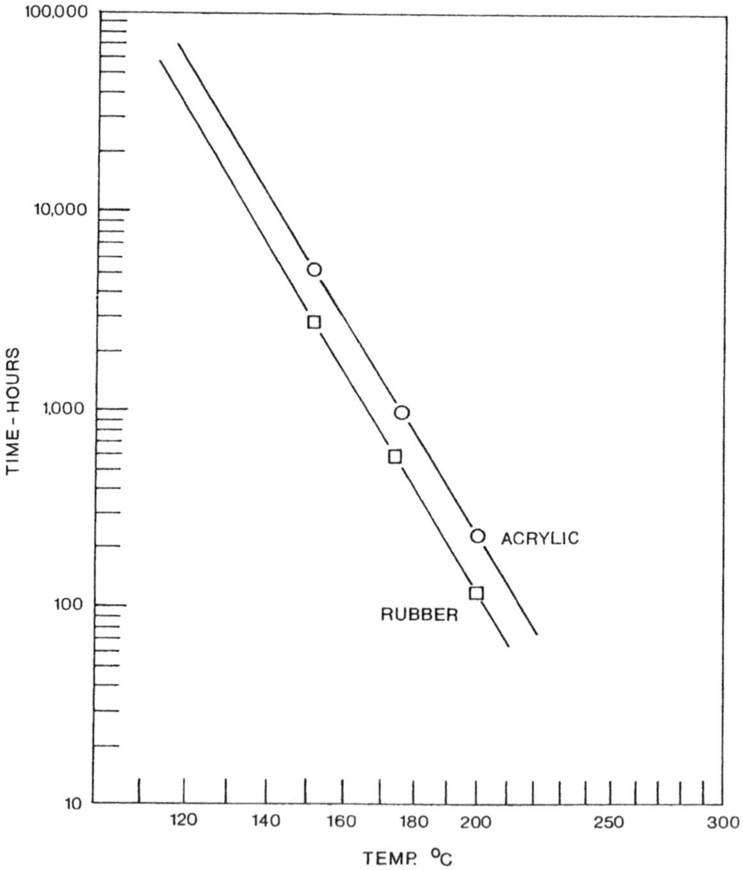

Fig. 10-15. Comparison of aging properties of acrylic and natural rubber adhesives.

11. Natural Rubber Adhesives

G. L. Butler
DRG Tapes and Adhesives
Borehamwood, England

The history of pressure sensitive adhesives is comparatively short; natural rubber was the elastomer used in the earliest of this type of adhesives, and it still remains today one of the major elastomers used in the manufacture of these adhesives, though the development of synthetic elastomers such as polyacrylates and S-B-S and S-I-S block copolymers is threatening the leading position that has been held by natural rubber.

The first use of natural rubber pressure sensitive adhesives was in surgical plasters. An early U.S. patent[1] taken out by Shecut and Day in 1845 describes a mixture of India rubber, turpentine extracts, Peruvian balsam, and pine gum.

However, the real development of the pressure sensitive adhesive industry began with the inventions of R. G. Drew in the late 1920s and early 1930s. Drew's inventions were primarily concerned with the development of crepe-paper based masking tapes and clear tapes based on regenerated cellulose film. He needed adhesives for these products that would adhere well under applicational pressure, and could be subsequently peeled off the surface without depositing the adhesive. A further necessary property of the adhesive was that the tape should be capable of being unwound without the adhesive offsetting to the reverse side of the tape. In these early days, primer coats to bond the adhesive to the web had not been developed, so the composition of the adhesive had to be such that it allowed satisfactory unwinding characteristics.

One of Drew's earliest patents[2] describes an adhesive composition for masking tape comprising plasticized plantation natural rubber, ZnO filler tackified with either coumarone-indene resin or a natural resin such as Burgundy Pitch. His early patent[3] for cellulose tape describes an adhesive based on masticated pale crepe latex, coumarone-indene tackifier, and liquid paraffin, claimed to increase the tack of the adhesive without causing the adhesive to offset, which would have occurred with added resin content.

those based on cellulose film, it is important that the color of the adhesive should be kept as light as possible. For this reason, it is normal to use a pale crepe grade of rubber. Where color is not important (e.g., dark colored tapes, certain masking tapes, cloth tapes, etc.), darker grades such as ribbed smoked sheet or SMR 5 are often used, as these grades are less expensive than pale crepe.

Natural rubber has to be masticated before it can be used in a pressure sensitive adhesive. This is necessary to break down the gel content and reduce the molecular weight of the natural material. Without mastication, it would be very difficult to obtain the desired level of tack in the adhesive and also, as most adhesive systems are coated from organic solvent solutions, the unmasticated rubber does not give a free-flowing adhesive suitable for subsequent coating without the use of excessive quantities of solvent.

Tackifying Resins

Pressure sensitive adhesives require a balance of three main properties: peel adhesion, cohesive holding power, and surface tack. Natural rubber alone has a very low tack and adhesion to surfaces. Consequently, it is necessary to add tackifying resins to the elastomer to produce the required balance of tack, peel adhesion, and resistance to shear forces.

Three main classes of tackifying resins have been used with natural rubber. These are:

1. Wood rosin and its derivatives
2. Terpene resins
3. Petroleum-based resins

Rosin and Derivatives. Apart from the wood rosin and its glycerol ester (gum ester), a number of other derivatives of this resin type is used with natural rubber. To vary the softening point, different esters are used, for example, pentaerythritol ester. The abietic acid molecule in wood rosin contains unsaturated double bonds that are susceptible to oxidation. Therefore, hydrogenated rosin derivatives are frequently used in adhesives requiring long-term aging properties. These include glycerol ester of partially hydrogenated rosin and the pentaerythritol ester.

Terpene Resins. Terpene resins, essentially consisting of polymerized β-pinene, started to be used in the late 1950s and early 1960s. They can be

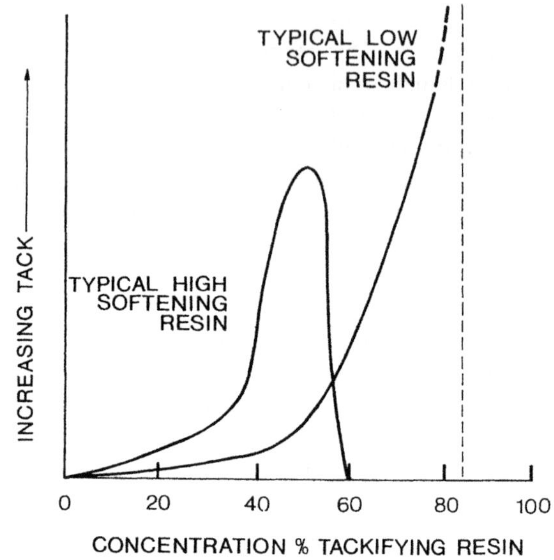

Fig. 11-1. Tack vs. concentration of tackifying resin in natural rubber for high and low softening point resins (after Wetzel).

resins of lower softening points (i.e., less than 95°C) provide higher maximum tack than resins of higher softening points.

Tack, however, is only one of the three main parameters that determine the properties of an adhesive. As we have seen, higher tack is generally achieved with resins of lower softening points, and with these resins generally, the higher

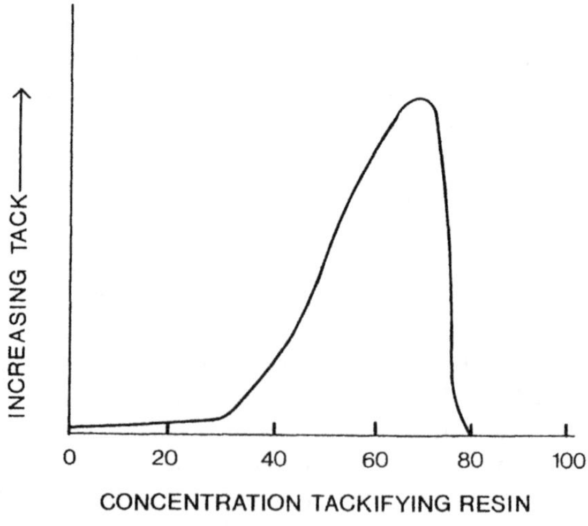

Fig. 11-2. Tack vs. concentration of tackifying resin in natural rubber for intermediate point resins (after Wetzel).

special factors that affect the choice of tackifying resins. If the adhesive is destined for application for a clear tape, then color is an important factor. In this respect, terpene resins are particularly useful, as are the hydrogenated resins, both rosin derived and also the hydrogenated polycyclic petroleum hydrocarbon resins. Not only do these resins show good initial color in a rubber-based adhesive, but they also give subsequent aging resistance due to the saturated nature of the resin molecule.

C. A. Dahlquist[10] has described a natural rubber adhesive tackified with alkylated polystyrene resins for a particularly good water-white color and good UV resistance. A typical resin is poly *tert*-butyl styrene with a molecular weight in the range 1000–2000, and can be used at a concentration of about 77 phr to give useful adhesives with milled natural rubber.

Most resins used to tackify natural rubber are compatible with the polymers. But it has been claimed[11] that the addition of small quantities of resins that either have a borderline compatibility or are incompatible with natural rubber, do lead to adhesives with considerably increased peel adhesion values. Examples of such resins include maleic modified ester of rosins and oil-soluble phenol aldehyde resins such as *p*-tertiary amyl phenol formaldehyde. These resins are added in addition to the normal tackifiers in amounts of less than 15% by weight of the rubber. The theory of this process is that the incompatible resins tend to separate from the body of the adhesive during application of the adhesive to the backing, and provide a resin-rich area at the surface, hence increasing adhesion, without affecting the cohesiveness of the adhesive layer as a whole.

Antioxidants

The *cis*-polyisoprene molecule of natural rubber contains unsaturated double bonds along the chain, and is subject to oxidation attack. This action on natural rubber has a degrading effect of chain scission with consequent loss of cohesive strength. This oxidation attack is normally increased by exposure to elevated temperatures or to UV light. Consequently, it is necessary to include antioxidants in the formulation of natural rubber based adhesives to ensure both adequate shelf life of the adhesive after it has been converted to tapes or labels, and also to afford a degree of protection to the adhesive in use.

In general, three main types of antioxidants have been utilized with natural rubber adhesives. First, there are the amine antioxidants, such as *N–N′* di-β-naphthyl-1,4-phenylenediamine (AgeRite D). This type of material, while acting as an efficient antioxidant, is a staining material, and can impart a slight stain to surfaces with which the adhesive comes into contact. The most widely used antioxidants have been the phenolic types; these include 2,5-di-(*tert* amyl) hydroquinone (Santovar A) and 2-2′-methylenebis(4-methyl-6-*tert* butyl phenol) (Antioxidant 2246). Phenolic antioxidants are nonstaining and are particularly effective

The addition of ZnO pigment tends to make the adhesive firmer. It has been claimed[14] that the addition of fine particle size silica improves the aging of natural rubber-based adhesives on fabric-based tapes. The addition of silica with particle sizes between 0.01 and 0.03 μm at levels between 2% and 8% of the adhesive is claimed to reduce the tendency of the roll of cloth tape to distort on aging and also to improve tack retention after aging.

Blends of Other Elastomers with Natural Rubber

Other elastomers can be added to natural rubber-based adhesives for either cost or performance reasons.

Synthetic *cis*-polyisoprene polymers can be added to natural rubber-based adhesive. The main problem with commercially available *cis*-polyisoprene polymers is that they do not possess the same green strength as natural rubber and consequently, in uncured adhesives, this leads to loss of cohesive strength. Quantities of the synthetic elastomers that can be added are limited.

Styrene-butadiene copolymers can be usefully blended with natural rubber systems. Such blends generally result in improved aging of the adhesive.[15] In the oxidation aging process, natural rubber undergoes chain scission and hence becomes softer, whereas SBR elastomers cross-link and become harder. Consequently, the two materials tend to balance each other. As well as improving aging resistance, blending with SBR can also increase the creep resistance.

MANUFACTURE

The vast majority of natural rubber pressure sensitive adhesives are applied to their substrates from solvent systems. The initial process in the manufacture is milling or breakdown of the rubber to achieve the required viscoelastic properties. This can be carried out on either a two-roll rubber mill or in an internal mixer. The degree of milling required depends both on the state of the raw rubber, which being a natural product, can vary from batch to batch, and on the viscoelastic properties required in the end adhesive. For most purposes, the rubber would be milled to a Mooney viscosity of between 60 and 75. The milling process can be accelerated by the addition on the mill of peptizers, such as pentachlorothiophenol, in order to reduce times and milling energy. The dissolution of the rubber in the solvent can be a lengthy process. In order to speed this up as much as possible, it is necessary to granulate the rubber in order to create the maximum surface area for the solvent to attack the rubber.

Natural rubber is a nonpolar material. Consequently, nonpolar hydrocarbon solvents are used. The choice of the solvent is normally determined by the

by Hoch[18] in 1963 when he stated that tack is largely dependent on the disperse phase that is effectively a viscous fluid that can wet and flow to contact a surface intimately. This viscous layer is so thin that failure cannot occur in it. Consequently, ultimate adhesive properties are dependent on the continuous elastomeric phase. Hence, the disperse phase contacts and wets the irregularities of the substrate, and the continuous phase carries the load.

The effect of solvent evaporation on the heterogeneity of rubber-resin films has been examined by Whitehouse, Counsell,and Lewis.[19] They investigated the chemical composition of natural rubber/rosin ester blends by ATR infrared spectroscopy, and found an excess concentration of resin at both surfaces of the adhesive layer, compared with the resin concentration in the bulk of the adhesive layer (Fig. 11-4).

Prolonged aging of the adhesive layer for about 500 days showed that the surface compositions of the films became identical to the bulk composition, thus indicating that the original heterogeneity of the adhesive films was due to

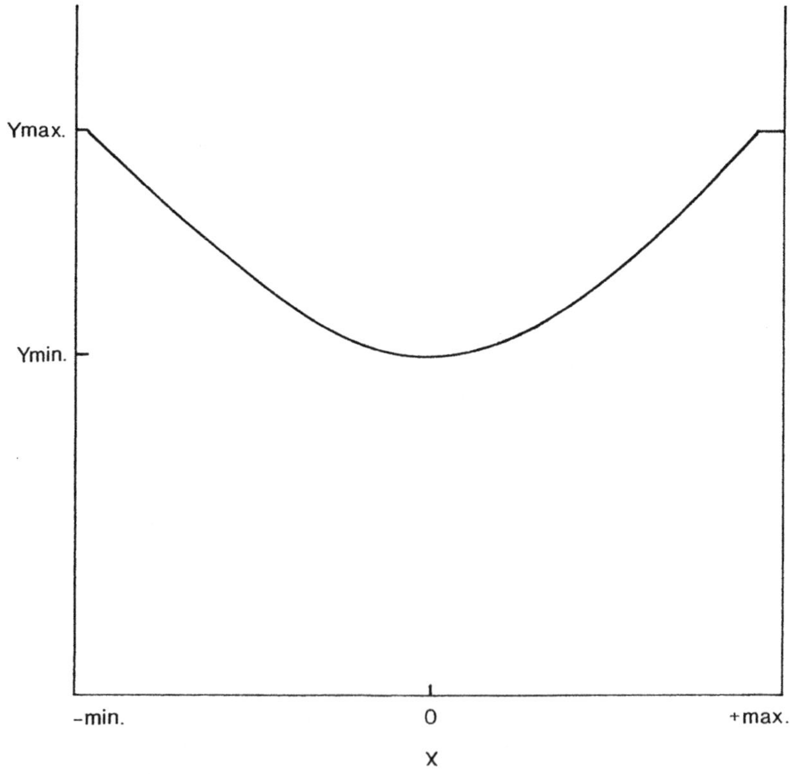

Fig. 11-4. Variation of volume fraction of resin Y with distance X from the central plane of the adhesive layer. (After Whitehouse, Counsell, Lewis.)

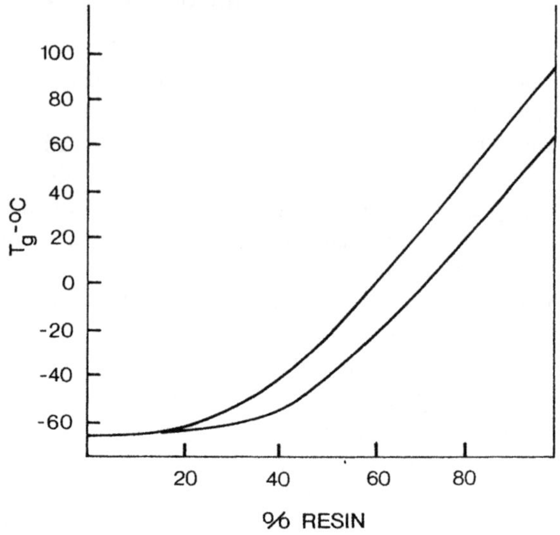

Fig. 11-5. Variation of transition temperatures with resin concentration with Piccolyte S70 (bottom curve) and S115 (top curve) in natural rubber. (After Sherriff, Knibbs, and Langley.)

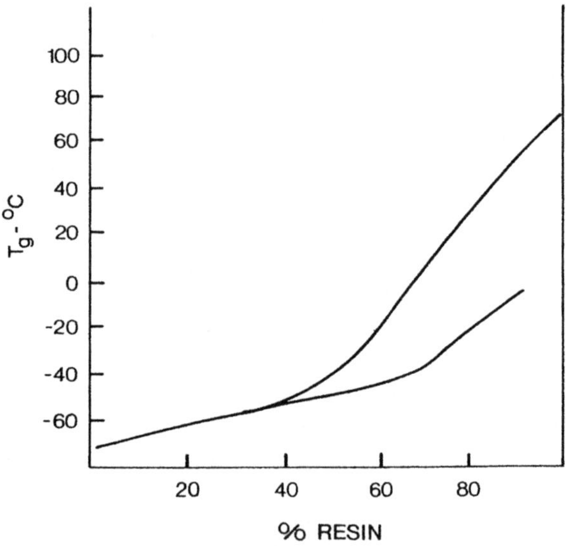

Fig. 11-6. Variation of transition temperatures with resin concentration with Pentalyn H in natural rubber. (After Sherriff, Knibbs, and Langley.)

Heat Resistance

The most important application for pressure sensitive adhesives where heat resistance is required is in the field of masking tapes. These tapes are used in paint spraying operations where the tape has to undergo the paint stoving process, with a typical baking cycle of 40 min up to 130°C. After undergoing this exposure, the masking tape has to be capable of being peeled off without leaving any adhesive residue, and thus peeling can take place while the tape is still warm.

The usual construction of a masking tape comprises an impregnated creped paper, coated with a pressure sensitive adhesive on one side and a paint-resistant backcoat on the other surface.

The conventional natural rubber adhesives previously described are quite obviously thermoplastic and do not have the required temperature resistance for a masking tape. Not only would the tape fall off in the oven, which could result in marking the uncured paint, but also it could not be removed without leaving adhesive residue behind. To impart the required degree of heat resistance to the adhesive, it is usual to effect a degree of cross-linking to the natural rubber. We have seen earlier that the tack and adhesion of pressure sensitive adhesives depend upon the viscoelastic properties of the rubber-resin blend. Consequently, the degree of cross-linking has to be carefully controlled; if the degree of curing is too high, then the adhesive will lose its ability to deform readily and hence wet the surface to which it is being applied, but it has to be sufficiently cured to have resistance to the paint solvent, and also not to be softened too much by the stoving cycle.

It is important, therefore, to be able to study the effectiveness of crosslinking on the thermal stability of the adhesive. One way of doing this is to measure the peel adhesion of the tape at elevated temperatures. This can be done by applying the tape to a polished steel panel, heating the panel to a required temperature (120°C), and then peeling the tape at an angle of 180 degrees at a fixed speed at the elevated temperature and measuring the pulling force required. Figure 11-8 illustrates the effect of cross-linking on peel adhesion measured at 120°C.

In this case, a curable adhesive has been coated on to paper backing, the solvent removed, and the adhesive cured for 1 min at various temperatures.

The adhesive studied in this case is based on natural rubber/rosin ester tackifier, ZnO pigment together with a sulfur plus accelerator curing system. The vertical axis shows the peel adhesion in grams per 25 mm, using a peeling speed of 15 cm/min. The effect of cross-linking can be easily seen on this graph. On the left, at low curing temperature, there is relatively little cross-linking, the adhesive is thermoplastic, peel adhesion is low, and failure is largely cohesive.

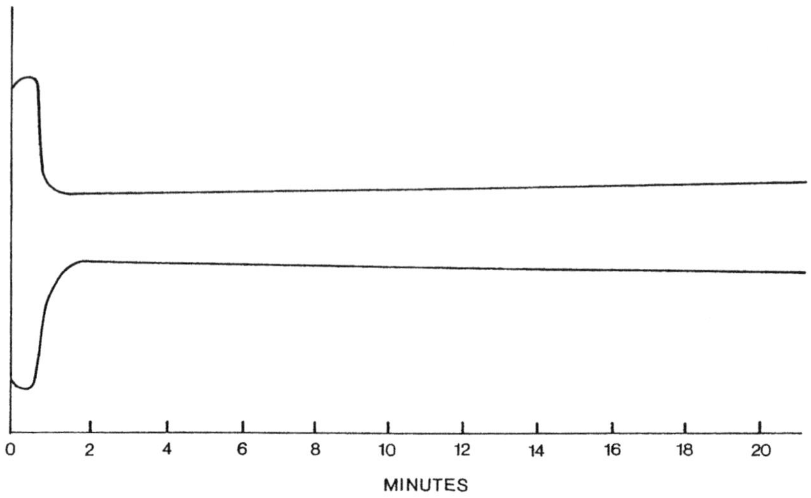

Fig. 11-9. Shawbury Curometer trace; curing temperature 200°C.

of the sample, which enables the stiffness of the sample and hence the state of cure to be examined. At the left side of the trace, it can be seen that the sample becomes less stiff for a short time. This is the period where the sample is heating up before curing starts, and on this trace, curing is taking place almost immediately after this induction period, as is shown by the rapidly increasing stiffness of the sample. As can be seen, the sample is fully cured within 1 min at 200°C.

There are several curing systems for natural rubber adhesives of this type. Perhaps the ones most commonly used are the phenol-formaldehyde resin cure systems. These have the advantage of imparting the required degree of cure at reasonable speeds, and they do not cause staining of any light-colored paints to which the tape may be applied. Phenolic resins of these kinds must be compatible with natural rubber and heat advancing. Bemmels[25] has described such a resin based on the alkaline condensation product of nonyl phenol with *para*-formaldehyde. Bemmels states that it is necessary that the alkyl group of the alkyl phenol should contain at least eight carbon atoms to maintain good compatibility with the natural rubber. A number of commercially available phenol-aldehyde resins can be used in this application.

As said earlier, the curing step is generally carried out in line with adhesive coating, so it is important that the curing reaction is carried out quickly to ensure economic production. Several methods have been proposed to accelerate the phenol-formaldehyde resin cure of pressure sensitive adhesive systems. One accelerator system[26] claimed is the use of organometallic derivatives of elements of Group IV in the periodic table, for example, zirconium tetraisopropoxide, or titanium tetrabutoxide.

subsequently dipped into varnish solutions. Typical varnish solvents are aromatic hydrocarbons (toluene) and alcohols (butanol).

Clearly, conventional uncured natural rubber-based adhesives have little resistance to hydrocarbon solvents. Degrees of curing described in this previous section, where the pressure sensitive properties are still maintained, have slightly improved solvent resistance in comparison with uncured adhesives, but the elastomers can still be readily swollen by a solvent such as toluene. This problem is overcome by providing a tape with a pressure sensitive adhesive that can be cured after the tape has been applied. After application, it is no longer necessary for the adhesive to remain pressure sensitive, but it still must remain well-adhered to the adherend. The adhesive is in essence a thermosetting one.

The curing systems are generally similar to the phenol-formaldehyde cure formulations described in the section on heat resistance. In this case, larger quantities of phenol-formaldehyde resin are used in the formulation to obtain a higher degree of cross-linking. In a typical heat-resistant masking tape, the amount of phenol-formaldehyde resin would be on the order of 10 parts per hundred rubber, whereas in a thermosetting adhesive, this would be increased to 20–30 parts per hundred rubber. The degree of solvent resistance does depend on the time and temperature of the curing cycle, but a typical cure cycle to obtain maximum solvent resistance is about 3 hr at 150°C. Phenol-formaldehyde curing resins are preferred for these thermosetting electrical applications, as they exhibit no curing of the rubber at room temperature, hence giving the tape a good shelf life. Also they cause no corrosion of copper electrical conductors, unlike sulfur curing agents.

Because of the high temperatures demanded in the curing cycle, rubber thermosetting pressure sensitive adhesives have to be coated on to heat-resistant backings. The most widely used backing is biaxially oriented polyester (e.g., DuPont Mylar). Others include glass cloth or specially impregnated papers. Methods of adhesive and tape manufacture are similar to those described in the section on manufacture of natural rubber pressure sensitive adhesives.

Improved Creep Resistance

As we have seen, uncured rubber-resin adhesives are perfectly adequate for many packaging and label applications. However, there are applications where it is necessary to improve the cohesive strength of the rubber-based adhesive. This is particularly the case with heavy-duty strapping tapes, where the adhesive can be subjected to considerable shear stress. Such tapes usually have a high tensile strength backing to provide the required strapping strength, and are based on films or paper with yarn or scrim reinforcement. More recently, these reinforced backings are being replaced by monoaxially oriented polypropylene films, which give the required high tensile strength at greatly reduced cost.

temperature of 49°C was significantly increased. Using irradiation doses of 1–3 megarads at a distance of 10 cm from the adhesive surface, shear resistance of the adhesive at 49°C was increased from failure after 6 min to no failure after 22–35 min. Not only did the adhesive show an increase of shear resistance, but other properties of the adhesive, peel adhesion, and tack, were not affected.

Improved Electrical Conductivity

For many purposes, pressure sensitive adhesives form a part of electrical insulating tapes. In this case, the conventional natural rubber adhesives that have been described show excellent electrical resistance. However, there are applications where it is desirable to have electrically conductive adhesive tapes; these tapes are generally based on a conductive metal foil backing (e.g., aluminum foil) coated on at least one side with a conductive adhesive layer. Such tapes can be used as solder replacements, pliable low voltage conductors, etc.

Electrical conductivity is imparted by incorporating conductive particles into the adhesive. One example[34] is to add milled aluminum spheres to a natural rubber/β-pinene blend. At a loading of 10.8 parts by weight per 100 of adhesive, the electrical resistance was lowered to 1.5 $\Omega/6.45\ cm^2$. However, the use of metallic particles does not always give very homogeneous results. An improved method[35] is to load the adhesive with conductive carbon black. Vulcan XC (Cabot Corporation) is such a carbon black, as is acetylene black. In order to obtain reasonable electrical conductivity, it is necessary to use heavy loadings of carbon black—up to 70 parts of black for 100 parts of base adhesive material. To offset the effects of such heavy loadings on adhesive properties, especially tack, it is necessary to use a very soft base adhesive, which can be obtained by excessive milling of the natural rubber. This system gives homogeneous electrical properties, with electrical resistance that can be below 0.05 $\Omega/6.45\ cm^2$.

EMULSION ADHESIVES*

Natural rubber adhesives are generally applied from either organic solvent solutions, or they are calendered as 100% solids compounds, but they are less frequently applied as latexes. Interest in emulsion adhesives has been growing as a result of legislation restricting solvent emissions and because of increasing

* This section was written by D. Satas.

latex stability rises rapidly over the first two weeks after centrifuging, especially in HA latex.

Metal ions, such as copper, iron, and zinc, have an adverse effect on latex properties. The latex should be kept away from contact with these metals.

Compounding

Latex is blended with tackifying resin dispersions in order to obtain desired tack and adhesion. Thickeners, antioxidants, and latex stabilizers may also be added. Potassium hydroxide has been used traditionally to stabilize latex. The merits of its use for pressure sensitive adhesives is questionable, especially if ZnO is not present in the adhesive formulation.[39] Surface active agents may be added to improve latex mechanical stability. Several surface active agents have been evaluated and their effects on adhesive properties determined.[39,40] The amount of surface active agent required to stabilize the latex is below 0.5 phr, which is small in comparison to the amount of surface active agent used to stabilize the tackifying resin emulsion.

Several antioxidants were evaluated and zinc dibutyldithiocarbamate was found to be the most effective with uncured adhesives. Cured adhesives performed better in aging tests.[39,40]

Natural rubber latex may be blended with synthetic rubber latex. Such blends often have a higher viscosity than either of the components due to the redistribution of surface active agents and the formulation of three-dimensional branched structures.

Tackifying resins are added to the latex in order to increase its tack and other adhesive properties. Resins are introduced as a predispersed aqueous emulsion or as a resin solution in an organic solvent. The solution easily disperses in the latex, but such coatings contain some solvent that might be objectionable. After coating, rubber and resin particles are fused by heating. The fusion takes place quickly. Oldack shows the tack development due to fusion as a function of heating time (Fig. 11-10).[39] It appears that tack decreases with increasing fusion temperature, for reasons that are not clear.

Properties of a latex adhesive tackified with glycerol ester of hydrogenated rosin are compared in Figure 11-11 to its solventborne counterpart.[39] Both adhesives were cast on polyester film. Properties are comparable, except that the latex adhesive exhibits a sharp increase in peel adhesion and quick stick at higher resin-rubber ratios. Polyterpene resins are also suitable for tackifying latex adhesives. Both of these resins yield a high tack adhesive, with rosin ester tackifiers generally giving higher quick stick values.[37] Aliphatic hydrocarbon resins are not as effective with latex adhesives. The effect of various tackifying resins on adhesive properties is shown in Table 11-1. The effect of resin blend

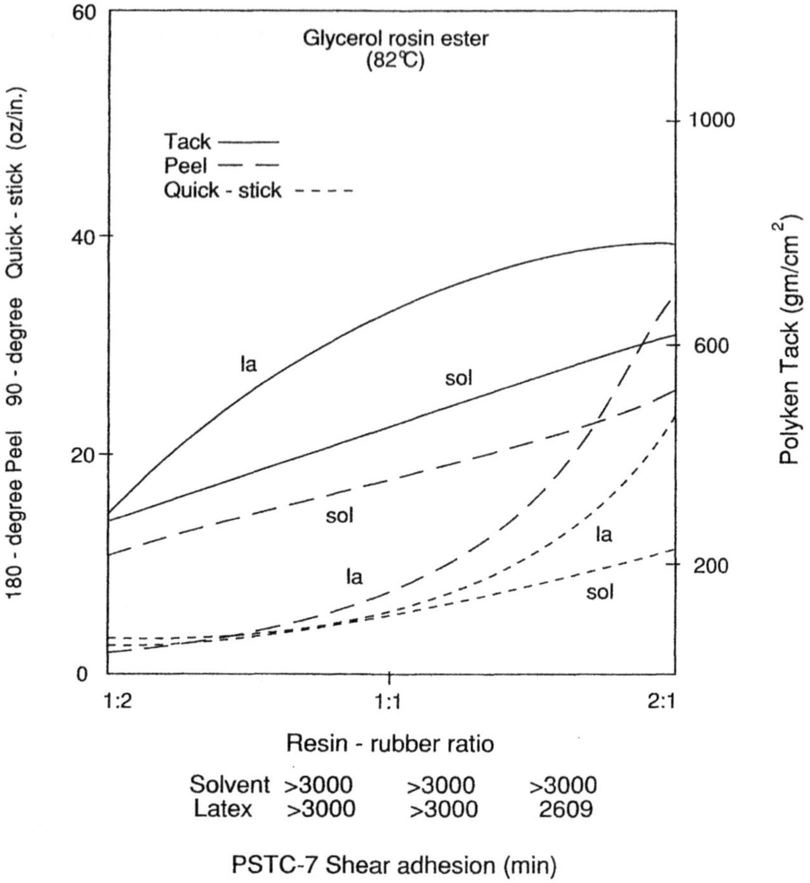

Fig. 11-11. Properties of a natural rubber latex adhesive compared to solution adhesive.[39]

Table 11-1. Effect of Various Tackifying Resins on Adhesive Properties

RESIN	SOFTENING POINT (°C)	RUBBER-TO-RESIN RATIO	BOND STRENGTH (N/cm) QUICK STICK	180-DEGREE PEEL	SHEAR HOLDING TIME (hr)
Rosin ester	75	60:40	0–29	0–85	14
		50:50	0–66	1–85	8
		40:60	1–68	4–79	3
Hydrocarbons					
Aliphatic	85	40:60	1–27	2–01	1
Aliphatic/aromatic	85	40:60	0	2–70	>150
Poly-β-pinene	85	50:50	0–50	1–60	30
		40:60	1–43	2–70	>150
	115	50:50	0–56	2–09	120
		40:60	0	0–19	80

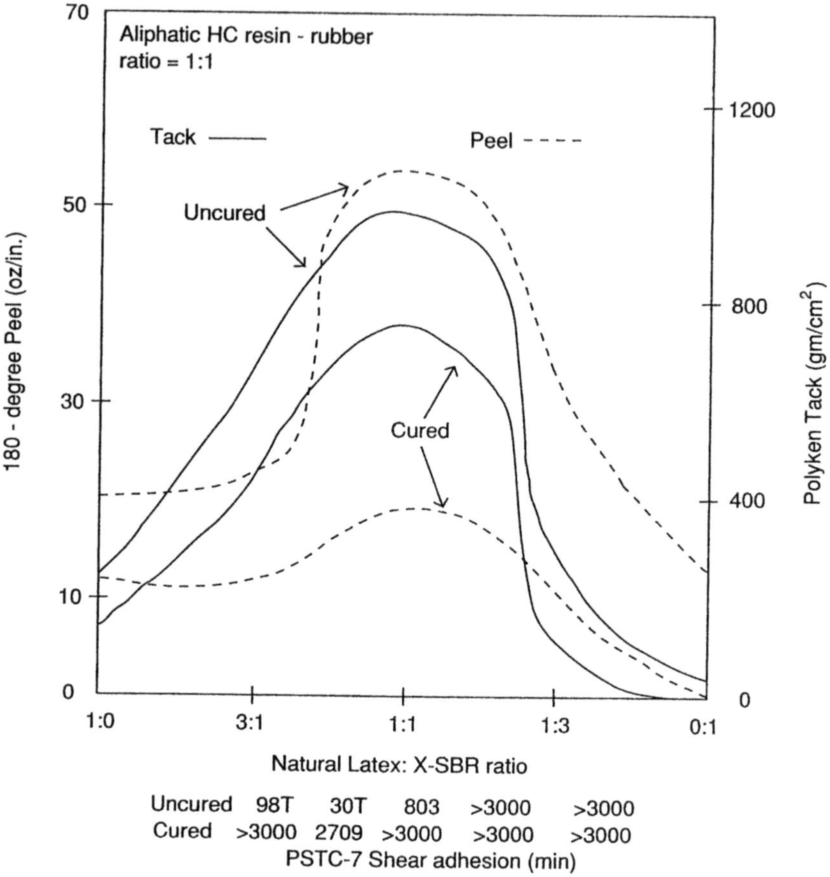

Fig. 11-14. Effect of cure on adhesive properties.

MAJOR APPLICATIONS

Packaging Tapes

Perhaps the largest single use of natural rubber adhesives is in the field of pressure sensitive packaging tapes. These products range from the well-known clear tapes used in the home to heavy-duty strapping tapes used in industry. For many years now, natural rubber has been the workhorse adhesive polymer on these tapes because of its unique balance of tack and cohesive strength combined with low cost.

The largest single use of packaging tapes today is the sealing of fiberboard cases. The penetration of the pressure sensitive tape with this market has been

Polypropylene film has been increasingly used in this application due to its low cost. In the United States, the polypropylene film tape has occupied a dominant position among packaging tapes. Packaging tapes are used both in the home and industry for hundreds of applications, applied to many different kinds of surfaces, and the natural rubber-based adhesive has the advantage of displaying good tack to a wide variety of substrates, which make it a preferred polymer in many of these tapes.

Masking Tapes

The most common and largest volume of masking tapes are those based on impregnated creped papers. These are mostly used in the holding of paper masks on vehicle bodies during spraying operations, both during the manufacture of vehicles, and perhaps more importantly, in vehicle repair shops. In the first case, paint stoving cycles of up to 40 min at 130°C can be encountered; in repair shops, lower temperatures of only 80°C are more common, and of course many paint spraying applications are air-dried. For these types of tapes, natural rubber adhesives are still the most commonly used. It is common to blend the natural rubber with a compatible SBR polymer to improve aging, and the adhesives are nearly always cured, usually with the phenol-formaldehyde resin systems discussed earlier. Figure 11-15 shows a masking tape being used to mask off the windows and stainless-steel fittings of a fire tender.

However, there are other types of masking tapes and applications where natural rubber adhesives are used. These include a lithographic masking tape, where a semitransparent tape is required for a removable application to a photographic plate to mask out an area of the plate. Such tapes are made from semitransparent red films, cellulose acetate, or unplasticized poly(vinyl chloride), coated with a low tack natural rubber pressure sensitive adhesive.

Double-face Tapes

One of the most rapidly expanding applications for pressure sensitive adhesives is in the field of double-face tapes. These products are finding an increasingly diverse range of uses in temporary or permanent mounting or splicing applications. These tapes usually comprise a carrier web, which can be either a film, paper, or fabric, coated on both sides with the pressure sensitive adhesive, with an interleave to keep the two adhesive layers apart. The interleave is commonly paper siliconized on both sides to facilitate easy removal in use.

Double-face tapes range from heavy-duty structures, using thick aggressive adhesive layers on fabric or paper webs to very thin products based on thin

Fig. 11-16. Double-face unplasticized poly(vinyl chloride) film tape holding insoles to the last in shoe manufacture.

Again, the all-around properties of natural rubber adhesive, which shows good adhesion to many surfaces, make it a preferred system for such a wide range of applications. Figure 11-16 shows a double-face thin unplasticized poly(vinyl chloride) film tape being used to hold the insole to the cast in the manufacture of ladies' shoes.

Specialty double-face tapes include those with thermosetting adhesives to give improved heat and solvent resistance. These have many uses in electrical component construction, both as a temporary or permanent holding medium, particularly in stick-wound coil production.

Electrical Tapes

Two types of natural rubber adhesives are used on electrical tapes. First, there are uncured, noncurable rubber-resin systems on conventional insulating tapes. In these cases, the adhesives are coated on either plasticized PVC or polyethylene films; these substrates are very flexible and conformable, have good electrical insulation resistance, and are ideal for general-purpose insulation tapes where operating temperatures do not exceed 80°C. Both films require a primer coat between film and adhesive; care must be taken with the plasticized PVC film

Protection Masks

Protection masks have also been a rapidly growing application for pressure sensitive adhesives in recent years. These materials comprise either a filmic or paper base web coated with a low tack adhesive. The protection masks are applied to a variety of substrates, such as sheet metal and plastic, to protect the metal or plastic surface against damage during fabrication, storage, transit, etc. The tape must have sufficient adhesion to adhere to the surface, but subsequently be capable of being easily removed, without leaving an adhesive residue on the metal or plastic surface. Protection mask materials are made in widths up to 2 m, so it is essential for the mask to have a low tack adhesive for easy removal.

Rubber-resin adhesives are used in some of these protection masks, generally coated on thin polyethylene or unplasticized poly(vinyl chloride) films. Low tack is achieved by the use of lower quantities of tackifier resin (15–30 parts per hundred rubber) and lower coating weights (10–15 g/m^2), compared with 20–25 g/m^2 for a packaging tape.

The use of natural rubber adhesives is restricted in this application by the comparatively poor aging properties of natural rubber in comparison with more saturated polymers, such as the acrylates and polyisobutylene. Consequently, protection masks with natural rubber adhesives are generally restricted to those applications where only short-term indoor protection is required, for example, protection of stainless steel during drawing fabrication into vegetable dishes, kitchen sinks, etc. Aging can be improved by the use of an opaque pigmented film that will give protection to the adhesive against light degradation.

Protection masks also offer the opportunity for the use of aqueous natural rubber systems, as only a low degree of tack is required on these products. This can be readily achieved with natural rubber latex, usually together with a minor amount of tacky polymers in emulsion form. These adhesives are usually coated on flat papers to make the protection mask.

Pressure Sensitive Labels

Natural rubber adhesives have been widely used in the manufacture of pressure sensitive label stocks; the largest volume of these have been labels based on unimpregnated paper, coated on one side with the pressure sensitive adhesive and the adhesive layer covered with a siliconized paper. For certain special applications, films of various kinds can replace the label paper. Although there are many variations in kind of adhesive labels, they mostly fall into two categories, removable and permanent. The former has a less aggressive adhesive which allows the label to be peeled off a surface after it has been applied; the latter

4. British Patent 425,343 (1933) (assigned to Minnesota Mining and Manufacturing Co.).
5. British Patent 425,159 (1933) (assigned to Minnesota Mining and Manufacturing Co.).
6. British Patent 514,402 (1938) (assigned to Minnesota Mining and Manufacturing Co.).
7. Wetzel, F. H. *Rubber Age* **82:** 291–295 (1957).
8. Wetzel, F. H., and Alexander, B. B. *Adhesives Age* **7** (1): 28–38 (1964).
9. Bull, R. F., *et al. Adhesives Age* **11:** 20 (1965).
10. Dahlquist, C. A. U.S. Patent 3,681,190 (1970) (assigned to Minnesota Mining and Manufacturing Co.).
11. British Patent 650,001 (1947) (assigned to Johnson & Johnson).
12. Bemmels, B. W. U.S. Patent 2,615,059 (assigned to Johnson & Johnson).
13. Samour, C. M. U.S. Patent 3,162,610 (1961) (assigned to Kendall Co.).
14. Blackford, B. B. U.S. Patent 2,909,278 (1957) (assigned to Johnson & Johnson).
15. British Patent 694,190 (1949) (assigned to Minnesota Mining and Manufacturing Co.).
16. Wetzel, F. H. *ASTM Bulletin* **221:** 64 (1957).
17. Hoch, C. W., and Abbott, A. N. *Rubber Age* **82:** 471–475 (1957).
18. Wittock, C. *J. Polymer Sci.* **C.** (3) 139–149 (1963).
19. Whitehouse, R. S., Counsell, P. J. C., and Lewis, C. *Polymer* **17:** 699–702 (1976).
20. Dahlquist, C. A. *Adhesives Age* **2:** 25 (1959).
21. Dahlquist, C. A. *Adhesives Fundamental & Practice.* Ministry of Technology. Maclaren, London (1969).
22. Sherriff, M., Knibbs, R. W., and Langley, P. G. *J. Appl. Polym. Sci.* **17:** 3423–3438 (1973).
23. Aubrey, D. W., and Sherriff, M. *J. Polymer Sci.* **C(16):** 2631 (1978).
24. Aubrey, D. W., and Sherriff, M. *J. Polymer Sci.* 2597–2608 (1980).
25. Bemmels, B. W. U.S. Patent 2,987,420 (1964) (assigned to Johnson & Johnson).
26. British Patent 960,509 (1963) (assigned to Johnson & Johnson).
27. British Patent 975,971 (1964) (assigned to Johnson & Johnson).
28. British Patent 1,172,670 (1969) (assigned to Johnson & Johnson).
29. Kisbany, F. N. U.S. Patent 2,881,096 (1959) (assigned to American Tape Co.)
30. British Patent 779,256 (1954) (assigned to Permacel Tape Corp.).
31. Crocker, G. J. British Patent 848,455 (1955) (assigned to Johnson & Johnson).
32. Klepetar, M., and Kucera, C. R. British Patent 1,234,860 (1969) (assigned to Johns-Manville Corp.).
33. Hendricks, J. O. U.S. Patent 2,956,904 (1960) (assigned to Minnesota Mining and Manufacturing Co.).
34. British Patent 1,169,946 (1966) (assigned to Minnesota Mining and Manufacturing Co.).
35. Stowe, R. H. U.S. Patent 3,778,306 (1971) (assigned to Minnesota Mining and Manufacturing Co.).
36. Eustis, W., and Orrill, G. R. U.S. Patent 2,429,223 (1940) (assigned to Kendall Co.).
37. Gazeley, K. F. Natural Latex and its Application in Adhesives. *1983 Proc. Industry Technical Conference.* Pressure-Sensitive Tape Council. Itasca, Ill. (April 5–7, 1983).
38. ASTM Standard D-1076, Part 37. American Society for Testing and Materials, Philadelphia, Pa., 1978.
39. Oldack, R. C. PSA's from Natural and Carboxylated SBR Latexes. Presented at the *Technical Meeting on Water Based Systems.* Pressure-Sensitive Tape Council. Chicago, Ill. (June 13–14, 1979).
40. Oldack, R. C., and Bloss, R. E. *Adhesives Age* **22**(4): 38–44 (1979).
41. Gazeley, K. F. *N.R. Technol.* **12,** Part 4: 85–92 (1981).

time, S/B latexes were also introduced to the carpet and textile industries as binders.

The first S/B latex for pressure sensitive adhesives was sold in the mid-1970s. The move of aqueous-based adhesives into this market was a natural one. Solvent-based products have been used for years as adhesive coatings, but they require both solvent removal and recovery. Aqueous-based adhesives eliminate this step. The S/B latexes offer a wide range of properties to the adhesives manufacturer, with the flexibility of determining adhesive properties through both copolymer composition and blending.

MONOMER AND COPOLYMER PROPERTIES OF STYRENE AND BUTADIENE

Table 12-1 lists some of the physical properties of styrene and butadiene. Butadiene is highly volatile and flammable, with a high heat of polymerization. Because of this, the safe, efficient production of S/B latexes requires sophisticated, highly instrumented process equipment.

The physical properties of S/B latexes reflect a combination of polystyrene and polybutadiene properties. Polystyrene latex contains a very hard polymer

Table 12-1. Properties of Styrene and Butadiene Monomers

	STYRENE	1,3-BUTADIENE
Empirical formula	C_8H_8	C_4H_6
Structural formula	$C_6H_5CH{=}CH_2$	$CH_2{=}CHCH{=}CH_2$
Molecular weight	104.14	54.09
Melting point (°C)	−30.6	−108.9
Boiling point (°C)	145.2	−4.4
Density at 25°C (g/ml)	0.902	0.650
Specific heat at 25°C (cal/g°C)	0.413	0.551
Heat of vaporization at 25°C (cal/g)	102.65	92.25
Heat of polymerization (cal/g)	160.2	326
Vapor pressure at 25°C (mm Hg)	6.9	2144
Explosive limits vol% in air)	1.1 to 6.1	2.0 to 11.5
Solubility in water (wt% at 15°C)	0.0063	0.13

more complicated recipe would be used to design a complete spectrum of specific end use properties.

Emulsion polymerization requires an emulsifier. The emulsifier contains both a hydrophobic (nonpolar) and a hydrophilic (polar) portion. When low levels of emulsifier are added to a well-mixed vessel of water, the emulsifier saturates the water phase. When the level is increased, the emulsifier molecules aggregate to form micelles. This level is the critical micelle concentration. At this point, the particle size of the latex is determined. Commonly used emulsifiers include anionic compounds such as sodium alkyl aryl sulfonates, cationic compounds such as quarternary ammonium compounds, and nonionic compounds such as polyethoxylated alkyl phenols.[1] For S/B latex production, anionic emulsifiers are most common.

As monomer is added, it accumulates within the micelles and the equilibrium is achieved in a swollen-micelle state. Polymerization begins when initiators are introduced. For S/B polymerizations, water-soluble initiators, such as potassium persulfate, are usually employed. These generate free radicals under the influence of heat or reducing agents.[2] The radicals react with monomer in the

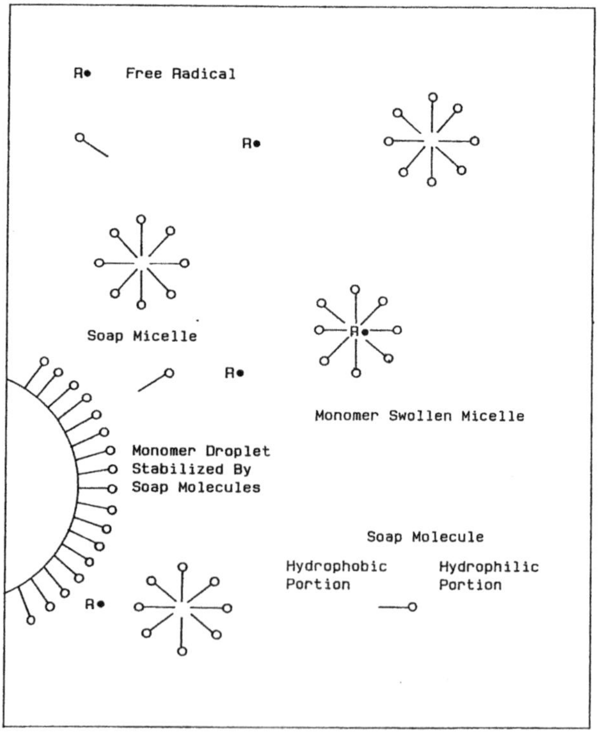

Fig. 12-1. Emulsion polymerization.

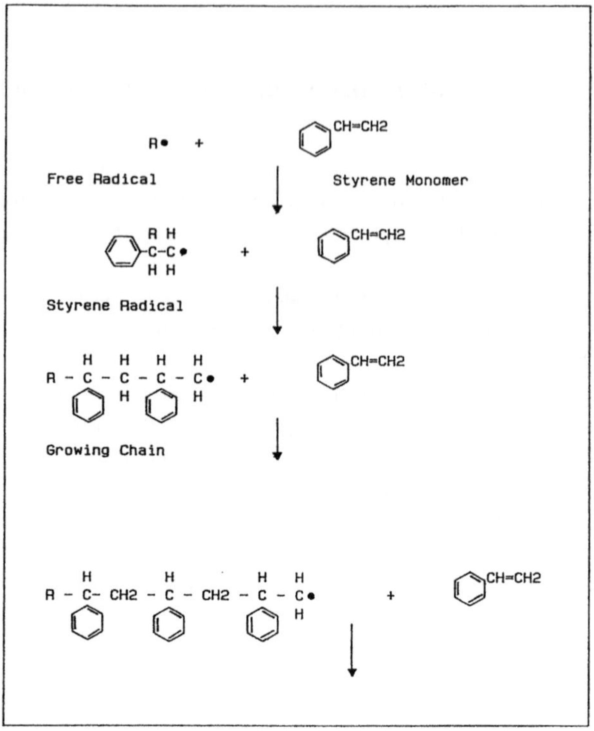

Fig. 12-3. Addition polymerization of styrene.

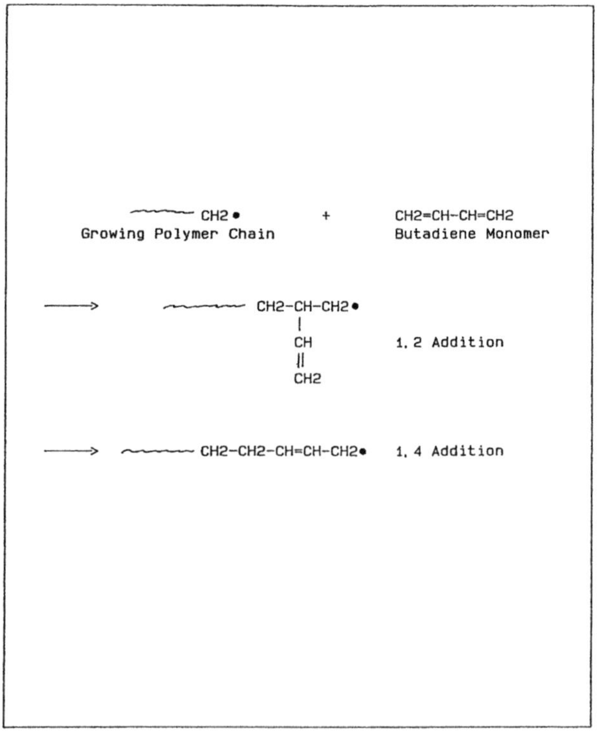

Fig. 12-4. Butadiene polymerization.

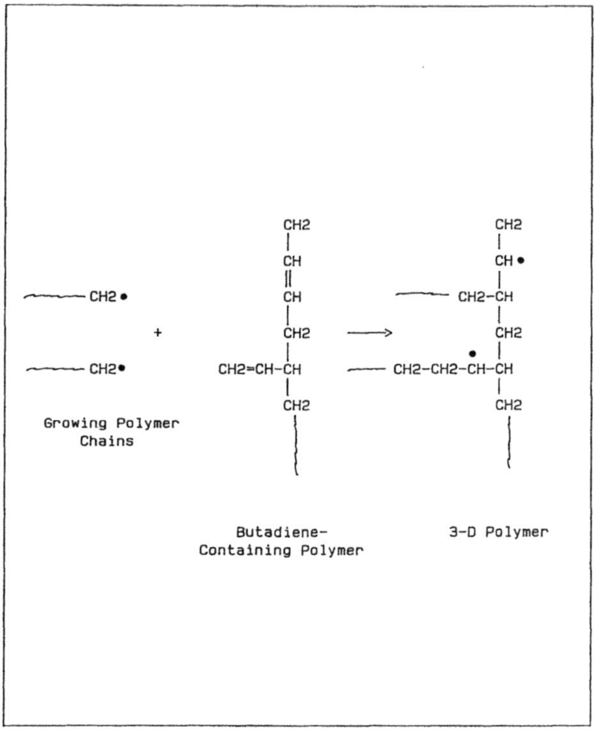

Fig. 12-7. Cross-linking—branching.

properties of an S/B polymer without a significant change in glass transition temperature.

POLYMER MODIFICATION

The addition of functional monomers can significantly change latex and adhesive performance. When added at low levels (generally less than 10%), they become part of the polymer backbone and can increase particle-to-particle bonding or cross-linking and improve polymer adhesion to the substrate.

The most widely used S/B modifiers are vinyl carboxylic acids (Fig. 12-8). These modifiers are usually part of the polymer network. Most of them are hydrophilic, and are bound near the surface. They often extend into the water phase (Fig. 12-9). These modifiers not only affect performance of the latex but also impart a degree of stability to the latex, similar to that provided by the adsorbed emulsifiers. Pressure sensitive adhesive coating operations are generally improved as a result of this added stability.

MOLECULAR STRUCTURE

Molecular structure, the combination of molecular weight and cross-linking, is controlled by polymerization temperature and initiator and chain transfer agent type. The degree of monomer conversion is also important. A low molecular weight polymer generally has low tensile strength, but high elongation (Fig. 12-10). An increase in elongation increases adhesive strength in an adhesive formulation as measured by peel adhesion (Fig. 12-11). A high molecular weight latex generally has high tensile strength, but low elongation (Fig. 12-10). An increase in tensile strength increases cohesive strength in an adhesive formulation as measured by shear bond (Fig. 12-12).

ADHESIVE FORMULATING

Pressure sensitive adhesive properties are dependent on the viscoelastic nature of the adhesive. The formulation of S/B latex and tackifier resin combines the elastic nature of the latex with the viscous nature of the tackifier. The low

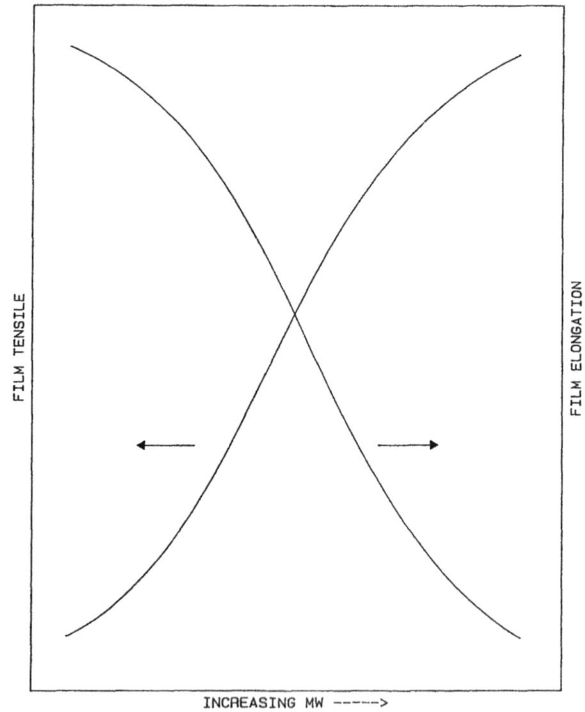

Fig. 12-10. The effect of molecular weight on film tensile strength and film elongation.

molecular weight, high T_g tackifier interacts with the high molecular weight, cross-linked, low T_g latex to form a system that exhibits the desired peel, tack, and shear properties.[4]

The choice of tackifier type and level depends on the balance of adhesive properties required and the effectiveness of the tackifier. Tackifiers are available as derivatives of natural products (rosins) and as synthetic hydrocarbon resins. Many of these are available commercially as dispersions, readily compatible with S/B latexes. Rosin-based tackifiers are the most commonly used tackifiers with S/B latexes. Table 12-4 lists some of the available rosin-based tackifier emulsions.

Tackifier concentration has a major effect on adhesive properties. As increased levels of tackifier are added, shear bond decreases (Fig. 12-13). However, peel adhesion and quick stick increase with increasing levels of tackifier (Fig. 12-14). At very high levels of tackifier, peel adhesion and quick stick usually begin to decrease. Table 12-5 lists typical test methods for measuring these properties.

Various theories on the physical process of tackification have been proposed. Some suggest a correlation between physical phase changes and performance, while others relate compatibility between polymer and tackifier to performance. In practice, the development of a formulation based on S/B latex requires routine and often extensive testing. Typically, when developing an adhesive formulation based on a particular S/B latex, different types and levels of tackifiers are screened (Table 12-6). A wide variety of properties can be designed by changing the ratio of tackifier to S/B latex, the tackifier type, or the tackifier softening point. The flexibility offered by blending can help reduce the number of adhesive raw materials inventoried.

Table 12-4. Rosin-Based Tackifier Dispersions

TRADEMARK AND GRADE	DESCRIPTION/ SOFTENING POINT	MANUFACTURER
Foral 85	Stabilized rosin ester (83°C)	Hercules
Staybelite Ester 10	Stabilized rosin ester (83°C)	Hercules
Snowtack 301CF	Gum rosin (66°C)	Tenneco
Snowtack SE80CF	Gum rosin (80°C)	Tenneco
Snowtack SE42CF	Gum rosin (42°C)	Tenneco
Snowtack SE25CF	Gum rosin (25°C)	Tenneco
Aquatac 5560	Tall oil rosin (60°C)	Sylvachem
Aquatac 5541	Tall oil rosin (41°C)	Sylvachem
Aquatac 5527	Tall oil rosin (27°C)	Sylvachem

Table 12-5. Adhesive Test Methods

Peel adhesion	PSTC-1, less than 30-sec dwell, stainless steel
Quick stick	PSTC-5, stainless steel
Shear bond	PSTC-7, ½ in. × ½ in. × 500 g, stainless steel
Williams plasticity	Two grams dry polymer, dried overnight at 23°C. Knead 3 min, condition 15 min, measure Williams plasticity after 10 min at 23°C under 5-kg load. Report as mm × 100.
Polyken[b] tack	Stainless-steel probe, 1-cm/sec rate, 1-sec dwell, 100-g/cm² force

Note: 1-mil adhesive on 2-mil Mylar[a] film (except where noted), equilibrated overnight, and tested at 23°C, 50% relative humidity.
[a] Trademark of E. I. DuPont deNemours & Co.
[b] Trademark of The Kendall Co.

Table 12-6. Pressure Sensitive Adhesive Formulated Properties[a] (Styrene/Butadiene Latex + Tackifier)

ADHESIVE FORMULATION (dry wt)		ADHESIVE PROPERTIES			
LATEX[b]	TACKIFIER	PEEL ADHESION (pli)	QUICK STICK (pli)	POLYKEN[f] TACK (%)	SHEAR BOND (hours)
100%		2.0	1.0	511	700
80%	20% Rosin Ester[c]	1.8	1.2	584	120
	20% Gum Rosin[d]	2.5	1.7	787	20
	20% Tall Oil[e]	1.7	1.1	643	120
60%	40% Rosin Ester[c]	3.0	2.0	1008	120
	40% Gum Rosin[d]	5.5(c)	2.9	1326	12
	40% Tall Oil[e]	3.1	2.1	1470	12
50%	50% Rosin Ester[c]	4.2	2.0	845	120
	50% Gum Rosin[d]	5.1(c)	3.1	1108	7
	50% Tall Oil[e]	5.5(c)	2.7	1727	9

[a] Adhesive dried 10 min at 115°C.
[b] Dow Latex PSA 578A, copolymer contains about 65% butadiene.
[c] 83°C softening point.
[d] 66°C softening point.
[e] 41°C softening point.
[f] Trademark of The Kendall Co.
(c) = cohesive failure.

POLYMER BLENDING

Styrene/butadiene latex is well suited to blending with other polymers, such as natural rubber latex. Natural rubber latex use in pressure sensitive adhesives has been limited, although natural rubber has been used in solvent adhesives

process, the aqueous adhesive is subjected to significant shear stress. The magnitude of this shear stress is dependent on the type of coating process and the coating speed. Viscosity, a measure of the adhesive's resistance to deformation under this stress, is determined through rheological measurements. It is measured as the slope of the tangent of the plot of shear stress versus shear rate. The non-Newtonian shear thinning or shear thickening behavior is also determined from these plots (Fig. 12-15).

For substrates such as silicone-coated release liners, an adhesive must have sufficient viscosity during coating to avoid crawl-back. Thickeners will aid the coating process by reducing the rate of crawl-back. However, streaks or ridges may be present where a thickener with poor leveling characteristics is used. For an adhesive to level, the product must have a low yield stress after all mechanical stress due to coating has been removed. Yield stress refers to the minimum stress necessary for a fluid to flow, and corresponds to the *y*-intercept in a plot of shear stress versus shear rate (Fig. 12-16). Coating these formulations onto silicone-coated release paper shows that the high viscosity, low yield stress combination provides the best wetting and leveling (Table 12-8).

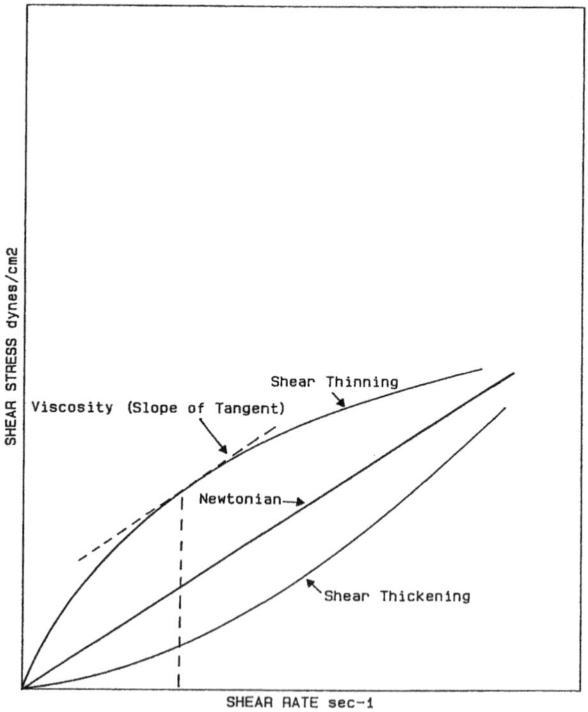

Fig. 12-15. Flow curves of time-independent fluids.

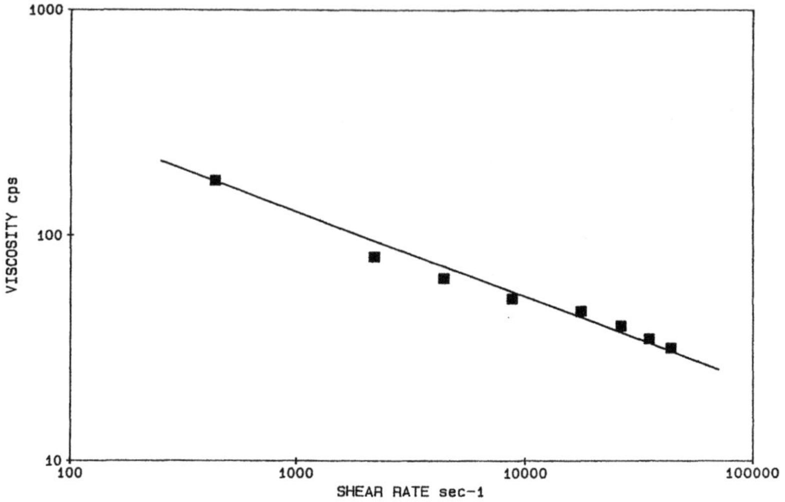

Fig. 12-17. High shear rheology (50% S/B latex, 50% rosin tackifier).

shear rate, the coating stability can be determined. If the adhesive exhibits good stability, a smooth profile of viscosity versus shear rate will be obtained. An unstable adhesive will show grit or coagulum formation, produce irregular plots, and ultimately cause the instrument to seize. Such an adhesive would be expected to coagulate during coating. Through these rheology studies, additives such as surfactants can be quickly screened for their effect on adhesive stability.

For the study of adhesive stability, and for wetting and leveling on specific substrates, pilot or production trials need to be run. The coating behavior is dependent on the adhesive, the coating method, the line speed, and the drying conditions.

ADHESIVE DRYING

When driving off water, skin formation on the surface of the coating should be avoided. Without sufficient channels through which the water can flow, the adhesive may blister. Two factors are important: the first is to initially heat the adhesive uniformly throughout, and the second is to keep from heating the adhesive surface too quickly. Through the use of infrared preheaters and zoned-forced-air ovens, S/B adhesive systems can be dried efficiently. Production systems have been designed to coat and dry emulsion adhesives at speeds of over 150 m/min.

Table 12-9. Dow Latex PSA 578A: Adhesion to Corrugated Board and Polyethylene[a]

PSA 578A	SNOWTACK[b] 301CF	SNOWTACK[c] SE25CF	QUICK STICK[d] C.B.	QUICK STICK[e] HDPE	SHEAR[f] BOND
50	50	—	1.2[g]	1.1[j]	34
	40	10	1.2[h]	1.6[k]	30
	30	20	1.2[i]	1.4	18
	20	30	1.2[i]	1.3	74
45	55	—	1.3[g]	0.6[j]	26
	50	5	1.4[g]	1.2[j]	30
	40	15	1.3[i]	1.5	19
	30	25	1.3[i]	1.7	16
40	60	—	1.3[g]	0.6[j]	18
	50	10	1.4[g]	1.5[j]	18
	40	20	1.4[h]	1.8	11
	30	30	1.4[h]	1.7	9

[a] Parts based on solids, 0.3 parts ASE-60 thickener added to all samples, transfer coated, approximately 1-mil dry adhesive.
[b] Tenneco Malros, Ltd. (63°C softening point).
[c] Tenneco Malros, Ltd. (25°C softening point).
[d] Quick stick on cardboard, labels pulled perpendicular to corrugations (pli).
[e] Quick stick on high-density polyethylene (pli).
[f] ½ in. × ½ in., 500 g (on stainless steel), min.
[g] Up to 80% fiber pickup.
[h] Up to 60% fiber pickup.
[i] Up to 40% fiber pickup.
[j] Spikes.
[k] Gaps.

Table 12-10. Dow Latex Blends in Freezer Grade Label Formulations

FORMULATION COMPONENTS (DRY BASIS)				ADHESIVE PROPERTIES			
)OW LATEX PSA 578A	NATURAL LATEX	HERCOLYN D TACKIFIER	PLASTI-CIZER	PEEL[a] ADHESION (pli)	FIBER[b] ADHESION (rating)	PE[b] ADHESION (rating)	R.T.[c] PEEL (pli)
20	20	60	—	tear	1.5	0	1.0
30	20	50	—	tear	2.0	0.5	1.1
40	20	40	—	2.0	1.5	0.5	1.0
20	20	60	2	tear	2.5	0.5	1.5
30	20	50	2	tear	2.5	0.5	1.7
40	20	40	2	tear	2.5	1.0	1.5
20	20	60	5	tear	3.0	1.5	1.2
30	20	50	5	4.0	3.0	1.0	1.3
40	20	40	5	tear	2.5	1.0	1.2

te: Samples transfer coated from silicone liner to facestock. Adhesive dried at 115°C for 3 min. Adhesive film is 1.0 mil y).
pplied at −20°F, immediately pulled using PSTC-1 method at 72°F.
pplied and tested at −20°F, 75% relative humidity; relative scale rating of 0–3, 3 is the best.
oom temperature peel using PSTC-1 method.

2. Blackley, P. C. *Emulsion Polymerization, Theory and Practice*, p. 155. Applied Science Publishers, London, 1975.
3. Ellseeva, V. I., Ivanchev, S. I., Kuchamov, and Levedev, A. V. *Emulsion Polymerization and Its Applications in Industry* (Trans. from the Russian by S. J. Teague), p. 153. Consultants Bureau, New York, 1981.
4. Satas, D. (ed.). *Handbook of Pressure-Sensitive Adhesive Technology*, pp. 364–369. Van Nostrand Reinhold, New York, 1982.

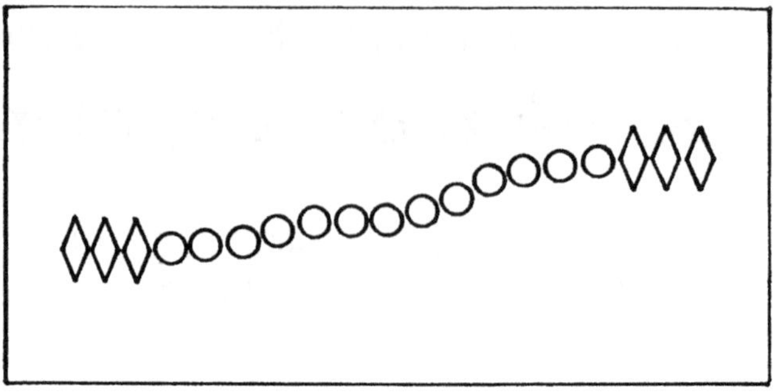

Fig. 13-1. Simplified representation of a thermoplastic rubber molecule.

rene-polybutadiene-polystyrene (S-B-S) polymers and the polystyrene-polyisoprene-polystyrene (S-I-S) polymers. This class of thermoplastic rubber was first marketed commercially in 1965 by the Shell Chemical Company. The second class of thermoplastic rubber consists of block polymers in which the elastomeric midblock is a saturated olefin rubber. The polymers in this class are the polystyrene-poly(ethylene/butylene)-polystyrene (S-EB-S) polymers and the polystyrene-poly(ethylene/propylene-polystyrene (S-EP-S) polymers. This class of thermoplastic rubber was first marketed commercially in 1972 by the Shell Chemical Company.

The usual method of preparation of thermoplastic rubbers is by anionic polymerization in solvent using an alkyl lithium catalyst such as butyl lithium. S-B-S– and S-I-S–type block polymers may be synthesized by two basic routes, sequential polymerization of all three blocks or sequential polymerization of two blocks, followed by coupling. These two processes are illustrated below, the sequential process in Equation (1) and the coupling process in Equation (2), for the preparation of an S-I-S–type polymer.

$$\text{BuLi} \xrightarrow{\text{Styrene}} \text{S-Li} \xrightarrow{\text{Isoprene}} \text{S-I-Li} \xrightarrow{\text{Styrene}} \text{S-I-S} \qquad (1)$$

$$\text{S-I-Li} \xrightarrow{\text{Coupling Agent}} \text{S-I-I-S} \qquad (2)$$

The sequential process results in what is called a linear S-I-S polymer. With the coupling process, use of a difunctional coupling agent gives a linear $(\text{S-I})_2$ polymer, while use of a multifunctional coupling agent gives a multiarmed $(\text{S-I})_n$ polymer. Usual coupling agents are esters or halogen-containing molecules. The literature reports catalytic hydrogenation of S-B-S and S-I-S polymers to produce saturated analogues of the S-EB-S and S-EP-S types, respectively.[1] A

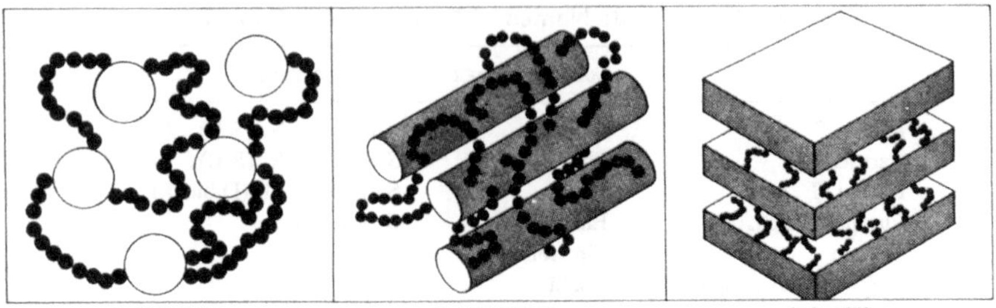

Fig. 13-3. Polystyrene domain configurations or morphologies.

Domains, the plastic endblock phase, are shown in spherical form Figure 13-2. These domains act as cross-links between the ends of many rubber chains, thereby locking the rubber chains and their inherent entanglements in place. This allows the thermoplastic rubber to behave like a conventionally vulcanized rubber that contains dispersed reactive filler particles. However, the thermoplastic rubber is physically cross-linked by the plastic endblock domains rather than being chemically cross-linked like a conventionally vulcanized rubber. These physical cross-link sites (domains) can be reversibly unlocked and reformed by various means, that is, solvation followed by solvent evaporation or through sufficient heating and shearing, then cooling.

The behavior of a thermoplastic rubber is highly dependent on the morphology or geometry of the endblock domains as they are dispersed in the rubber matrix. In commercial thermoplastic rubbers useful for pressure sensitive adhesives, the endblock phase is present in the smaller proportion and is dispersed in a continuous rubber matrix, as suggested by Figure 13-3a, b or c. The uniform dispersion of spherical endblock domains shown in this figure (a, spheroids) is approached only in carefully prepared laboratory samples with low endblock phase concentration. Depending on the endblock phase concentration and on actual processing conditions used to prepare a given sample, the geometry or morphology of the dispersed phase may be spherelike (a), rodlike (b), or lamellar (c) as depicted in Fig. 13-3. The existence of these different morphologies has been confirmed by electron micrographs.[2,3]

In the latter two cases, the endblock phase may extend as a continuous plastic, rodlike or lamellar, network throughout the rubber matrix. This tends to be the case when the endblock phase concentration is above about 20%w for neat (uncompounded) S-B-S thermoplastic rubbers. In this situation, as the sample is stretched, the initial stress is borne by the plastic network, and the stress-strain properties are greatly affected as discussed below.

Under all conditions the dimensions of the dispersed phase are restricted. Since the junctures between the endblocks and midblocks are located at the phase boundary, the domain thickness is limited to the distance that can be

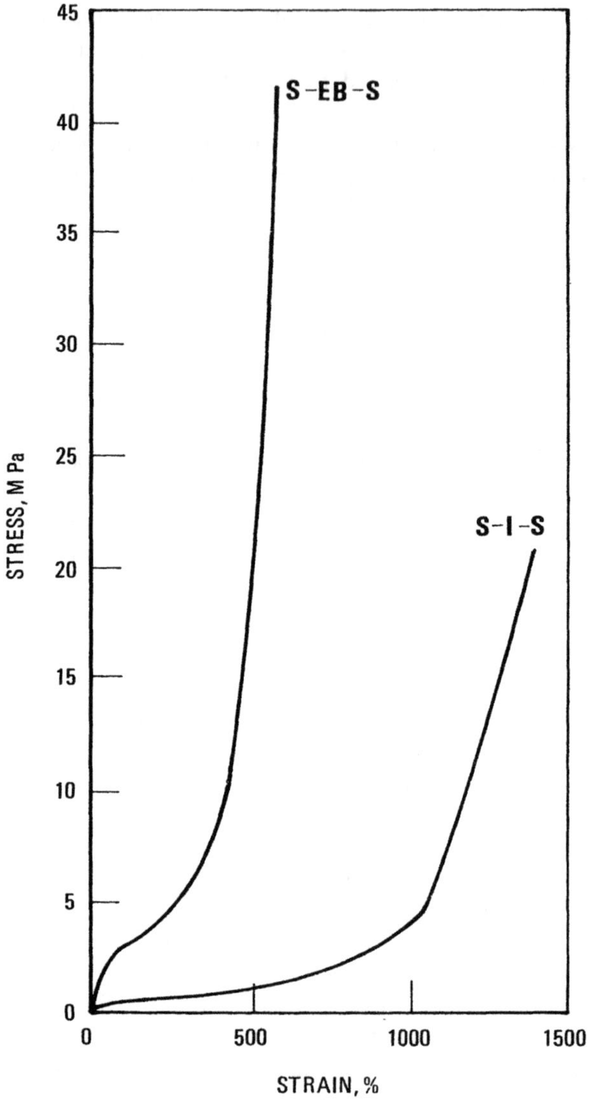

Fig. 13-4. Typical stress-strain curves for thermoplastic rubbers.

and the rubber midblock, solution viscosities are relatively low. The solution viscosity of a 25%w solution of most thermoplastic rubbers in toluene will be from about 0.2 to 20 Pa·s at 25°C.

Hot melt viscosities of neat thermoplastic rubbers can be extremely high at low shear rates. Typically, the hot melt viscosities are 10^4–10^5 Pa·s at about

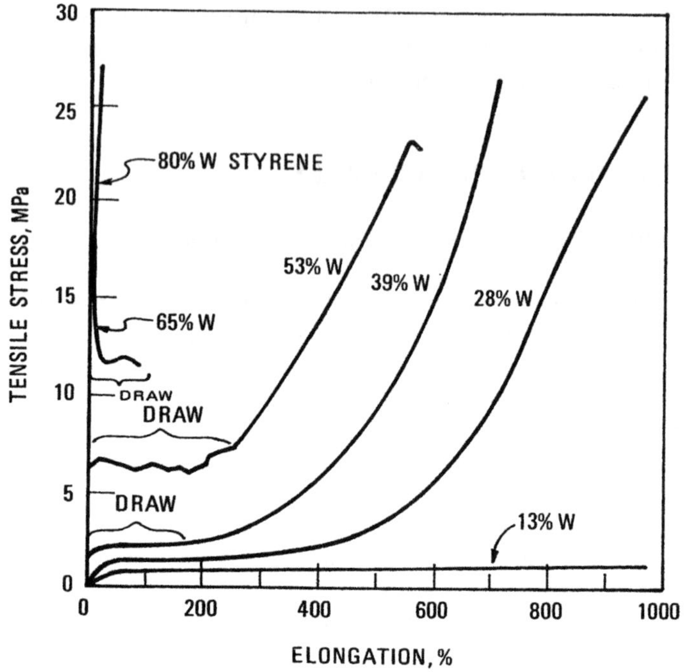

Fig. 13-5. Effects of polystyrene endblock concentration on linear S-B-S polymer tensile properties.

At polystyrene endblock concentrations of 20–30%w, the stress-strain curve resembles that of a vulcanized rubber. At concentrations above about 33%w, the phenomenon of "drawing" commonly exhibited by thermoplastics occurs. This happens when a continuous rodlike or lamellar endblock network exists as suggested in Figure 13-3. When such a sample is stretched, an initial yield stress is observed. Then, as the relatively weak plastic structure is disrupted by further elongation, drawing occurs. When the stress is released, the plastic network will gradually reform. Higher temperatures will speed reforming of the plastic network. At even higher endblock concentrations, the plastic phase is continuous and the midblock phase dispersed to give a system resembling high impact polystyrene.

Effects of Additives. The ratio of endblock phase to midblock phase can be varied by adding materials that associate preferentially with one phase or the other. Coumarone-indene resins, for example, associate with the endblock phase in S-B-S and S-I-S polymers. Figure 13-6 shows how the initial portion of the stress-strain curve of the neat polymer (curve B) is shifted (curve A) when the endblock phase concentration is increased by adding an endblock

from different solvents. Curve A is for a film cast from a 60–40%v blend of *n*-hexane/MEK. In this case, the last component to evaporate is MEK, which dissolves the polystyrene phase only. Thus, the rubber midblock phase precipitates first and the polystyrene phase last, so that a continuous polystyrene phase tends to be formed rather than discrete polystyrene domains. The curve shows an initial yield, followed by cold drawing, and a final rise in stress to break point.

Curve B shows the stress-strain behavior for the film cast from solution in toluene, a relatively good solvent for both phases. This curve represents a structure close to an equilibrium state because the solvent is a good solvent for both phases.

Curve C is for a film cast from a 16–20–64%v blend of ethylbenzene/*n*-butyl acetate/Shell Sol 340EC. In this case, the last solvent to evaporate is Shell Sol 340EC, a very poor solvent for polystyrene, and the polystyrene is precipitated first, forming more discrete domains and fewer styrene continuous regions.

Properties intermediate to those shown by the three curves in Figure 13-7 can be achieved by proper choice of solvents and solvent blends. Annealing the films represented by curves A and C will tend to shift both curves toward the intermediate equilibrium position of curve B. The rate at which the films anneal is temperature dependent. Annealing effects may be encountered in processes where the film passes through a high temperature drier.

Choice of Compounding Ingredients

Resins. Resin selection is probably the most important aspect in the development of pressure sensitive adhesive formulations based on thermoplastic rubbers. When considering the effects of resins on adhesive properties, the distinction must be made between the effects of midblock-compatible resins and the effects of endblock-compatible resins, since their effect on tack is widely different.

Rubber Phase Associating Resins. Aliphatic olefin-derived resins, rosin esters, polyterpenes, and terpene phenolic resins derived from petroleum or terpentine sources and having relatively low solubility parameters tend to associate with the rubber midblocks and not with the polystyrene endblocks. As the molecular weights or softening points of these resins are lowered, their solubility in the endblocks increases. A partial list of commercial resins in this category is presented in Table 13-2.

The primary function of a midblock-compatible resin is to develop tack in the thermoplastic rubber. This type of resin can also serve to improve the

tures (T_g) greater than those of the rubber midblock phase will increase the glass transition temperature of the rubber-resin blend. This will result in reduction of low-temperature tack and flexibility. Partial compensation for this reduction can be obtained through selection of midblock plasticizers or midblock compatible resins that have very low T_g. However, using low molecular weight plasticizer may also have an effect on upper service temperature and adhesion to certain substrates because of some unavoidable plasticizing of the endblocks. The importance and application of altering the T_g of thermoplastic rubber compositions is discussed more fully under low-temperature labels and tapes in the latter part of this chapter.

Endblock Phase Associating Resins. Polyaromatics, coumarone-indene resins, and other high-solubility parameter resins derived from coal tar or petroleum and having softening points above about 85°C tend to associate with the polystyrene endblocks and not with the rubber midblocks. As the molecular weights or softening points of these resins are lowered, their solubility in the midblocks increases. A partial list of typical and commercial endblock-associating resins is presented in Table 13-3.

The addition of nonplasticizing endblock-associating resins tends to stiffen the compound by increasing the modulus and increases the tendency to draw as discussed previously. This occurs as the fraction of endblock phase increases and its morphology becomes more rodlike or lamellar. In pressure sensitive adhesives, only a limited amount of high softening point endblock-compatible resin can be included in the formulation before the stiffness increases and the aggressive tack is reduced to unacceptable levels. Somewhat higher levels of

Table 13-3. Endblock Phase Associating Resins for Thermoplastic Rubbers

TRADE NAME	SOFTENING POINT (°C)	CHEMICAL TYPE	SUPPLIER
Amoco 18 Series	100, 115, 145	Alphamethyl styrene	Amoco
Kristalex Series	85-140	Alphamethyl styrene	Hercules
Piccotex Series	75-120	Alphamethyl styrene/ vinyl toluene	Hercules
Nevchem Series	100-140	Aromatic hydrocarbons	Neville
Picco 6000 Series	70-140	Aromatic hydrocarbons	Hercules
Nevindene LX Series	100-150	Heat reactive hydrocarbon	Neville
Cumar Series	10-130	Coumarone indene	Neville
Cumar LX-509	155	Coumarone indene	Neville
Piccovar AP Series	Liquid[a]	Alkylaryl resin	Hercules
Piccovar 130	130	Alkyl aromatic polyindene	Hercules

[a] Because of the low molecular weight, these resins are soluble in both endblock and midblock phases of thermoplastic rubbers.

Table 13-4. Properties of Plasticizing Oils

TRADE NAME	SUPPLIER	SOLUBILITY PARAMETER, δ^a HILDEBRANDS	AVERAGE MOLE WEIGHT[b]	SPECIFIC GRAVITY (15.6°C)	VOLATILITY LOSS WEIGHT[c]	OIL ABSORBED[d] (phr)
Polypropene D-60	Amoco	6.55	800	0.86	0.1	25
Polybutene-18	Chevron	6.95	600	0.88	0.1	39
Tufflo 6206	Arco	7.06	660	0.88	0.05	31
Polybutene-12	Chevron	(7.04)	530	0.88	—	(53)
Tufflo 6056	Arco	7.18	550	0.87	0.3	56
Polybutene-8	Chevron	(7.18)	440	0.86	—	(75)
Kaydol	Witco	7.34	480	0.89	—	82
Tufflo 6026	Arco	7.29	410	0.86	1.0	76
Polybutene-6	Chevron	7.34	315	0.84	10.0	103
Tufflo 6016	Arco	7.51	390	0.85	2.0	106
Tufflo 6204	Arco	7.60	440	0.92	0.5	96
Shellflex 371	Shell	7.60	410	0.90	0.9	112
Tufflo 6094	Arco	7.60	410	0.92	0.8	95
Tufflo 6054	Arco	(7.66)	380	0.92	1.3	(127)
Tufflo 6014	Arco	7.73	320	0.89	12.0	214

[a] Calculated from experimentally determined surface tension, average mole weight, and specific gravity. Values in parentheses were interpolated.

[b] By ebullioscopic methods of Mechrolah osmometer.

[c] Twenty-two hours and 107°C.

[d] Grams of oil absorbed by 500 to 1000 micron films of Kraton D1101 rubber at room temperature per 100 grams original weight after soaking 100 hours. Films were prepared by casting from toluene solution on mercury and drying very slowly. Values in parentheses were interpolated or extrapolated from related data.

Fillers. Nonreinforcing fillers such as clay, talc, and whiting, can be used in pressure sensitive adhesives to pigment the adhesive and to lower raw material costs. The amount of filler that can be used is limited because fillers increase the stiffness and reduce cohesive strength and aggressive tack, although tack reduction may be useful in certain applications. Carbon black, which functions as a pigment and screening ultraviolet stabilizer, forms permanent gel structures with the unsaturated midblocks. This reduces solubility and thermoplasticity.

Stabilizers. Unsaturated thermoplastic rubbers are susceptible to attack by oxygen, ozone, and ultraviolet radiation, especially when stressed. Therefore, stabilizers should be incorporated in the formulation to provide resistance to attack during processing and to protect the finished adhesive during its service life.

Although the stability of the saturated thermoplastic polymers is substantially better, it is a recommended practice to include stabilizers in formulated products containing these more stable polymers.

In choosing a stabilizer package, it should be noted that the rubber midblock is more susceptible to attack than the polystyrene domains. It is therefore more advantageous to use stabilizers that associate primarily with the rubber midblock and protect double bonds along the midblock chain.

The two types of rubber midblock, polybutadiene and polyisoprene, in the unsaturated thermoplastic rubbers behave differently when attacked by oxygen, ozone, or ultraviolet radiation. S-B-S polymers tend to cross-link, with the films becoming hard and brittle. S-I-S polymers, on the other hand, tend to undergo chain scission, whereby the films become softer and tackier and loose cohesive strength.

Protection Against Oxygen Attack. Oxidative attack occurs in two ways. One is under normal end use conditions; the other occurs at high temperature under shear. The latter instance would be typical during the mixing of hot melts, especially if the operation takes place in air over an extended period.

A list of some of the available antioxidants found to be effective in thermoplastic rubber formulations is given in Table 13-5. Their relative physical effectiveness will vary with aging conditions (temperature, amount of contact with air, etc.) and with the criteria used to judge the amount of degradation. Combinations of antioxidants may be more effective than one alone. It is essential to minimize the amount of contact with air by using an inert gas blanket when preparing hot melt adhesives.

Table 13-6. Antiozonants for Thermoplastic Rubbers

TRADE NAME	CHEMICAL TYPE	SUPPLIER
NBC	Nickel dibutyldithiocarbamate	duPont
Pennzone B	Dibutylchiourea	Pennwalt
Ozone Protector 80	Undisclosed	Reichhold

Protection Against Ozone Attack. As mentioned previously, unsaturated thermoplastic rubbers are susceptible to degradation by ozone, particularly when under stress. Degradation is evidenced by surface crazing and hardening.

Table 13-6 lists some antiozonants for use in thermoplastic rubber formulations. Antiozonants such as nickel dibutyldithiocarbamate (NBC) and Pennzone B improve the resistance of thermoplastic rubbers to ozone, but also produce discoloration and staining. Improvement is also provided by Ozone Protector 80, a nonstaining antiozonant, but considerably higher loadings are required than with NBC or Pennzone B. It should be noted that Pennzone B is not suitable for hot melt formulations, as it accelerates the cross-linking of the rubber segments.

Protection Against Ultraviolet Radiation. Degradation by exposure to ultraviolet light is denoted by discoloration and embrittlement or strength loss of thermoplastic rubber compounds. In the vast majority of indoor applications, this type of degradation is not a major consideration. If, however, direct exposure to sunlight is expected, unsaturated thermoplastic rubbers will be susceptible

Table 13-7. Ultraviolet Inhibitors for Thermoplastic Rubbers

TRADE NAME	CHEMICAL TYPE	SUPPLIER
Eastman OPS	Octylphenyl salicylate	Eastman
Eastman RMB	Resorcinol monobenzoate	Eastman
Permasord MA	2-hydroxy-4-(2-hydroxy-3-methacrylyloxy) propiobenzo-phenone	National Starch
Tinuvin P	Substituted benzotriazole	Ciba-Geigy
Tinuvin 326 Tinuvin 327	Substituted hydroxyphenyl benzotriazole	Ciba-Geigy
Tinuvin 770	Hindered amine	Ciba-Geigy
Antioxidant 330	1,3,5-trimethyl-2,4,6-tris(3,5-di-tert-butyl-4-hydroxy benzyl)benzene	Ethyl Corp.
Irganox 1076	Octadecyl 3-(3′,5′-*di-tert*-butyl-4′-hydroxy-phenyl) propionate	Ciba-Geigy
Irganox 1010	Tetrakis[methylene 3-(3′,5′-*di-tert*-butyl-4′-hydroxyphenyl)propionate]methane	Ciba-Geigy

in this chapter and specific examples of tape and label formulations discussed at its conclusion.

Basic Formulation Development Technique

A final formulation developed for a specific end use will contain some combination of the various ingredients just discussed. Since adhesive properties depend not only on the types of ingredients but also on their concentration and their interaction with each other, finding the proper combination for a particular adhesive can be a large task. In this section, the use of adhesive property contour diagrams is suggested as a rapid and reliable approach to formulation development.

An adhesive property contour diagram can be thought of as a topographic map. The *X* and *Y* axes of the map are the concentrations of two of the compounding ingredients. The lines on this rectangular concentration grid are contours at which the particular adhesive property has a constant value. Thus, at any point on the grid representing a particular combination of the two ingredients, the value of the adhesive property is immediately known.

The construction and use of adhesive property contour diagrams can be illustrated by the following practical example on the development of a formulation for a hot melt pressure sensitive adhesive for a general-purpose masking tape.

To begin, the adhesive formulator must select the critical property requirements that must be met. In this example, the adhesive must have a low value for rolling ball tack, moderate holding power to kraft paper, and low melt viscosity. Using the formulating principles presented, the formulator must then select the two ingredients most likely to affect these critical properties. In this example, the properties will be dominated by the concentration of midblock resin and midblock plasticizer. Adhesives are then prepared at compositions corresponding to selected points on the concentration grid covering the range of interest. In this example, sixteen adhesives were initially prepared containing Kraton® D1107 rubber, midblock resin (Wingtack 95) at 75, 100, 150, and 200 parts per hundred rubber (phr), each containing 0, 25, 50, and 100 phr of midblock plasticizer (Shellflex® 371 oil). The performance of these adhesives in the critical property tests is then determined and the results are written at the appropriate point on the concentration grid. From these data, contour lines are drawn through points at which it is estimated that the adhesive property will have a constant value. If necessary, regions of special interest may be further refined by preparing and testing more samples in the area. Computer contouring is also possible with some of the software programs now on the market.

There are many advantages of developing adhesive formulations using contour diagrams. One major advantage is that they clearly show how properties

particular ingredients. If the requirements of the general-purpose masking tape adhesive used in this example were that rolling ball tack must be one centimeter or less, holding power must be at least 200 minutes and melt viscosity must be less than 40 Pa·s, superposition of the contour diagram shows that the only practical formulation will contain about 125 phr of midblock resin and 25 phr of midblock plasticizer. The contour diagrams also show how sensitive the properties are to changes in concentration of the ingredients, and thus how closely ingredient concentrations must be controlled during manufacturing to assure satisfactory and consistent product performance.

Contour diagrams can be constructed using data obtained on samples cast from solutions, even if the intended use of the adhesive is for hot melts. Adhesive properties will depend upon the characteristics of the solvent from which the adhesive was cast. However, if the solvent used in preparing the adhesive is a good solvent for all ingredients in the formulation, the measured properties will be similar to properties of the same adhesive prepared as a hot melt, assuming the hot melt has not been degraded. Reagent-grade toluene is usually a satisfactory solvent. There are two advantages of doing the formulation development in solvent rather than in hot melt. First, the personnel required to mix the adhesive and prepare test samples is substantially less and thus the time required to develop contour diagrams will be less. Second, heat history or degradation of the adhesive during processing, a very important variable that can significantly affect adhesive properties, is eliminated. Data from the contour diagrams generated from the solution cast adhesives will indicate the most appropriate combination of such ingredients for use in the hot melt adhesive. However, if the hot melt adhesive does not have exactly the required properties, the contour diagrams also show trends that indicate how to adjust the concentration of the ingredients of the hot melt adhesive to reach the target performance properties.

APPLICATION OF FORMULATING PRINCIPLES

Understanding of the thermoplastic rubber molecular structure, neat polymer properties, and general formulating principles forms the basis for the logical next step: To make useful and competitive pressure sensitive adhesive finished products. What added considerations must now be taken into account to reach this goal? Three general areas for consideration are identified:

1. Pressure sensitive adhesive product performance
2. Adhesive manufacture
3. Adhesive product manufacture (finished tape or label)

boundary range established for acceptable performance based on (a) end use performance need, or (b) a property that characterizes the adhesive polymer type or its variability rather than finished product performance?

4. What other real performance characteristics of the established product or adhesive polymer are not defined or measured in the specifications or its tests, yet are known to be important in this product?
5. What new test method, criteria, and guidelines should be established for a thermoplastic rubber adhesive that are relevant to the end use performance, and will allow assessment of its performance on a sound basis? Is there a correlation with performance?
6. What structural differences between the current adhesive polymer and the block copolymer candidate may significantly alter results via the test method employed? Where can the structural differences in block copolymers be used to advantage?
7. What technical data need to be developed in order to verify performance via new test methods that reflect actual performance?

With any new or alternative adhesive system product, answers to these questions form a sound technical basis for judging the finished product, assuring quality in manufacture, and developing product information that allows its introduction and successful use in the marketplace.

Adhesive Manufacture

Adhesive formulations are most frequently prepared from crumb or pelletized thermoplastic rubber by solvent solution or hot melt mixing techniques. Commercial thermoplastic rubbers are not offered as emulsions.

Thermoplastic rubbers have two distinct characteristics that differentiate them as a group from conventional rubbers, and are important in considering adhesive manufacturing via solvent or hot melt mixing techniques.

Styrene endblock-type thermoplastic rubbers produce low solution viscosities in a good polystyrene solvent, for example, toluene. Thermoplastic rubbers without premastication have lower solution viscosities at comparable concentrations than do elastomers such as SBR, natural rubber, and polychloroprene after milling. This is because thermoplastic rubbers have relatively low, precisely controlled molecular weights. Molecular weights are allowed to be relatively low because the ABA structure and polystyrene endblock domain network afford high cohesive strength at these comparatively lower molecular weights. Unvulcanized conventional rubbers rely primarily on chain entanglements for strength, thus molecular weights must be high, resulting in higher solution viscosities.

The following subsections serve as a guide to preparing solvent solution adhesives and hot melt adhesives with thermoplastic rubbers in greater detail.

Solvent Behavior of Thermoplastic Rubbers. The behavior of thermoplastic rubber toward solvents is unique because it is comprised of two different polymeric chains present in every molecule. Each midblock and endblock portion of the molecule has a characteristic solubility parameter.* Thus, the ability of the neat thermoplastic rubber molecule to be solvated is a strong function of the solvent or solvent blend chosen. As such, solution viscosities can be varied over a wide range as a function of solvent choice or blend at a given solids level concentration. This characteristic can be used advantageously in developing formulations where adhesive solids level and viscosity must be closely balanced, or optimized, for given solvent coating equipment, as well as a means to alter drying rate characteristics.

In addition, a judicious choice of resins and resin concentrations can lead to enhanced solution viscosities, and potentially higher solids systems with less costly or more environmentally acceptable solvent blends.

For A-B-A the thermoplastic rubbers, two solubility parameters are involved, one for the polystyrene endblock and one for the rubber midblock. A "good" solvent for a thermoplastic rubber must therefore be one that dissolves both endblock domains and midblock segments.

To illustrate, Figure 13-10a indicates the approximate range of solubility parameter of solvents for rubber midblocks. The boundaries are not sharp near the extremes because the polymer molecules tend to collapse while still solvated. Figure 13-10b indicates, in a similar way, the range of solubility parameter of solvents for polystyrene endblocks. Figure 13-10c shows what happens when both blocks are combined in a single block copolymer molecule. A central range of solvents readily dissolves both blocks and gives low viscosity solutions. With solvents at the lower solubility parameter end, the polystyrene endblocks tend to remain associated and form what might be termed solvated endblock domains. In this region, solution viscosity increases rapidly with decreasing

* Hildebrand's solubility parameter (represented by δ) is a measure of the total forces holding the molecules of a solid or a liquid together. It has the units of $(cal/cm^3)^{1/2}$. Every compound is characterized by a specific value of solubility parameter, although this value may not always be recorded in convenient literature references. Materials having the same solubility parameter tend to form homogeneous mixtures or to be miscible. Those with different solubility parameters tend to form separate layers or to be mutually insoluble. Other indexes of solvent compatibility (Kauri-Butanol value, Aniline Point, Dilution Ratio, etc.) can often be correlated with solubility parameter. For example, Kauri-Butanol (K-B) values in the range of 20 to 100 for hydrocarbon solvents can be used to predict thermoplastic rubber solubilities. A solvent can be judged good or poor by comparison of its K-B value with those of known good or poor solvents. (Discussions of solubility parameter concepts are presented in references 4 and 5. K-B values and solubility parameters for a range of commercial solvents are listed in reference 6.

solubility parameter until rigid cross-linked-like gels are formed. With solvents at the high solubility parameter end, the situation is somewhat different. First, the midblocks become less and less soluble and tend to associate together, while the endblocks are highly solvated. Opalescent solutions with time-dependent viscosities are formed. Thermoplastic rubber crumb can be dispersed in this sort of borderline solvent (for example, methyl ethyl ketone).

An example of these principles is shown in Figure 13-11; these curves represent the viscosities of 15%w solutions of S-B-S thermoplastic rubbers in *n*-hexane/toluene solvent blends over a zero to 100% toluene concentration range. When an aliphatic solvent, such as *n*-hexane, or other pure solvents with solubility parameters below 7.4 are chosen, the S-B-S solution viscosity is manyfold higher than that shown for a pure toluene solution of the same polymer. This is because *n*-hexane is a nonsolvent for the polystyrene endblock, while toluene dissolves both the midblock and endblock segments of the thermoplastic rubber molecule.

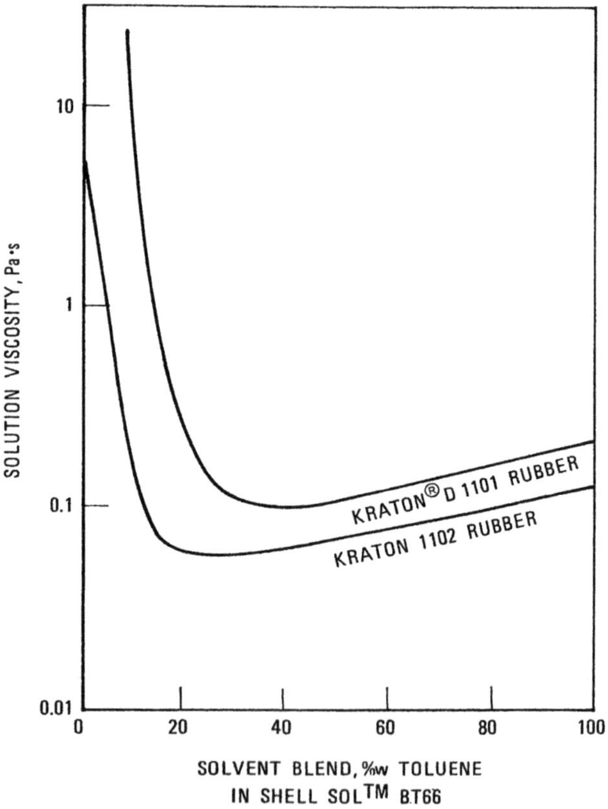

Fig. 13-11. Effect of toluene concentration on the solution viscosity of 15%W S-B-S thermoplastic rubbers in aliphatic hydrocarbon/toluene blends at 23°C.

solvent blends of *n*-hexane and toluene for 15%w solids S-B-S solution (plot similar to Fig. 13-11). The net effect is that the addition of this resin has shifted the minimum solution viscosity position of the rubber-resin blend to a point where the concentration of a good endblock solvent, toluene, is lowered when compared to the neat polymer thermoplastic rubber solution minimum viscosity point. This suggests that the solution viscosity as a function of solvent blend can be perturbed by the addition of resins that can act to alter the net solubility parameter of the solvent system, and shift the minimum viscosity position. In this case, the resin is acting more like toluene solvent in the rubber-resin solution, allowing a lower concentration of this good endblock solvent to achieve a minimum viscosity.

Contours of such interactions could be plotted for various resin types, resin concentrations, solids levels, polymer types in order to map and optimize solution viscosities and solvent blends.

Table 13-8A shows results for a series of solution viscosity measurements made on S-I-S linear thermoplastic rubbers when blended with midblock and endblock resins at a *constant solids level* (60%) in toluene solvent. Table 13-8B shows the formulations of these blends.

Comparing examples 1 and 2 or 6 and 7, the expected change in viscosity is noted when the polymer content of the 60% total solids polymer-resin blend is reduced by substitution with resin. In these cases, the viscosity at 60% solids decreases with this substitution of polymer by resin.

As endblock-associating resin is substituted for midblock resin (examples 3 and 4; 8 and 9 of Table 13-8B), increase in solution viscosities is noted. In

Table 13-8A. Solution Viscosity vs. Resin Level and Type: Constant 60%w in Toluene

SAMPLE NUMBER[a]	RESIN BLEND	RATIO OF S-I-S[bc] TO RESIN	SOLUTION VISCOSITY[d] (Pa·s)
1[b]	100% midblock	1 to 1	21.8
2	100% midblock	5 to 7	11.4
3	6 to 1 (midblock:endblock)	5 to 7	14.6
4	5 to 2 (midblock:endblock)	5 to 7	20.0
5	5 to 2 (midblock:plasticizing endblock)	5 to 7	9.2
6[c]	100% midblock	1 to 1	110.0
7	100% midblock	5 to 7	33.4
8	6 to 1 (midblock:endblock)	5 to 7	47.5
9	5 to 2 (midblock:endblock)	5 to 7	88.4
10	5 to 2 (midblock:plasticizing endblock)	5 to 7	10.9

[a] See Table 13-8B for detailed formulation of S-I-S rubber-resin blends.
[b] Samples 1 to 5 based on Kraton D1107 rubber.
[c] Samples 6 to 10 based on Kraton D1111 rubber.
[d] Measured at 25.0°C.

this case the endblock resin substitution is increasing the integrity and extent of the endblock domain structure and concentration in solution, and the resultant effect of increasing viscosity is expected. The trend is internally consistent for the two linear S-I-S polymers used in generating these data, Kraton D1111 rubber, a 22% styrene polymer, and Kraton D1107 rubber, which is a 15% styrene polymer. As might be expected, however, the higher styrene content polymer shows higher solution viscosities than the lower styrene content polymer as endblock resin is added and solids level is held constant. Finally, the choice of a low softening, plasticizing endblock resin (examples 5 and 10 of Table 13-8B) demonstrates its utility to act as a "solvent," and produce the lowest solution viscosities for the examples shown.

A method by which pressure sensitive formulations can be compared is in examining the interaction of solvent blends and weight percent adhesive solids possible at a constant solution viscosity, where data much like those generated in Table 13-8A are used to develop the curves in Figure 13-13. Higher solids content, that is, less solvent, and the use of less costly and more available solvents in solvent blends, is the driving force for such analyses. While formulation variables are not controlled in the series of curves shown in Figure 13-13, some general comments can be made on the method and its

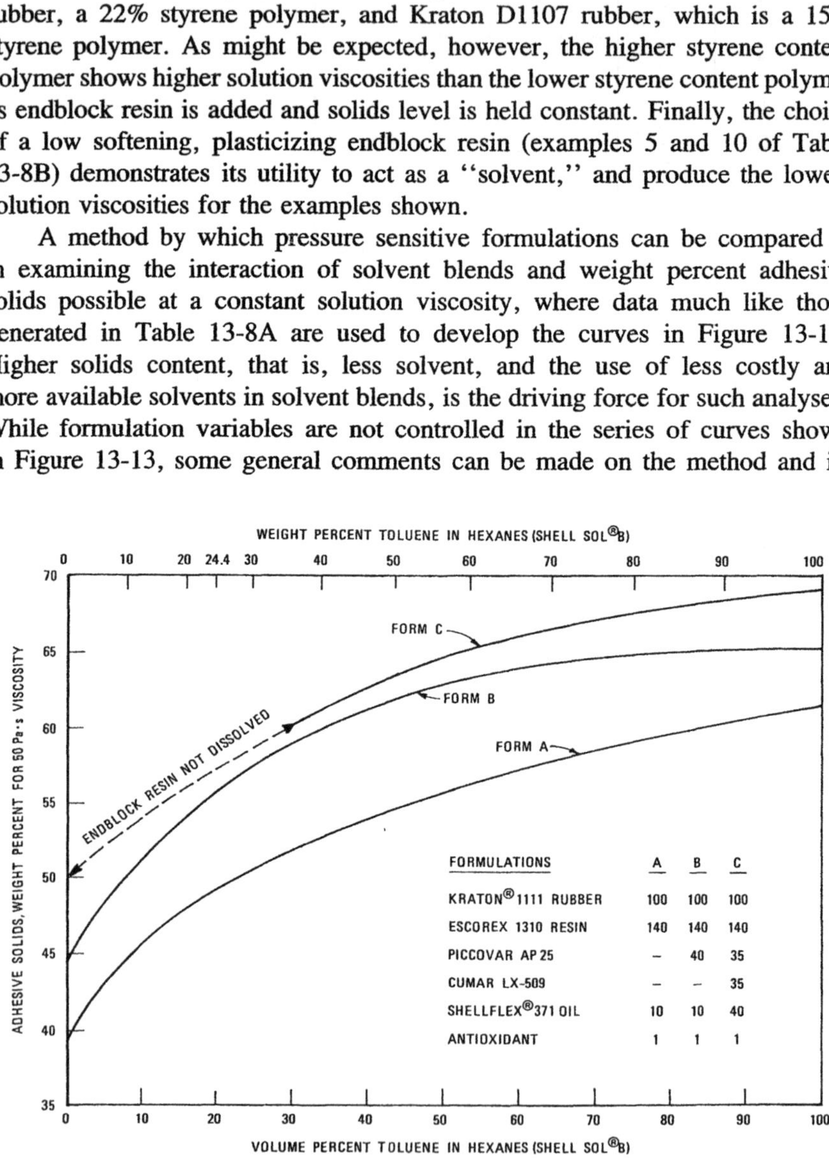

Fig. 13-13. Effect of toluene concentration on pressure sensitive adhesive solids and composition in solvent blends to produce constant 50 Pa·s solution viscosities.

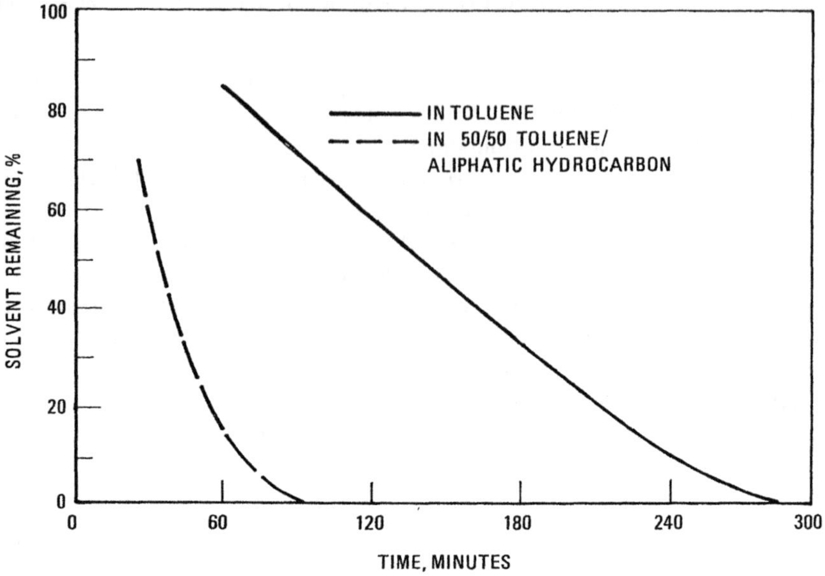

Fig. 13-14. Solvent evaporation rates for 10%W solutions of an S-I-S thermoplastic rubber (Kraton D1107 rubber) in toluene or aliphatic hydrocarbon (Shell Sol BT66) solvent.

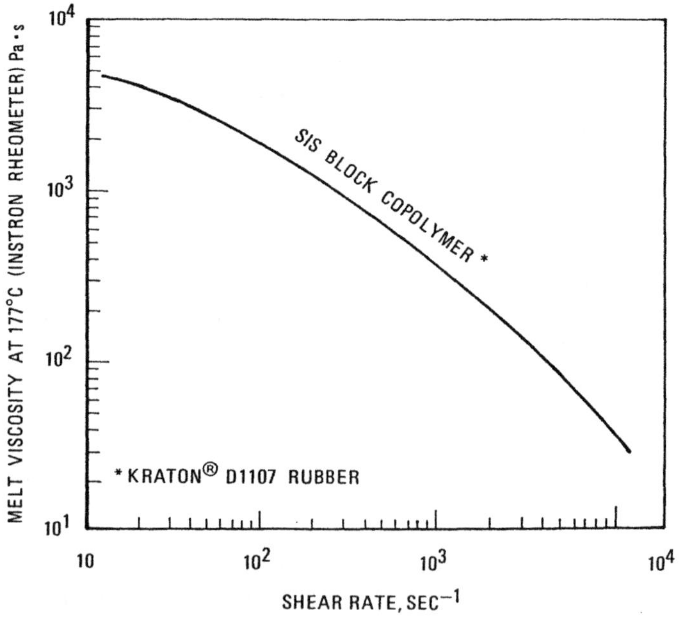

Fig. 13-15. Melt viscosity vs. shear rate for a neat S-I-S polymer.

High Shear Melt Mixing Precautions. Thermoplastic rubbers, particularly those with the highest melt viscosities, may build up excessive temperatures in high shear internal mixers (e.g., Banbury mixers) if exposed to the high shear conditions in the absence of plasticizers and stabilizers. Certain precautions should be noted. While mass temperatures up to 180°C are normal in the mixing of thermoplastic rubber formulations, higher temperatures may indicate too high a melt viscosity or too vigorous a mixing action. If the temperature were allowed to rise uncontrolled in such a situation, vaporization of plasticizer light ends and polymer decomposition could occur, creating a potential fire hazard. Excessive temperatures can be minimized during the mixing operation by adding a resin or plasticizer or by reducing mixer speed. It is often desirable to investigate the mixing characteristics of a new composition on a small scale before undertaking large-scale runs.

Saturated thermoplastic rubbers, because they have an olefin midblock, are substantially more resistant to degradation. They can be held at high temperature for long periods of time with negligible oxidative degradation. In the absence of oxygen they begin rapid thermal degradation at temperatures above about 275°C; 280°C is considered the maximum temperature for processing the saturated midblock thermoplastic rubber polymers.

Resins used in hot melt mixing can be premelted and in effect, become solvents for the polymer. A resin that associates with the polystyrene endblock will therefore act as a solvent for this phase during hot melt mixing and enhance polymer dissolution. As mixing progresses, less mechanical energy is required to cause flow.

Successful mixing of hot melt pressure sensitive adhesive based on unsaturated midblock thermoplastic rubbers thus depends on minimizing oxidative degradation by balancing the following variables in all types of mixing equipment and mixing processes:

1. Assure adequate and uniform heating and shearing to soften and disperse polymers, resins, oils, and other ingredients to a homogeneous pressure sensitive adhesive.
2. Minimize heat history in the process, that is, time and temperature reduced to a minimum in accomplishing dispersion.
3. Minimize exposure of the hot melt to air (oxygen) by using an inert gas blanket.

Manifestations of poor control and balance include gel and viscosity increase with S-B-S thermoplastic rubbers that tend to cross-link, and loss of network strength, that is, lower viscosity, poorer holding power, and lower service temperature in finished adhesives through chain scission with S-I-S thermoplastic rubbers.

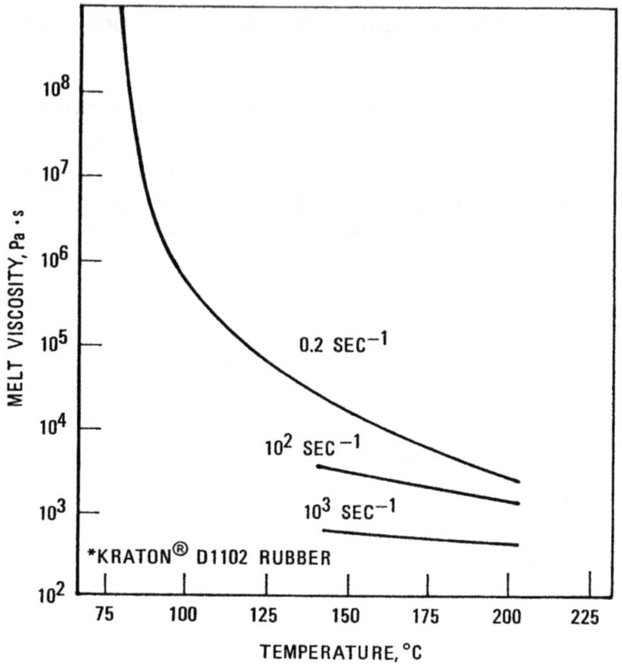

Fig. 13-17. Temperature vs. melt viscosity at constant shear rates for a neat S-B-S polymer.

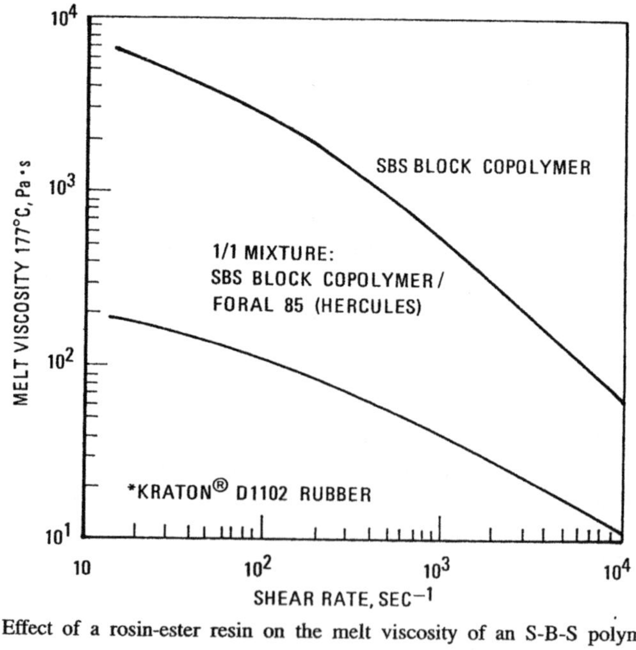

Fig. 13-18. Effect of a rosin-ester resin on the melt viscosity of an S-B-S polymer at various shear rates.

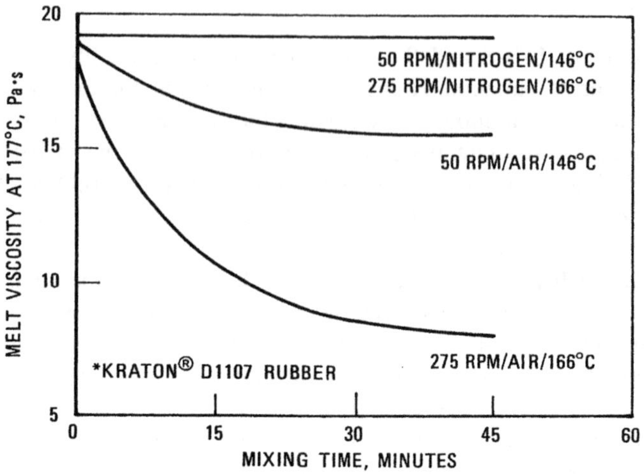

Fig. 13-19. Effect of mixing conditions on melt viscosity retention of a pressure sensitive adhesive based on an S-I-S polymer.

Figures 13-20a and 12-20b demonstrate the beneficial effect nitrogen blanketing has in preventing degradation of an S-I-S–polymer-based pressure sensitive adhesive when held in a melt holding tank for extended periods of time at elevated temperature. Severe decreases in melt viscosity and adhesive holding power are shown after less than 24 hours when the adhesive is in contact with air (Fig. 13-20a); the same adhesive with nitrogen blanketing shows little reduc-

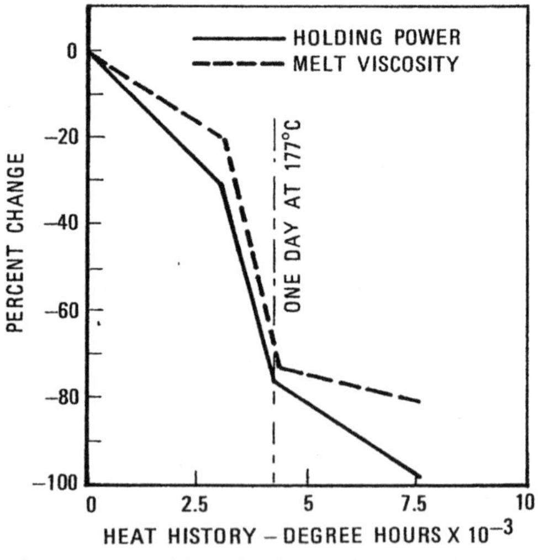

Fig. 13-20a. Hot melt pressure sensitive adhesive pot life study in contact with air at 177°C.

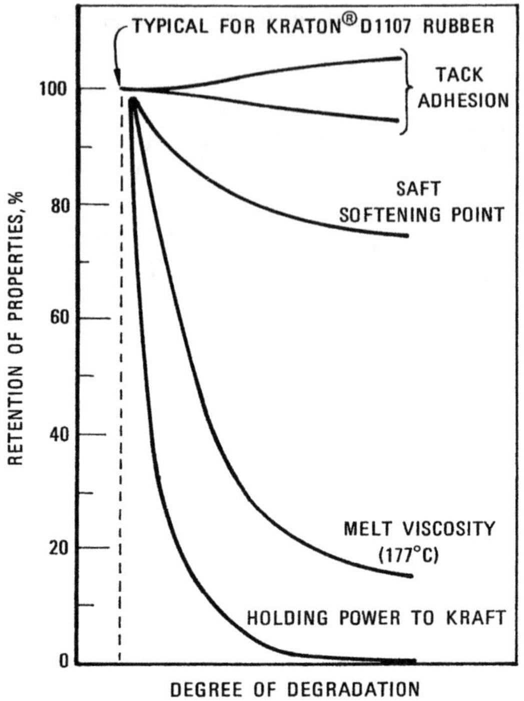

Fig. 13-21. Effect of degradation on retention of pressure sensitive adhesive based on an S-I-S polymer.

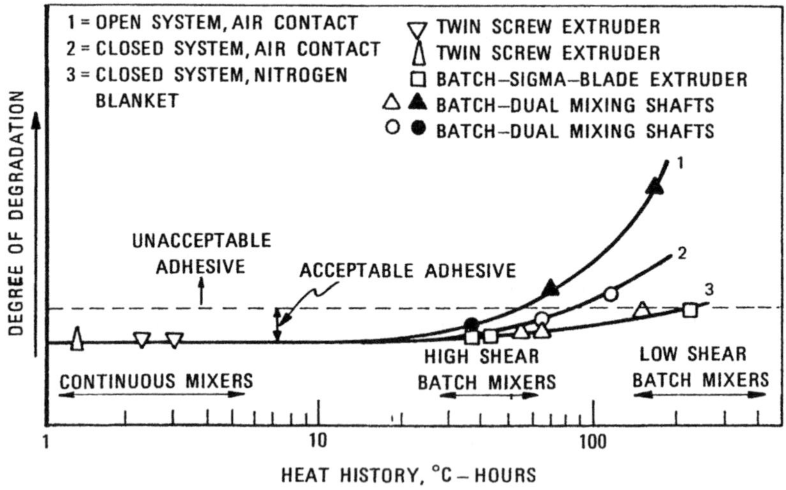

Fig. 13-22. Effect of heat history on degree of degradation for a pressure sensitive adhesive based on an S-I-S polymer using various hot melt mixing equipment.

Table 13-9. General-purpose Tape Formulation and Properties Based on an S-I-S Thermoplastic Rubber

FORMULATION 7A	PARTS BY WEIGHT
S-I-S polymer[a]	100.0
Hydrocarbon tackifying resin[b]	140.0
Paraffinic/naphthenic oil[c]	10.0
Antioxidant[d]	2.0

	HOT MELT COATED TAPES			
	CREPE PAPER		POLYESTER	
ADHESIVE PROPERTIES	TARGET	FORMULATION 7A	TARGET	FORMULATION 7A
Coat weight (g/cm^2)	—	43.0	—	36.0
Rolling ball tack (cm)	1.0	1.3	1.5	1.3
Polyken probe tack (kg)	0.3	0.6	1.0	1.3
180° peel strength (N/m)	400	550	500	800
SAFT to Mylar (°C)	60	71	93	85
SAFT to kraft (°C)	54	61	65	67
Holding power (shear)				
to steel (min)[e]	200	500	2000	2000
to kraft (min)[e]	175	300	2000	2000
Box loop test (days)	10	>30	>30	>30

[a] Kraton D1107 Rubber; Shell Chemical.
[b] Such as: Escorez 1310; Exxon Chemical or Wingtack 95; Goodyear Chemical, or other 100°C softening point C_5 resins.
[c] Shellflex 371 Oil,; Shell Chemical.
[d] Such as: Irganox 1010; Ciba-Geigy or Antioxidant 330; Ethyl Corp., Irganox 1330; Ciba-Geigy.
[e] 12.7 × 12.7-mm contact, 1-kg load.

label formulations to help lower the cohesive strength and reduce formulation costs.

New polymers, such as Kraton D1112X and D1117X rubber, have been developed specifically for adhesives where low cohesive strength and lower viscosity is desired. The formulation shown in Table 13-10 demonstrates a better balance of pressure sensitive adhesive properties and die cuttability than Kraton D1107 rubber when using either Kraton D1112X and D1117X rubber in this formulation.

Duct Tape

Duct tape is the familiar silver colored polyethylene/cloth laminate pressure sensitive product. Its name is derived from the major intended use it serves—to wrap air-conveying ducts, keeping insulation positioned and in place. It is

Table 13-11. Thermoplastic Rubber Pressure Sensitive Adhesive Duct Tape vs. Commercial Duct Tape Products

	HOT MELT S-I-S[a]		COMMERCIAL DUCT TAPE			
IDENTIFICATION	LOW-COAT WEIGHT	HIGH-COAT WEIGHT	A	B	C	D
PROPERTIES SUBSTRATE	JOANNA WESTERN[b]		POLYETHYLENE/CLOTH SCRIM			
Adhesive coat weight (g/m^2)	124	210	—	—	204	228
Rolling ball tack (cm)	1.2	0.5	0.9	0.8	1.7	1.2
Polyken probe tack (kg)	0.6	0.9	0.6	0.5	0.8	0.7
SAFT to Mylar (°C)	74	81	75	42	100	114
SAFT to kraft (°C)	77	77	61	40	100	102
180° peel to steel (N/m)	1460	2100	370	390	520	580
180° peel to backing (N/m)	790	1320	230	300	840	860
Quickstick to steel (N/m)	820	1030	230	300	320	370
Holding power to kraft (min)						
12.7 × 12.7 mm, 2-kg load	165	160	10	1	—	—
12.7 × 12.7 mm, 1-kg load	4450	5450	90	10	30	70
Holding power to steel (min)						
12.7 × 12.7 mm, 2-kg load	120	55	40	5	—	—
12.7 × 12.7 mm, 1-kg load	300	190	—	—	30	70
Box loop test (days)						
25.4-mm test strip	>35	>35	80	<1	—	—
12.7-mm test strip	>35	>35	<1	<1	—	—

[a] Formulation: 100 parts Kraton D1107 Rubber, 140 parts Wingtack Plus, 10 parts Shellflex 371, 5 parts Butazate, 5 parts TiO^2.
[b] Prelaminated polyethylene/cloth scrim made by Joanna Western Mills Co.

applications. However, as the temperature at which the product will be used approaches the glass transition temperature of the endblock phase of the rubber (usually about 92°C), the polystyrene endblock domains begin to soften and unlock the physical cross-links. When this occurs, the rubber midblocks are no longer cross-linked and the adhesive shows a drastic reduction in cohesive strength. Therefore, adhesives based on thermoplastic rubber cannot be used, for example, on high-temperature masking tapes that must have some solvent resistance and withstand a significant load at temperatures in paint baking ovens of up to 170°C. This limitation can be overcome by cross-linking the rubber in the adhesive so that it will no longer be thermoplastic.

The double bonds in the midblocks of the unsaturated S-B-S and S-I-S polymers provide the sites for cross-links. These unsaturated thermoplastic rubbers can be cross-linked with the usual sulfur, sulfur donor, or peroxide curing systems traditionally used for cross-linking unsaturated rubbers. There are many

Table 13-13. Pressure Sensitive Adhesive Formulations Cross-linked by Exposure to Ultraviolet Radiation

FORMULATION	A	B
S-I-S polymer[a]	100	—
Experimental $(S\text{-}I)_n$ rubbers[b]	—	100
Wingtack 95[c]	80	80
TMPTA[d]	7.5	7.5
Irgacure 651[e]	1.0	1.0

ADHESIVE PROPERTIES	A		B	
ULTRAVIOLET CURE	BEFORE	AFTER	BEFORE	AFTER
Belt speed (m/min)	—	3.0	—	12.0
Gel content (%)[f]	0	92	0	89
Rolling ball tack (cm)	1.4	14.3	2.8	4.1
Polyken probe tack (kg)	0.9	0.8	1.0	0.9
180° peel (N/m)	720	530	630	540
Holding power to steel (hr)[g]	>70	>44	>88	>88
Holding power to kraft (hr)[g]	>70	0.3	>88	>88
70°C holding power to kraft (hr)[h]	1.5	0.5	5	27

[a] Kraton D1107 rubber; Shell Chemical.
[b] Shell Chemical.
[c] Goodyear.
[d] Trimethylolpropane triacrylate, SR-351; Sartomer.
[e] Ciba-Geigy.
[f] Percent polymer fraction not soluble in toluene.
[g] 12.7 × 12.7 cm with 2-kg load.
[h] 25.4 × 25.4 cm with 1-kg load.

addition, a photoinitiator, Irgacure 651, is needed to generate free radicals by dissociating rapidly under ultraviolet exposure, which initiate the cross-linking reaction. The formulations can be cross-linked by exposure, under a nitrogen blanket, to ultraviolet light by passing the adhesive under a high-pressure mercury lamp. However, it is required that the light be filtered to remove the portion of the spectrum whose wavelength is less than 310 nanometers in order to prevent degradation of the surface of the adhesive. Formulation A, in Table 13-13, is based on a conventional S-I-S polymer and requires a slow conveyor belt speed of 3.0 meters/minute to obtain sufficient cross-linking. Formulation B, on the other hand, is based on an experimental $(S\text{-}I)_n$ polymer designed for radiation cross-linking and can be cross-linked at a belt speed of 12.0 meters/minute.

Adhesives can also be cross-linked by exposure, under a nitrogen blanket, to electron-beam radiation at a dose of 1 to 10 megarads. Table 13-14 demonstrates the effectiveness of Kraton D1320X rubber, Formulation B, in obtaining a cross-link network at 5 megarads without the use of a reactive monomer. Formula-

Table 13-15. Effect of Resins on Electron Beam Cross-linking of $(S\text{-}I)_n$ Thermoplastic Rubber Pressure Sensitive Adhesives

FORMULATIONS	1	2	3	4
Kraton D1320X Rubber; Shell	100.0	100.0	100.0	100.0
Escorez 5380; Exxon	60.8	—	—	—
Regalrez 1018; Hercules	61.4	—	—	—
Wingtack 95; Goodyear	—	69.4	50.1	—
Escorez 1310; Exxon	—	—	—	74.2
Adtac B10; Hercules	—	52.8	—	48.0
Wingtack 10; Goodyear	—	—	72.1	—
Antioxidant 330; Ethyl Corp.	0.4	0.4	0.4	0.4
BHT	0.3	0.3	0.3	0.3

IDENTIFICATION NUMBER	ELECTRON BEAM DOSE (Mrad)	GEL CONTENT[a] (%)	ROLLING BALL TACK (cm)	SAFT–MYLAR[b] (° C)	95°C HOLDING POWER[c] (min)	180° PEEL (N/m)
ESCOREZ 5380/REGALREZ 1018						
1	0	0	1.7	74	6	540
	1.7	45	1.4	104	50	470
	3.8	90	1.6	112	670	470
	6.5	93	3.2	113	800	420
	8.0	95	4.0	114	>1000	390
WINGTACK 95/ADTAC B10						
2	0	0	2.5	92	23	840
	1.7	4	2.5	97	50	720
	3.8	69	1.9	115	165	720
	6.5	89	4.3	122	290	680
	8.0	95	4.3	114	500	700
WINGTACK 95/WINGTACK 10						
3	0	0	1.5	88	11	630
	1.7	0	1.3	93	26	610
	3.8	62	1.3	100	125	530
	6.5	77	3.0	116	130	600
	8.0	90	1.8	114	140	560
ESCOREZ 1310/ADTAC B10						
4	0	0	2.9	91	12	770
	1.7	3	1.8	96	33	770
	3.8	74	1.7	110	180	630
	6.5	75	3.1	111	190	650
	8.0	82	4.6	115	650	670

[a] Percent polymer fraction not soluble in toluene.
[b] 25.4 × 25.4 cm with 1-kg load, temperature at failure of adhesive bond.
[c] 25.4 × 25.4 cm with 1-kg load.

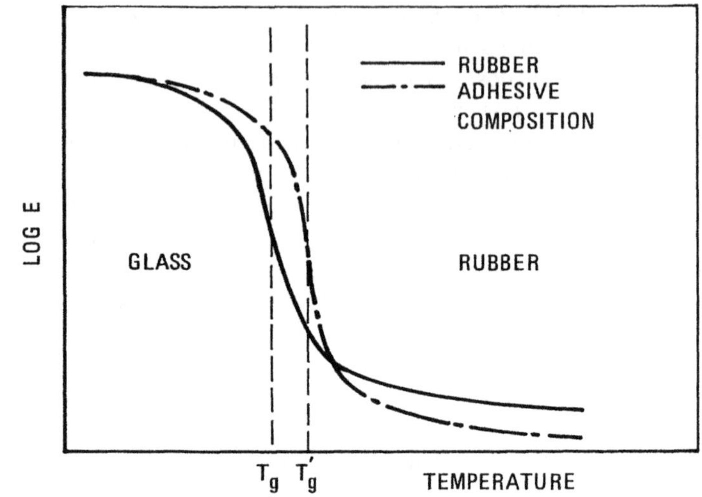

Fig. 13-24. Schematic presentation of modulus, *E*, as a function of temperature and pressure sensitive adhesive glass transition temperature T_g.

deformed and flow and make a good contact with the adherent. The rubber component in a pressure sensitive adhesive alone has practically no tack. However, decreasing the modulus by addition of sufficient amounts of rubber-compatible resins and plasticizers raises the aggressive tack at room temperature.

The modulus value of the adhesive and thus the tack at any given temperature is highly dependent on the glass transition temperature of the rubber component (T_g) and the glass transition temperature of the adhesive composition (T'_g). These effects are shown schematically in Figure 13-24. As the temperature decreases, the modulus increases. The adhesive becomes less able to flow and make contact with the substrate. At or below the glass transition temperature (T'_g) the adhesive behaves as a rigid solid and the tack is nil. It is clear that the lower the T'_g of an adhesive, the longer the tack properties will be maintained with a decrease in temperature.

The T_g of elastomers is rather low; for polyisoprene and polybutadiene, midblock components in unsaturated thermoplastic rubbers, T'_g are $-58°$ and $-91°C$, respectively. However, when midblock-associating resins are added, to obtain tack at ambient temperature, the T_g of the composition increases rapidly with increasing resin content (T_g to T'_g shift illustrated in Figure 13-24). The extent of this shift depends on the type of resin and plasticizers incorporated, and the resultant T_g value of the ultimate adhesive blend.

By using materials with relatively low T_g and low rate of change in modulus with temperature, adhesives can be formulated with good low-temperature tack.

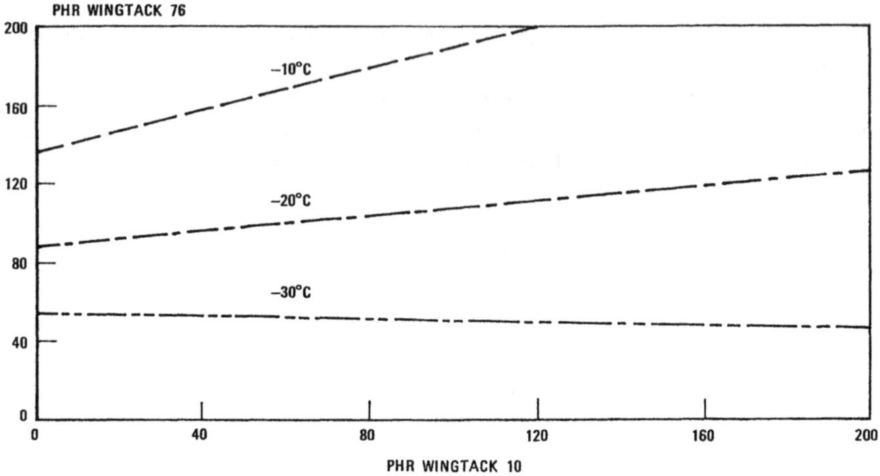

Fig. 13-25. Contour isopleths for calculated glass transition temperature, T_g, for two resins in an S-I-S–polymer (Kraton D1107 rubber) -based system.

Because the low-temperature performance is determined by the latter, this value is used in the calculation.

A probe tack test was used to measure the tack at the low temperatures. In this test, a flat-ended stainless-steel probe was contacted under a light weight with the adhesive surface for a very short time. The force required to withdraw the probe was measured. From the table it can be seen that measurable tack is developed as long as the test temperature is above the T_g' of the given adhesive formulation.

Contour diagrams presenting T_g' for adhesive formulations as a function of type and amount of the ingredients can be very helpful in predicting formulations with good low-temperature properties. In Figure 13-25 an example is given for a combination of linear S-I-S polymer, Wingtack 76 as tackifying resin and Wingtack 10 as plasticizer. From this figure it is easy to find formulations in which tack could be expected down to −30°C.

However, other property requirements, such as melt viscosity, ambient tack properties, peel adhesion, and holding power, limit the area from which an adhesive for low-temperature tack can be chosen. Formulations with a relatively low amount of high softening resin and a relatively high amount of plasticizer may fulfill both the requirement for low-temperature tack and low melt viscosity; however, 180° peel adhesion and holding power will be unacceptably low; also there is the possibility of having the adhesive bleed through the backing.

By superpositioning of the contour diagram for T_g' with those of the other critical properties, it may be possible to choose a proper formulation with a satisfactory balance in properties, including low-temperature performance.

Table 13-18. Weather-resistant Pressure Sensitive Adhesives Based on Saturated S-EB-S Thermoplastic Rubbers

FORMULATIONS (Dry weight, phr)	A	B	C
Kraton G1657X rubber[a]	100	100	—
Kraton G1652 rubber	—	—	100
Arkon P90[b]	91	—	—
Regalrez 1078[c]	—	68	58
Regalrez 1018	—	45	54
Indopol H100[d]	46	—	—
Irganox 1010[e]	1.0	1.0	1.0
Tinuvin 327[e]	0.25	0.25	0.25
Tinuvin 770	0.25	0.25	0.25
ADHESIVE PROPERTIES			
Calculated T_g (°C)	−16	−16	−16
Rolling ball tack (cm)	3.0	1.5	1.7
Polyken probe tack (kg)	0.39	0.84	0.49
180° peel (N/m)	560	540	370
SAFT to Mylar (°C)[f]	88	75	87
Holding power to steel (hr)[g]	20	33	13
Hot melt viscosity @ 177°C (Pa·s)	83	45	100

[a] Shell Chemical.
[b] Arakawa.
[c] Hercules.
[d] Amoco.
[e] Ciba-Geigy.
[f] 25.4 × 25.4 cm with 1-kg load, temperature at failure of bond.
[g] 12.7 × 12.7 cm with 2-kg load.

of an adhesive based on an unsaturated S-I-S rubber. The 180° peel strength of the adhesive cast on polyester backing and adhered to glass was measured as a function of exposure time in a xenon lamp weatherometer. As shown in Table 13-19, the performance of the adhesive based on the unsaturated polymer, Formulation C, is poor. After less than six days exposure, the adhesive was so badly degraded that it failed cohesively when peeled from the glass. As exposure is continued, the peel sample continues to fail cohesively, but at lower and lower peel strength as the cohesive strength of the adhesive is further reduced by degradation.

The ultraviolet resistance of the adhesives based on saturated polymers and ingredients, Formulations A and B, is much better than that of Formulation C in Table 13-19. In fact, these adhesives appear to weather better than the polyester backing since, after extended exposure, the backing tears during during the peel-strength measurement.

14. Butyl Rubber and Polyisobutylene

J. J. Higgins
Exxon Chemical Company
Polymers Group-Americas
Florham Park, New Jersey

F. C. Jagisch
Exxon Chemical Company
Polymers Group-Americas
Baton Rouge, Louisiana

N. E. Stucker
Exxon Chemical Company
Polymers Group-Americas
Baytown, Texas

Butyl rubber and polyisobutylene are elastometric polymers[1] used quite widely in pressure sensitive adhesives both as primary elastomers and as tackifiers and modifiers. Butyl rubber is a copolymer of isobutylene with a minor amount of isoprene, while polyisobutylene is a homopolymer.

As shown in Figure 14-1, the carbon-hydrogen backbone of butyl rubber is relatively long and straight, containing between 47,000 and 60,000 units. This regular structure, with few double bonds or reactive sites, renders butyl very stable and quite inert to the effects of weathering, age, and heat. It has good resistance to vegetable and animal oils and to attack by chemicals. Being an all-hydrocarbon material, the butyl polymer has very low water absorption and is soluble in typical hydrocarbon solvents. The many side groups attached to the polymer chain produce a high degree of damping. Since these side groups are not large in size and are regularly spaced, close, unstrained molecular packing results. To this is attributed the unique low air, moisture, and gas permeability of this polymer.

Note: There is no liability accepted for any loss or damage that may occur from use of this information, nor is there any warranty against patent infringement.

Table 14-1. Commercial Grades of Exxon Butyl Rubber and Vistanex® Polyisobutylene

GRADE[a]	APPROXIMATE VISCOSITY AVERAGE MOLECULAR WEIGHT	APPROXIMATE ISOPRENE UNITS/100 MONOMER UNITS (MOLE % UNSATURATION)	COMMENTS
Vistanex LM-MS	44,000	0	Semiliquid, tacky polymers used mainly as tackifiers
LM-MH	53,000	0	
LM-H	63,000	0	
Exxon Butyl 065	350,000	0.8	Low MW grades with high age stability
165	350,000	1.2	
268	450,000	1.6	Used for high cohesive strength
269	450,000	1.6	Highest viscosity/cohesive strength grade
365	350,000	2.1	
077	425,000	0.8	Used in curing applications
			Special BHT-stabilized FDA grade
Chlorobutyl 1065	350,000	1.9	Contain approximately 1.2 wt % chlorine
1066	400,000	1.9	
1068	450,000	1.9	
Bromobutyl 2222	375,000	1.6	Contain approximately 2.0 wt % bromine
2233	400,000	1.6	
2244	450,000	1.6	
2255	450,000	1.6	
Vistanex MM L-80	900,000	0	Lowest viscosity of MM grades
MM L-100	1,250,000	0	Widely used in pressure-sensitive adhesives
MM L-120	1,660,000	0	
MM L-140	2,110,000	0	Widely used in pressure-sensitive adhesives
			Highest viscosity of MM grades

[a] Specific gravity: 0.93: Bromobutyl.
0.92: All other grades.

provide improved flexibility and impact resistance, particularly at low temperatures.

In addition to the Vistanex grades described, BASF (Badische Anilin und Soda Fabrik Co.) also supplies polyisobutylene homopolymers under the trade name Oppanol.

Halogenated Butyl Rubber

Chlorobutyl rubber[6] is prepared by chlorinating the regular butyl polymer under controlled conditions so that reaction is primarily by substitution and little of the unsaturation originally present in the macromolecule is lost. The chlorine is believed to enter the molecule at the highly reactive allylic position, one carbon removed from a double bond. Approximately 1.2 weight percent chlorine is present in the commercial grades of Exxon Chlorobutyl. The chlorine tends to enhance the reactivity of the double bonds as well as supply additional reactive sites for cross-linking. As a result, a wide variety of different type cure systems can be used to vulcanize this elastomer.

Chlorobutyl can be readily blended and cured with more highly unsaturated elastomers. In the adhesives area, it has been blended with both regular butyl and natural rubber and then preferentially cured through the chlorine to improve strength. The reactive chlorine will also tend to increase adhesion to many polar substrates.

A brominated butyl polymer[7] is also available. Bromobutyl is structurally similar to Chlorobutyl, but provides an additional level of cross-linking activity. These halogenated butyl polymers are marketed by both Exxon Chemical Company and Polysar Limited.

Butyl Rubber Latex

Butyl rubber can be emulsified to give a latex that might typically have a solids content of approximately 60 weight percent, a pH of 5.5, and a Brookfield viscosity of 2500 cps. An anionic emulsifier is used to prepare butyl latex having an average particle size of 0.3 μm. BHT can be used as an antioxidant, with a small amount of preservative sometimes added. One supplier of butyl latex is Burke-Palmason Chemical Company (Pompano Beach, Florida). Vistanex polyisobutylenes have also been produced in latex form.

Butyl latex has excellent mechanical, chemical, and freeze thaw stability, which allows for a wide latitude in compounding and blending with other ingredients. Because of its stability, it cannot be coagulated in a controlled manner

1. For strength, choose the highest molecular weight polyisobutylene (PIB) or butyl rubber grade consistent with application or processing requirements.
2. Blend butyl rubber with polyisobutylene and cure the butyl portion. Since the PIB does not cure, the extent of cure in the system and thus the cohesive strength is governed by the butyl content.
4. Partially cure butyl rubber. This can best be done by adding a carefully measured amount of curative and fully curing to the extent of the curative present, or by using one of the available precured grades.
4. Blend either PIB or butyl with halogenated butyl and cure only the halogenated butyl portion by utilizing the halogen functionality. As an example, a curative that will react with Chlorobutyl and not with butyl is zinc oxide. This technique is also of interest in blends of halogenated butyl with either SBR or natural rubber for pressure sensitive adhesives.

Another example of the compounding latitude afforded by polymer selection in the polyisobutylene family is tack. Tack is increased with Vistanex LM, butyl rubber can be depolymerized to any desired tacky state, or a wide range of compatible tackifiers can be incorporated.

In terms of the butyl polymers, the variable of greatest interest to the adhesives compounder is usually molecular weight. All grades of butyl rubber are of such low unsaturation that this characteristic does not affect most performance properties in uncured or partially cured formulations. However, in applications not requiring cure, the lowest unsaturation grades are usually chosen for optimum aging.

Pigments and Fillers

The same pigments and fillers commonly used with other rubbers can be compounded with butyl rubber and polyisobutylene, and the general principles of selection are the same. Very fine pigments increase cohesive strength and stiffness, reduce cold flow, and also reduce tack. Platy pigments such as mica, graphite, and talc are preferred for acid and chemical resistance and low gas permeability. Some of the coarser pigments increase tack. Zinc oxide increases tack and cohesive strength (and also plays an important chemical role in the vulcanization of butyl). Aluminum hydrate, lithopone, whiting, and the coarser carbon blacks such as thermal blacks also increase tack with moderate increase in cohesivity. Clays, hydrated silicas, calcium silicates, silico-aluminates, and the fine furnace and thermal blacks increase cohesive strength and stiffness. Stiffness can also be increased by use of very fine silica pigment and magnesium oxide or carbonate.

an epoxy silane like A-187. There is evidence that amino silanes like A-1100 and methacryoxy silanes like A-174 may also be of value. In certain textile applications, isocyanate primers or adhesion promoters can be used with butyl cements.

Antioxidants are often used in butyl and PIB adhesives to protect against severe environmental aging caused by ultraviolet light or heat. Materials such as Ethanox 702, Butyl Zimate, AgeRite White, Irganox 1010, BHT, and even sulfur have been used.

Curing Systems

There are four curing systems that are of general interest in cements, adhesives, sealants, and coatings based on butyl and halogenated butyl.[12] These are the (1) "quinoid" cure, (2) cure with sulfur or sulfur donor compounds, (3) resin cure, and (4) for halogenated butyl only, the zinc oxide cure. The introduction of cross-linking increases cohesive strength as stated previously but reduces tack and solubility in hydrocarbon solvents and thus finds only limited use in pressure sensitive type adhesives. Therefore, only summary information on these cure systems will be provided here.

The quinoid cure is an old, easily controlled system that typically is formulated as a split batch, two-part cement designed to cure at or even below room temperature. The resultant network is tightly cured, highly resistant to ozone, heat, chemicals and other environmental attack and provides good electrical properties. Dark color bodies are formed during the reaction, so quinoid cures cannot be used to make white products.

A commonly used system makes use of *p*-quinone dioxime (commercially known as QDO) or dibenzoyl *p*-quinone dioxime (DBQDO) combined with an oxidizing agent such as MnO_2, PbO_2, Pb_3O_4, or benzothiazyl disulfide. The DBQDO system is less active, and thus is easier to control at elevated temperatures.

Sulfur curing systems for butyl include elemental sulfur, thiuram or dithiocarbamate accelerators, and thiazole or thiazyldisulfide activators. Zinc oxide or other metallic oxides are necessary to attain satisfactory cure. The relatively minor differences in unsaturation between grades of butyl are significant in regard to sulfur cures, the rate of cure, and number of attainable cross-links increasing with increasing polymer unsaturation. Sulfur cures find little commercial value in adhesives, however, due to the generally elevated temperatures required to generate cross-links.

The resin cure of butyl normally uses one of a series of active brominated phenolic resins such as Schenectady's SP-1055. The resin cure of butyl can be made to proceed at temperatures from ambient to very high, depending on the specific resin chosen and on the concentration and type of cure activator.

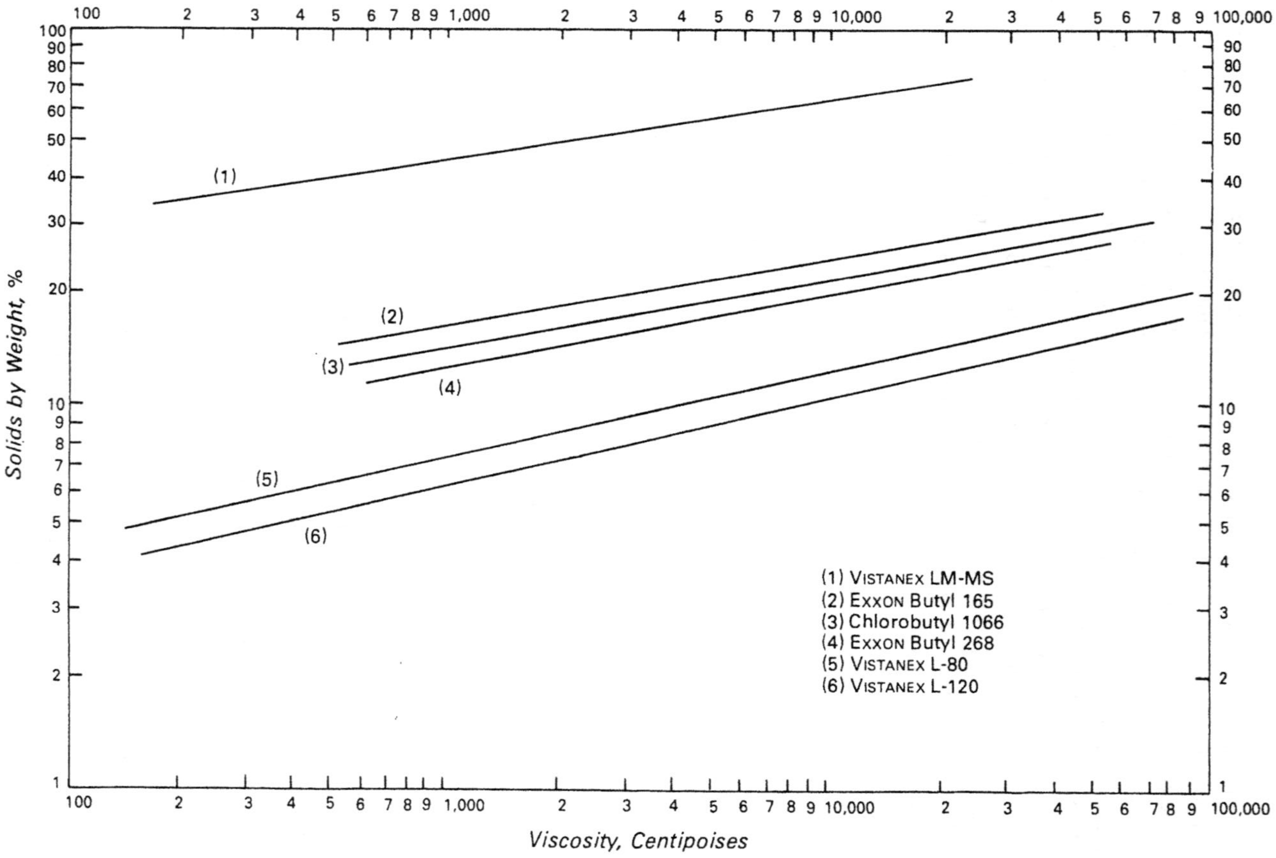

Fig. 14-2. Solution viscosities of isobutylene polymers in toluene.

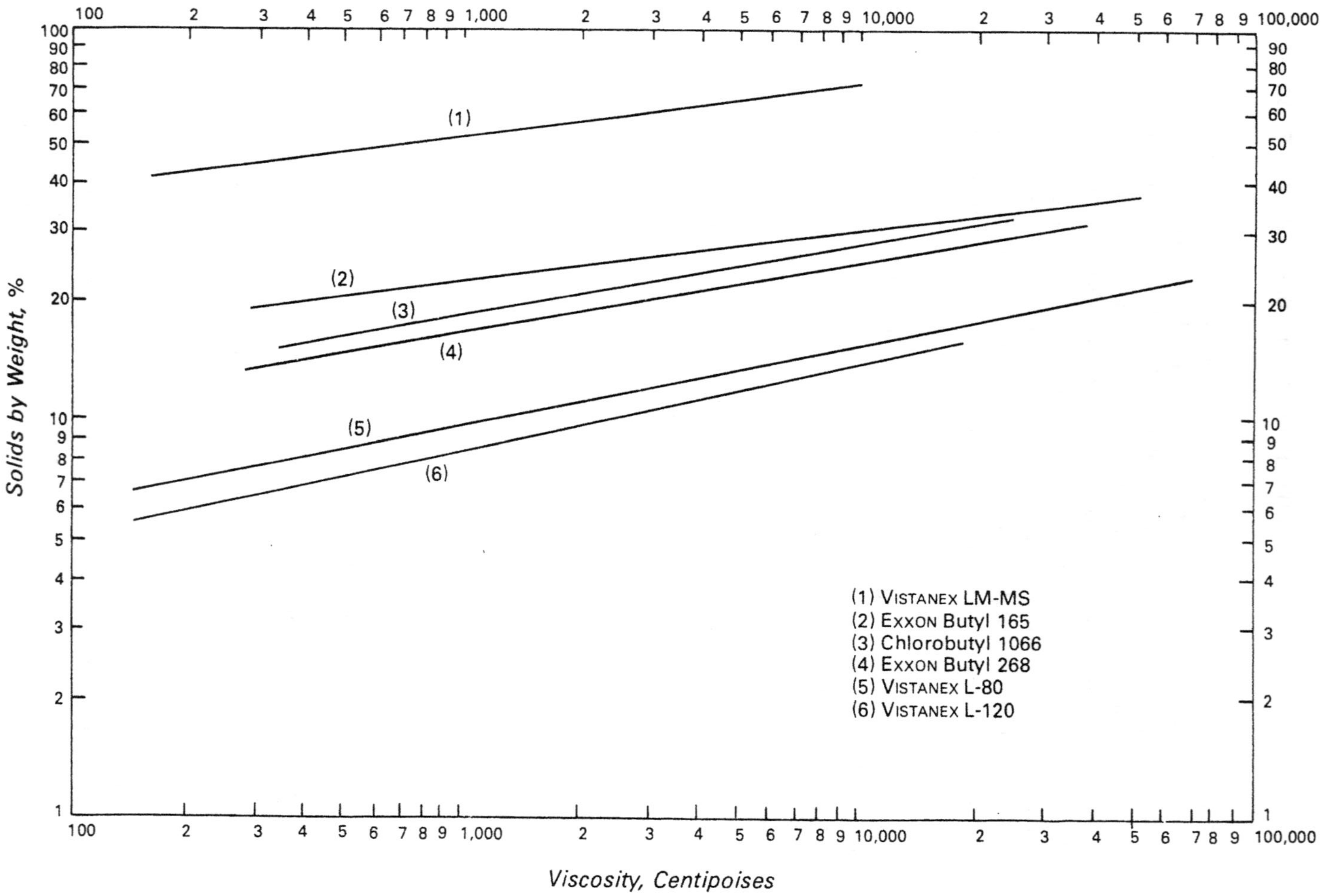

Fig. 14-3. Solution viscosities of isobutylene polymers in hexane.

are tack and hold, which is the ability to resist creep under deadload. As noted, butyl and polyisobutylene are inherently tacky polymers. This tack can be increased with a wide variety of resins and other tackifiers. The cohesive strength is low compared to some other pressure sensitive adhesive polymers, such as natural rubber, but can be increased if required by (1) incorporation of high molecular weight PIB or natural rubber, (2) the choice of the other ingredients, particularly resins and fillers, and (3) the partial or preferential curing techniques noted previously. The polyisobutylene polymers are primarily used in label pressure sensitive adhesives and in certain tapes where high cohesive strength is not necessary.

The age resistance of polyisobutylene polymers manifests itself as lack of hardening and permanency of tack, and this has led to their use where long service life or aged adhesive integrity is required, as, for example, in removable label pressure sensitive adhesives. A removable label adhesive that ages poorly may lose cohesive strength and consequently will not peel as cleanly as desired. The higher molecular weight grades when used as the base polymer also die cut well, an important requirement for use in labels. Butyl and polyisobutylene are thus preferred materials in removable pressure sensitive label and protective tape adhesives.

The low-temperature flexibility and tack of these polymers results in a service temperature range extending below ambient. Vistanex LM low molecular weight PIB is a particularly useful low-temperature tackifier. The polyisobutylene polymers find application in solventborne, low-temperature pressure sensitive adhesives such as freezer label adhesives.

The low order of toxicity, favorable FDA status, and light color result in a variety of medical pressure sensitive adhesive uses for selected polyisobutylene polymer grades. Applications include surgical tape, oral bandages, and ostomy appliances.

Adhesives formulated with butyl and polyisobutylene are used for adhesion to a variety of substrates including nonporous, hard to adhere to materials such as polyethylene. Other common substrates include paper, polyvinyl chloride, and polyester film.

Several typical or starting point solution pressure sensitive adhesive formulations illustrating some of the uses previously described are given in Table 14-2. They are provided for general guidance on materials selection and quantity, but likely require modification to suit specific ingredients, manufacturing, and finished property requirements. All concentrations in formulations are expressed in parts by weight. Suppliers of the ingredients are listed in Table 14-7.

Hot Melt Pressure Sensitive Adhesives

The polyisobutylene family of polymers can also be formulated in hot melt compositions and can impart many of the same property benefits noted for solvent-based systems. They improve flexibility, particularly at low temperatures, provide good aging and chemical resistance, and serve to contribute toughness and strength in thermoplastic mixtures. Increases in hot tack, adhesion, and overall seal strength have been widely documented.[19–23] Polyisobutylene grade choice and concentration are important considerations, however, in that these elastomeric products become more difficult to melt mix, significantly increasing melt viscosity as PIB molecular weight increases. Vistanex MM high molecular weight grades, in fact, are somewhat limited in hot melt applications.

For hot melt pressure sensitive adhesives[24] butyl rubber is usually blended with significant amounts of melt materials, for example, tackifying resin, petrolatum, and amorphous polypropylene, to reduce viscosity to desirable levels. Low volatility hydrocarbon oils, polybutenes, and paraffin and microcrystalline waxes also depress melt viscosity, but excessive wax can adversely affect tack. Vistanex LM low molecular weight PIB is an effective butyl plasticizer and enhances tack. Applications for butyl hot melt pressure sensitive adhesives often involve long service life, low temperature, and polyolefin plastic film adhesion, capitalizing on the inherent features of this polymer noted previously. An illustrative starting point formulation for adhesive-backed carpet tiles is shown in Table 14-3.

The Vistanex LM grades are perhaps the most functional polyisobutylene polymers for hot melt pressure sensitive adhesives. They are easiest to melt

Table 14-3. Butyl Rubber Hot Melt Pressure Sensitive Adhesive

Exxon Butyl 065	100
Escorez 1304 resin	100
Petrolatum[a]	50
Amber microwax[b]	150
Antioxidant[c]	1
Brookfield viscosity at 121°C	
Spindle SC4-29 (cps)	80,000
Surface	Tacky, firm
Adhesion[d]	
Vinyl foam to plywood	Good–excellent
Rubber foam to plywood	Good–excellent

[a] 57°C M.P.
[b] Be Square 175.
[c] Ethanox 702 or Irganox 1010.
[d] Adhesives coated on foam and subsequently manually pressed onto plywood.

Table 14.4. Low Molecular Weight Polyisobutylene Hot Melt Pressure Sensitive Adhesives

	1	2	3
A. General Recipes			
Vistanex LM-MS	25	35	45
Escorez 1310 resin	50	50	15
Petrolatum[a]	15	—	40
High molecular weight LDPE[b]	10	—	—
Paraffinic process oil	—	15	—
Brookfield Viscosity at 177°C			
Spindle SC4-27 (cps)	1,900	1,200	850
Surface	Tacky, firm	Very tacky, soft	Very tacky, soft
Softening Point			
Ring and Ball (°C)	101	49	55
B. Low-temperature Label Pressure Sensitive Adhesive			
Kraton 1107[c]	50		
Finaprene 1205[d]	50		
Vistanex LM-MS	90		
Escorez 2393 resin	130		
Naphthenic process oil	58		
Antioxidant[e]	2		
Rolling Ball Tack (cm)	2.5		
180-degree Peel (g/cm)			
Stainless steel	880[g]		
Treated LDPE	570		
180-degree Peel at 0°F (g/cm)[f]			
Stainless steel	1,250[g]		
Treated LDPE	90		
Loop Tack at 0°F (g/cm)[f]			
Stainless steel	950		
Corrugated board	290[h]		
Waxed board	410[g]		
C. EVA-based Pressure Sensitive Adhesive			
Elvax 40	100		
Vistanex LM-MS	20		
Escorez 2393 resin	80		
Antioxidant[e]	1		

[a] 57°C M.P.

[b] A low MW LDPE grade such as Epolene C-10 would give lower viscosity and softening point.

[c] Styrene-isoprene-styrene block copolymer.

[d] Styrene-butadiene block copolymer; improves polyisobutylene compatibility with Kraton 1107.

[e] Irganox 1010.

[f] Adhesive and adherend conditioned at −18°C before contact; adhesion determined at −18°C.

[g] Some adhesive transfer from polyester backing noted.

[h] Fiber tearing noted.

have a fabric or film backing) and are typically available in roll form on a release liner.

There are numerous pressure sensitive tapes with various adhesive and backing combinations for specific electrical applications. The nonpolarity of the polyisobutylene polymers makes them quite resistant to attack and swelling by the commonly used vinyl plasticizers. Thus, they are less susceptible to softening when used on vinyl film backings.

Typical nonconductive splicing tape and pressure sensitive electrical tape formulations are given in Table 14-6. The splicing tape mix is subjected to brief heat treatment in a Banbury mixer to maximize cohesive strength and electrical properties.

Table 14-7. Suppliers of Trade Named Formulating Ingredients

MATERIALS	SUPPLIER
A-Fax 600	Hercules Inc.
AgeRite Resin D	R. T. Vanderbilt Co.
AgeRite White	R. T. Vanderbilt Co.
Be Square 175	Petrolite Specialty Polymers Group
Bromobutyl	Exxon Chemical Co., Polysar Ltd.
Butyl Zimate	R. T. Vanderbilt Co.
Chlorobutyl	Exxon Chemical Co., Polysar Ltd.
DBQDO	Lord Corp.
Elvax 40	E. I. DuPont, Inc.
Exxon Butyl	Exxon Chemical Co.
Epolene	Eastman Chemical Products Inc.
Escorez® Resins	Exxon Chemical Co.
Ethanox 702	Ethyl Corp.
Finaprene 1205	Fina Oil & Chemical Co.
Hercolyn D	Hercules Inc.
Irganox 1010	Ciba-Geigy Corp.
Kalar 5214, 5245, 5263, 5264	Hardman Inc.
Kalene 800 and 1300	Hardman Inc.
Kraton 1107	Shell Chemical Co.
Laktane	Exxon Corp.
Mistron Vapor Talc	Cyprus Industrial Minerals
Oppanol	BASF
Parapol® 950 and 1300	Exxon Chemical Co.
Pentalyn Resins	Hercules Inc.
Piccolyte S115	Hercules Inc.
Poly DNB	Lord Corp.
Polysar Butyl	Polysar Ltd.
QDO	Lord Corp.
Satinton W/Whitex	Engelhard Corp.
Schenectady Resins	Schenectady Chemicals, Inc.
Silanes	Union Carbide Corp.
Vistanex®	Exxon Chemical Co.

15. Acrylic Adhesives

Donatas Satas
Satas & Associates
Warwick, Rhode Island

Acrylic polymers have been known for a long time, but their utilization as pressure sensitive adhesives is relatively recent.

Acrylic acid was first synthesized in 1843. By 1901, research was being carried out on acrylic acid esters.[1] The first commercial production of poly(methyl methacrylate) took place in 1927 by Roehm and Haas AG in Germany. Acrylic dispersions have been produced by Badische Anilin und Soda Fabrik since 1929. The suitability of polyacrylates for pressure sensitive adhesives was recognized as early as 1928.[2] Despite all this early work, polyacrylates found extensive use for pressure sensitive adhesives only in the 1950s and attained their current importance in the 1960s.

Polyacrylates of a certain monomer composition are inherently pressure sensitive without any additional compounding. Besides polyacrylates, only polyvinyl ethers and some ethylene/vinyl acetate copolymers have this property. Such single-component adhesives have some advantages over compounded products.[3] Low molecular weight ingredients that can migrate to the surface of an adhesive coating are absent. The elimination of compositional differences in the adhesive coating is highly desirable; the presence of low molecular additives at the adhesive–adherend interface can affect the bond. Such variation is difficult to avoid in multiphase systems, while uniformity is more easily achieved in single-component adhesives.

Polyacrylates possess some inherent properties that are superior to many other polymers used for pressure sensitive adhesives. The polymer is saturated and resistant to oxidation. It is transparent, colorless, and does not yellow upon exposure to sunlight. The resistance to oxidation surpasses that of most polymers used for pressure sensitive adhesives, except silicones.

Acrylic adhesives are available as solutions, aqueous emulsions, hot melts, and 100% reactive solids. Unlike most of the other polymers used for pressure

acrylic esters and a small amount of polymerizable surfactant and zwitterionic monomers has been prepared that does not exhibit adhesion buildup.[5]

An opposite effect is obtained by the incorporation of azlactone and carboxylic acid groups carrying monomers. Such copolymers exhibit a substantial increase of adhesion after aging, amounting to as much as a 150% increase in peel adhesion over the initial value.[6] Development of the ultimate bond may take as long as three weeks at room temperature, or less than one hour at 90–100°C. This is attributed to the chemical reaction between azlactone groups and nucleophilic groups on the substrate, resulting in the formation of covalent bonds, to the reaction of azlactone groups with water that is adsorbed on the substrate surface, and to the ring-opening hydrolysis of the azlactone ring, resulting in a net expansion of volume.

The use of various secondary and functional comonomers has been described in the patent literature and many are listed in the appendix at the end of the chapter. Almost any conceivable monomer that is capable of undergoing vinyl polymerization has been mentioned in the voluminous patent literature. Some of these monomers modify the adhesive properties significantly to meet various application requirements, while some only provide a convenient way to change the polymer composition. Because of the early disclosure of the suitability of polyacrylates for pressure sensitive adhesives, it was not possible for any single patent to have a dominant position in this field. This condition was responsible for the growth of a large number of compositional patents different in the use of some comonomer, sometimes of questionable importance, in obtaining the specific adhesive properties. The patent by Ulrich[7] was as close to a dominant position as possible under the circumstances. This patent claims the use of small quantities of various functional monomers, such as acrylic, methacrylic, or itaconic acids and their amides.

The generalized pressure sensitive acrylic polymer composition is as follows:

Main monomer: 50–90%
Modifying monomer: 10–40%
Monomer with functional groups: 2–20%

In order to possess a low modulus of elasticity and an acceptable tack, the polymer must have a sufficiently low T_g. As shown in Table 15-1, of the commonly available monomers only 2-ethylhexyl acrylate ($T_g = -70°C$) and *n*-butyl acrylate ($T_g = -54°C$) have sufficiently low T_g. Monomers that give homopolymers of a low T_g plasticize the adhesive, while monomers that give homopolymers of a high T_g increase the adhesive stiffness. Soft polymers have a higher tack, as shown in Figure 15-1.[8] The lower the glass transition temperature, the higher the tack. The modulus of elasticity follows the same pattern (Fig. 15-2). Glass transition temperature is a useful indicator in choosing como-

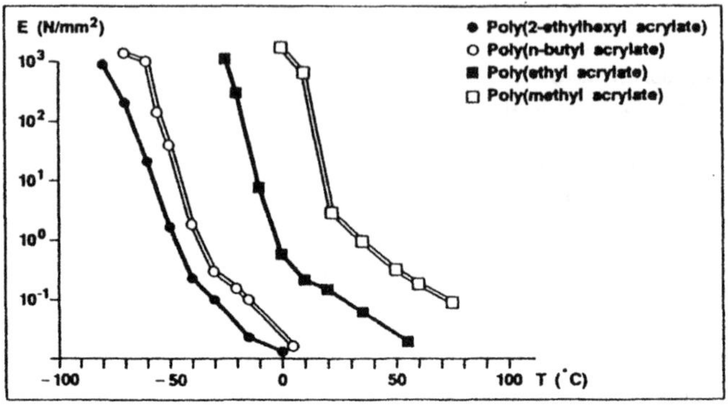

Fig. 15-2. Modulus of elasticity of various polyacrylates as a function of glass transition temperature. (Courtesy of BASF AG, Ludwigshafen.[8])

nomers, but its usefulness does not extend into the fine adjustment of adhesive properties.

It is necessary to raise the T_g of soft acrylic homopolymers in order to increase the peel adhesion at room temperature at normal peel rates. This is accomplished by copolymerization with a monomer that raises T_g.[9] Such monomers as vinyl acetate, ethyl acrylate and methacrylate, methyl methacrylate, or acrylic acid are useful for this purpose.

The polymer flexibility and tackiness increases with increasing side-chain length until a certain chain length is exceeded, and the chains start to form crystalline regions that cause stiffening of the polymer. In the case of straight side-chain polymers, this maximum softness occurs in *n*-octyl acrylate (T_g = −80°C) for the acrylic polymer series and in *n*-decyl or *n*-dodecyl methacrylates in the methacrylic polymer series.[10,11] The maximum softness occurs higher in the methacrylic series because of the steric hindrance of the methacrylic group.[12] The softness and tackiness maximum is reached at a higher chain length if these properties are tested at temperatures above 20°C.

Branched side chains stiffen the polymer in the range from propyl to hexyl. In case of longer chains, branching decreases the tendency to form crystalline regions and therefore branching makes the polymer softer.

A similar effect of branch chain length was observed in the case of pressure sensitive vinyl acetate/dialkyl maleate copolymer.[13] Increase of the branch chain length from C_9 to C_{10} to C_{12} shifted the transition from cohesive to adhesive failure to higher peel rates, but also caused a decrease in the peel force in the region of the adhesive failure (see Fig. 15-3).

The effect of introducing a comonomer of a different chain length is not dissimilar from the effect of branching. The physical properties of a copolymer may be roughly intermediate of those of the homopolymers and might not be drastically different from the properties of corresponding homopolymer blends.

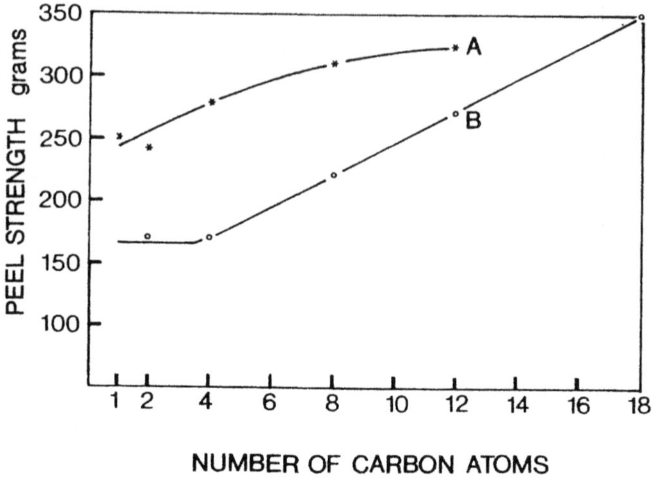

Fig. 15-4. Peel strength of methyl methacrylate copolymerized with various alkyl methacrylates and acrylates. Mole ratio 9:1. Curve A shows acrylic comonomer; curve B shows methacrylic comonomer.

It is claimed that addition of a small amount of monoethyl acid phosphate to a tackified acrylic adhesive improves its adhesion to plastic surfaces.[19]

Ethylene (10–30 weight percent) was copolymerized with acrylic esters and vinyl acetate to give a pressure sensitive adhesive.[20] Free radical initiated emulsion polymerization was carried out in a pressurized autoclave.

VINYL ACETATE/ACRYLIC COPOLYMERS

Vinyl acetate/acrylic ester copolymers constitute a special group of acrylic pressure sensitive adhesives. Because of the high amount of vinyl acetate used, such copolymers are usually referred to as vinyl acrylics. They are somewhat different: such adhesives are less polar and adhere better to polyolefins and other nonpolar plastics, they usually exhibit a high tack, and may have a lower shear resistance as well as somewhat poorer aging properties than all acrylic adhesives. Vinyl acetate is also a less expensive monomer than acrylic esters.

Aliphatic hydrocarbon solvent-compatible pressure sensitive adhesive solution polymers were prepared from 2-ethylhexyl acrylate/vinyl acetate/acrylic acid.[21] Solution and emulsion copolymers of 2-ethylhexyl acrylate and vinyl acetate were compared. Solution polymers exhibited a higher tack because of a lower molecular weight.[22]

Improved adhesion to polyolefin film and other nonpolar surfaces was also obtained by blending an acrylic pressure sensitive adhesive with an ethylene-vinyl acetate copolymer instead of copolymerization.[23]

Table 15-2. Effect of Molecular Weight on Some of the Adhesive Properties

POLYMER	INTRINSIC VISCOSITY	NUMBER AVERAGE MOL. WT.	PROBE TACK (g)[a] AT PRESSURE			CREEP (hr)		SPLITTING TEMPERATURE (°C)
			10 g	100g	500 g	24°C	71°C	
A	4.50	1,500,000	163	493	545	>200	16	>93
B	3.04	770,000	133	513	557	120	2	88
C	2.65	251,000	180	537	637	10	0.2	37
D	2.17	250,000	157	550	670	9	0.2	29
E	1.71	353,000	143	453	473	1.8	0.04	20
F	1.67	276,000	170	310	373	1.3	0.03	21
G	1.62	360,000	260	563	593	1.2	0.03	<20

[a] Contact time 1 sec, rate of separation 1 cm/sec, temperature 23°C.

weight should be considerably higher to develop a sufficiently high resistance to creep, unless the level of secondary bonding is high.

Table 15-2 shows the effect of molecular weight on the properties of pressure sensitive adhesive polymers. The polymer series A through G are of identical monomer composition.[32] The molecular weight was varied by varying the rate of polymerization; lower molecular weight polymers are obtained at higher reaction rates. The intrinsic viscosity was determined by regular viscometric techniques and the number-average molecular weight by an osmometer. Adhesive properties correlate better with intrinsic viscosity than with number-average molecular weight. Number-average molecular weight is sensitive to the presence of low molecular weight materials, while mechanical properties are mainly determined by the higher molecular weight fraction. Tack was determined by the Polyken Probe Tack Tester. Resistance to creep was measured by applying a shear force to an adhesive bond and observing the time required for a complete failure. The splitting temperature is defined as the temperature at which the transition of mode failure in the 180-degree peel test takes place from adhesive to cohesive.[33]

Figure 15-5 shows the dependence of peel force on peel rate, and Figure 15-6 shows the dependence of peel force on the temperature for polymers of various molecular weights, as shown in Table 15-2. The solid lines indicate an adhesive failure; the broken lines, a cohesive failure, where a visually detectable layer of adhesive remains on the steel panel after the test.

From Figure 15-5 it is evident that the molecular weight has an important effect on the behavior of these adhesives. High molecular weight adhesives A and B show an adhesive failure at the rates tested. Intermediate molecular weight polymers C and D show a transition from adhesive to cohesive failure. Low molecular weight adhesives E, F, and G show a cohesive failure throughout the peel-rate range tested.[32] Extension of the testing to higher peel rates should also show a transition to adhesive failure in these polymers.

temperature. Force–temperature and force–time curves are qualitatively similar and superimposable. Adhesives E and F exhibit the transition of mode of failure at the lower end of the temperature scale. Similar behavior would be expected by extension of the test conditions to higher peel rates.

The effect of molecular weight on adhesive properties has been discussed by Druschke.[8] Figure 15-7 shows tack and initial peel adhesion as a function of *K* value. The *K* value is obtained by measuring the relative viscosity of dilute polymer solutions and is related to the molecular weight. At *K* values above 50 the mode of failure in peel tests changes to adhesive failure, as required for functional pressure sensitive adhesives. As the *K* value increases, tack decreases, indicating poorer wettability by higher molecular weight adhesives.

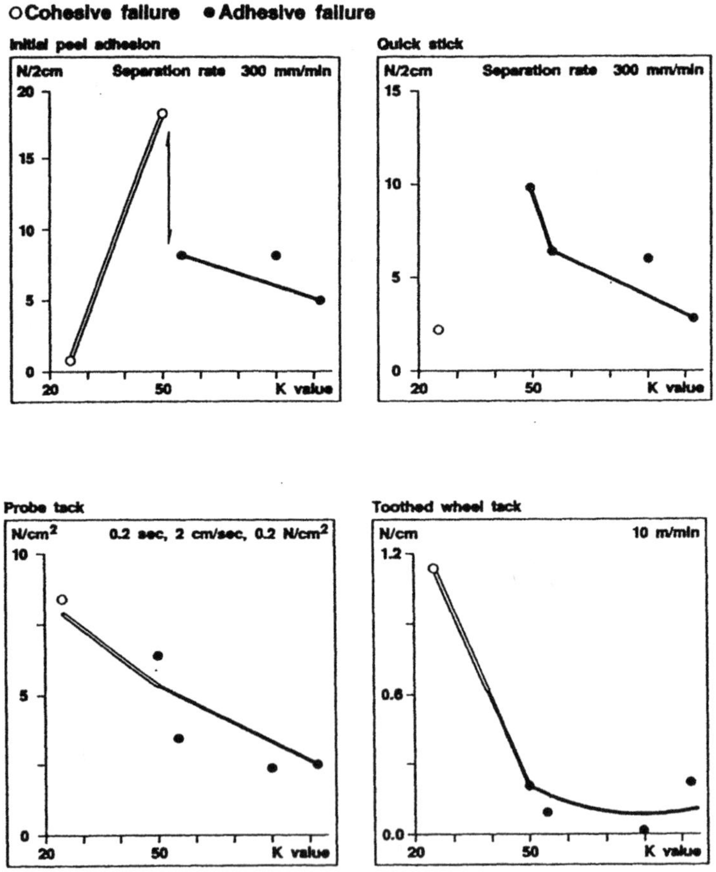

Fig. 15-7. Tack as a function of the *K* value. (*Courtesy of BASF AG, Ludwigshafen.*[8])

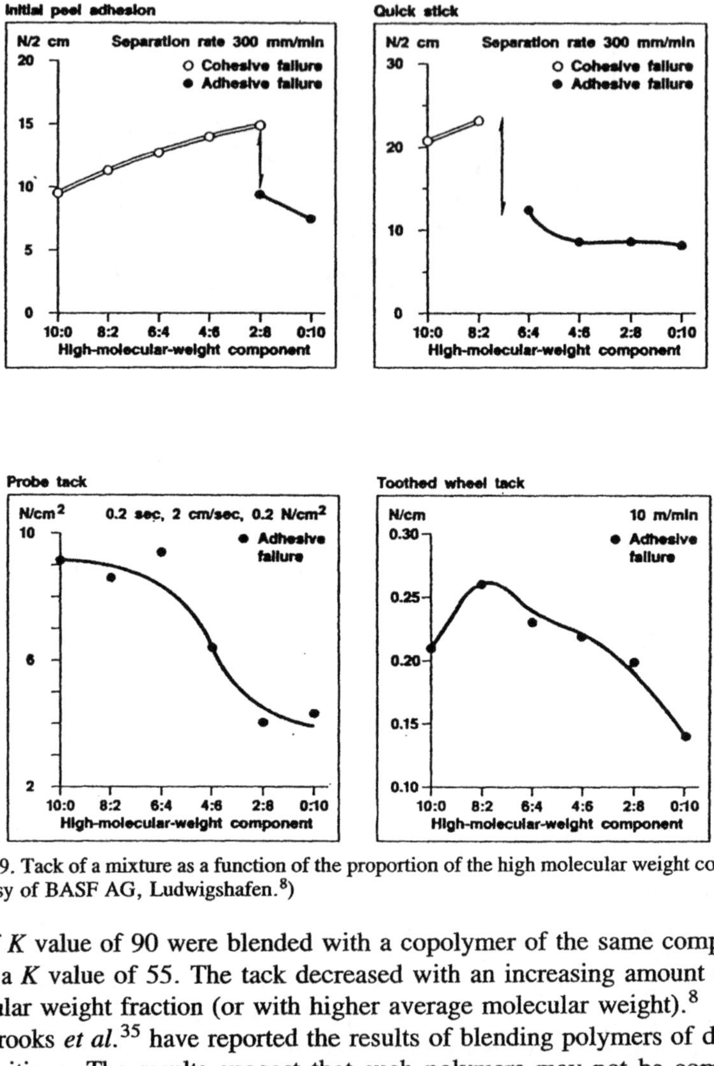

Fig. 15-9. Tack of a mixture as a function of the proportion of the high molecular weight component. (Courtesy of BASF AG, Ludwigshafen.[8])

mer of *K* value of 90 were blended with a copolymer of the same composition but of a *K* value of 55. The tack decreased with an increasing amount of high molecular weight fraction (or with higher average molecular weight).[8]

Brooks *et al.*[35] have reported the results of blending polymers of different compositions. The results suggest that such polymers may not be compatible and that a two-phase system is formed. The more similar the composition, the better such polymers will blend. Figure 15-10 shows the properties of polymer C and D blends. Peel adhesion and loop tack exhibit maxima, suggesting phase separation in the blend. Figure 15-11 shows the properties of polymer E and F blends. These polymers are of similar composition and are apparently compatible. The change of properties is additive, and such results could be achieved just as easily by tailoring the polymer properties during polymerization, rather than by blending. The composition of these polymers is given in Table 15-3.[35]

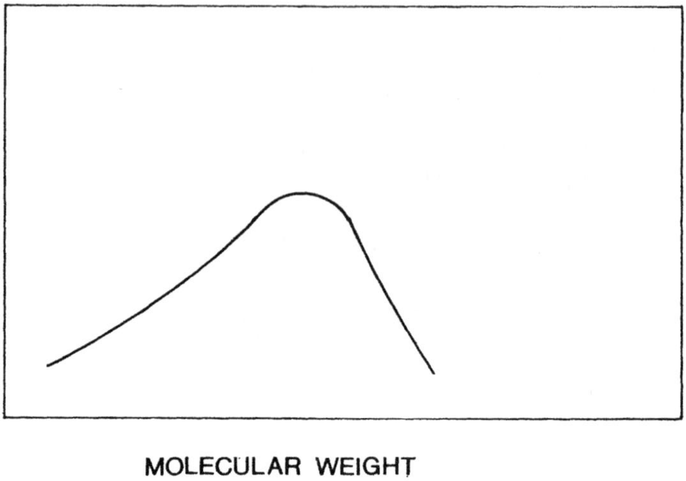

Fig. 15-12. Molecular weight distribution of a typical acrylic pressure sensitive adhesive.

The molecular weight distribution of acrylic pressure sensitive adhesives is quite broad, containing a considerable amount of the low molecular weight fraction. A typical molecular weight distribution is shown in Figure 15-12. In some cases, it has been found that blending two polymers of different molecular weight gives improved properties. Such blends exhibit two peaks on the molecular weight distribution curve, as illustrated in Figure 15-13.

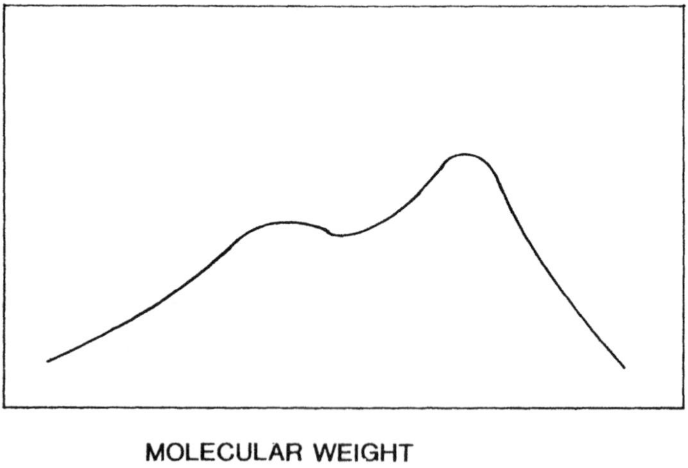

Fig. 15-13. Molecular weight distribution of a typical blend of two acrylic pressure sensitive adhesives.

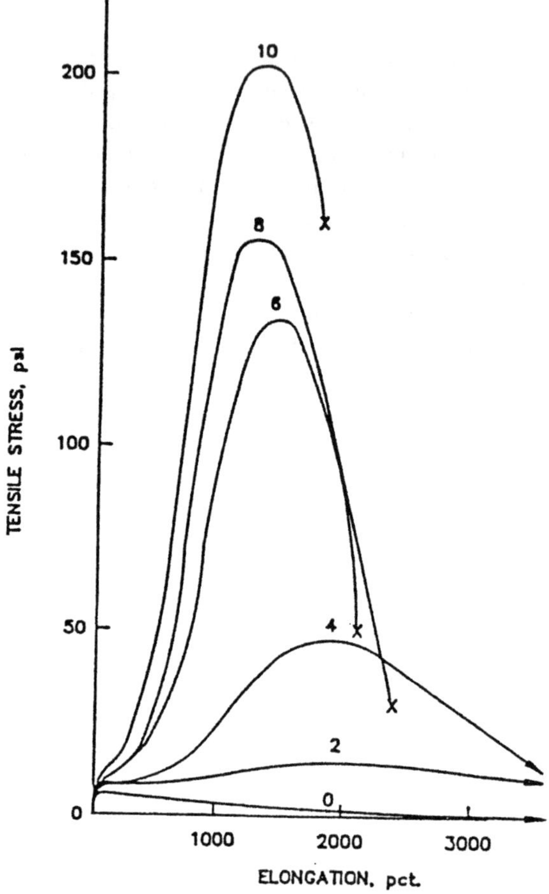

Fig. 15-15. The effect of cross-linking on tensile and elongation of a pressure sensitive acrylic polymer. The number on each curve is an index representing the amount of cross-linker. (Courtesy Pressure Sensitive Tape Council.)[36]

The effect of cross-linking on the mechanical properties of acrylic adhesives is reported by Dale *et al.*[36] Tensile strength increased by as much as a factor of 20 and the elongation decreased by a factor of at least 2 (see Figure 15-15). The results of dynamic testing show a large increase in the storage modulus with an increasing amount of cross-linker.

The transition from a cohesive to an adhesive failure during peeling advances to a lower peel rate and higher temperature with increasing cross-linking density. Cross-linked adhesive may not exhibit a noticeable transition at all, but fail adhesively under all conditions.

The cross-linking density should be low for pressure sensitive adhesives—

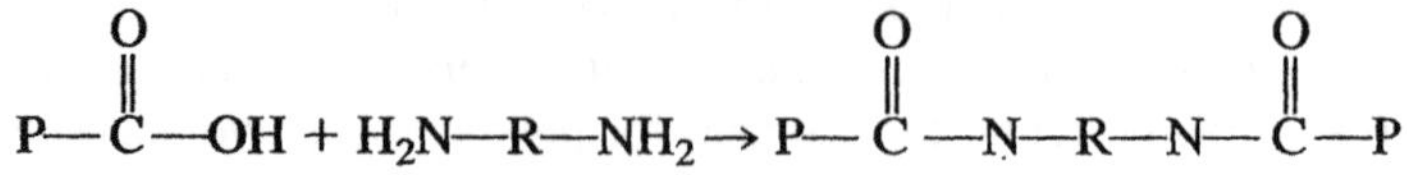
(1)

3. Ester group formation with polyepoxides.

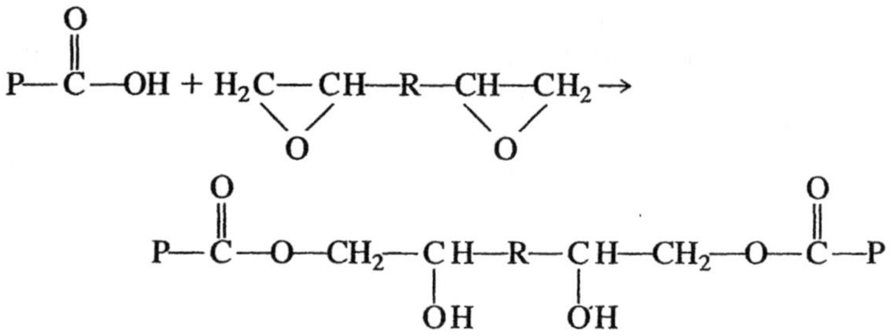
(2)

4. Ester formation with polyols. Ester formation is a slower reaction and requires higher temperature and removal of water.

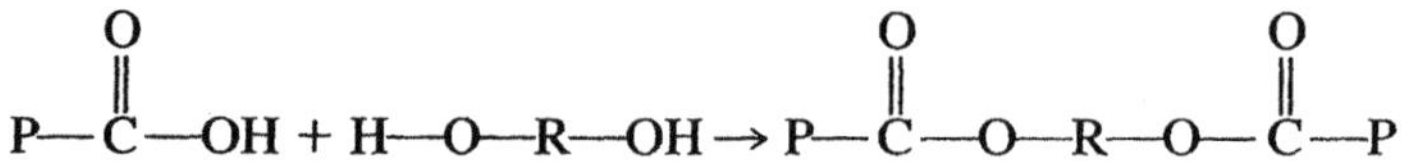
(3)

5. Ester formation with polyimides.

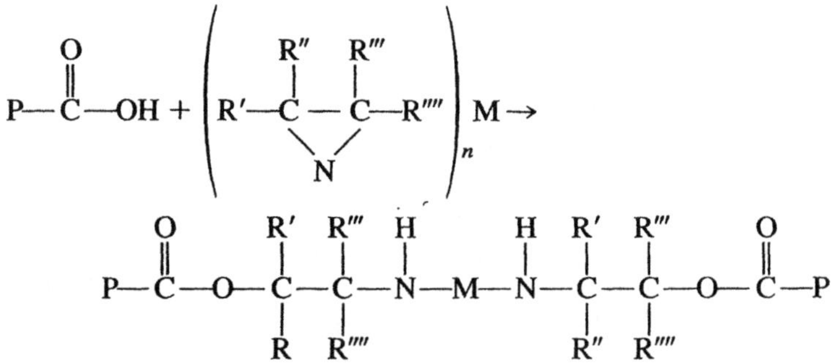
(4)

6. Reaction with mono and polycarbodimides.
7. Amide formation with polyisocyanates.
8. Anhydride formation.
9. Radiation decarboxylation leading to radical formation and cross-linking by polymerization and recombination.

Self-curable adhesives based on fumaric acid esters have been prepared by incorporating carboxylic groups bearing monomers and compounds that contain hydroxy, epoxy, *N*-methyloamide, aldehyde, or isocyanate groups.[51]

A pressure sensitive adhesive latex containing carboxylic groups was reacted with maleic hydrazide (two parts by weight) at room temperature. Adhesives for applications to human skin were cross-linked by a cross-linking agent that has tetrahydro-4*H*-12,3,5-oxadiazine-4-one.[52] A carboxylic group containing adhesive was cured with 2-oxazoline.[53] Polyfunctional aziridines have been used as room temperature cross-linking agents for carboxylic group containing latexes.

Hydroxyl groups are also useful for cross-linking purposes, although their reactivity is usually lower than that of carboxylic groups. Hydroxyl groups react with epoxy, urea, and formaldehyde condensation products, and with acidic groups at elevated temperatures.[54] Glyoxal may be added to the adhesive emulsion. It reacts with the dried film by acetal formation with polymers containing hydroxyl groups.

Epoxy groups react with acids, anhydrides, and amines. Glycidyl group-containing monomers, especially glycidyl methacrylate, are used as cross-linking agents. The epoxy group remains largely unchanged during the emulsion polymerization—only 7% of the epoxide will be converted to dihydroxide, which might have some reactivity with carboxyl groups.[39] The inclusion of an epoxide or hydroxide monomer in conjunction with a carboxyl-containing monomer lowers the cross-linking temperature. Self-reaction with an epoxide may occur when an aminomethacrylate (dimethylaminoethyl methacrylate) is present. It acts as a catalyst in polymerizing the epoxy group, resulting in cross-linking. A room temperature self-curing pressure sensitive adhesive has been prepared by incorporating five parts of acrylic acid and one part of glycidyl methacrylate into the polymer.[55]

Acrylic adhesive containing glycidyl acrylate or methacrylate has been cross-linked by heating at 60–100°C in the presence of zinc chloride catalyst,[56] and at 100°C without the catalyst, but in the presence of carboxylic groups.[57]

Acrylamide and its derivatives are frequently used as cross-linking monomers for the preparation of self-cross-linking emulsions that cross-link on drying.[39] The monomer is water soluble and has a high reactivity and therefore it is not easily copolymerized; it tends to form water-soluble homopolymer. A pressure sensitive adhesive consisting of 2-ethylhexyl acrylate/butyl acrylate/ethyl methacrylate/acrylic acid/acrylamide (58:30:8:2:0.8) was prepared containing very high solids (73%).[58] Methylolacrylamide and methoxymethylacrylamide have a lower polymerization reactivity and therefore are more easily copolymerized. Both of these monomers are water soluble and have a similar cross-linking efficiency to that of acrylamide. Methylolacrylamide is a useful cross-linking agent with epoxy or carboxylic groups containing polymers. Iso-

Table 15-4. Effect of Polyethylene Glycol Dimethacrylate and Acrylic Acid on Adhesive Properties

FORMULATION	PEEL STRENGTH (g/cm)	COHESIVE STRENGTH (g/cm)	SOLVENT RESISTANCE (sec)		
			ACETONE	TOLUENE	ISOPROPANOL
1	130	1800	450	46	480
2	40	3400	92	37	195
3	70	3300	1000	56	1800
4	86	4500	>1000	54	>1800
5	130	6800	>1000	>100	>1800

FORMULATIONS (mole percent)

NUMBER	2-ETHYLHEXYL ACRYLATE	METHACRYLIC ACID	ACRYLIC ACID	METHYL ACRYLATE	POLYETHYLENE GLYCOL DIMETHACRYLATE
1	94.8	5.2	—	—	—
2	89.9	10.1	—	—	—
3	94.8	5.2	—	—	0.02
4	81.0	4.8	—	14.2	0.02
5	95.0	—	5.0	—	0.01

min, required to shear the adhesive. Solvent resistance was measured by bonding the tape to a thin sheet of paper and by immersing the laminate in a solvent. Time to bond failure is recorded as solvent resistance. Other observations indicate that a small amount of polyethylene dimethacrylate is an effective cross-linker in the presence of acrylic or methacrylic acid, but not alone. The use of a larger quantity of the cross-linking monomer resulted in a gel in the case of a solution polymer or in a grainy adhesive in the case of an emulsion polymer.[68] Generally, only a few percent of difunctional cross-linking agent can be used, as a larger amount causes the precipitation of latex particles.[69]

Latent cross-linking can be introduced into the polymers by incorporating ethylene glycol dimethacrylate or its low molecular weight polymers. Ethylene glycol dimethacrylate, and especially its dimer, trimer, or low molecular weight polymer, consists of a long, flexible chain with the methacrylate groups at the ends. The reactivities of both methacrylate groups are equal until one group becomes attached to the polymer chain. At this time, the reactivity of the pendant group is decreased, because of the restriction of its movement. Thus, at the end of the reaction, a number of pendant methacrylate groups are left unreacted. These are available for cross-linking among themselves at temperatures higher than the polymerization temperature, because their mobility, as well as the mobility of the main polymer chain, increases with increasing temperature. These groups then can be brought into contact to accomplish the cross-linking.[70]

Acrylic adhesives have been cross-linked by exposure to accelerated elec-

acid/glycidyl methacrylate.[86] The copolymer was of sufficiently low molecular weight to be coatable with the addition of only a small amount of solvent. When mixed with zinc chloride solution and heated to 100°C for 10 minutes, the coating cured to a solid tacky adhesive coating.

Acrylic hot melt adhesives have been difficult to prepare and many attempts resulted in a polymer of insufficient cohesive strength at room temperature. It has not been commercially feasible, so far, to prepare acrylic block copolymers equivalent to styrene/isoprene/styrene block copolymers. Competing side reactions occur during anionic polymerization reactions with acrylate monomers, lowering the yield of the desired product.[87]

Various methods have been used to obtain acrylic hot melts. Attempts have been made to blend high and low molecular weight polymers and to incorporate a tertiary amine-containing monomer, which is reacted with a heavy-metal-containing compound. The use of methacrylate-terminated polystyrene macromolecular monomers is proposed for the preparation of acrylic hot melt adhesives.[88] These monomers have been prepared and copolymerized with acrylic monomers in solution polymerization yielding a graft copolymer.[89] The interaction of the grafted groups improves the cohesive strength at room temperature, but the bonds are weakened at elevated temperatures, resulting in a polymer of sufficiently low viscosity to be applied by hot melt technique.

Hot melt acrylic adhesive consisting of acrylic copolymers containing NCO-reactive groups, blocked isocyanate cross-linker, and a catalyst is suggested.[90]

Hot melt adhesives have also been prepared by the addition of a small amount of monomers containing carboxylic, hydroxyl, and amino groups.[91]

SECONDARY BONDING

The cohesive strength of an adhesive is also influenced by other bonds besides the covalent cross-linking. The most common bonds are the ones caused by dispersion forces that act between all atoms and are largely responsible for intermolecular adhesion. These forces act over a short distance; the interaction energy decreases with the sixth power of the distance.

Cohesive properties of a polymer can be affected by introduction of polar groups that interact forming secondary bonding. Such interactions are classified as hydrogen bonding, dipole–dipole and dipole-induced dipole bonding. It is not always possible to distinguish among various types of bonds, since the same groups might be involved in different types of bonding. While the effect of such bonding is similar to covalent cross-linking, such polymers are soluble, and the strength of such bonds decreases rapidly with increasing temperature because the bond strength decreases with increasing distance between atoms.

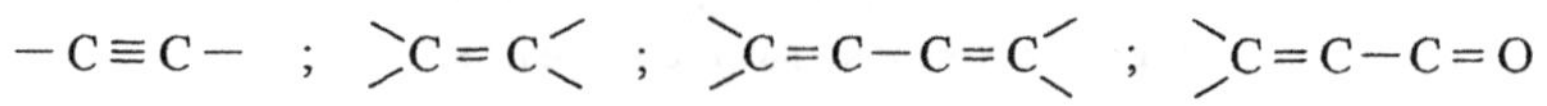
(7)

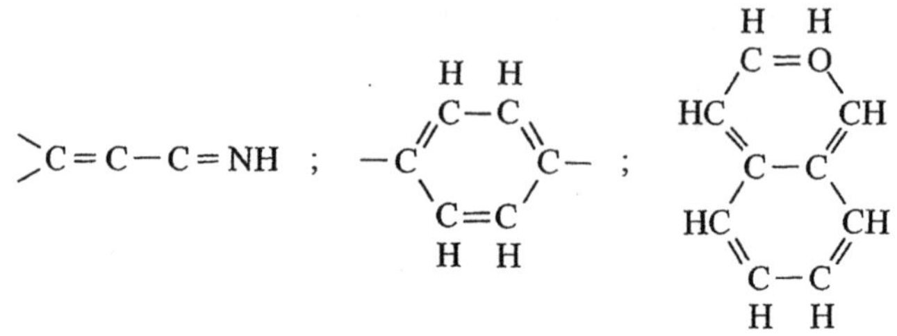
(8)

Hydrogen bonding takes place between the negative pole of a dipole and the positively charged H atom of another dipole as shown below.

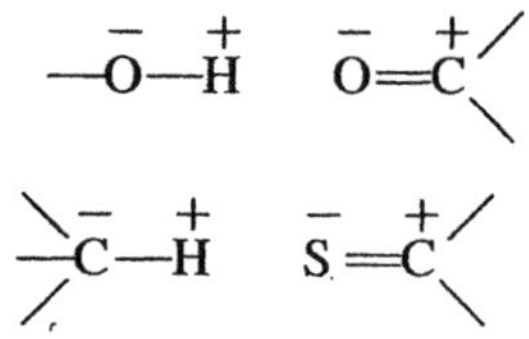
(9)

The range of hydrogen bonding is 2.6–3.0 Å, and the bond is very strong.

Acrylonitrile has a very strong dipole —C≡N capable of forming hydrogen bonds, or interacting with other dipoles. Introduction of acrylonitrile as a comonomer, even in small amounts, has a great effect on the pressure sensitive properties of acrylic polymers. It serves well as an example of the importance of secondary bonding. Other polar groups generally have a lesser effect. Table 15-6 shows the effect of acrylonitrile.

The glass transition temperature increases with an increasing amount of acrylonitrile. The tack remains constant, except at higher concentrations of the polar monomer. Sample L, containing 10% acrylonitrile, shows a decrease in tack. The creep resistance increases with increasing acrylonitrile content. The increase is less pronounced at elevated temperatures, but considerable at room temperature.

Figures 15-16 and 15-17 show the peel force as a function of peel rate and temperature. The transition from cohesive to adhesive failure is observed in some polymers (H, I, and J in Fig. 15-16). In the range of adhesive failures, the peel force decreased with increasing acrylonitrile content. The decreasing compliance of the adhesive distributes the stress over a smaller area that accounts

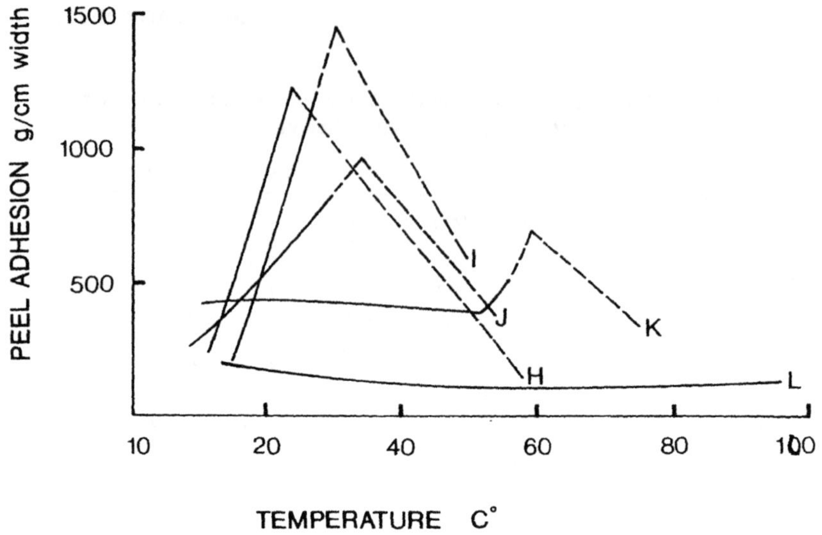

Fig. 15-17. Dependence of peel adhesion on temperature for various acrylonitrile-containing polymers shown in Table 15-6.

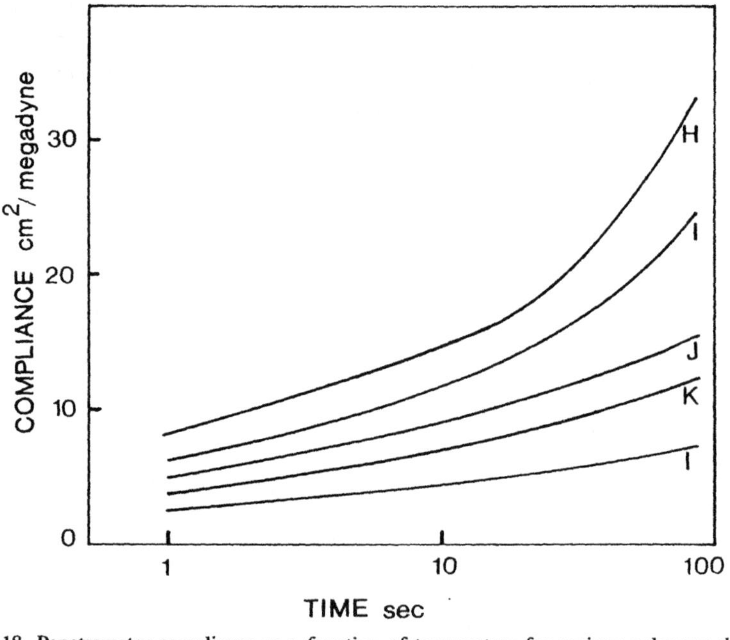

Fig. 15-18. Penetrometer compliance as a function of temperature for various polymers shown in Table 15-6.

Table 15-7. Effect of Bound Sodium on Pressure Sensitive Adhesive Properties of the Polymer

SAMPLE	BOUND SODIUM (wt %)	TACK (g/cm^2 at 10 g CONTACT PRESSURE)	PEEL ADHESION (g/cm width)	RESISTANCE TO CREEP (hr)
1	0.0	163	524	0.6
	0.06	123	312	4.5
	0.08	77	480	1.7
	0.32	40	368	>100
2	0.0	607	971	0.2
	0.34	247	513	4.2
	0.50	142	502	7.5
	0.52	133	558	2.9
	0.59	92	279	68
	0.69	0	279	64

The effect of ionic binding on splitting temperature is shown in Table 15-8. Splitting temperature is defined as the temperature at which the mode of failure changes from adhesive to cohesive. Instead of a clean peel, as obtained in the case of adhesive failure, the polymer splits at or above this temperature, leaving a residue on the steel plate. The data shown are obtained from the

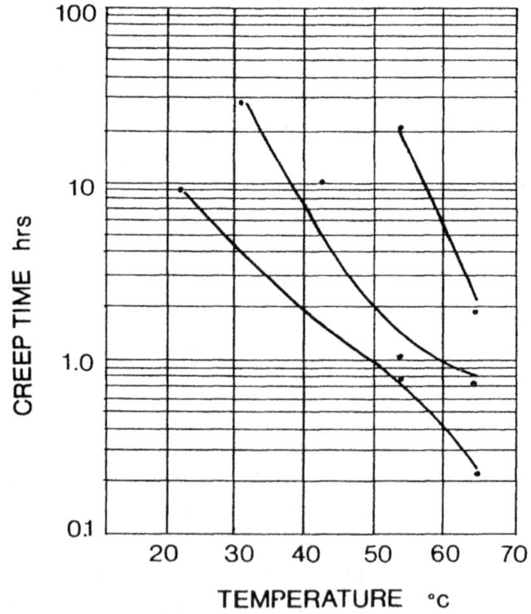

Fig. 15-19. Creep resistance as a function of temperature. Sodium content: Top curve 0.8%; middle curve 0.08%; bottom curve 0%.

properties of many polymers, including pressure sensitive adhesives. The effect could be interfacial or it could be dependent on the change of adhesive bulk properties; copolymerization with carboxylic acids yields polymers of higher plasticity.[21] The effect of carboxylic groups on the peel adhesion in poly(butyl acrylate) pressure sensitive adhesives has been studied by Aubrey and Ginosatis and is discussed in several publications.[9,96,97] This work is briefly mentioned in the chapter on peel adhesion. An attempt was made to separate the interfacial from the bulk effects, leading to the conclusion that the carboxylic groups have indeed a beneficial effect on peel adhesion attributable to the interfacial changes, rather than to a change in adhesive bulk properties.

Mayo and Reegan reported their findings on the effect of copolymerizing methyl methacrylate with acrylic acid, acrylamide, methacrylic acid, and methacrylamide.[14] The peel adhesion of these hard non-pressure sensitive copolymers is shown in Table 15-9. Acrylic acid was the most effective comonomer. The decrease of peel adhesion in the case of a higher concentration of acrylamide is attributed to the increase of crystallinity. The authors explain the difference

Table 15-9. Effect of Various Comonomers on the Peel Adhesion

	PEEL ADHESION (g)			
COMONOMER (MOLE %)	ACRYLIC ACID	ACRYLAMIDE	METHACRYLIC ACID	METHACRYLAMIDE
0	160	160	160	160
2	220	200	170	140
5	250	270	170	170
10	310	280	100	100
15	—	140	—	—
20	420	—	—	—

COMONOMER (10 mole %)	PEEL ADHESION (g)
$CH_2 = C(CH_3)CO_2CH_2CF_3$	90
$CH_2 = C(CH_3)CO_2CH_2CH_3$	170
$CH_2 = C(CH_3)CO_2CH_2CH_2CN$	170
$CH_2 = C(CH_3)CO_2CH_2Si(CH_3)_2OSi(CH_3)_3$	210
$CH_2 = C(CH_3)CO_2CH_2CH_2Cl$	220
$CH_2 = C(CH_3)CO_2CH_2CH_2Br$	260
$CH_2 = C(CH_3)CO_2CH_2CH(OH)CH_3$	260
$CH_2 = C(CH_3)CO_2CH_2CH_2OH$	270
$CH_2 = C(CH_3)CO_2CH_2CH_2$— 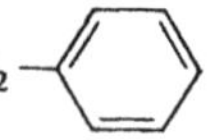	290
$CH_2 = C(CH_3)CO_2CH_2$— 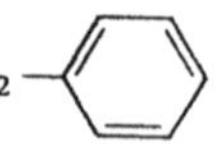	350
$CH_2 = C(CH_3)CO_2CH_2CH_2OCH_2CH_3$	360

do not yield films with good low-temperature properties, and still might cause loss of adhesion, since only a small amount of plasticizer accumulation on the adhesive surface is required to produce a drastic reduction of tack.

Plasticizer–adhesive compatibility determines the rate of plasticizer migration and the amount of the plasticizer that will migrate from the film into the adhesive. If at equilibrium conditions the amount of plasticizer in the adhesive is large, its effect on the adhesive properties, especially on shear resistance, will be extensive. If the amount of the plasticizer in the adhesive is low, the adhesive properties might be affected only slightly.

Plasticizer, in addition to its effect on bulk adhesive properties, may also accumulate at the film–adhesive interface, causing loss of adhesive coating anchorage, or more often, at the adhesive surface, causing a loss of tack and adhesion. The latter condition develops when the adhesive is not compatible with the plasticizer and does not form a barrier to its migration. Plasticizer on the adhesive surface may also be picked up by the tape when the adhesive is in contact with the film in a roll. Infrared techniques for surface analysis are useful in determining the presence of plasticizer on the adhesive surface.

While most pressure sensitive adhesives are affected by the migratory plasticizers, some adhesives exhibit only a low and tolerable level of reduction of adhesive properties. Lower polarity, as influenced by some monomers of the acrylic adhesive, decreases its compatibility with generally polar vinyl plasticizers, and such adhesives are less susceptible to a reduction of adhesive properties. Cross-linking decreases plasticizer solubility in the adhesive and substantially improves its performance.

COMPOUNDING

Acrylic pressure sensitive adhesives do not require compounding in order to develop the optimum combination of properties. The adhesive properties, how-

Table 15-10. Effect of Tackifying Resin on Adhesive Properties

RESIN CONTENT (phr)	PEEL ADHESION (g/cm width)	CREEP RESISTANCE (hr)	TACK[a] (g/cm^2) AT A PRESSURE OF		
			10 g	100 g	500 g
0	275	32	117	530	515
5	319	25	183	540	548
10	341	11	200	540	555
20	352	5	322	548	555
30	462	2.7	315	570	622

[a] Polyken Probe Tack Tester, contact time 1 sec, rate of separation 1 cm/sec, temperature 23°C.

Larger amounts of nonreactive pigments such as clay and calcium carbonate can be tolerated before a noticeable effect on peel adhesion is observed. Zinc oxide and colloidal silica are active pigments, and much smaller amounts can be tolerated before the adhesive properties are lost.

Flame-retardant properties are incorporated into an acrylic pressure sensitive adhesive by compounding with phosphate plasticizers and antimony oxide.[29]

Acrylic pressure sensitive adhesives have been compounded with other polymers for special effects. Compounding with a noncompatible synthetic elastomer gave an adhesive especially suitable for vibration dampening over a wide temperature range.[91]

COATING METHODS

Pressure sensitive acrylic adhesives are available in various forms for coating. Solutions in hydrocarbon solvents are commonly used, although the method is under pressure to be replaced by other adhesive coating techniques, for environmental and economical reasons. Many large coaters are choosing to remove the solvent vapor from the exhaust by activated carbon adsorption and then to recover the adsorbed solvent. Despite a large investment, this method is economically feasible for controlling vapor emissions to the atmosphere and conserving on the solvent usage.

Adhesive polymer solutions are usually made by solution polymerization, but sometimes the polymer is emulsion polymerized, coagulated, washed, and then dissolved. The normal range of polymer solids in solution coatings is 25–40%. Some high solid content (50–70%) nonaqueous dispersions have also been reported.[98]

Liquid acrylic oligomers or monomers have been polymerized *in situ* to pressure sensitive adhesives. The reaction can be initiated by radiation or by thermal decomposition of a catalyst.

Liquid acrylic polymers containing hydroxyl groups and isocyanate-terminated prepolymer are combined and reacted to form a pressure sensitive adhesive. Such a system is on the market[19] and is processed as a high solid solution (80% solids) or as a 100% solid coating. At 150°C, complete cure, using dibutyl tin dilaurate catalyst, is obtained in 30–45 sec; at 93°C, in 2.5–4 min. The adhesive properties depend on the OH/NCO ratio, which is adjusted prior to coating.

Acrylic latexes are widely used for pressure sensitive applications. They are compounded with thickeners, antifoaming agents, wetting agents, and other ingredients to render the latexes easily coatable. The additives must be carefully chosen and evaluated because of their possible effect on the adhesive properties.

SOLUTION POLYMERIZATION

Solution polymerization of pressure sensitive acrylates is a homogeneous free radical polymerization process. The monomers are soluble in the solvent and the polymer is also usually soluble in the same solvent or solvent blend. The main advantage of the solution polymerization is that a pure polymer is obtained in a ready-to-coat condition. Unfortunately, the solvents used also act as chain transfer agents and cause a decrease in the molecular weight that has a negative effect on the physical properties of the adhesive. Ethyl acetate, cyclohexane, toluene, and *n*-heptane are commonly used solvents. The two former ones give a higher molecular weight polymer. The lower molecular weight must be compensated by cross-linking, hydrogen bonds, or other means. Various multivalent metal ions are used to cross-link solution polymers.[46]

The lower the chain transfer constant, the higher the molecular weight. The chain transfer constant is defined as the ratio of the rate constant for transfer with the solvent (termination) and for propagation. Impurities in the solvent increase the chain transfer.

The molecular weight also depends on the monomer concentration. Lower molecular weight polymers are obtained at low monomer concentrations. High polymer concentrations in the solution increase branching and cross-linking due to chain transfer with a dead polymer.

Various chemical initiators are used to start the polymerization. Organic, oil-soluble peroxides are commonly used. Their main disadvantages are occasional formation of peroxide cross-links and the dependence of the decomposition rate on the type of solvent used and other variables. Azo compounds, such as 2,2′-azobisisobutyronitrile, do not possess some of the disadvantages of peroxides and are especially useful for solution polymerization. Initiator concentrations of 0.05–2% of the monomer weight are used. A lower amount of initiator favors the higher molecular weight. The rate of the polymerization reaction is directly proportional to the monomer concentration, and it is related to the square root of the initiator concentration.

One of the disadvantages of solution polymerization is the increase of viscosity with an increasing degree of conversion, increasing molecular weight, and solids content. Formation of difficult-to-handle gels can occur easily if any cross-linking takes place.

The most common way to carry out the solution polymerization of acrylic pressure sensitive adhesives is to start with a high monomer concentration (as high as 70%) using a minimum amount of catalyst, choosing a solvent of a low chain transfer constant, and running the reaction at the reflux conditions that help to remove the heat of reaction. Remaining monomer is added as the reaction proceeds, solvent is added to regulate the viscosity, and additional catalyst keeps the reaction going. The rate of reaction in solution polymerization

$$N = \chi\left(\frac{\rho}{\mu}\right)^{2/5}(aS)^{3/5}, \tag{10}$$

where

N = number of particles
ρ = rate of radical production
μ = rate of volume increase of a particle
a = area occupied by one soap molecule
S = total amount of soap present in the system
χ = a constant which varies between 0.37–0.53

The constant's value is 0.53 if an assumption is made that only micelles can capture radicals. Since it is clear that polymer particles can capture free radicals as well, the constant's value of 0.53 constitutes the upper limit. The value of 0.37 for the constant χ is obtained if an assumption is made that the number of radicals entering a particle is independent of the particle radius. This constitutes the lower limit. The emulsion particle size can be controlled by several variables. An increased amount of surfactant decreases the particle size simply by providing a larger number of micelles. The delay in monomer and surfactant addition increases the particle size as compared to adding all the surfactant at once. An increase in reaction temperature increases the particle size, as well as the presence of electrolytes. Acrylic acid and other water-soluble monomers can reduce the latex particle size.

In practice deviations from theoretically predictable behavior are observed quite often. This is mainly due to the water solubility of some acrylic monomers used for acrylic adhesives. Polymerization of water-soluble polymers takes place in the aqueous phase. Some water-soluble monomers, such as methacrylamide, are difficult to copolymerize because of the formation of homopolymer formed by solution polymerization before the methacrylamide monomer can copolymerize with other monomers. It has been shown that butyl acrylate emulsion polymerization does not obey the Smith-Ewart theory.[108] Butyl acrylate is, of course, a very important monomer for pressure sensitive acrylic adhesives. The reasons for the deviation are slow termination and probably radical transfer out of particles.

EMULSIFIERS

Emulsifiers perform several important functions in emulsion polymerization: they form micelles, where the polymerization takes place, and they protect the polymer particles from coalescence. Protection by emulsifiers is especially important during the critical stage when all the monomer has migrated to polymer–monomer particles that are soft and can coalesce upon contact.

latexes without migratory surfactants is to employ polymerizable surfactants. Such monomers contain functional groups that are capable of latex stabilization and also vinyl or other groups that can copolymerize with other monomers. Such compounds are incorporated into the polymer chain and cannot migrate. Patent literature contains numerous examples of such compounds and their use in emulsion polymerization.

Carboxylic groups containing monomers are very important in the formation and stabilization of emulsion particles. Acrylic, methacrylic, and itaconic acids and their salts have been used for these purposes. Carboxylic groups orient themselves at the particle surface, providing the stabilizing effect. Emulsion stability achieved by the use of carboxylic groups is superior to that obtained by conventional surfactants, because a higher concentration of carboxylic groups can be placed on the particle surface by the polymerizable surfactants. Stable pressure sensitive adhesive emulsions have been prepared by employing acrylic acid neutralized with morpholine and also vinyl pyrrolidone as a comonomer.[112] Pressure sensitive adhesives have been prepared employing amic salt monomers.[113]

Sodium vinyl sulfonate is used in vinyl acetate, styrene, and vinylidene chloride polymerization, but not often in the polymerization of acrylic monomers, because of the unfavorable reactivity ratio. Quaternary ammonium salt of vinyl sulfonic acid was used in the emulsion polymerization of acrylic esters.[114] Sodium sulfoethyl methacrylate is a useful polymerizable surfactant and has been employed for the preparation of removable pressure sensitive adhesives.[115] Sulfonic groups containing monomers were first copolymerized with a small quantity of another monomer and then the remaining comonomers were added. A thin envelope of ionic groups containing copolymer was formed around the nonionic core by this procedure. Similar results were achieved by introducing the polymerizable surfactants at the end of their polymerization reaction.[116]

Polymerizable cationic compounds are useful as surfactants. Dimethylaminoethyl methacrylate was reacted with ethyl acrylate and used to stabilize acrylic emulsions, including pressure sensitive adhesives.[117] Quaternary ammonium salts with a lipophilic group are described for uses as polymerizable surfactants.[118]

INITIATORS

Chemical free radical producers are most often used to initiate the emulsion polymerization. The initiators are classified into dissociative and redox systems. The former yields free radicals upon dissociation, and the latter, upon reaction between a reducing and an oxidizing agent.

The persulfate/ferrous ion system can be used for initiation of emulsion polymerization and proceeds according to the following reaction:

$$S_2O_8^{2-} + Fe \xrightarrow{2+} SO_4^{-}\cdot + SO_4^{2-} + Fe^{3+} \quad (13)$$

The persulfate/bisulfite system is a popular redox system for initiation at 50°C and higher temperatures. Free radicals are produced according to the reaction:

$$S_2O_8^{2-} + HSO_3^{-} \rightarrow SO_4^{2-} + SO_4^{-}\cdot + HSO_3\cdot \quad (14)$$

The sulfate radical is more important in the initiation reaction than the bisulfite radical. To polymerize 10 kg of monomer, 25 g of potassium persulfate dissolved in 1 liter of water and 5 g of sodium bisulfite in 500 ml of water were sufficient. Persulfate solution was added to the emulsion and the bisulfite was added gradually. The reaction was carried out at 70°C. Persulfates give a higher conversion of monomer to polymer than peroxides.

The polymerization reaction does not start immediately upon addition of the catalyst. Usually a delay of 5 to 30 min is observed. This induction period is attributed to the presence of inhibitor in the monomers and to the oxygen in the air. The length of the induction period varies depending on these variables, initiator concentration, pH, and the rate of mixing. Fast mixing tends to inhibit the reaction. For commercial manufacturing of acrylic pressure sensitive adhesives, it is not necessary to remove the inhibitors from the monomers or to carry out the reaction under an inert atmosphere.

After addition of the initiator and after a suitable induction period, a temperature rise is observed. The reaction can be carried out at approximately constant temperature by applying cooling and withholding the addition of the initiator until the temperature starts decreasing. Further addition of the initiator causes a rise again and the reaction is allowed to proceed at some control temperature. Sometimes the control temperature is chosen to be the boiling point of the lowest boiling monomer. The removal of the heat of reaction is much easier if the reaction is run at reflux conditions. Vinyl acetate containing pressure sensitive adhesives are often polymerized at 72–73°C, the boiling point of vinyl acetate.

The monomer emulsion can be prepared separately in a disperser and introduced pre-emulsified into the reactor, or the monomer mixture can be added to the reactor containing water and emulsifiers and dispersed by mixing. A portion of the monomers can be withheld and added gradually as the reaction proceeds. This technique allows an easier control of the exotherm. Likewise, some of the surfactant can be withheld in order to control the number of particles and thus the particle size. The withheld surfactant does not participate in micelle formation and is used for protection of polymer particles only.

Monomer is pumped from the storage to the reactor, preblending tank, or a disperser, if used in the process, via a volumetric measuring device. The reactor is a jacketed, baffled vessel, constructed from 316 SS, from glass-lined steel, or other suitable material. It is equipped with a condenser, or at least a cooled vapor pipe, and a mixer. A turbine-type mixer with variable speed control or at least two speeds is preferred. Faster speed is used to prepare the emulsion and slower speed is used during the polymerization. Agitation is quite important. Fast mixing can extend the induction period and cause polymer coagulation. The emulsion particles are very soft at the beginning and can easily coagulate if the shear forces are too high. The temperature is controlled by cooling and by withholding initiator addition that slows down the reaction. Figure 15-21 shows a schematic flow diagram of a batch polymerization process.

The polymerization can also be carried out in a continuous manner. The advantages gained in decreasing the size of the equipment are minor, and continuous polymerization is not important for polymerization of pressure sensitive adhesives.

The polymer emulsion is filtered, preferably by a rotating screen, and stored for further use. Stainless steel, aluminum alloy, reinforced polyester, or epoxy resin-lined tanks are used for emulsion storage. The tanks should be

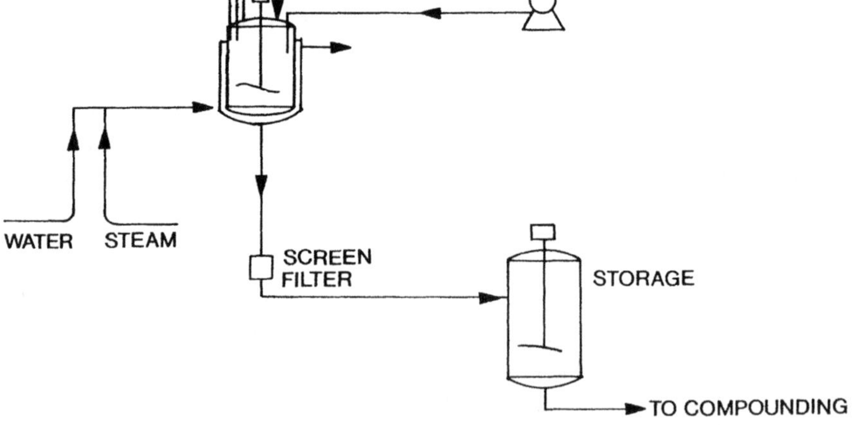

Fig. 15-21. Schematic diagram of batch polymerization equipment.

Appendix 15A. Monomers Used for Pressure Sensitive Acrylic Adhesives

COMONOMER	REFERENCE
Acrylates	
acrylic acid	many references
acrylic acid formal	17
methyl	31,48,55,69,104,119,120, 121,122,123,176
ethyl	many references
n-propyl	60,104
butyl	many references
pentyl	125,126
isopentyl	104,127
2-methylbutyl	128
amyl	21,59
n-hexyl	60,128,129
2-ethylbutyl	60,127,129
methylpentyl	43,130,131,178
heptyl	43
octyl	21,22,27,28,59,122,128,131, 132,133,178
2-ethylhexyl	many references
n-decyl	9,129,134,135,178
isobornyl	92
n-undecyl	125,128
n-dodecyl	28,60,125,129
tridecyl	134
lauryl	21,55,60,135
stearyl	123,136
10-cyclohexylundecyl	137
fusel oil	59,131
6-methoxy	43,178
methoxyethyl	16,25,126,179

Appendix 15A (*cont.*)

COMONOMER	REFERENCE
n-dodecyl	43,60,125,129,134
lauryl	127
isobornyl	92,124,126
hydroxyethyl	50,78,131,150
hydroxypropyl	50,71,78,140,151
methoxymethyl	179
methoxyethyl	16,179
ethoxyethyl	16
butoxyethyl	16,179
cyanoethyl	65,150
dimethylaminoethyl	44,92,104,120,125,132,152, 153,175
t-butylaminoethyl	44,92,104
glycidyl	48,49,51,55,56,57,86,120,140,141, 144,145,154
benzophenoneglycidyl	140
3-(3,4-dicholorophenoxy)-2-hydroxypropyl	121
3-(2,4,6-trichlorophenoxy)-2-hydroxypropyl	121
3-(2,3,4,5-tetrachlorophenoxy)-2-hydroxypropyl	121
3-(pentachlorophenoxy)-2-hydroxypropyl	121
10-chlorodecyl	43,178
2-sulfoethyl	104
sodium-2-sulfoethyl	5,149
3-(trimethoxysilyl)propyl	63,133
Dimethacrylates	
ethylene glycol	66,67,128,140
1,3-butylene glycol	140
triethylene glycol	71,140,153
polyethylene glycol	65,66,67,140
dimethylaminoethyl	140
Trimethacrylates	
trimethylolethane	66,67
trimethylolpropane	146,177
Tetramethacrylates	
pentaerythritol	66,67,140
Acrylamides	
acrylamide	7,20,21,46,59,65,129,130,133
N-methylol	54,128,155,156
N-ethanol	173
N,N-dimethyl	144
N-(1,1-dimethyl-3-oxobutyl)	155
N-[2-(methyl-4-oxopentyl)]	155
N-*t*-butyl	63,119,142,152,156,157
N,N-diacetonyl	42,50
octyl	121,156
N-*t*-C_9	134
N-*t*-C_{12}	134

Appendix 15A (*cont.*)

COMONOMER	REFERENCE
fumaric acid diamide	129
fumaric acid nitrile	129
fumaric acid dinitrile	129
Crotonic acid derivatives	
crotonic acid	145,161,166
glycidyl crotonate	145
Itaconic acid derivatives	
itaconic acid	7,29,31,44,92,120,124,128,134,145
itaconic anhydride	44
half esters of itaconic acid	167
Citraconic acid derivatives	
citraconic acid	44
citraconic acid anhydride	44
half esters of citraconic acid	167
Succinic acid derivatives	
β-methacrylooxyloxyethyl acid succinate	120
Maleic acid derivatives	
maleic acid	44
maleic anhydride	31,44,53,55,119,133,136,140,144, 148,163,164
methyl maleate	140
butyl maleate	69,134
dibutyl maleate	122,134
t-octyl maleate	134
di(2-ethylhexyl) maleate	134,165
bis(2-hydroxyethyl) maleate	50
maleic acid amide	129
maleic acid diamide	129
N-methylol maleamide	173
maleic nitrile	129
maleic dinitrile	129
Maleamic acid derivatives	
Primene maleamic acid	65,134,168,169
N-methylol maleamic acid	173
N-*t*-C_{12} maleamic acid	174
Vinyl compounds	
vinyl chloride	60,135,158
vinylidene chloride	60,122
vinyl acetate	20,21,44,47,51,52,57,69,86,112, 119,120,121,122,124,131,132, 133,134,135,136,139,141,143,145, 146,147,149,150,151,154,158, 163,164,165,170,173,178,180,181,182
vinyl propionate	21,60,112,122
vinyl butyrate	21,141
vinyl valerate	21
methyl vinyl ketone	122

Appendix 15A (*cont.*)

COMONOMER	REFERENCE
3-[(2-methacryloxyethyl)dimethyl ammonium]propionate betaine	43
Isocyanates	
methylenebisphenyl-4-4′-diisocyanate	71
acrylic acid-2-isocyanate ester	144
polyisocyanate prepolymers	151
Silanes	
3-methacryloxypropyltrimethoxysilane	71,119,124

REFERENCES

1. Peckmann, H. V., and Rohm, O. *Berichte der Deutschen Chemischen Gesellschaft* **34:** 429 (1901).
2. Bauer, W. German Patent 575,327 (1933) (assigned to Roehm and Haas AG).
3. Satas, D. *Adhesives Age* **15**(10): 19–23 (1972).
4. Winslow, L. E., Bennett, R. E., and Overstreet, T. S. U.S. Patent 4,645,711 (1987) (assigned to Minnesota Mining & Manufacturing Co.).
5. Silver, S. F., Winslow, L. E., and Zigman, A. R. U.S. Patent 3,922,464 (1975) (assigned to Minnesota Mining & Manufacturing Co.).
6. Charbonneau, R. R., Heilman, S. M., Rasmussen, J. K., and Tumey, M. L. U.S. Patent 4,619,867 (1986) (assigned to Minnesota Mining & Manufacturing Co.).
7. Ulrich, E. W. U.S. Patent 2,884,126 (1959) and Re 24,906 (assigned to Minnesota Mining & Manufacturing Co.).
8. Druschke, W. Adhesion and Tack of Pressure-Sensitive Adhesives, presented at the AFERA Meeting, Edinburgh, Scotland (October 1–4, 1986).
9. Aubrey, D. W. Pressure-Sensitive Adhesives—Principles of Formulation. *Developments in Adhesives-1,* pp. 127–156 (William C. Wake, ed.). Applied Science Publishers, Barking, England, 1977.
10. Rehberg, C. E., and Fisher, C. H. *J. Am. Chem. Soc.* **66:** 1203 (1944).
11. Rehberg, C. E., and Fisher, C. H. *Ind. Eng. Chem.* **40:** 1429 (1948).
12. Nuessle, A. C. *Chem. Ind. (London)* 1580–1594 (1966).
13. Gerard, C., Marco, C., and Mayence, G. Rheological and Adhesive Properties of Copolymers Based on Dialkylmaleate, in *Proc. 7th Int. Congress of Technology.* Gotheborg, Sweden, 1976.
14. Mayo, T. J., and Reegan, S. L. *Proceedings of the Symposium on Adhesion and Cohesion* pp. 209–217 (Philip Weiss, ed.) Elsevier, Amsterdam, 1962.
15. Ames, W. A. U.S. Patent 4,164,514 (1979) (assigned to Eastman Kodak Co.).
16. Sanderson, F. T., Kowalski, A., and Kottke, R. H. U.S. Patent 4,077,926 (1978) (assigned to Rohm and Haas Co.).
17. Young, D. W. U.S. Patent 4,254,251 (1981).
18. Pietsch, H., and Curts, J. U.S. Patent 3,728,148 (1973) (assigned to Beiersdorf AG).
19. Koeble, E. A. U.S. Patent 4,217,434 (1980) (assigned to Stauffer Chemical Co.).

48. McKenna, L. W. Jr., Finkelstein, R. S., and Haynes, J. K. Jr. U.S. Patent 4,234,660 (1980) (assigned to Monsanto Co.).
49. Lehmann, G. W. H., and Curts, H. A. J. U.S. Patent 3,563,953 (1971) (assigned to Beiersdorf AG).
50. McKenna, L. W. Jr. U.S. Patent 3,886,126 (1975) (assigned to Monsanto Co.).
51. Guse, G., and Pietsch, H. G. U.S. Patent 3,923,752 (1975) (assigned to Beiersdorf AG).
52. Takanen, L. W., Palmius, K. and Myrthil, G. U.S. Patent 4,079,030 (1978) (assigned to Salve SA).
53. Goel, A. B. U.S. Patent 4,626,575 (1986) (assigned to Ashland Oil, Inc.).
54. Coffman, A. M. U.S. Patent 3,738,971 (1973) (assigned to B. F. Goodrich Co.).
55. Knapp, E. C. U.S. Patent 3,284,423 (1966) (assigned to Monsanto Co.).
56. Lehmann, G. W. H., and Curts, H. A. J. U.S. Patent 4,038,454 (1977) (assigned to Beiersdorf AG).
57. Kordzinski, S., and Horn, M. B. U.S. Patent 3,579,490 (1971) (assigned to Ashland Oil & Refining Co.).
58. Japan Kokai Tokkyo Koko JP 59, 152 971. Daicel Chemical Industries. Chemical Abstracts 1984, **101:** 231678.
59. Ulrich, E. W. U.S. Patent 2,973,286 (1961) (assigned to Minnesota Mining & Manufacturing Co.).
60. British Patent 930,761 (1963) (assigned to Farbenfabriken Bayer AG).
61. Mendelsohn, M. A. *Ind. Eng. Chem., Prod. Res. Dev.* **3:** 67–72 (1964).
62. Fravel, H. G. Jr., and Cranley, P. E. *Adhesives Age* **27**(11): 18–20 (1984).
63. Brenner, W. *Exploratory Research on Novel Ambient Temperature Curing Techniques for Adhesives, Sealants and Laminates.* NYU DAS 79-07, report April 1979, Naval Air Systems Contract No. N00019-78-C-0169, New York University.
64. Mark, H. *Proceedings of the Symposium on Adhesion and Cohesion,* pp. 240–269 (P. Weiss, ed.) Elsevier, Amsterdam, 1962.
65. Samour, C. M., and Satas, D. U.S. Patent 3,492,260 (1970) (assigned to Kendall Co.).
66. Ono, T., and Matsuguma, Y. U.S. Patent 3,971,766 (1976) (assigned to Teijin Ltd.).
67. Ono, T., and Matsuguma, Y. U.S. Patent 3,983,297 (1976) (assigned to Teijin Ltd.).
68. D. Satas. Unpublished data.
69. Warson, H., and Kinsler, D. British Patent 1,174,914 (1969) (assigned to Dunlop Co., Ltd.).
70. Loshaek, S., and Fox, T. G. *J. Am. Chem. Soc.* **75:** 3544 (1953).
71. Douek, M., Schmidt, G. A., Malofsky, B. M., and Hauser, M. U.S. Patent 3,996,308 (1976) (assigned to Avery Products Corp. and Loctite Corp.).
72. Gehman, S. D., and Gregson, T. C. *Rubber Chem. Technol.* **33**(6): 1375–1437 (1960).
73. Hendricks, J. O. U.S. Patent 2,956,904 (1960) (assigned to Minnesota Mining & Manufacturing Co.).
74. Kasper, A. A. U.S. Patent 3,328,194 (1967) (assigned to Kendall Co.).
75. Christenson, R. M., and Anderson, C. C. U.S. Patent 3,725,115 (1973) (assigned to PPG Industries, Inc.).
76. Malloy, D. E. Jr. *Plast. Eng.* **11**(12): 33–36 (1984).
77. Dabroski, W. C. U.S. Patent 3,285,771 (1966) (assigned to Johnson & Johnson Co.).
78. Horn, M. B., and Capone, V. P. U.S. Patent 3,269,994 (1966) (assigned to Catalin Corp. of America).
79. Kealy, J. P., and Zenk, R. E. U.S. Patent 4,418,120 (1983) (assigned to Minnesota Mining & Manufacturing Co.).
80. McCarthy, W. J. U.S. Patent 3,465,058 (1969) (assigned to B. F. Goodrich Co.).
81. Schenk, W. N. U.S. Patent 4,305,996 (1981) (assigned to B. F. Goodrich Co.).

F. G o.).
Goo .
Patent 4 (1979)
F. G o.).
(assig iersdorf
e Se hesives.
ake B a, Fla. S
to At hfield C
74) (a CPC Int
5) (a Sanyo
Patent (1979)
Polys
977.
1981
ng 2 1986);
eedi , pp.
ncyc Polyme
64.
is Polyme
or .
R. Patent 4
).
3, p
r Co.
o.).
l Soda
75) (
igned

117. Samour, C. M. U.S. Patent 3,399,159 (1968) (assigned to The Kendall Co.).
118. Samour, C. M., and Richards, M. C. U.S. Patent 3,751,451 (1973) (assigned to The Kendall Co.).
119. Doehnert, D. F. U.S. Patent 3,558,574 (1971) (assigned to Johnson & Johnson Co.)
120. Gardner, D. M., and McKenna, L. W. Jr. U.S. Patent 3,893,982 (1975) (assigned to Monsanto Co.).
121. Pastor, S. D., and Skoultchi, M. M. U.S. Patent 4,052,527 (1977) (assigned to National Starch and Chemical Co.).
122. Minnesota Mining & Manufacturing Co. British Patent 624,764 (1949).
123. Dahlquist, C. A., and Zenk, R. E. British Patent 870,022 (1961) (assigned to Minnesota Mining & Manufacturing Co.).
124. Crocker, G. J. and Doehnert, D. F. U.S. Patent 3,551,391 (1970) (assigned to Johnson & Johnson Co.).
125. Davis, I., Skoultchi, M. M., and Fries, J. A. U.S. Patent 3,925,282 (1975) (assigned to National Starch and Chemical Co.)
126. Guerin, J. D., Hutton, T. W., Miller, J. J., and Zdanowski, R. E. U.S. Patent 4,045,517 (1977) (assigned to Rohm and Haas Co.).
127. Kugler, J. H. U.S. Patent 2,544,692 (1951) (assigned to Minnesota Mining & Manufacturing Co.).
128. Reinhard, H., Mueller, A., and Dotbauer, B. U.S. Patent 3,725,122 (1973) (assigned to Badische Anilin und Soda Fabrik AG).
129. Dalibor, H., Gutet, R., and Stenzel, G. U.S. Patent 3,842,133 (1974) (assigned to Reichhold-Albert Chemie AG).
130. Tomita, J., and Strahan, T. W. U.S. Patent 3,515,578 (1970) (assigned to Minnesota Mining & Manufacturing Co.).
131. Aranyi, C., Gutfreud, K., Hawrylewicz, E. J., and Wall, J. S. U.S. Patent 3,607,370 (1971) (assigned to U.S. Secretary of Agriculture).
132. Christenson, R. M., and Anderson, C. C. U.S. Patent 3,725,115 (1973) (assigned to PPG Industries, Inc.).
133. Fournier, A. A. U.S. Patent 3,725,121 (1973) (assigned to Johnson & Johnson Co.).
134. Samour, C. M. U.S. Patent 3,299,010 (1967) (assigned to The Kendall Co.).
135. Davis, I. J., and Sirots, J. U.S. Patent 3,535,295 (1970) (assigned to National Starch and Chemical Co.).
136. Kosaka, Y., Uemura, M., Fujiki, T., and Saito, M. U.S. Patent 3,838,079 (1974) (assigned to Toyo Soda Mfg. Co., Ltd.).
137. Gobran, R., Knoepfel, H., and Silver, S. F. U.S. Patent 3,924,044 (1975) (assigned to Minnesota Mining & Manufacturing Co.).
138. Dowbenko, R., and Christenson, R. M. U.S. Patent 3,897,295 (1975) (assigned to PPG Industries, Inc.).
139. McKenna, L. W. Jr., and Gardner, D. M. U.S. Patent 3,763,117 (1973) (assigned to Monsanto Co.).
140. Mowdood, S. K. and Given, D. A. U.S. Patent 3,998,997 (1976) (assigned to Goodyear Tire and Rubber Co.).
141. Guse, G., and Pietsch, H. G. U.S. Patent 3,690,937 (1972) (assigned to Beiersdorf AG).
142. Brown, F. W., and Keck, F. L. Canadian Patent 769,780 (1967) (assigned to Minnesota Mining & Manufacturing Co.).
143. Maska, R. U.S. Patent 3,701,758 (1972) (assigned to National Starch and Chemical Corp.).
144. Lehmann, H., and Curts, J. U.S. Patent 3,729,338 (1973) (assigned to Beiersdorf AG).
145. Yamada, F., Fukuda, Y., and Uraya, T. U.S. Patent 3,849,840 (1974) (assigned to Kanebo Ltd.).

176. McKenna, L. W. Jr., Finkelstein, R. S., and Haynes, J. K. Jr. U.S. Patent 4,185,051 (1980) (assigned to Monsanto Co.).
177. Ryrfors, L. O., and Hassander, M. S. O. U.S. Patent 4,234,467 (1980) (assigned to Hoechst-Perstorp AB).
178. Knoepfel, H., and Silver, S. F. U.S. Patent 3,770,708 (1973) (assigned to Minnesota Mining & Manufacturing Co.).
179. Sanderson, F. T., Kowalski, A., and Kettke, R. H. U.S. Patent 4,077,926 (1978) (assigned to Rohm and Haas Co.).
180. Kealy, J. P., and Zenk, R. E. U.S. Patent 4,418,120 (1983) (assigned to Minnesota Mining & Manufacturing Co.).
181. Lazear, N. R., and Stackman, R. W. U.S. Patent 4,185,050 (1980) (assigned to Celanese Corporation of America).
182. Grunewalder, V. J. and Koeble, E. A. U.S. Patent 3,697,618 (1972) (assigned to National Starch and Chemical Co.).

adding resins and plasticizers. In fact, the proportion of additives often exceeds that of the elastomers. The form of a typical formulation (by %) is as follows:

Elastomers: 30–50
Resins: 20–40
Plasticizers: 0–40
Fillers: 0–10
Stabilizers/antioxidants: 1–2

The elastomers and resins, and often the plasticizers as well, are of two to three different types whose properties have to be mutually reconciled.[3] Obviously, the formulation must allow for fluctuations in quality and properties that are inevitable in natural products, in this case rubber.

Pressure sensitive adhesives formulated with these elastomers must be dissolved in an organic solvent before they can be applied in the form of coatings. Preparation of rubber solutions is very time-consuming and requires sophisticated equipment.

First, the rubber is mechanically degraded on two roll mills or in intensive mixers until it acquires the necessary intrinsic viscosity.[4] After that, it is granulated in order to allow the solution to be prepared more rapidly. Even then, several hours are required. High-speed dissolvers must be used, and the entire dissolving process is very energy intensive.[5] The other additives are also incorporated in the dissolvers.

Hot melt pressure sensitive adhesives are formulated in a similar manner. As a rule, the elastomer is a thermoplastic rubber or, for certain applications, EVA copolymers with a high vinyl acetate content. Naturally, these adhesives dispense with a dissolving stage, but mixing in compounders and melting vessels consumes even more time and energy.

Another aggravating factor is what is known as the heat history. The processing properties, for example, the viscosity, and the performance characteristics, for example, the tack and color, may change to an extent, depending on the temperature and the duration of exposure to heat during production and ancillary processing. For instance, utmost care must be taken in proportioning antioxidants to ensure that the polymer chains do not break apart either during formulation of the adhesive or during subsequent coating, because the final product, after all, should also remain resistant to aging.

The tack of certain products, such as polyisobutenes and poly(vinyl ethers), depends on their molecular weight. Here, too, modification is an absolute necessity for attaining optimum tack and cohesion by combining products of various molecular weights. Once again, resins are usually added to increase the tack. These polymers are often combined with others.

In contrast to the products that have been mentioned up to now, polyacrylates

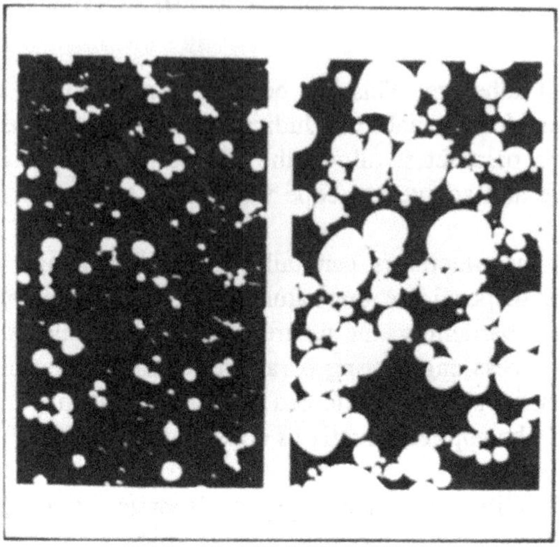

Fig. 16-1. Electron microscope microphotograph of an acrylic emulsion. (*Courtesy of BASF AG, Ludwigshafen.*)

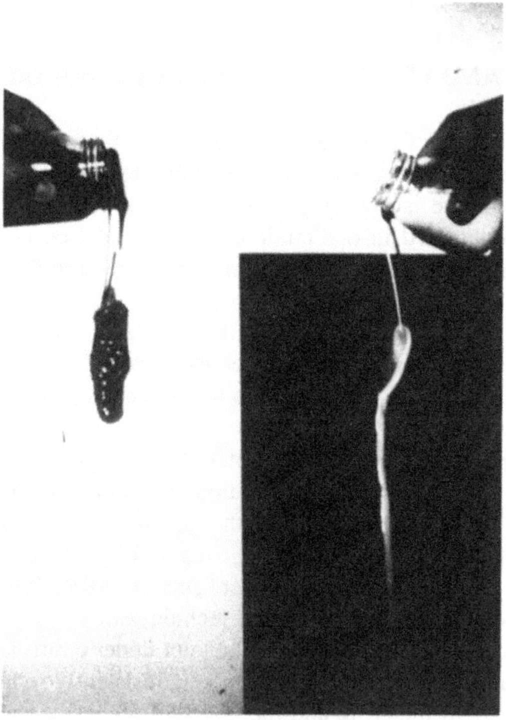

Fig. 16-2. Differences in viscosity of an adhesive solution and a dispersion.

Fig. 16-3. Schematic diagram of the mechanism of film formation from a dispersion.

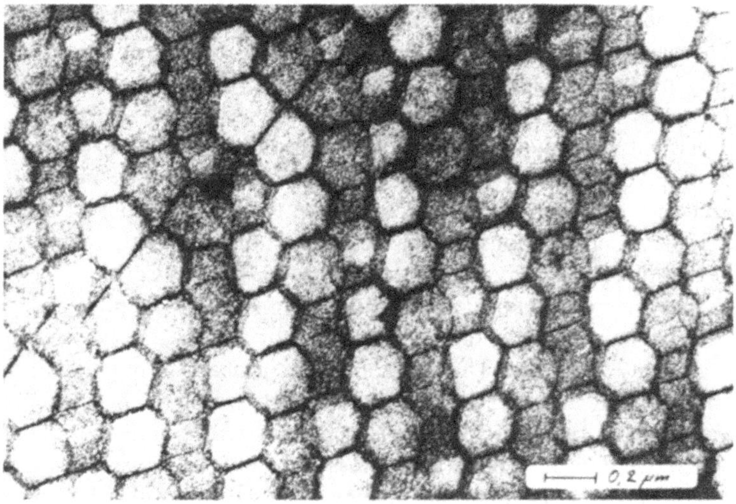

Fig. 16-4. Electron microscope microphotograph illustrating the formation of a thin pressure sensitive adhesive film from a dispersion. (*Courtesy of BASF AG, Ludwigshafen.*)

As a rule, these aims can be achieved by the following means.

- By mixing two or more acrylic dispersions together
- By mixing the acrylic dispersion with resins and/or plasticizers
- After the desired adhesive properties have been obtained, by adapting the adhesive to the existing coating parameters by thinning with water and/or by adding auxiliaries, for example, wetting agents or thickeners[9]
- By adding fillers and/or pigments to color the adhesive compounds (brown adhesive tapes for packaging), to label them, or to achieve special effects (e.g., zinc oxide for medical plaster)

Modification by the Adhesives Manufacturer

Coaters may obtain their adhesives ready-to-coat from manufacturers who have developed them from the raw materials at their disposal and have ensured that their properties and coating characteristics meet the user's specifications. Thus, processors can apply the adhesives direcly on their coaters without any further adjustment.

Modification by Coaters

If coaters prepare the adhesives themselves, they must have facilities for developing formulations in the laboratory, for mixing, and for quality assurance. Frequently, they can more than offset the costs involved by producing numerous adhesives from small inventories of raw materials or by great flexibility in the event of a changeover. Countermeasures can be immediately adopted if difficulties occur, and even mixtures of poor stability can be processed without difficulty. Since they devise their own formulations, the coaters acquire a specific know-how and thus often gain a clear lead over competitors.

There is no doubt that the procedures described here and the associated expenditures generally cannot be justified unless large amounts of dispersions are processed, that is, the amounts concerned can be delivered and stored economically in bulk.

Modification by the Dispersions Manufacturer

A few dispersions manufacturers offer pressure sensitive adhesives that are ready for application. Under these circumstances, they can also be regarded as adhesives manufacturers. If the demand is large enough, custom-made, ready-

for-use dispersions can be produced; for instance, by controlling the process parameters in the polymerization reactor accordingly, by adding auxiliaries to the formulation, or by mixing different polymer dispersions.

The various possibilities and stages are shown schematically in Figure 16-5.

MODIFICATION EXAMPLES

Aqueous acrylic dispersions are much easier to modify than rubber solutions. They flow readily, and thus do not present any problems in pumping and mixing. Since they are neither flammable nor injurious to health, there is no need to introduce many time-consuming safety measures. The only precaution necessary is to ensure that the place of work is well ventilated in conformance with the usual industrial standards.

Small-scale processors produce the adhesives in drums. The ingredients are added by hand, and weight control is absolutely essential. It is also important that the formulation be homogeneously mixed by a low shear propeller stirrer. High-speed dissolver disks with pegs or serrations are unsuitable, because they may impair the shear stability of polymer dispersions.

It is more economical to have the dispersions delivered in tank trucks than in drums.[5] This is shown schematically in Figure 16-6.

Storage tanks, pipelines, and equipment for polymer dispersions must be of a corrosion-resistant material, e.g., glass-reinforced polyester resins or stainless steel. Water-soluble compounds formed by contact with base metals impair the dispersions' stability. Thus, agglomerates may be formed initially at the tank walls and lead to coagulation.[14]

Combination of Various Dispersions

The simplest means of obtaining optimum properties for a given application is to mix two or more acrylic dispersions together in the appropriate ratio.

The pH value of the dispersions to be mixed should be raised to about 6–8 in order to ensure mutual compatibility and to lengthen the shelf life of the mixture.

The best means of raising the pH is to add concentrated ammonia or 10% sodium hydroxide to the dispersion in a low-speed propeller mixer. The same technique is subsequently adopted to mix the dispersions together. Care must be taken to avoid entrapped air, because it causes foam that is very difficult to destroy.

Many acrylics have excellent cohesion and good adhesion properties, but

their tack does not quite suffice for a given application. The remedy in this case is to add tackifiers, which are mostly resins. Another means of improving the initial tack of base dispersions, such as Acronal 85 D or 80 D, to the desired level in the production of adhesive tapes for packing or household is to add 10–20% of acrylics of a low molecular weight, such as Acronal 7 D.

The same principle applies to the production of permanent adhesive paper labels. In this case, the base dispersions are products with greater adhesion and less cohesion, such as Acronal V 205/V 207, and the initial tack is increased still further by adding small amounts of an acrylic tackifier.

Different requirements are imposed on some other self-adhesive articles. An example is protective film, which must have an adhesion that remains weak over long contact periods. Special self-cross-linking acrylics, for example, Acronal 35 D and more recent developments, are used as the base products for this purpose. In a very few applications, the adhesive must display greater tack but allow the film to be peeled off without leaving a residue, even after it has been attached for a long period. Consequently, low molecular weight acrylics would not be considered, and highly cohesive products with good initial tack are required. Thus, in this case, the base dispersions would be tackified.

The situation is similar for removable plasticized PVC film. Butyl acrylate homopolymers have excellent adhesion to plasticized PVC labels, which allows them to be peeled off very smoothly. A suitable product for this application is Acronal 4 D, which has good tack, medium cohesion, and high resistance to plasticizers. Since traces of plasticizer obviously reduce the tack, these adhesive labels must have a higher initial tack, which can be achieved very easily by adding 10–20% of a very adhesive, highly cohesive dispersion, such as Acronal 80 D.

Very often, the cohesion and peel characteristics of the base dispersions used for adhesive packaging tapes do not suffice for masking paper and crepe adhesive tapes. This problem is solved by highly cohesive products that allow the tapes to be readily stripped. However, the dispersion added must not reduce the tack excessively. An example of a suitable product of medium tack is Acronal 50 D, which is partially pre-cross-linked and is fully cross-linked by heat.

Similar demands are imposed on sound-deadening materials. The cohesion of a product eminently suitable for polyurethane foams would be far too low for heavy bitumen roofing and would lead to rapid failure. Here, adhesives with very high cohesion must be used, or highly cohesive polymers must be mixed with the base dispersion.

Certain applications, such as special spray tests performed by the automobile industry, impose severe demands on the water resistance. Consequently, this is another property that must be taken into consideration in selecting acrylic dispersions. Here, too, raw materials manufacturers offer suitable dispersions

sive, and can be readily modified: ratios of dissolved resin to polymer of up to 1:1 are feasible. It is very important that the mixture be given time to mature. In other words, it must be allowed to stand for one to two days before it is briefly stirred and applied on a coater. During this time, the resin, which is initially present in the form of separate particles, gradually diffuses into the acrylic polymer. As a result, a homogeneous distribution is obtained, and the swelling effect exerted by the solvent causes the mixture to thicken somewhat.

Pressure sensitive adhesives of great shear strength but comparatively hard tack are obtained from resins with softening points of the order of 70°C, as determined by the ball-and-ring method. If these resins are blended with plasticizing resins, the tack will generally be much increased. This applies particularly to the tack (and the adhesion) on polyolefin adherends.[8] A drawback of adding resins is that it impairs the adhesive's resistance to aging. Important applications for formulations of this nature are paper labels and double-sided adhesive tapes, for example, for laying carpets.

If one of the great advantages of polymer dispersions, namely, the absence of solvents, is to be retained, resin dispersions should be given preference to resin solutions. One property of resin dispersions that is generally observed is that they reduce the cohesion more than resin solutions.

The effects that can be achieved by adding resin dispersions are evident from Table 16-1. The example taken has been a 70% acrylic dispersion for paper labels.

Wet techniques are adopted for removable PVC labels, particularly large stickers. For this purpose, the adherend is moistened with a wet sponge in order to ensure that the position of the label can be better adjusted. An unpleasant consequence, particularly if the PVC film is transparent, is that coatings thus produced from polymer dispersions become opaque. Although Acronal 4 D coatings absorb very little water (about 6% after 24 hours), temporary opalescence cannot be avoided. This cloudiness is not brought about by the effective water content, but by the differences in light refraction within the microscopic zones formed by water absorption. The size of these zones ranges from one-tenth to a multiple of the wavelength of light. For example, the interspaces with a high emulsifier concentration absorb more water than the actual polymer particles.

The Addition of Plasticizers

Plasticizers allow the adhesive film to flow more easily and are responsible for more rapid wetting of the adherend. They thus increase the initial peel strength, though obviously at the cost of the cohesion or resistance to heat deformation. For instance, the following combination has better adhesion on

Other types of plasticizers are used for specific tasks. Thus, chlorinated paraffins or organic bromophosphoric compounds are employed if greater flame retardance is needed.

There is one aspect of modifying with plasticizers that deserves mentioning. The PVC film for the production of insulating tapes or labels contains large amounts of plasticizers that can migrate into the adhesive layer and alter its properties in the course of time until equilibrium is attained. In cases of this nature, the properties of the adhesive can be kept constant if some plasticizer, preferably the same as that in the PVC film, is included in the formulation.

Modifying with Poly(vinyl ethers)

Poly(vinyl ethers), mainly poly(vinylmethyl ether) and poly(vinylisobutyl ether), act as tackifiers in a similar manner to that of resins or low molecular weight polyacrylates. However, they are decidedly more expensive than the usual resins and are thus used solely for achieving special effects. Lutonal M 40, a poly(vinyl methyl ether) supplied by BASF, is a water-soluble plasticizing resin that increases the adhesive's affinity for water and solubility in water. It is usually stirred directly into the dispersion in the form of an aqueous solution. However, it must be noted that this product thickens the adhesive and that its solubility is reversible. In other words, it becomes insoluble at temperatures above 28°C and can thus cause the mixture of dispersions to coagulate.

Poly(vinylisobutyl ethers) are being offered as 55% secondary dispersions that do not present any problems if they are combined with acrylic dispersions. Examples are Lutonal I 60 D and I 65 D supplied by BASF. In common with poly(vinylmethyl ethers), these two products readily absorb plasticizers without an attendant loss in tack. Consequently, they are eminently suitable as tackifiers for acrylic dispersions for coating decorative laminates or PVC labels. Many acrylic dispersions absorb very little plasticizer, but rapidly lose their tack when they come in contact with a plasticized PVC film. Under these circumstances, the plasticizer migrates through the interface or interstices between the polymer particles and forms a kind of release film on the surface. On the addition of 20–30% of a poly(vinyl ether), the plasticizer is fixed in the polymer and thus cannot migrate, as is evident from Table 16-2.

Plasticized PVC adhesive tapes can be produced from dispersions with good adhesion and very high cohesion by adding poly(vinylisobutyl ethers). Tape of this description has been produced without a primer and release coat from a commercial grade of plasticized PVC film and an adhesive of the following composition: 80% Acronal 80 D plus 20% Lutonal I 65 D. In view of the differences that exist between various types of PVC film (i.e., diverse plasticizers, PVC qualities, lubricants, and stabilizers) and the manifold interactions, a typical

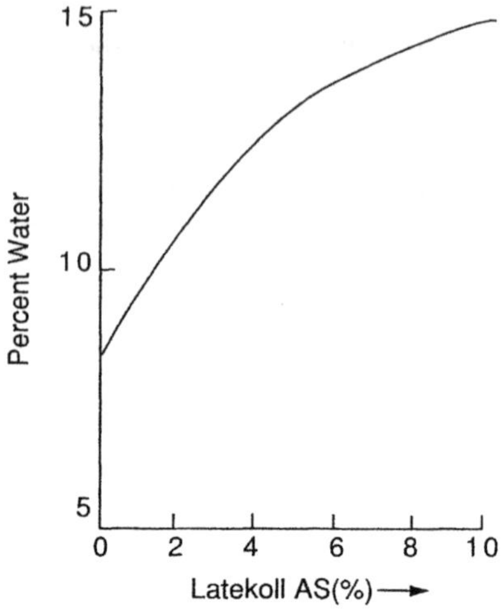

Fig. 16-7. Dependence of water absorbed by an adhesive film on the amount of thickener used.

The proportion of a thickener should be kept as small as possible, because an increase entails a decrease in the resistance to water offered by coatings formed from polymer dispersions. The amount of water absorbed after 96 hours immersion by a coating formed from a vinyl chloride/*n*-butyl acrylate copolymer dispersion is shown as a function of the proportion of ammonium polyacrylate (Latekoll AS) added in Figure 16-7.[15]

Large amounts of thickener may also exert an adverse effect on the properties of the adhesive. In most cases, they somewhat reduce the tack and the peel strength. This detrimental effect is not exhibited by Collacral SK, a specially developed thickener.

The effect of various thickeners on the peel strength of some pressure sensitive adhesive dispersions is shown in Table 16-3.

In various application techniques, such as, knife-over-roll or reverse roll, the product fed directly into the nip, die, etc., must have a comparatively high viscosity. The efficiency of various thickeners in raising the viscosity of several acrylic dispersions at various pH values was the subject of a series of tests. Typical representatives were chosen from the numerous types of thickener. They included a vinylpyrrolidone copolymer (Collacral VL), a polyacrylate (Latekoll D), a sodium carboxymethylcellulose (in the form of a 2% 1000 mPa·s aqueous solution), and a polyacrylate-diurethane (Collacral SK). Figures 16-8 through 16-11 show that the efficiency of the individual products depends on the pH value. The carboxymethylcellulose and the diurethane thickeners display good efficiency over a comparatively wide range of pH values. As opposed to this, the polyvinylpyrrolidone is most efficient in the slightly acid

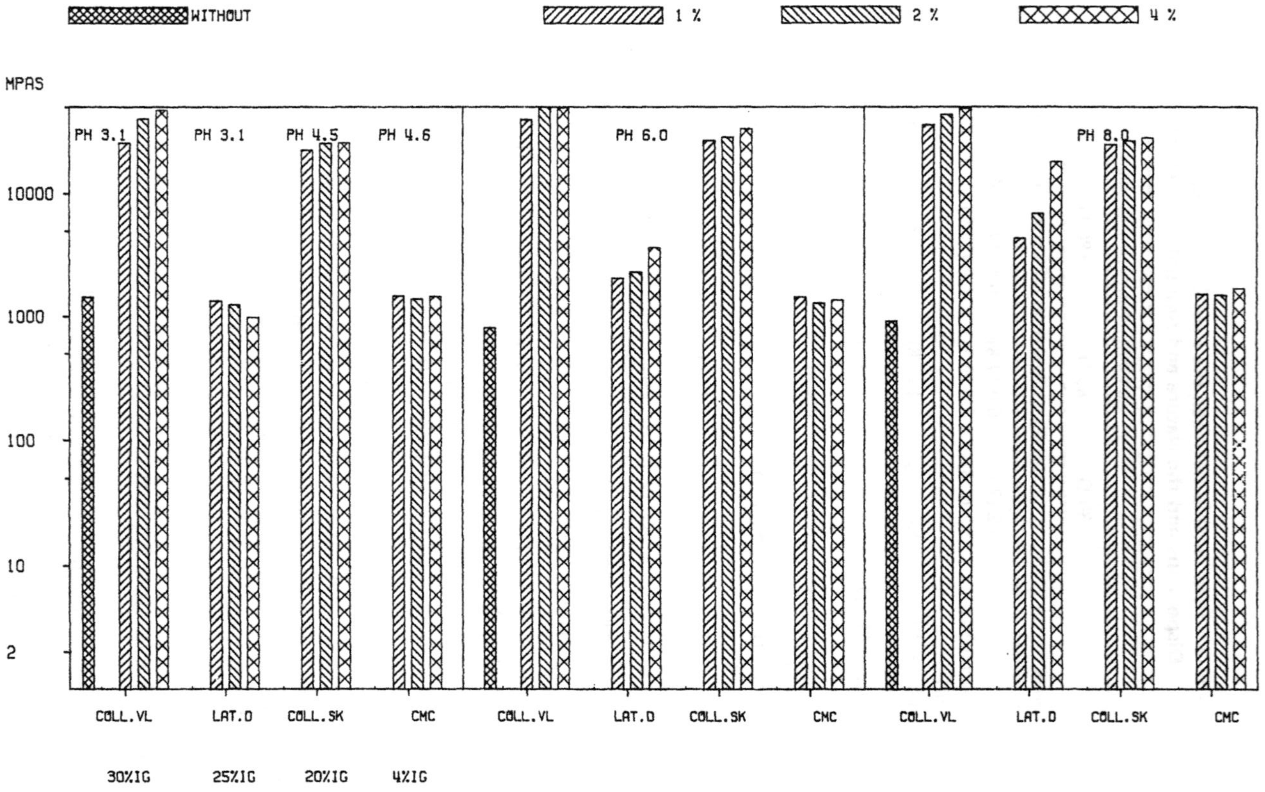

Fig. 16-8. Dependence of Acronal *V* 205 dispersion viscosity on pH and amount of thickener.

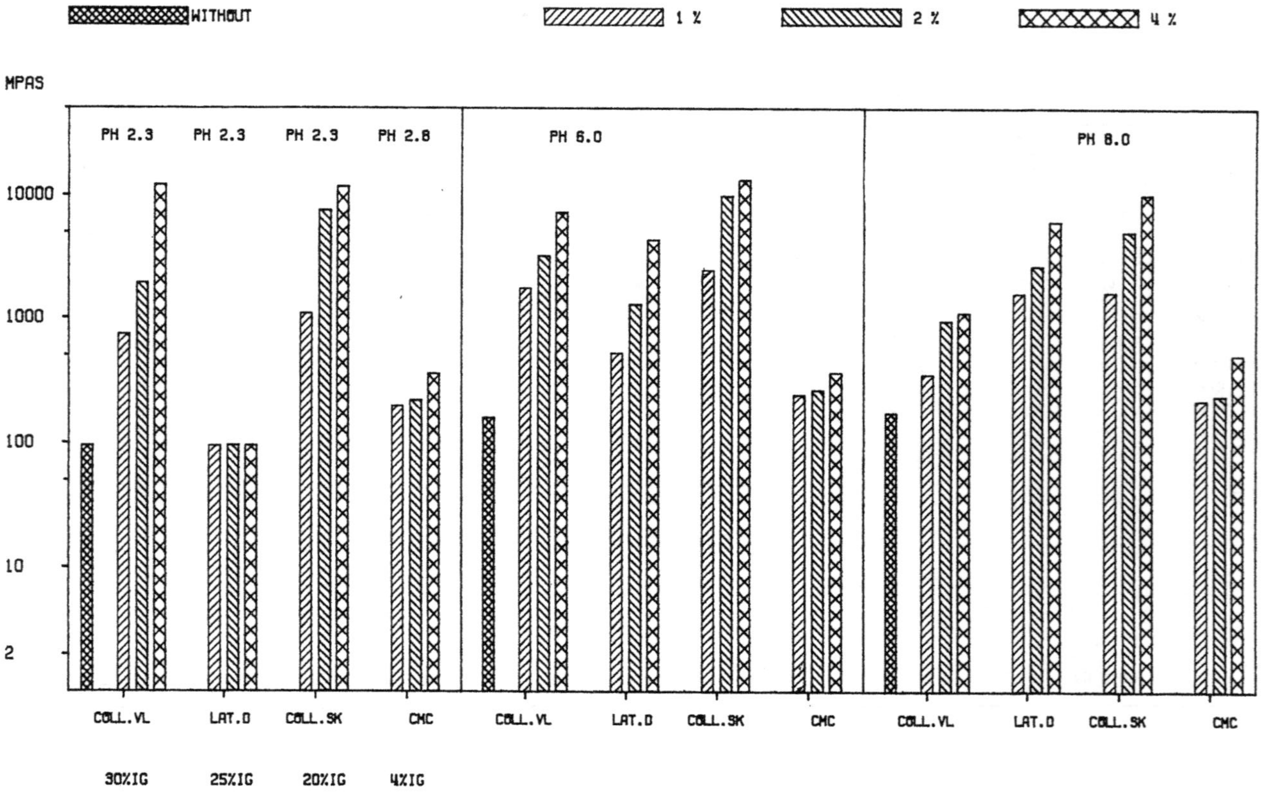

Fig. 16-10. Dependence of Acronal 850 dispersion viscosity on pH and amount of thickener.

A different situation exists if the coating technique necessitates low viscosities, for example, in modern reverse gravure, Meyer bar, or air-knife coating. In this case, there is a low resistance to flow, and the coating of dispersion, which is initially distributed evenly over the surface, contracts comparatively rapidly if the surface tension is too high. Under these circumstances, the surface tension of the pressure sensitive adhesive dispersion must be brought into line with that of the adherend, such as the film or silicone paper, by adding wetting agents. Obviously, these surfactants increase the risk of foaming and, in some cases, the amount of water absorbed by the adhesive coating. For this reason, the proportions in which they are added should be kept as small as possible.

It can be seen from Table 16-4 that an increase in the pH value suffices to reduce the surface tension and improve wetting. The table also shows the effect exerted by various wetting agents on different dispersions. The comparatively low foaming surfactants tested in this study were the sodium salt of a sulphosuccinic acid ester (Lumiten I-RA/anionic), the sodium salt of a fatty acid condensation product (Lumiten I-AFK/anionic), a sodium alkane sulphonate (Emulsifier K 30 supplied by Bayer AG), and the nonionic Emulan P.

An important point to observe in the application and evaluation of wetting agents is that the mixture must be allowed to mature. It should stand at least overnight to permit equilibrium between the emulsifying agent already present in the dispersion and the surfactant subsequently added. This maturing period improves the efficiency. Often, no effects are observed if the dispersion is tested immediately after the wetting agent has been added. In fact, the properties are occasionally impaired in the initial stages.

Unfortunately, no laboratory methods have yet emerged for determining the minimum amount of wetting agent required. The reason for this is that the amounts required in the laboratory are always higher than those on the coater, which achieves better wetting because of the high shear at the instant of application on the substrate. The higher the coating speed, that is, the greater the shear, the better the wetting. No exact reason can be given for this phenomenon, which is related to the limiting dynamic viscosity.

High-speed coaters impose an additional restraint on contraction of the adhesive, because the coated article passes rapidly through the drying tunnel, with the result that drying begins at a very early stage.[5]

Defoamers

The emulsifying and wetting agents and protective colloids in the dispersion are responsible for a tendency to foaming. Foam may cause trouble if the coater is run at high speeds, if there is a large excess of dispersion, or if the dispersion is very strongly agitated by the roll or recirculated. Foam suppression

Table 16-4. Effect of Surface Active Agents

ADDITION EMULSIFIER	WETTING FAULTS	SURFACE TENSION WITH TENSIOMETER	FOAM HEIGHT (FRITTE METHOD) BUILDUP 1′	2′	3′	4′	5′	DECREASE 1′	2′	3′	4′	5′	6′	7′	8′	9′	10′	WATER UPTAKE IN % AFTER 24 HR	96 HR
Acronal 85 D without	Drastic	34.2	95	170	230	265	295	235	140	105	65	50	35	25	15	10	0	11	27
pH 8 with ammonia	Severe	30.1	105	195	280	360	435	430	430	420	415	410	405	400	400	395	390	14	29
pH 8 + 1% Lumiten I-AFK	None	31.5	120	235	355	400	400	398	390	375	345	320	305	280	255	245	240	28	45
pH 8 + 1% Emulsifier K 30	None	31.4	120	250	365	400	400	399	398	397	396	396	395	395	394	394	393	15	53
pH 8 + 0.5% Lumiten I-RA	Light	28.1	115	210	305	405	500	500	500	495	495	495	495	490	490	490	485		
pH 8 + 1% Lumiten I-RA	None	30.4	105	200	300	395	490	490	490	490	490	485	485	480	480	480	480	26	72
pH 8 + 1% Lumiten I-RA + 0.1% Surfinol 104 H	None	27.5	100	195	290	375	470	470	470	470	465	465	465	465	465	460	460		
pH 8 + 1% Emulan P	None	34.5	115	235	340	400	400	390	360	250	160	75	30	5	5	0	0	20	41

Table 16-5. Coloring Pressure Sensitive Adhesive Dispersions

ADHESIVE	COAT WEIGHT (g/m^2)	CARRIER	PEEL STRENGTH IN *N*			SHEAR STRENGTH UNDER 1-kg LOAD, 5-cm^2 BOND	
			TACK	CONTACT TIME (1 min)	CONTACT TIME (24 hr)	ROOM TEMPERATURE (24 hr)	50°C (24 hr)
Acronal V 205	25	Host	6.8 A[a]	6.2 A[a]	15.9 C[b]	>24 hr	[d]F 5 hr 29 min
+ 2% Chalk P 5	25	Host	3.8 A[a]	3.8 A[a]	6.7 A[a]	>24 hr	>24 hr
+ 5% Chalk P 5	25	Host	2.3 A[a]	2.8 A[a]	5.5 A[a]	>24 hr	>24 hr
+ 10% Chalk P 5	25	Host	1.9 A[a]	2.7 A[a]	4.5 A[a]	>24 hr	>24 hr
+ 20% Chalk P 5	25	Host	1.4 A[a]	1.3 A[a]	4.7 A[a]	>24 hr	>24 hr
+ 2% Titanium dioxide	25	Host	5.4 A[a]	4.8 A[a]	11.3 A[a]	>24 hr	[d]F 3 hr 02 min
+ 5% Titanium dioxide	25	Host	4.5 A[a]	4.4 A[a]	8.8 A[a]	>24 hr	[d]F 3 hr 54 min
+ 10% Titanium dioxide	25	Host	3.8 A[a]	4.5 A[a]	8.9 A[a]	>24 hr	[d]F 4 hr 38 min
+ 20% Titanium dioxide	25	Host	2.9 A[a]	3.9 A[a]	8.3 A[a]	>24 hr	[d]F 3 hr 07 min
+ 2% Ground Quartz S 100	25	Host	2.3 A[a]	4.5 A[a]	11.2 C[b]	[b]C 1 hr 55 min	[d]F 13 min
+ 5% Ground Quartz S 100	25	Host	0 A[a]	2.1 A[a]	5.8 LF[c]	[d]F 3 min	[d]F immediately
+ 10% Ground Quartz S 100	25	Host	0 A[a]	0.3 A[a]	0.5 LF[c]	[d]F immediately	[d]F immediately
+ 20% Ground Quartz S 100	25	Host	0 A[a]	0 A[a]	0 A[a]	[d]F immediately	[d]F immediately

Note: Laboratory atmosphere: 23°C and 50% relative humidity; width of specimen strips: 20 mm; rate of peeling: 300 mm/min; adherend: mirror-finish, chromium-plated brass panel.

[a] A = Adhesion failure.

[b] C = Cohesion failure.

[c] LF = Light film (trace of adhesive on metal panel).

[d] F = Film.

Preserving the Dispersions

Preservatives are usually added to polymer dispersions to ensure that they are protected from bacteria and fungi during transportation and subsequent brief storage periods. If the acrylic dispersion is turned around rapidly and the formulated pressure sensitive adhesive is processed within a few days, there is no need for any further preservative. This generally applies to coaters who formulate their own adhesives.

Adhesives manufacturers who supply coaters must add fungicides and bactericides to their pressure sensitive adhesives. The pH values of the dispersions during formulation should be in the neutral to faintly basic range, in which the ingredients are most compatible with one another. But it is precisely in this pH range that the dispersions are most open to attack by bacteria and fungi. The preservatives used are the same as those that have proved successful in emulsion paints, such as formaldehyde and its derivatives, benzimidazoles, isothiazolones, and chloracetamides.

The choice of preservative and its concentration depend on the application.[17] It is common practice to add two different preservatives in different amounts. Their efficiency should be continuously checked, because—even if they are present in adequate amounts—certain strains of bacteria and fungi may acquire immunity in the course of time.

Microorganisms can enter a system at any point, for example, through the vents. The measures adopted to combat them must include routine cleaning of tanks, pipelines, and mixing equipment, and the subsequent treatment with formaldehyde. After thorough mechanical cleaning, the system should be filled with an aqueous solution consisting of one liter of 30% formaldehyde per cubic meter. This solution should be allowed to act for 48 hours to ensure that all the microorganisms are eradicated. After the tanks, etc., have been thoroughly rinsed with water, they can again be filled with the dispersion.[14]

Metering

The individual components of a formulation are usually weighed out in transportable containers. Special designs of gear-type meters with ball bearings are suitable for measurements in pipelines. Since they can only measure volume, air bubbles may be embraced by the measurements. Filters or screens should be installed upstream from these instruments in order to protect them. Various types of pumps may be resorted to for the polymer dispersions, such as eccentric single-rotor screw, centrifugal, rotary piston, and diaphragm pumps, but they must not be allowed to run dry under any circumstances.[14]

Auxiliaries that are added in small amounts, for example, thickeners, wetting

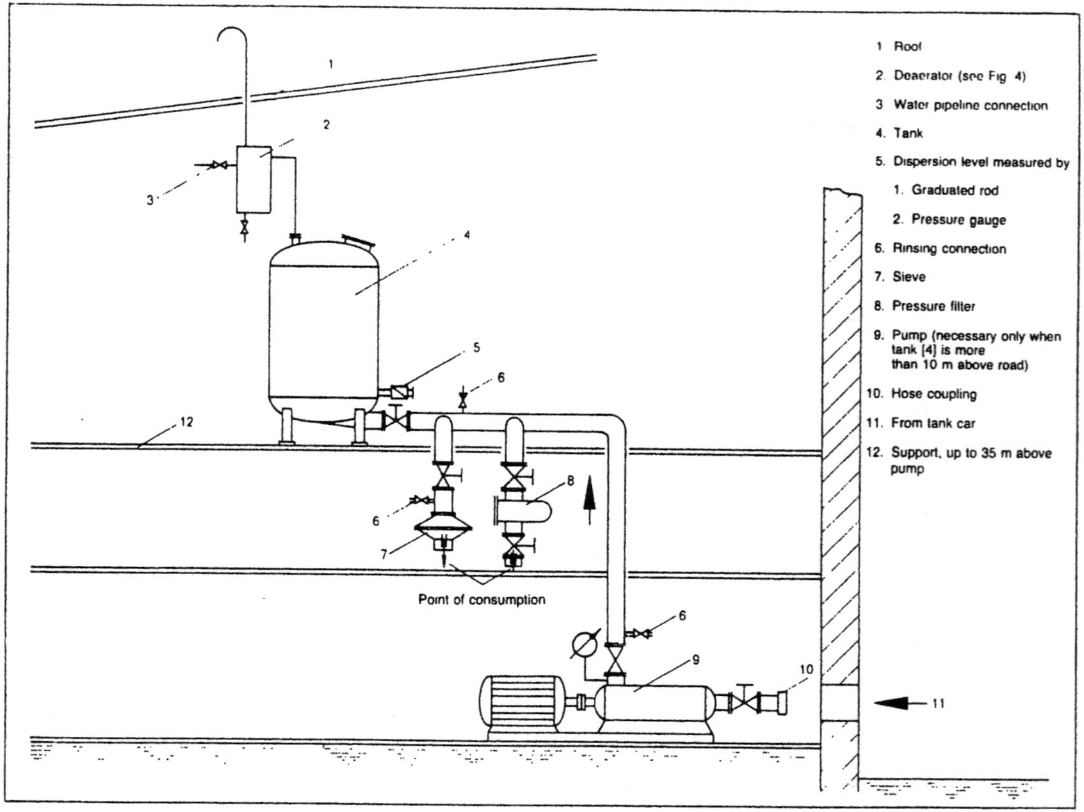

Fig. 16-12. Aboveground storage tank filled by pump.

have a higher specific gravity than that of water. Nevertheless, average shelf lives of at least six months to one year without sedimentation are expected nowadays, and most acrylic dispersions meet these expectations.

Drying

Drying is a very complex mechanism that is governed by many factors. One of the most important is heat transfer from the air into the coating, and it mainly depends on the pressure difference and the area of contact. The mass transfer involved in evaporation is related to the difference in vapor pressure. Both heat and mass transfer depend on the current of air that passes over the area of contact, and thus on the efficiency of the drier or air jets. If the drying conditions are too severe, a skin may form at the surface of the coating and restrict the release of water vapor. As a result, blisters may occur in the coating.[18]

The great advantage of dispersions during the drying stage is that the aqueous phase is nonflammable, nonpolluting, and odorless. The main drawback is that the evaporation enthalpy is almost five times greater than that of the solvents otherwise used for pressure sensitive adhesives. This disadvantage would, at first glance, appear to be insuperable, but the high energy consumption involved in evaporation is more than offset by the following two facts.

1. The solids content is decidedly higher, namely, 60%, 65%, and even 70%.
2. Air can flow at much lower rates over the surface of the coatings.

Adhesive solutions mostly have a solids content of the order of 25%, but acrylic dispersions may contain as much as 70% solids. Thus, even if it is assumed that the concentration of solids in a dispersion is only 65%, the amount of water to be evaporated would still be less than one-half of the amount of solvent to be evaporated from an adhesive solution. The example can be better illustrated by actual figures: the mass of solvent to be evaporated from each kilogram of adhesive is 750 grams, but the corresponding mass of water is only 350 grams.

A more important consideration is that safety regulations require much higher airflow rates for drying coatings obtained from solvent-type adhesives. For instance, the lower explosion limit for toluene is 1.2% by volume, while the concentration in practice must be restricted to 30–40% of this limit in order to allow a factor of safety. This implies that large volumes of air must be heated solely for diluting the solvent vapors, and the heat thus lost cannot be recovered.

It can be demonstrated mathematically that if the production rate is the

therefore be designed to cope with pressure sensitive adhesive dispersions of both flexibile and inflexibile rheology.

An important requirement imposed on pressure sensitive adhesive dispersions is that their runnability must be just as good as that of solvent-type and hot melt adhesives. In other words, they must permit current coating speeds of 100–150 m/min and, in the near future, coating speeds of 200–300 m/min. This is a requirement that is not easy to fulfill, and the situation is aggravated by the fact that polymer dispersions have a greater tendency to foam, owing to the presence of the emulsifying agents required in their production.

Leading dispersion manufacturers allow for this situation by testing new products in the development stages on a laboratory coater under conditions closely simulating those encountered in practice. Nevertheless, the fact remains that the situations change from one case to another and that a pressure sensitive adhesive dispersion must be adapted to cope with each and every coating task. As a result, liaison must be maintained between the supplier's technologists and the formulators and coaters.

The formulations devised in experiments on full-scale laboratory coaters permit time to be saved and reduce the losses in materials entailed in starting up an adhesives coating line.[9]

Acrylic dispersions and their modification permit chemists great latitude in optimizing pressure sensitive adhesives. From the economic point of view, inventories of basic acrylics for pressure sensitive adhesives should be kept as small as possible. Since acrylic dispersions can be readily pumped, no great outlay is required for their storage, metering, filtering, and mixing. It is advisable to store at least the main products in tanks. Recently developed acrylic dispersions permit just as much scope for modification as the well-established types, but have superior tack and adhesion. Thus they can also be adopted as base products. It need hardly be mentioned that they also have high cohesion.

REFERENCES

1. Ward, M. A. L. *Pressure-Sensitive Materials in the European Market.* Labelex Europe, 1985.
2. Jaeger, H. *Acrylic Dispersions for Pressure-Sensitive Adhesives.* AFERA, Taormina, 1984.
3. Wieder, H. *Solvent-Type Pressure-Sensitive Adhesives.* Presented at the 5th Klebstoff-und Veredelungsseminar, Munich, West Germany, 1980.
4. Butler, G. L. Natural Rubber Adhesives. *Handbook of Pressure-Sensitive Adhesive Technology* (D. Satas, ed.). Van Nostrand Reinhold, New York, 1982.
5. Tuerk, J. *Coating with Pressure-Sensitive Adhesive Dispersions.* Convertex, Blackpool, England, 1984.
6. Rym, F. Solvent-Base Acrylic Adhesives. Presented at the FINAT Seminar, Düsseldorf, West Germany, 1981.
7. Eisentraeger, K., and Druschke, W. Acrylic Adhesives and Sealants. *Handbook of Adhesives,* 2nd edition (I. Skeist, ed.). Van Nostrand Reinhold, New York, 1977.

17. Vinyl Ether Polymers

Helmut W. J. Mueller
BASF AG, Ludwigshafen/Rhein

Polymers and copolymers of vinyl ethers are used for various applications. Homopolymers with the general formula

$$-[CH_2-\underset{\displaystyle OR}{\underset{|}{C}H}]_n- \qquad (1)$$

where

$$\begin{array}{ll} \text{R is} -CH_3 & \text{(methyl)} \\ \text{or} -CH_2-CH_3 & \text{(ethyl)} \\ \text{or} -CH_2-\underset{\displaystyle CH_3}{\underset{|}{C}H}-CH_3 & \text{(isobutyl)} \end{array}$$

and copolymers with acrylates are used as raw materials for producing pressure sensitive adhesive compounds. Homopolymers of the aforementioned vinyl ethers are most widely used. These polymers may be of oily to rubberlike consistency.

Vinyl ether was first converted to a balsamlike polymer more than 100 years ago.[1] Between 1920 and 1930, Reppe chemistry made vinyl ethers readily accessible so that they became technically attractive. Systematic research work was carried out to investigate their polymerizability. Production of polymers on an industrial scale commenced in 1938 in the Ludwigshafen works of the then IG-Farbenindustrie (now the main works of BASF AG). After 1940, production was taken up by GAF Corporation, and later by Union Carbide Corporation, which discontinued their production in 1976.

Table 17-1. Monomers Used for Producing Raw Materials for Adhesives

PRODUCT	STRUCTURE	MOLECULAR WEIGHT	DENSITY AT 20°C (g/cm³)	FLASH POINT (°C)	MELT POINT (°C)	BOILING POINT (°C)
Vinyl methyl ether	$CH_2{=}CH{-}O{-}CH_3$	58.1	0.747	−60	−122	~6
Vinyl ethyl ether	$CH_2{=}CH{-}O{-}CH_2{-}CH_3$	72.1	0.754	−45	−115	~36
Vinyl iso-butyl ether	$CH_2{=}CH{-}O{-}CH_2{-}CH(CH_3){-}CH_3$	100.2	0.769	−15	−112	~83

brown discoloration. This usually has no influence on the pressure sensitive adhesive properties. Vinyl ether polymers are very resistant to hydrolysis and are tasteless and odorless. Products with low to medium molecular weight (K value < 65) may have a weak odor that is caused by residual monomers and oligomers. Tests for determining the primary skin irritation caused by vinyl ether polymers on the dorsal skin of white rabbits showed no or very slight irritation. The instructions of the manufacturers must be observed in using the products for medical applications. The solubility of the vinyl ether polymers is indicated in Table 17-3.

The solubility is influenced by the alkyl group. The solubility of polyvinyl methyl ether in water is due to hydrogen bonding between the water molecules and ether oxygen. The intrinsically insoluble polyvinyl methyl ether is surrounded by water molecules and is thus solvated. The bridge bonds are broken on application of energy and the nonhydrated polymer precipitates. In the case of Lutonal M 40 solutions, for instance, this process takes place at approximately 28°C. The precipitation point is influenced by the concentration of the aqueous solution and by any solvents present.

STABILIZATION

The influence of oxygen, heat, and light causes chain cleavage, cross-linking, and reactions with oxygen. These change the molecular weight, molecular weight distribution, adhesiveness, color, etc., of the polymer. The changes are usually the result of free radical reactions. The first step is cleavage of a carbon-hydrogen bond by uptake of energy, leading to the formation of free radicals. The energy can be introduced as heat, light (UV radiation), or mechanical action. By oxidation, new radicals are formed via peroxide intermediates. The reaction thus initiated is continued either by chain cleavage (causing softening of material) or cross-linking (causing hardening of material).[5]

These aging processes affect vinyl ether polymers in an undesirable way as well. Antioxidants must therefore be used in order to obtain adhesive compounds with constant properties. Numerous products are available on the market. Products that have proved very effective include Ionox 330 (Shell), Irganox 1010 (Ciba-Geigy), and Antioxidant ZKF (Bayer Leverkusen).

Figure 17-1 shows the aging behavior of a Lutonal I pressure sensitive adhesive applied from a hot melt. The adhesive compound was tested in unstabilized form and with an addition of approximately 1.5% of Antioxidant ZKF. The bonding strength was measured in the original condition, after 2-, 8-, and 24-hr aging at 150°C in a circulated air oven. Within a few hours, the nonstabilized adhesive degraded to such an extent that it became completely unusable. The stabilized adhesive, however, changed insignificantly for practical application.

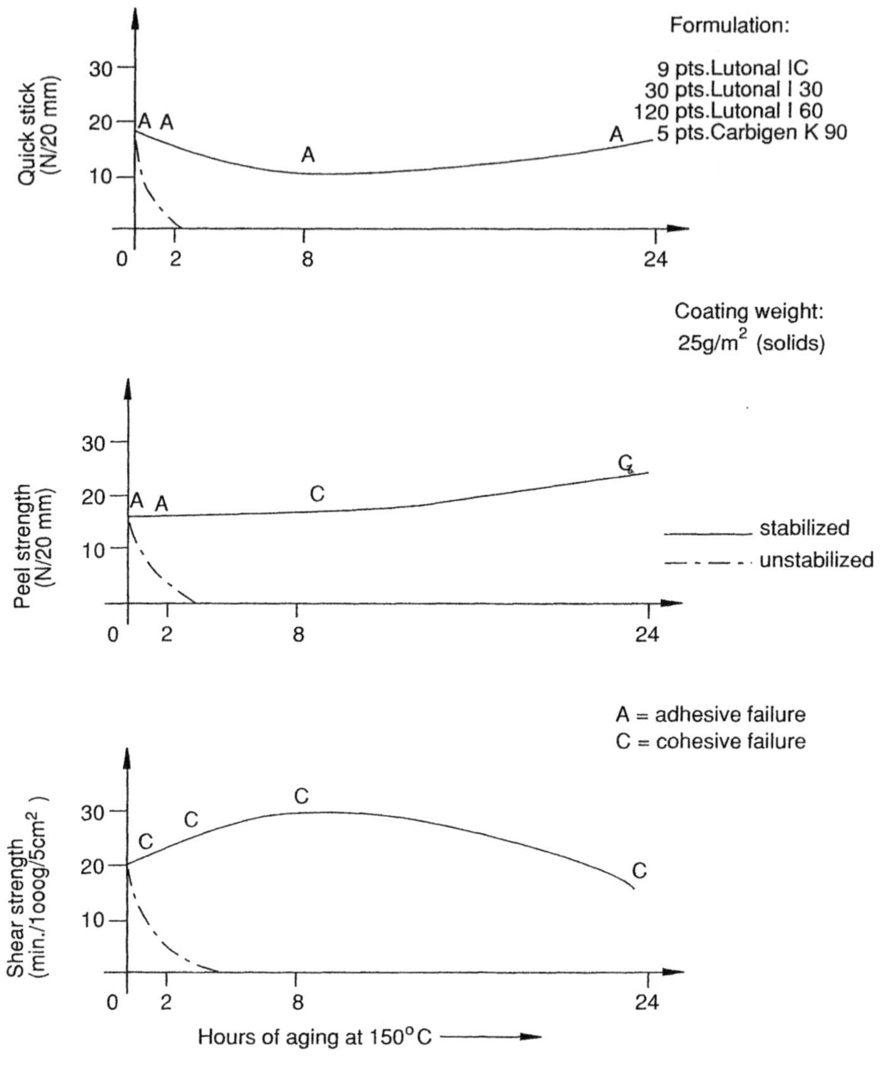

Fig. 17-1. Adhesion values of Lutonal I adhesive with and without stabilizing agend after exposure to heat.

First, a slight hardening took place, and a negligible tendency to chain fracture was observed thereafter. The results of this extremely stringent short-term test confirm the good experience that has been gained in practice with polyvinyl ethers over the course of decades. The adhesion values were determined by internal BASF methods with due regard to the PSTC standards.

The diagrams in Figure 17-2 show the viscosity behavior of stabilized

allows the production of pressure sensitive articles with excellently reproducible properties. The commercial products are available either in solvent-free form or as solutions in conventional solvents. When the base materials for pressure sensitive adhesives must be transported over long distances, it is advisable to use products free from solvent. In such cases, it is advantageous that polyvinyl ether solutions can be made up not only in solution kneaders and planetary mixers but also in dissolvers with a high shearing effect.

Adhesive compounds based on vinyl ether polymers can be processed on all conventional coaters. These compounds were considered already in the early stage of developing solvent-free coating processes. Apart from other substances, vinyl ether polymers were suggested for use as raw materials for hot melt adhesives in 1953, for processing on conventional coaters[6,7] in 1957, and for the extrusion process[8] in 1960. The extrusion process allows short exposure of the adhesive to heat and in-line melting and coating. The formulation of adhesive compounds that can be applied as hot melts is subject to certain limitations. To achieve the required coating behavior, it is often necessary to subordinate quality requirements to the processing requirements.

The formulations suggested in the paragraphs below should be applied at a rate of approximately 25 g/m^2. The optimum coating weight depends on the purpose of the finished article and the nature of the adhesive compound. A coating weight of 20–30 g/m^2 is usually applied for articles used on smooth surfaces, such as glass, lacquered wood, or metal surfaces. Articles for use on uneven surfaces, such as pressure sensitive tapes for laying carpets, require coating weights of up to 100 g/m^2. Moreover, there is a certain relationship between adhesion and coating weight, as shown in Figure 17-3. With low coating weights, the adhesion normally increases with the coating weight, provided that adhesive failure takes place on separation. The adhesion is largely independent of the coating weight when the bond separates in the adhesive layers, that is, when cohesion failure occurs. The optimum coating weight must be determined by trials in each individual case.

Vinyl Methyl Ether Polymers

Vinyl methyl ether polymers are applied as blending components for pressure sensitive adhesive compounds based on other raw materials—usually aqueous dispersion-type adhesives. They are used primarily for achieving defined hydrophilic properties and increasing the adhesiveness.

Tapes with adhesive compounds containing polyvinyl methyl ether are used, for instance, for bonding paper roll to paper roll in continuous printing of paper webs. The bonded areas are cut out after printing. They must be dispersible in water so that the paper can be repulped.[9–11]

70 parts Acronal V 205
30 parts Lutonal M 40, 50% in water
40 parts water
5 parts toluene

In the following typical formulation for plasticized PVC decorative films, Lutonal M 40 is used for increasing the adhesion.

70 parts Acronal 80 D
30 parts Lutonal M 40 ~ 50% in water
5 parts ethyl acetate

At temperatures above 28°C, mixtures of Lutonal M 40 and polymer dispersions are unstable on account of the solubility inversion of the vinyl methyl-ether polymers. A slightly gritty to completely coagulated mixture is formed, depending on the temperature and duration of exposure. These mixtures can be rendered stable even at temperatures above the precipitation point of Lutonal M 40 by adding small quantities of solvent, if necessary, in combination with an emulsifier. Acronal V 205 has a solids content of approximately 70%. Mixtures of this polymer dispersion, of exceptionally high solids content, and Lutonal M 40 cannot be stabilized solely by the means previously described. The solids content of the typical formulation containing Acronal V 205 was therefore reduced to approximately 50% by adding 40 parts water.

When used in adhesive compounds for coating plasticized PVC, vinyl ether polymers fulfill another function. They reduce the decrease in adhesive strength caused by plasticizer migration. This phenomenon is probably due to the good compatibility of the vinyl ether polymers with a great variety of solvents and plasticizers. Vinyl ether polymers take up the plasticizer migrating from the carrier film and form a stable mixture. The usual accumulation of plasticizer on the surface of the adhesive compound and the resultant decrease in adhesive strength can thus be diminished. However, it was found that this effect is largely dependent on the nature of the film, so that investigations should be carried out in each individual case.

Vinyl Ethyl Ether Polymers

Vinyl ethyl ether polymers are very similar to the substantially more widely used vinyl isobutyl ether polymers with regard to properties and processibility. In principle, the formulations and application data given below are also applicable to polyvinyl ethyl ethers. One difference is that the water vapor permeability of these polymers is about equivalent to the average loss of water through the

This hot melt pressure sensitive adhesive was already mentioned. It can be used, for instance, for double-faced tapes with high coating weight on a textile carrier, for example, for securing edges (carpet laying tape).

Widely differing requirements are imposed on the adhesive behavior of pressure sensitive medical articles. Formulations must therefore be worked out in each individual case with due regard to the required properties. The following formulation for an adhesive compound applied from a solution may serve as a guide. The developed adhesive should in any case be examined for its skin irritating effect.

10	parts Lutonal IC
60	parts Lutonal I 60
5	parts Lutonal I 30
0.5	part Antioxidant ZKF
25	parts zinc oxide
~80	parts mineral spirit (b.p. 65–95°C)

Pressure sensitive tapes and films produced with hot melt adhesive compounds consisting of 40–60 parts Lutonal I 60 and 60–40 parts bitumen can be used as "bituminous adhesive off the roll" for a great variety of applications. They can, for instance, be used on building sites where cookers for hot bituminous adhesives are undesirable because of the danger of fire and accidents, or for other reasons.

Vinyl isobutyl ether polymers of a medium degree of polymerization are also used as blending components for adhesive compounds based on other raw materials. They are mainly applied as secondary dispersions in admixture with aqueous acrylic polymer dispersions and improve the tack, which could also be achieved with resins and other polymers of a medium degree of polymerization. In addition, blending of acrylic emulsions with a vinyl isobutyl ether emulsion yields adhesives with improved adhesion as in the case of vinyl film products made with monomeric plasticizers, as already mentioned in the discussion of vinyl methyl ethers. Furthermore, the anchorage of the adhesive compound to the substrate and the desired adhesive legging can be built in the adhesives used for flexibile vinyl film products. These effects are provided by Lutanol I 65 D, as shown in the following adhesive formulation, which is suitable for flexible vinyl decorative films, tapes, and label stock: Acronal 80 D, 80 parts; Lutanol I 65 D, 20 parts.

CONCLUSION

The significance of vinyl ether polymers has changed over the course of time. However, these polymers still have a firm foothold as products for special

18. Silicone Pressure Sensitive Adhesives

Loretta A. Sobieski and Thomas J. Tangney
Dow Corning Corporation
Midland, Michigan

Silicone pressure sensitive adhesives are useful in many specialty applications whose requirements preclude the use of conventional organic materials. They are often chosen for a particular performance characteristic that is dictated by the application. Silicones should be considered when exposure to harsh environments, including temperature extremes, chemicals, moisture, and radiation, is expected. They are known for their durability and long service life. In pressure sensitive adhesives, this is evidenced by the clean removability of silicone coated pressure sensitive tapes after long-term service, even at high temperatures.

FUNDAMENTAL TECHNOLOGY

Silicone pressure sensitive adhesives are similar to their organic counterparts in that they are based on two major components: a polymer, or gum, and a tackifying resin. The polymer is typically a high molecular weight polydimethylsiloxane or polydimethyldiphenylsiloxane that contains residual silanol functionality (SiOH) on the ends of the polymer chains. The resin is a three-dimensional silicate structure that is endcapped with trimethylsiloxy groups ($OSiMe_3$). However, it too contains some residual silanol functionality. During processing, the resin and the polymer are dissolved together in a hydrocarbon solvent. Processing conditions, and levels of silanol functionality, are selected so that a condensation reaction takes place between the resin and the polymer (Fig. 18-1). This condensation results in a network of polymer chains that are crosslinked with resin molecules. More specifically, this network has a two-phase morphology consisting of a continuous polymer-rich phase and a discontinuous resin-rich phase covalently bound together at the interface through siloxane

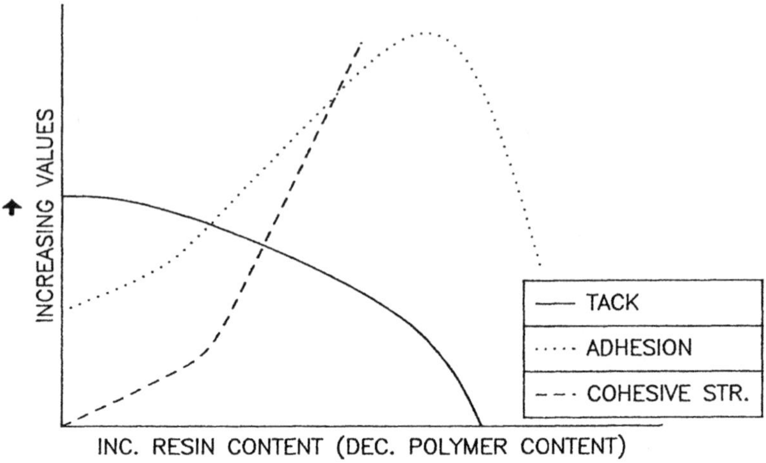

Fig. 18-2. The effect of resin and polymer content on performance properties.

ally affected by the level of silanol functionality present on both the resin and the polymer.[3] In the limiting case, adhesives formulated from resins and polymers that have no silanol functionality cannot form a cross-linked network through silanol condensation. These adhesives are very tacky and exhibit moderate peel adhesion values, but are very low in cohesive strength. If the silanol functionality is present only on the polymer, adhesives with somewhat lower tack and higher peel and cohesive strengths are produced. From a practical standpoint, adhesive properties are most easily controlled by controlling the level of functionality on the resin. As this level of functionality is increased, the cross-link density of the network is increased, resulting in a predictable effect on performance properties. Tack and peel adhesion decrease slightly while cohesive strength increases significantly (Fig. 18-3). In addition, these performance properties, as well as bulk physical properties, can also be affected by changes in the molecular weight of either the polymer or the resin.

CURE CHEMISTRY

Silicone pressure sensitive adhesives are usually supplied as solutions in a hydrocarbon solvent. Upon solvent evaporation, they immediately exhibit pressure sensitive properties. Simple solvent removal results in high tack, low cohesive strength materials that are useful in some applications. More typically, a catalyst is added to the adhesive to reinforce the siloxane network and increase cohesive strength. The three types of catalysts most commonly employed are organic peroxides, amino silanes, and metal salts of organic acids. Peroxides, such as

Finally, metal salts of organic acids are also used as condensation catalysts to increase cross-link densities of some silicone pressure senstive adhesives.

PRIMERS

As is characteristic of polydimethylsiloxanes, silicone pressure sensitive adhesives have very low surface energies[5] (18–22mN/m) and are good film formers. This allows them to adhere to most backing materials, including some of the traditionally difficult to wet, low-energy surfaces such as tetrafluoroethylene and polymide films, silicone rubber, and silicone coated fiberglass. However, for some tape applications, even greater anchorage to the tape backing is required to prevent legging (adhesive delamination from the tape backing). Anchorage can be increased through the use of primers or adhesion promoters that are cured on to the substrate prior to the application of the pressure sensitive adhesive. Polydimethylsiloxane paper release coatings are widely used as primers. Due to their similar surface energies and chemical compositions to the adhesives themselves they facilitate adhesion to the substrate. These materials are usually applied in very thin films and can be cured at low temperatures. If application of a primer coat is not possible, adhesion to the backing can be improved by addition of a silane coupling agent to the pressure sensitive adhesive prior to curing.

TYPICAL PROPERTIES

Silicone pressure sensitive adhesives display numerous advantages over their organic counterparts. They have very low glass transition temperatures (about −120°C) and are quite thermally stable. This allows them to be used over a much wider temperature range than even the best organic pressure sensitive adhesives.[6] Table 18-1 illustrates how a silicone pressure sensitive adhesive retains peel strength between −73°C and 250°C without becoming brittle at either temperature extreme.

Silicone adhesives also display very good electrical properties, maintaining high dielectrical strength, good arc resistance, and low loss factors (Table 18-2), even after extended exposure in adverse environments. Furthermore, they resist breakdown upon exposure to an electrical current and do not yield carbon residues or current leakage. Along with low surface energies, their ability to conform to irregular surfaces makes them the adhesives of choice for many applications. They are also very resistant to many environmental factors, such as moisture, ultraviolet radiation, ozone, biological attack (fungus and mildew), and chemicals like oils, acids, and bases. In addition, silicone pressure sensitive

the point where other constructions can be fabricated. When adhesive strength is high enough to warrant a release system, mechanical liners are frequently used.

The largest application for silicone pressure sensitive adhesives is in the manufacture of printed circuit boards. The circuit is covered with a polyester-backed, silicone pressure sensitive adhesive platers tape to mask the board leaving the contact fingers exposed. Tin–lead solder is stripped from these fingers to expose bare copper. Nickel, followed by gold, is then plated on the copper.[7] During these processes, the taped board is in contact with water and a variety of chemical solutions, including solutions of strong inorganic acids. The platers tape is used to prevent the leakage of any of these solutions back to the circuitry. Such leakage can result in corrosion of the exposed copper and board rejection. Silicone pressure sensitive adhesives are formulated to optimize their conformability to the irregular surface of printed circuit boards. Few organic materials approach the low surface energy of silicones that is required for this application.

Other applications during the processing of printed circuit boards also utilize silicone pressure sensitive adhesive tapes. Polyimide-backed tapes are used to mask various areas of the board during wave soldering and hot air leveling. In these operations, the silicone pressure sensitive adhesive must remain in place during immersion in aqueous flux solutions and 260°C solder, and while hot air is blown across the board.

Various electrical applications comprise the next largest market segment for silicone pressure senstive adhesive tapes. Because of their excellent electrical propeties, silicone pressure sensitive adhesives are used on tetrafluoroethylene (TFE), glass cloth, and TFE coated glass cloth backings to make insulating tapes. These tapes are used to wrap wires and coils in appliance motors and aircraft engines to maintain electrical performance over wide temperature ranges.

A unique application for silicone pressure sensitive adhesives is the bonding of other silicone fabrications. Very low surface energy silicone rubber and silicone release coatings are virtually impossible to adhere to with organic pressure sensitive adhesives. Silicone splicing tape, made with a polyester backing, is used to join rolls of silicone coated release liner as they pass through adhesive coaters, curing ovens, and other web handling operations. The pressure sensitive adhesive requires enough cohesive strength to hold the web under tension at temperatures up to 150°C. Similarly, silicone pressure sensitive adhesive tapes are used to splice TFE, TFE coated fabrics, and silicone coated fabrics.

A broad range of applications for silicone pressure sensitive adhesive tapes is their use as process aids. Glass cloth, or silicone coated glass cloth backed tapes are used as plasma or flame spray tapes. Double-sided cloth tapes are also common. This type of tape is used to mask areas of metal surfaces in flame spray and grit blasting operations during the repair of heavy machinery

Table 18-3. Performance Properties of Medical vs. Industrial Pressure Sensitive Adhesives

	MEDICAL-GRADE ADHESIVE	INDUSTRIAL-GRADE ADHESIVE
Release (g/cm)[a]	1.4	79
Tack (g/cm^2)	56	1141
Peel adhesion (g/cm)	328	555

[a] From BIO-RELEASE® X7-2751 medical-grade silicone release coating cured on 2-mil polyester film. Adhesive solvent cast on liner, cured, and laminated to 2-mil polyester film. Laminates aged one week at room temperature.

rates of drug permeability, allowing a controlled rate of delivery to the patient over the wear time of the patch.[10]

Since the drug can be loaded directly into the adhesive, chemical resistance is important. Some drugs, especially amine-containing compounds, can react with the adhesive resulting in a loss of pressure sensitive properties. Recently, silicone pressure sensitive adhesives have been developed that resist the effects of these drugs and retain their adhesive properties.[11] This is shown in Table 18-4, where we see that a drug-resistant medical-grade adhesive has much better tack retention than a standard medical adhesive after loading with Progesterone and Indomethacin.

Table 18-4. Adhesive Properties of Drug-loaded Silicone Pressure Sensitive Adhesives

DRUG-RESISTANT MEDICAL ADHESIVE			
DRUG	RELEASE (g/cm)	PEEL ADHESION (g/cm)	TACK (g/cm^2)
Control (no drug)	1.2	568	64
Progesterone	2.2	518	70
Indomethacin	1.6	517	63
STANDARD MEDICAL ADHESIVE			
DRUG	RELEASE (g/cm)	PEEL ADHESION (g/cm)	TACK (g/cm^2)
Control (no drug)	1.4	328	56
Progesterone	0.8	250	0
Indomethacin	1.3	469	42

Note: Aged 8 weeks at room temperature; 1.0 weight percent drug loading.

19. Miscellaneous Polymers

Donatas Satas
Satas & Associates
Warwick, Rhode Island

Pressure sensitive adhesives can be compounded by employing various elastomers. Pressure sensitive properties depend on the adhesive's rheology, which can be changed by compounding with various materials of lower molecular weight, such as resins and plasticizers, and not on the properties of the elastomer alone. The elastomer portion of the adhesive determines the cohesive strength and the shear resistance. The adhesive modulus must usually be decreased by additives to yield a soft and tacky material. Various polymers may be used as the basis for pressure sensitive adhesives, but only a small number of elastomers is used in large volume. Many other polymers discussed in this chapter are of interest for products with special properties and their use may be small. Styrene-butadiene rubber is the only exception, since it is used quite widely and for many diverse applications.

STYRENE-BUTADIENE RUBBER

Random styrene-butadiene copolymers (SBR) are used for pressure sensitive adhesives in formulations with natural rubber and other elastomers, or as the sole elastomer. Butadiene-styrene latexes are discussed separately in Chapter 12. The use of adhesive solutions and of solid adhesive compounds for calendering is, however, much more widely spread than that of emulsions. Low styrene content copolymers are used for pressure sensitive adhesives because of their lower modulus. Such copolymers are compounded with tackifying resins and other additives. SBR is not compatible with β-pinene resins, which are the preferred tackifiers for natural rubber, but it is compatible with rosin-type tackifiers and with α-pinene resins. Lanolin improves the compatibility of SBR with

conditions. Sulfur cure by the addition of elemental sulfur is rarely used. The most common curing formulations, the so-called efficient vulcanization systems, employ a sulfur donor compound, zinc oxide, a fatty acid activator, and an organic accelerator.[5] There are some chemical compounds (thiuram disulfides) that act as both sulfur donor and accelerator. An efficient vulcanization system requires a low sulfur concentration. The exact curing mechanism is not completely understood but is known to involve several steps.

A typical accelerated rubber vulcanization is shown below.

Rubber	100
Zinc oxide	1–40
Fatty acid	1–4
Sulfur	0.5–4
Accelerator	0.5–2

Tackifying resins and other compounding ingredients are added to the cured rubber. The formulation of a thiurads-cured SBR adhesive for a calendering application is shown below.

SBR	100
Tackifying resin	125
Zinc oxide	86
Titania	30
Lanolin	20
Mineral oil	13
Thiurads	1

Vulcanized SBR solution adhesives are prepared by compounding the adhesive on a mill and then dissolving it. The adhesive may be precured on the mill—by masticating at 90–120°C for 20 minutes to several hours; by allowing it to cure in an autoclave at elevated temperature; or by curing it in the drying oven after it is coated. Curing times of well-compounded adhesives may take 2–5 minutes at 150–180°C, or longer at lower temperatures. It is therefore, not surprising that it is difficult to complete the cure during the drying operation. The cure may continue during tape storage, causing some changes in the adhesive propeties. A sulfur-cured adhesive formulation suitable for masking tapes and other general-purpose applications is shown below.[6]

SBR (50 Mooney value)	100
Cycloparaffin tackifier	85
Alicyclic petroleum oil	40
Titanium dioxide	50

deficiency. The formulation of such an adhesive is shown below.[10] The adhesive was cured at 188°C for 20 seconds.

Polyisoprene (slightly milled)	100
Polyterpene resin	55
Amberol ST-137	15
Zinc resinate	5

Polyisoprene blends with other elastomers: natural rubber,[11] block copolymers,[12] and others have been used for pressure sensitive adhesives. Very low molecular weight liquid polyisoprene was also used to formulate pressure sensitive adhesives[13] and to improve the tack and adhesion of natural-rubber-based adhesives.[14] A polyisoprene-styrene copolymer was used to prepare a pressure sensitive adhesive.[10] An isoprene piperylene copolymer was synthesized and used for pressure sensitive adhesives.[15]

POLYBUTADIENE

Polybutadiene-based adhesives have good water resistance and form good bonds at low temperatures. A formulation of such an adhesive is shown below.[16]

1,4-polybutadiene	100
Polyterpene resin, S-1010	50
Polyterpene resin, S-115	20
Phenol formaldehyde resin	35
Dioctyl phthalate	10
Heptane	500
Ethanol	10

Polybutadiene is also used for hot melt adhesive formulations.[17]

BUTADIENE-ACRYLONITRILE RUBBER

Butadiene-acrylonitrile rubbers have been used for adhesives with improved solvent resistance. They were used for vinyl film tapes[18] and other electrical tapes.[19] A simple formulation of such an adhesive is shown below.

Butadiene-acrylonitrile copolymer (70:30)	100
Diphenyl cresyl phosphate	50
Tackifying resin	40

reacting polyol with diisocyanate and using pinene resin as a tackifier.[25] The tack level was controlled by varying the ratio of diisocyanate to polyol. This adhesive is suitable for manufacturing a double-face tape that can be wound on itself without an interliner. A low peel adhesive has been prepared by polymerizing a low molecular weight acrylic solution polymer containing hydroxyl groups with an isocyanate-terminated prepolymer.[26] An adhesive for cleanly removable labels, even after aging at elevated temperature, has been described.[27] Pressure sensitive adhesive coatings were prepared *in situ* by reacting polyols with diisocyanates after coating.[28,29] Capped diisocyanates are described for use with *in situ* polymerizable adhesives.[30]

ETHYLENE-VINYL ACETATE COPOLYMERS

Ethylene-vinyl acetate copolymers of high vinyl acetate content were used for compounding the early hot melt pressure sensitive adhesives. These adhesives had one main deficiency: their shear resistance was low, especially at higher temperatures. The properties of such adhesives depend largely on the amount of vinyl acetate and on the molecular weight.[31] A generalized formulation of a typical ethylene-vinyl acetate copolymer based hot melt pressure sensitive adhesive is shown below.[32]

EVA copolymer	30–50%
Tackifying resin	30–50%
Plasticizer	0–20%
Filler	0–5%
Antioxidant	0.5%

Such compounded EVA copolymers may have good tack and peel adhesion initially, but these might drastically decrease on aging. This is usually caused by the incompatibility of tackifying resins with the EVA copolymer. Incompatibility can be revealed by microscopic observation, as evidenced by the presence of a discontinuous phase.[33] The incompatibility could also be detected by differential scanning calorimetric measurements.[34] Endothermic transitions due to crystalline sites in the EVA copolymer are disrupted by compounding with a tackifying resin. These can reappear later on aging.

The introduction of block copolymers increased the application of hot melt pressure sensitive adhesives, and these polymers replaced EVA copolymers completely.

Inherently tacky ethylene-vinyl acetate copolymer emulsions are available. They are useful especially for the lamination of difficult-to-adhere films. Pressure sensitive uses of such adhesives beyond lamination applications are expected to grow.

5. Hertz, D. L. Jr. *Elastomerics* **116**(11): 17–21 (1984).
6. Pike, C. O., and Bemmels, C. W. U.S. Patent 2,576,968 (1951) (assigned to Industrial Tape Corp.).
7. Joesting, E. O., and Ethier, D. O. U.S. Patent 2,708,192 (1955) (assigned to Minnesota Mining & Manufacturing Co.).
8. Korpman, R. U.S. Patent 3,231,419 (1966) (assigned to Johnson & Johnson Co.).
9. Korpman, R. U.S. Patent 3,535,152 (1970) (assigned to Johnson & Johnson Co.).
10. Weidner, C. L. U.S. Patent 3,725,322 (1973) (assigned to Johnson & Johnson Co.).
11. Takamatsu, H., Minatono, S., Terao, K., and Yamauchi, J. U.S. Patent 4,181,635 (1980) (assigned to Kuraray Co., Ltd.).
12. Korpman, R. U.S. Patent 3,625,752 (1971) (assigned to Johnson & Johnson Co.).
13. Minatono, S., Takamatsu, H., Yamauchi, J., and Terao, K. U.S. Patent 4,204,046 (1980) (assigned to Kuraray Co., Ltd.).
14. *Rubber World* **179**(5): 37–40 (1979).
15. Lakshmanan, P. R., Swift, H. E., and Wu, C. Y. U.S. Patent 4,096,103 (1978) (assigned to Gulf Research & Development Co.).
16. Freeman, C. H., and Sauer, C. I. U.S. Patent 3,451,537 (1969) (assigned to Minnesota Mining & Manufacturing Co.).
17. Jurrens, L. D. U.S. Patent 4,025, 478 (1977) (assigned to Phillips Petroleum Co.).
18. Waldman, H. L. U.S. Patent 3,374,134 (1968) (assigned to Johnson & Johnson Co.).
19. Hendricks, J. D., and Dahlquist, C. A. U.S. Patent 2,601,016 (1952) (assigned to Minnesota Mining & Manufacturing Co.).
20. Feeney, G. W., and Burroway, G. L. U.S. Patent 3,157,319 (1979) (assigned to The Goodyear Tire & Rubber Co.).
21. Fitch, J. C., and Snow, A. M. Jr. *Adhesives Age* **20**(10): 23–26 (1977).
22. Matulewicz, C. M., and Snow, A. M. Jr. *Adhesives Age* **24**(3): 40–43 (1981).
23. Dahl, R. U.S. Patent 3,437,622 (1969) (assigned to Anchor Continental).
24. Dahl, R. U.S. Patent 3,515,773 (1970) (assigned to Continental Tapes Inc.).
25. Webber, C. S. U.S. Patent 3,246,049 (1966) (assigned to Norton Co.).
26. Schenk, W. N. U.S. Patent 4,306,996 (1981) (assigned to B. F. Goodrich Co.).
27. Mueller, H., and Szonn, B. U.S. Patent 3,930,102 (1975) (assigned to Beiersdorf AG).
28. Dahl, R. U.S. Patent 3,802,988 (1974) (assigned to Continental Tapes, Inc.).
29. Kest, D. O. U.S. Patent 3,879,248 (1975).
30. Hartmann, H., Druschke, W., Eisentraeger, K., and Mueller, H. U.S. Patent 4,087,392 (1978) (BASF AG)
31. Zimmerman, A. B., and Fischer, J. *Adhesives Age* **18**(9): 29–32 (1975).
32. Litz, R. J. *Adhesives Age* **14**(7): (1971).
33. Powers, P. O. Growth and Change of Hot Melts. *Adhesion Science and Technology* (L-H. Lee, ed.). Plenum, New York, 1975.
34. Rifi, M. R. Pressure Sensitive Adhesives Structure-Performance Relationship. *Course Notes 1977, Cavalcade of Hot Melts, Their Future, Their Problems,* pp. 17–24. Technical Association of the Pulp and Paper Industry, Hilton Head Island, S.C. April 18–21, 1977.
35. Park, V. K. U.S. Patent 3,850,858 (1974) (assigned to Eastman Kodak Co.).
36. Hoh, G. L. K., and Reardon, J. E. U.S. Patent 3,932,326 (1976) (assigned to E. I. du Pont de Nemours & Co.).
37. Isayama, K. Sealants Adhesives and Pressure Sensitive Adhesives Prepared from Polyether Oligomers with Terminal Silyl Function Groups. *Abstracts of Poster Papers. 1987 International Meeting. The Adhesion Society.* Williamsburg, Va., February 22–27, 1987.
38. De Groot von Arx, E. U.S. Patent 3,096,202 (1963) (assigned to Johnson & Johnson Co.).

the tree and collecting the oleoresin that exudes from the wound. Sulfuric acid is usually used to enhance exudation and increase the yield of oleoresin. More recently, other materials, such as the herbicide paraquat, have been found to further increase the yield of oleoresin as well as the resinous content of the wood in the lower portion of the tree.[1] The oleoresin obtained from this process is then distilled to give gum rosin and turpentine.

Wood rosin is obtained from aged pine stumps that are pulled up, washed, and mechanically reduced to chip form. These chips are then extracted with solvent followed by distillation and further refining to give wood rosin.

Tall oil rosin is the third type of rosin, and it is obtained from tall oil, which is a by-product of the paper industry. During the pulping operation, wood chips are cooked under heat and pressure in alkaline medium. The pulp is removed by filtration and washing, and the liquid portion is concentrated resulting in a precipitate that is removed and acidified to give crude tall oil. This crude tall oil consists mainly of fatty acids and rosin acids, which are separated by distillation.

Rosin from any of these three sources is a mixture of organic acids called rosin acids. Minor components consist of rosin esters and anhydrides, unsaponifiable matter, and fatty acids.

The rosin acids can be divided into two different types. These are the abietic acid type and the pimaric acid type, and unmodified rosin consists primarily of abietic-type acids. Figure 20-1 shows some of the more common members of these two types. It can be seen that the difference in the two types is the substitution pattern at the C_{13} position. Abietic acids have an isopropyl group, while the pimaric acids have a methyl and a vinyl group.

Rosin by itself is generally unsuited for use in modern adhesive systems because it is subject to oxidation and other reactions as well as crystallization. Because of this, most rosins for adhesive use have been subjected to one or more of a series of reactions to form modified rosins or rosin derivatives. Among the more important reactions for the production of materials for use in adhesives are disproportionation, polymerization, hydrogenation, and esterification.

Disproportionation takes place when rosin is heated to about 270°C. The reaction may be accelerated by the use of various catalysts such as platinum or iodine. The result is a hydrogen exchange between the rosin acids where, for instance, two abietic acids are converted into a dihydroabietic acid and a dehydroabietic acid.

Polymerization is affected by treating a solution of rosin in an inert solvent with an acid catalyst such as boron trifluoride or sulfuric acid. Removal of the catalyst and solvent results in a product composed largely of rosin dimer and having a higher softening point than unmodified rosin.

Hydrogenation using Raney nickel or a similar catalyst can be tailored to

purpose. When special stability is desired rosin esters are produced from hydrogenated rosins.

A few examples of rosin-based tackifier resins used in pressure sensitive adhesives are shown in Table 20-1.

HYDROCARBON TACKIFIER RESINS

Hydrocarbon tackifier resins are low molecular weight polymers derived from monomers obtained from petroleum, coal, and wood. Unlike the rosin derivatives, hydrocarbon resins are true polymers and their performance characteristics are determined by monomer composition, polymerization method, and final product molecular weight.

Chemically, hydrocarbon resin streams may be classified as containing primarily aromatic, aliphatic, and diene (cyclic olefin) monomers. These are also referred to as C-9, C-5, and $(C\text{-}5)_2$, respectively, corresponding to the average number of carbon atoms per monomer molecule. Polymerization of these streams is carried out using a Lewis acid catalyst or by a free radical process using heat and pressure. These streams, polymerization methods, and the general designations of the resins produced are shown in Figure 20-2.

Because of the variety of feed streams available and the various polymerization techniques that are employed, a wide variety of hydrocarbon resins are available with a broad range of properties and prices. Furthermore, it is this diversity that has made hydrocarbon resins useful in so many tackifier applications, especially those involving pressure sensitive adhesives. The following discussion briefly treats each of the main types of hydrocarbon resins. The typical range of properties available will be outlined. Tables 20-2 through 20-7 contain representative lists of the materials available from various manufacturers. Each trade name used here represents only one resin out of a series or family that may be available under that general name.

Table 20-1. Rosin Tackifiers

TRADE NAME	DESCRIPTION	MANUFACTURER
Floral 85	Glycerine rosin ester	Hercules, Inc.
Pentalyn H	Hydrogenated pentaerythritol ester	Hercules, Inc.
Staybelite Ester 10	Hydrogenated glycerine ester	Hercules, Inc.
Sylvatac RX	Modified tall oil rosin	Sylvachem Corp.
Sylvatac 95	Polymerized rosin	Sylvachem Corp.
Zonester 85	Rosin ester	Arizona Chemical Co.

Table 20-2. Aromatic Resins

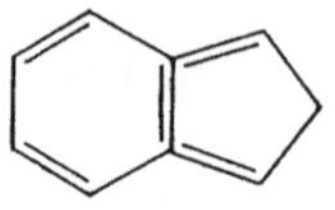

Indene

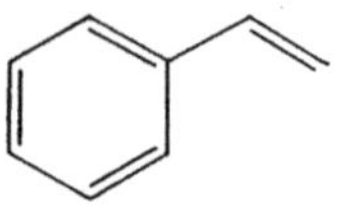

Styrene

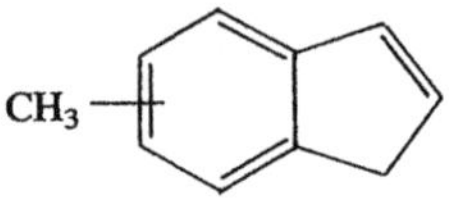

Methylindene(s)

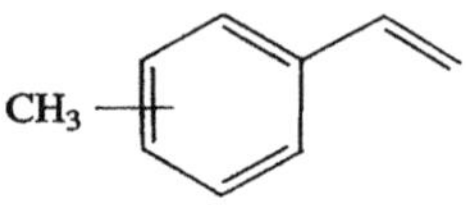

Methylstyrene(s)

TRADE NAME	MANUFACTURER
Amoco Resin 18	Amoco Chemicals
Escorez 7105	Exxon Chemicals
Nevchem 100	Neville Chemical Co.
Nevex 100	Neville Chemical Co.
Pentaprene 100	Chemfax, Inc.
Petrorez 100	Lawter International
Picco 6100	Hercules, Inc.
Resinall 711	Resinall Corp.

adhesives. The high softening point grades are finding increased use in adhesives based on the thermoplastic elastomers. They associate with the styrene end blocks, and can thus be used to modify the strength, viscosity, and upper service temperature of the adhesive blend.

Aromatic resins are also used in the production of inks and various coatings as well as in rubber compounding, where they serve as process aids, plasticizers, and tackifiers.

Aliphatic Hydrocarbon Resins

Aliphatic resins (Table 20-3) are produced from light, so-called C-5 petroleum fractions. The principal monomers are *cis-* and *trans*-piperylene. Also found in these streams are varying amounts of isoprene, 2-methylbutene-2, and, in some cases, dicyclopentadiene. These streams are polymerized using aluminum chloride, and the softening point is controlled by adjusting the composition of the diluent or monomers. This differs from the production of aromatic resins using BF_3 where the softening point is controlled by changing the reaction conditions or the addition of plasticizer oils.

Mixed Aliphatic/Aromatic Resins

The mixed C-5/C-9 resins (Table 20-4) are hybrid resins produced by blending C-5 and C-9 streams in varying proportions and then polymerizing with either $AlCl_3$ or BF_3 as required. These resins are usually produced in only one or two softening point grades, and the properties obtained are generally intermediate between those of either straight aromatic or aliphatic resins. These resins are the result of research efforts by various manufacturers to fine-tune the compatibility and tackifying characteristics of their C-5 resins to produce products with outstanding performance in the hot melt and pressure sensitive adhesives.

Heat Reactive Hydrocarbon Resins

The next general class of hydrocarbon resins is the so-called heat reactive or diene resins (Table 20-5). These are petroleum resins with a mixed aromatic and aliphatic composition. The streams used to produce these resins contain indene and styrene, mentioned before, along with substantial amounts of dicyclopentadiene, dimethyl-dicyclopentadiene, and various other C-5 dimers, better known as diene dimers. These resins are generally polymerized by the action of heat and pressure in a free radical process. This method of polymerization is generally less costly than the use of a Friedel-Crafts catalyst, and the resins are, therefore, usually less expensive than the catalytic resins.

Colors of the heat reactive resins can vary from light pale yellow to dark amber. The iodine numbers of these reactive resins are quite high, ranging from about 120 to 220, depending on the feed blend and the actual polymerization conditions employed. Specific gravities are similar to those of aromatic resins, ranging from 1.08 to 1.10, and molecular weights generally fall between 500 and 800. Softening points of various grades usually fall between 95° and 140°C, although some resins with higher values are available for special applications.

Because of the lower cost and excellent compatibility characteristics of

Table 20-4. Mixed Aliphatic/Aromatic Resins

TRADE NAME	MANUFACTURER
Escorez 2101	Exxon Chemicals
Hercotac AD-1100	Hercules, Inc.
Nevpene 9500	Neville Chemical Co.
Super Nevtac 99	Neville Chemical Co.
Super Sta-Tac 100	Reichhold Chemicals, Inc.
Wingtack Extra	The Goodyear Tire & Rubber Co.
Wingtack 86	The Goodyear Tire & Rubber Co.

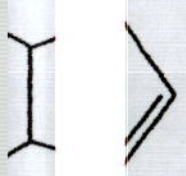

Table 20-6. Terpene Resins

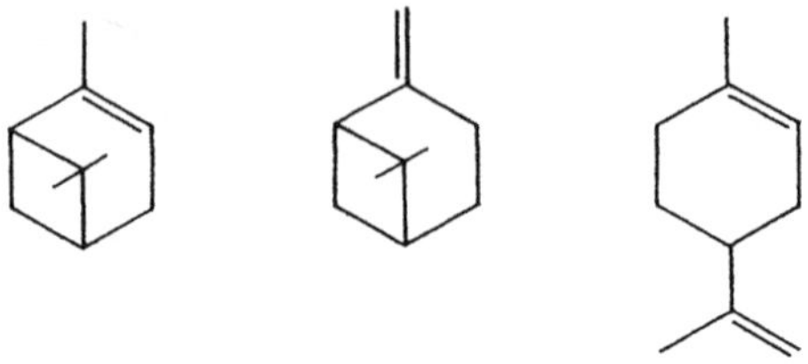

α-Pinene β-Pinene Dipentene

TRADE NAME	MANUFACTURER
Nirez 1100	Reichhold Chemicals, Inc.
Piccolyte S-100	Hercules, Inc.
Zonarez 7100	Arizona Chemical Co.
Zonatac 105	Arizona Chemical Co.

softening points, β-pinene resins have the highest molecular weights and lowest gravities, while dipentene resins have the highest gravities and lowest corresponding molecular weights.

Terpene resins, especially β-pinene resins, were the resins of choice in the production of pressure sensitive adhesives for many years. These resins have outstanding tackifying abilities when used with natural rubber, a wide range of softening points to choose from, and extremely broad Food and Drug Administration clearances. More recently, limonene and limonene-styrene copolymer resins have been developed that are almost water white or colorless. They are excellent tackifiers for butadiene-based elastomers and have become very popular for use in adhesives for the disposable products markets. Over the last ten years, however, due to rising costs and uncertain availability, the terpene resins and rosin derivatives have lost considerable ground to the growing number of less expensive synthetic hydrocarbon resins designed specifically to replace them.

Modified or Special Resins

The final group of resins to be discussed are the so-called special or modified resins. This is a catchall group used for resins having special properties that do not quite fit into the general classes of resins already mentioned. These resins are prepared by various techniques and by blending selected feed streams or pure monomers to obtain resins with special solubility, compatibility, or

performance characteristics are not too difficult if a systematic approach is taken.

Table 20-8 groups the more common base polymers used in pressure sensitive adhesives along with the types of tackifiers that can normally be used with them. Using this as a starting point, the first step is to contact the suppliers of these types of tackifiers for information on the specific resins they supply, their properties and prices, and any specific recommendations they might make. This information is then evaluated to eliminate resins with properties and prices that do not lend themselves to the application being studied.

The second step in choosing a tackifier is to screen the selected candidates to determine their suitability as tackifiers for the system being studied. This screening test can be quite simple, and may only involve dissolving roughly equal quantities of the base polymer and tackifier in a suitable solvent and casting a film that may be evaluated, upon drying, by touching it and observing its clarity.

Other important preliminary tests may involve simple stability tests such as heating a sample of resin in an oven for a few hours to observe the formation of foaming, skinning, or dark colors.

Table 20-8. General Tackifier Recommendations

BASE POLYMER	TACKIFIERS
1. Natural rubber (NR) *cis*-Polyisoprene (IR) Styrene/isoprene block copolymers	Aliphatic resins Mixed C-5/C-9 resins Rosin derivatives Special resins Terpene resins
2. Styrene-butadiene rubber (SBR) Styrene/butadiene block copolymers	Aromatic resins Heat reactive resins Mixed C-5/C-9 resins Rosin derivatives Special resins Terpene resins
3. Ethylene-vinylacetate copolymers	Aromatic resins Rosin derivatives Special resins
4. Atactic polypropylene	Aliphatic resins Mixed C-5/C-9 resins Terpene resins
5. Miscellaneous elastomers	Various depending on the polarity of the system

ture is probably the most important. Shear adhesion or time-to-failure values in this test can then be expressed as a function of the steady flow viscosity (V_O) as shown below.[4]

$$T = \frac{L^2 W V_O}{2MDG} \tag{2}$$

where

T = time to fail
L = overlap
W = width
D = adhesive thickness
M = load
G = gravity constant.

From this it can be seen that shear is essentially a viscosity measurement and may also be expressed by the log mixing rule just shown.

Peel adhesion is a widely used and studied test for the characterization of pressure sensitive adhesives. The values obtained by any of the various peel methods are primarily related to the stiffness of the adhesive. Indeed, Aubrey and Sherriff have shown that peel values for rubber-resin adhesives are directly related to dynamic storage modulus values and the glass transition temperature of the adhesive. For a mutually compatible rubber-resin blend, the glass transition temperature may be related to composition using the simple weighted average or the more complicated Flory–Fox equation.[5]

Tack is a fundamental property of pressure sensitive adhesives. While tack is easily defined in a general sense, it is not easy to test in a reliable, straightforward manner. Generally, rolling ball tack and quick stick may be considered as variations of peel adhesion tests. The probe tack test is considered the best measure of tack from both a theoretical and aesthetic point of view.[6,7] Probe tack is related to the wetting ability of the adhesive along with its compressibility and stiffness. These are the same variables that determine peel adhesion values and, as such, tack measured by almost any means except rolling ball can also be satisfactorily compared to adhesive T_g.

Tackifier Evaluation

Based on these considerations, it is possible to relate the performance values of most of the common adhesive tests to the composition of the formulation using a simple linear expression of the weight fraction of the various components. This can be of tremendous utility in the evaluation of tackifier resins and subse-

(r), a tackifier (t), and a plasticizer (p). Using the log mixing rule to define the behavior observed in a shear adhesion test, we write the following equation.

$$\log S = X_r \log S_r + X_t \log S_t + X_p \log S_p. \tag{3}$$

Because we are using weight fractions of the components, $X_r = 1 - (X_t + X_p)$. Substituting for X_r in the Equation (3) gives a new equation:

$$\log S = \log S_r + X_t \log (S_t/S_r) + X_p \log (S_p/S_r). \tag{4}$$

Recognizing that S_r, S_t, and S_p are constants allows us to express log S in the form of a simplified linear model containing only the weight fractions of the two components of interest:

$$\log S = B_0 + B_t X_t + B_p X_p. \tag{5}$$

Using this same technique for viscosity, peel, and tack, we obtain a family of general models that can be used to relate the various adhesive test values (Y) to the weight fractions (X_i) of the components in the formulation. These are shown in Table 20-10.

It will be noted that a second model has been shown for the tack and peel tests. This comes from a consideration of the fact that the more complex expression for T_g is generally more precise. As a result, addition of the cross product term, $B_{tp} X_t X_p$, can sometimes be used to improve the fit of the adhesive data to the model, as will be shown.

A detailed treatment of statistics and experimental design are far beyond the scope of this chapter or book. However, the results obtained using a simple two-variable two-level adhesive design to evaluate a solid tackifier resin and liquid resin plasticizer combination are shown in Table 20-11 along with the

Table 20-10. General Performance Models

Viscosity and Shear
$\log Y = B_0 + B_t X_t + B_p X_p$
Peel and Tack
$Y = B_0 + B_t X_t + B_p X_p$
or
$Y = B_0 + B_t X_t + B_p X_p + B_{tp} X_t X_p$

model coefficients obtained and some statistics associated with the models. Wingtack 95 and Wingtack 10 resins were evaluated in a hot melt pressure sensitive adhesive formulation based on a styrene-isoprene block copolymer, Kraton D-1107. A simple two-level/two-factor design consisting of four "corner" points and one "center" point was used. Because the polystyrene blocks in the copolymer are insoluble in the polyisoprene adhesive matrix at room temperature, they do not contribute to the development of the adhesive properties related to T_g. Therefore, in developing this design only the polyisoprene portion of the Kraton D-1107 was considered in calculating the weight fractions of the various components.

The R values in Table 20-11 are a statistical measure of the fit of the data to the model. A perfect fit gives an R^2 value of 1. Except for tack, the fit of the data to the two-term models is excellent, and simple addition of the cross product to the tack model results in a significant improvement in this model as well.

Results from this type of evaluation can be extremely valuable. They can be used to predict the adhesive performance for any desired composition as long as the T_g of the blend does not exceed about 0°C. Substitution of desired adhesive values into the model equations gives a new set of equations that can be solved to obtain the composition necessary to produce those values. Also, the coefficients obtained can be used to compare the relative cost versus performance properties of a number of tackifier resins and plasticizers. Finally, because these equations are derived from fundamental relationships regarding the rheological behavior of polymer blends, they can be extended to provide information on the effects of basic changes in tackifier structure, composition, and molecular weight on the basic performance properties of pressure sensitive adhesives.

REFERENCES

1. Rothrock, C. W. Jr., and Rhyne, J. B. *TAPPI* **60**(6): 54–57 (1977).
2. Aubrey, D. W., and Sherriff, M. *J. Polym. Sci., Polym. Chem. Ed.* **18:** 2597 (1980).
3. Nielsen, L. E. *Polymer Rheology,* Chaps. 4 and 6. Dekker, New York (1977).
4. Dahlquist, C. A. *Handbook of Pressure-Sensitive Adhesive Technology,* Chap. 5 (D. Satas, ed.). Van Nostrand Reinhold, New York (1982).
5. Nielsen, L. E. *Polymer Rheology,* Chap. 5. Dekker, New York (1977).
6. Johnston, J. *Adhesives Age* **26**(12): 34–38 (1983).
7. Johnston, J. *Adhesives Age* **26**(13): 24–28 (1983).

or internal phase and the bulk substance the continuous or external phase.[2] A dispersion typically consists of a solid disperse phase and a liquid continuous phase. An emulsion consists of both a liquid disperse phase and a liquid continuous phase. Oil-in-water emulsions have oil as the disperse phase and water as the continuous phase. Water-in-oil emulsions have water as the disperse phase and oil as the continuous phase. Emulsifiers or surfactants are used to facilitate emulsification and to promote emulsion or dispersion stability.

Surfactants are classified as anionic, cationic, or nonionic according to the charge carried by the surface active part of the molecule.[3] An anionic surfactant will ionize in solution to yield a large negatively charged ion (anion) and a small positively charged ion.[4] The anionic surfactant derives its name from the surface active anion. Anionic surfactants can be further subdivided into general categories based upon chemical structure: sulfates, alkyl sulfonic acids and their salts, carboxylic acids and their salts, alkylaryl sulfonic acids and their salts, and miscellaneous.

Cationic emulsifiers will ionize to form a large positively charged ion (cation) and a small negatively charged ion. They can be subdivided into primary, secondary, or tertiary amine salts, quaternary ammonium compounds, and miscellaneous.

A nonionic surfactant will not ionize in solution. Nonionic surfactants are compatible with either anionic or cationic surfactants. They can be classified as having alkoxy ether, alkoxy ester, or amide linkages to the solubilizing groups.

Amphoteric (or ampholytic) surfactants assume either an anionic or cationic character. The ionic character of an amphoteric surfactant is a function of the pH of the system.

Resins can be emulsified or dispersed in two general ways, direct and inversion.[5] In the direct method, if the material is a liquid at room temperature, it can be dispersed directly into water using an emulsifier and good agitation. If it is a solid at room temperature, the resin can be dispersed at an elevated temperature where it is molten. Alternatively, the resin can be dissolved in a solvent; the resulting solution is then emulsified. In this case, the solvent must be removed after emulsification in order to have a solvent-free or low-solvent dispersion.

In the inversion method, a water-in-oil emulsion is first formed by slowly adding water to the oil phase in the presence of an emulsifier. As more water is added, the mixture eventually becomes thick at the inversion point where the water becomes the continuous phase. The fine particle size is developed by shearing the viscous mass at the inversion point. After inversion, additional water lowers the percent total solids of the dispersion.

migrate to the adhesive–substrate interface and cause loss of adhesion. An emulsifier could possibly migrate into a porous substrate, such as label stock. These properties should be kept in mind and tested for with a specific adhesive system in a specific application. An emulsifier could also cause foam generation during application. Care must be taken if defoamers are used, in order to avoid any detrimental effect by the defoamer on adhesive properties.

Another concern specific to resin dispersion/polymer latex systems is that of mixing and particle fusion. In solution adhesives, the adhesive components are mixed on a molecular basis.[5] Therefore, when the solvent is evaporated the adhesive components are as finely mixed as possible in the dried film. In an aqueous adhesive mixture of resin dispersion and polymer latex, each component is present in individual particles composed of many molecules. On evaporation of the water, the particles come together. However, some combination of additional time and temperature is needed to fuse these particles to the same intimate molecular mixture as obtained from a solution (Fig. 21-1). The ease of achieving this intimate fusion of particles is dependent upon the softening point of the resin, the mutual solubility of the resin and polymer, and the relative particle size of the resin dispersion and polymer latex.

COMMERCIAL RESIN DISPERSIONS

Various resin suppliers have taken different approaches to answering the need for emulsion adhesive raw materials. The following discussion is not meant to be all-encompassing either in terms of resin-dispersion suppliers or of all their products, but rather to show the variety of resins available in dispersion form. Resin suppliers continue to introduce resins in dispersion form for aqueous adhesive applications.

High solids, water-dispersible resins are one approach to the development of tackifying resin products for emulsion adhesive systems. Aquatac tall oil rosin-based water-dispersible resins are offered by Arizona-Sylvachem. These high solids resins are especially suitable for adhesive formulators who have facilities to handle and mix viscous materials.[6] Typical properties of three of these products are given in Table 21-1.

Hercules offers Hydrotac 114, a water-dispersible form of Hercolyn D (hydrogenated methyl ester of rosin).[7] Typical properties are as follows:

Total solids (%)	100
Brookfield viscosity, 35°C (cps)	1190
Viscosity, 25°C Gardner-Holdt	$Z_1 - Z_2$
Gardner Color	3+

Table 21-1. Aquatac High Solids, Dispersible Tackifying Resins Typical Properties[6]

	PRODUCT DESIGNATION		
	AQUATAC 8005	AQUATAC 9027	AQUATAC 9041
Total solids (%)	80	90	90
Softening point ring and ball (°C)	5	27	41
Brookfield viscosity, 135 (57°C) (cps)	1000	6000	10,500
Gardner color maximum	12	13	15
Specific Gravity at 25°C	1.03	1.03	1.03

The high solids products available from this supplier offers cost-efficient transportation and a means of adjusting the final solids of the adhesive. It is dispersed in water with mild heat and agitation.

Aqueous dispersions of tackifying resins, typically about 50–60% solids, are generally easier to handle as they allow direct mixing with polymer latexes without use of heat. A number of companies provide a wide variety of aqueous resin dispersions that offer a range of properties when combined with various polymer latexes. A sampling of dispersions available from several suppliers can be found in Tables 21-2 through 21-5.[8–10] Both rosin-resin dispersions, as typified by Staybelite 10 Ester-55WKX (glycerol ester of partially hydrogenated rosin, from Hercules), Snowtack 301CF (disproportionated rosin, from Tenneco), and Aquatac 5527 (tall oil rosin, from Arizona-Sylvachem), and hydrocarbon resin dispersions, such as Piccopale 85-55WKX (petroleum aliphatic resin, from Hercules) and Escorez ECR-324X (aliphatic/aromatic resin, from Exxon) are available. The choice of a specific resin dispersion will be dependent upon the degree of compatibility with the polymer latex being used and the adhesive performance requirements of the specific applications, as discussed in subsequent sections.

SELECTION OF MODIFYING RESINS

The major factors in determining the type of resin to use with a particular polymer latex are resin composition or chemical structure and molecular weight. Using the concept of a resin functioning as a solid solvent for a portion of the polymer, similar solubility characteristics are necessary if the resin is to act as a tackifier.[11] (This concept is discussed further in the section on polymer latexes.) Low molecular weight resins will typically be better solvents for a given polymer

Table 21-3. Tenneco Snowtack Resin Dispersions Typical Properties[9]

	PRODUCT DESIGNATION						
	SNOWTACK 301CF	SNOWTACK 52CF	SNOWTACK 42CF	SNOWTACK SE80CF	SNOWTACK SE45CF	SNOWTACK SE62CF	SNOWTACK SE25CF
Resin type	Disproportionated rosin	Stabilized rosin acid	Disproportionated rosin	Disproportionated rosin ester	Disproportionated rosin ester	Disproportionated rosin ester	Disproportionated rosin ester
Resin softening point ring and ball (°C)	66	52	42–47	85	45	62	25
Total solids (%)	50 ± 1	50 ± 1	50 ± 1	50 ± 1	50 ± 1	50 ± 1	50 ± 1
Resin Gardner color[a]	11	11	11	9	11	9	11
Resin acid value (mg KOH/gm)	160	155	100	14	10	16	12

[a] Fifty percent weight-to-weight ratio in toluene.

NATURAL RUBBER
- PETROLEUM ALIPHATICS
- PETROLEUM AROMATICS
- TERPENES
- ROSIN ESTERS

CARBOXYLATED STYRENE-BUTADIENE RUBBER
- ROSIN ESTERS
- PETROLEUM AROMATICS
- LOW MOLECULAR WEIGHT PURE MONOMER AROMATICS
- ALPHA-PINENE TERPENES

ACRYLIC POLYMERS
- ROSIN ESTERS
- PURE MONOMER AROMATICS
  - LOW MOLECULAR WEIGHT POLYSTYRENES AND COPOLYMERS OF ALPHA-METHYLSTYRENE OR VINYLTOLUENE

Fig. 21-2. Tackifying resins for various elastomers.

3. Glass transition temperature
4. For styrene-butadiene rubber (SBR) elastomers, the level of bound styrene

In SBR systems, the styrene imparts cohesive strength to the polymer and thus to the finished adhesive. Too much styrene in the polymer, however, will increase the glass transition temperature of the elastomer to a point where the flow of the formulated adhesive at room temperature and subsequent wetting of the substrate will be inhibited.

From previous viscoelastic studies, we know that pressure sensitive adhesive performance is dependent upon having the glass transition temperature of the resin-rubber blend within a specified temperature range. For example, room temperature, pressure sensitive adhesives will typically have a glass transition temperature of approximately $-15°$ to $+5°C$. The desired glass transition temperature range is obtained by blending a high glass transition temperature hard resin with a low glass transition temperature elastomer.

Figure 21-3 shows curves obtained from a dynamic mechanical study of an acrylic-resin blend. The maximum in each curve corresponds to the glass transition temperature of the blend. The ability to modify the glass transition temperature of the blend by varying resin level is apparent. A low glass transition temperature elastomer is desirable to give formulating latitude.

Molecular weight distribution is also an important consideration in developing a pressure sensitive adhesive latex. Elastomers with a broad molecular

weight distribution typically show better performance than materials with narrow molecular weight distributions. The concept of tackifying resins functioning as solid solvents for portions of the elastomer has been used for many years to explain the need for a broad molecular weight distribution. According to this theory, resins will dissolve the lower molecular weight fractions of the elastomer and create tack. The undissolved higher molecular weight fractions are needed to maintain cohesive strength. Without a proper molecular weight distribution, the desired balance of tack and strength cannot be obtained.

Figure 21-4 shows this concept graphically. If there is a large amount of high molecular weight material, as shown in the top drawing, only a small amount of polymer will dissolve in the tackifying resin and only a low level of tack will be obtained. Even with low tack, however, cohesive strength will be high once a bond is established. If, on the other hand, there is a large amount of low molecular weight material but not much high molecular weight material, as shown in the middle drawing, the resin will dissolve a significant portion of the polymer to give a very tacky product, but there will be insufficient high molecular weight material to give strength to the adhesive. A broad, nearly symmetrical molecular weight distribution, as in the bottom drawing, would

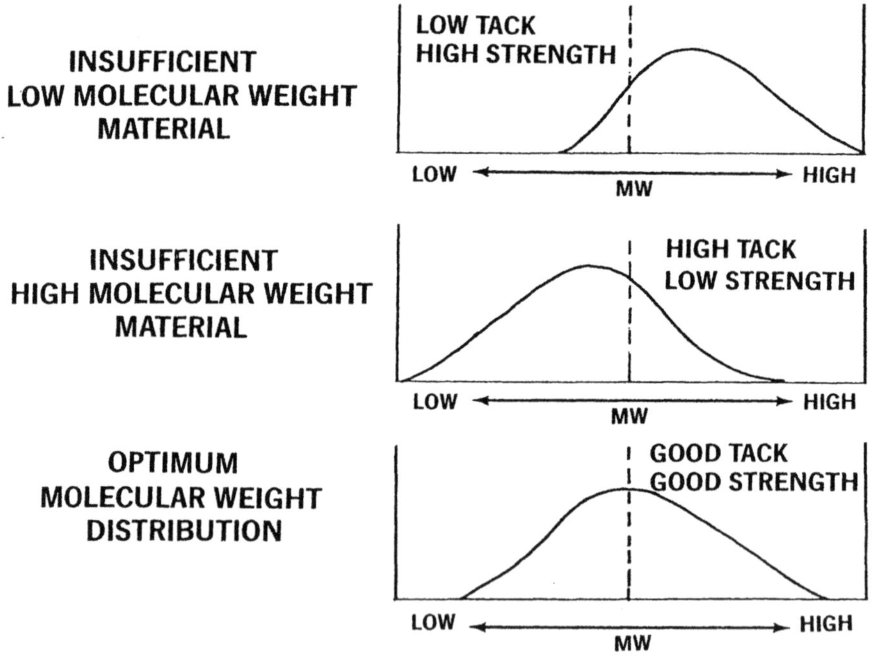

Fig. 21-4. Effect of molecular weight distribution on adhesive properties.

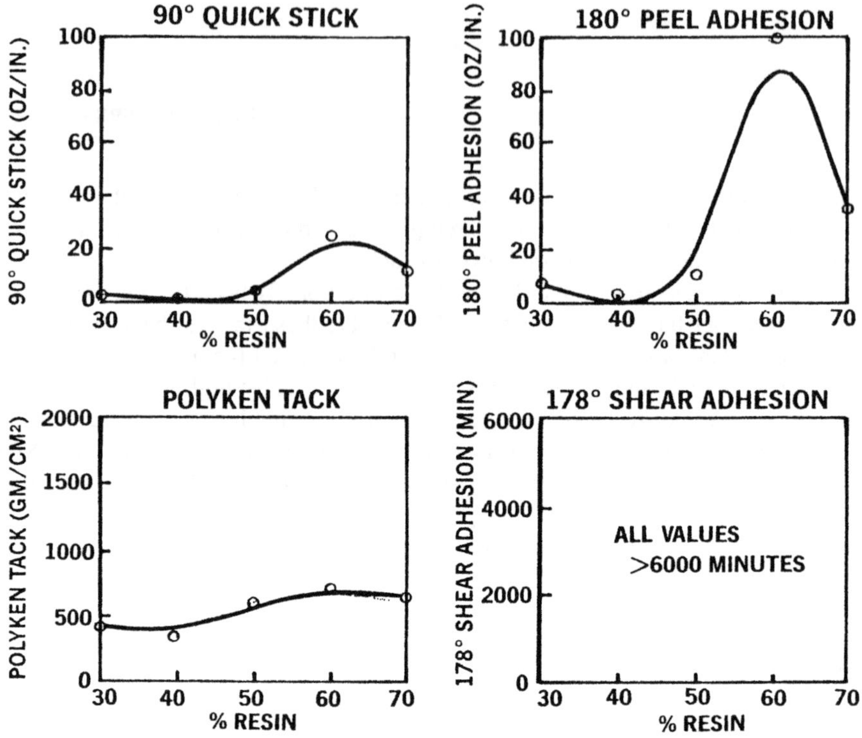

Fig. 21-5. Pressure sensitive adhesive performance of Foral 85/Carboxylated SBR Latex A.

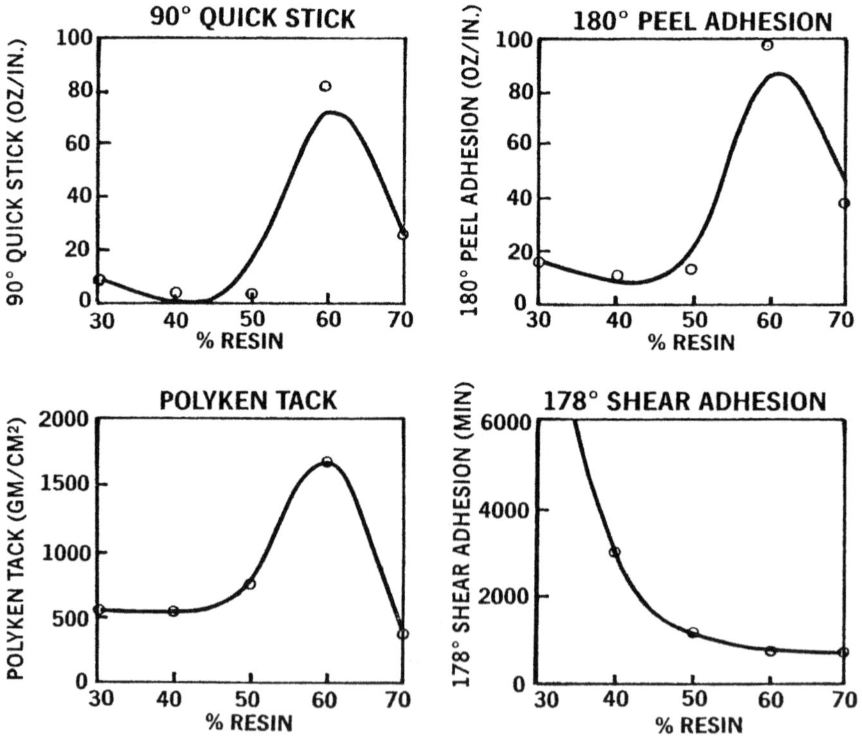

Fig. 21-6. Pressure sensitive adhesive performance of Foral 85/Carboxylated SBR Latex B.

Latex D (Fig. 21-8), a high molecular weight latex, contains a high level of styrene (46%) and therefore has a high glass transition temperature (−17°C). This latex does not respond well to resin addition. Tack values are low but, as we saw in the previous data, shear adhesion is good.

Latex E (Fig. 21-9) is also high in styrene content (44%) and in glass transition temperature (−18°C). Its molecular weight and molecular weight distribution are such that it has a substantial amount of low molecular weight material, but an insufficient amount of higher polymer to provide strength. Therefore, as tack begins to increase upon resin addition, there is a decrease in shear adhesion. In addition, the optimum level of tack is reached at a lower resin level than found in latexes A, B, or C. This is due to the smaller proportion of butadiene. The tack results from the association of resin and butadiene. In the region of optimum tack, the maximum amount of resin that the butadiene can hold has been added. Once this level is reached, the resin acts only as a filler, and tack begins to fall off. Latexes with increased levels of butadiene can incorporate more resin before this maximum is reached.

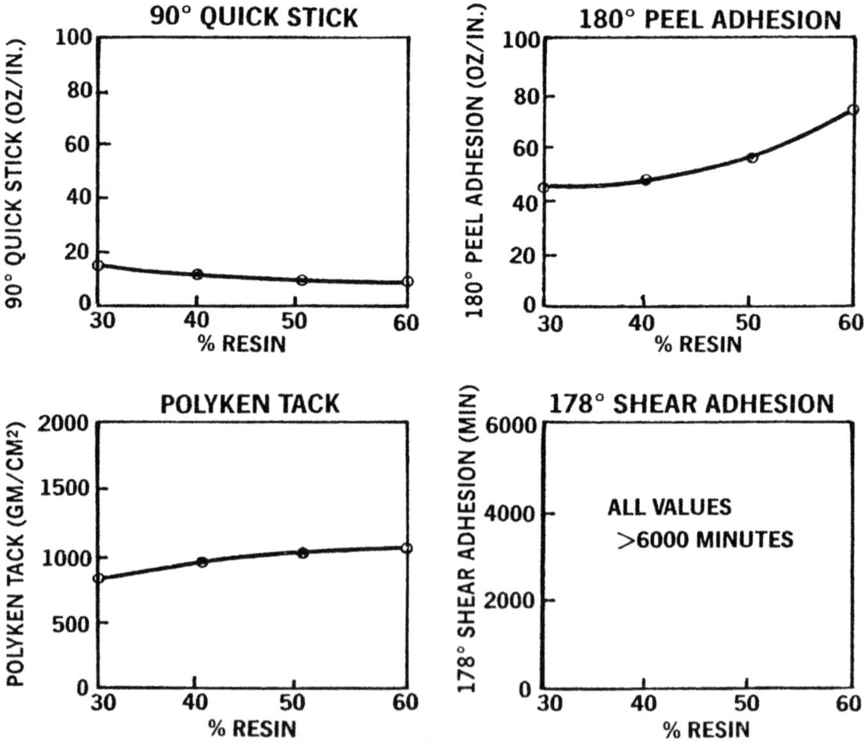

Fig. 21-8. Pressure sensitive adhesive performance of Foral 85/Carboxylated SBR Latex D.

From the standpoint of formulating pressure sensitive adhesive properties, Latex F (Fig. 21-10) represents an improvement over Latexes D and E. The lower level of styrene (25%), lower glass transition temperature (−46°C), and broader molecular weight distribution allow for the development of good tack properties without a sacrifice of shear adhesion.

The adhesive properties discussed thus far have all been developed with carboxylated SBR latexes. Acrylic latexes will also respond to resin tackification. Acrylics have typically been used in solvent systems with very little or no tackifying resin. This is because a certain level of pressure sensitive adhesive properties can be built into an acrylic polymer. However, acrylic performance can usually be enhanced by resin modification. In addition, with so many different pressure sensitive adhesive applications, it would be impractical to design a polymer for each one. It is more reasonable to design a polymer or several polymers that can be modified for various applications through resin modification.

Figure 21-11 shows the response of one particular acrylic latex (Latex G) to resin tackification by Staybelite Ester 10. This polymer was designed to give high shear adhesion. Upon resin addition, there is a significant improvement

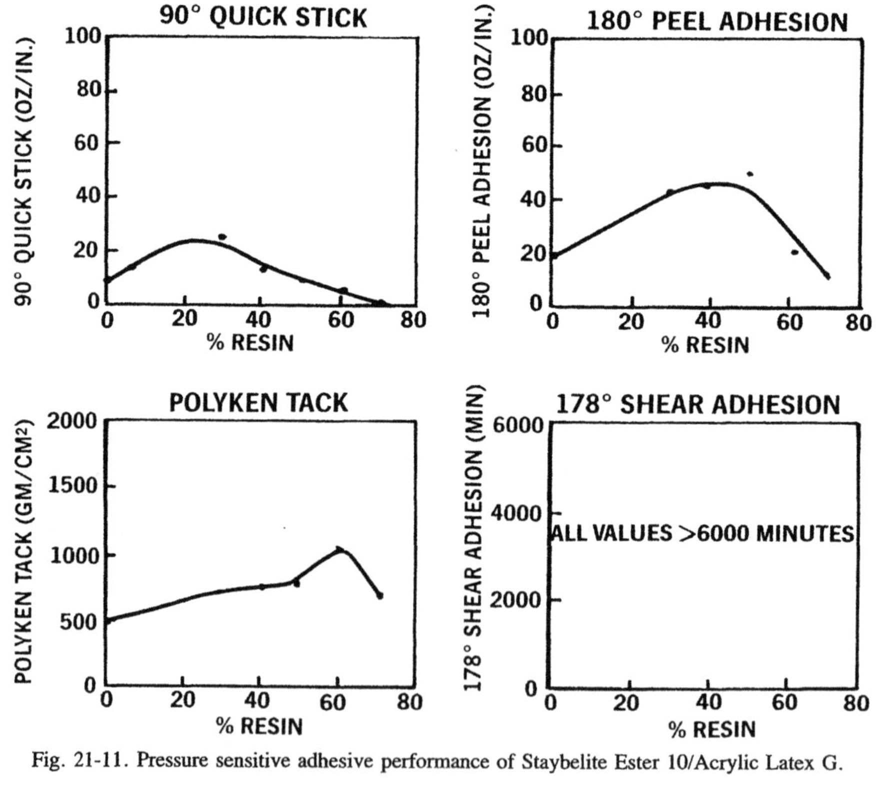

Fig. 21-11. Pressure sensitive adhesive performance of Staybelite Ester 10/Acrylic Latex G.

in 90-degree quickstick and 180-degree peel. The improvement in tack is accomplished with no loss of shear adhesion.

Figures 21-12 and 21-13 show the response of two other acrylic latexes to resin tackification. In both cases, we see an increase in pressure sensitive adhesive properties over the unmodified latex. Through resin modification, the range of pressure sensitive adhesive properties that can be obtained with the polymer latex has been extended.

APPLICATIONS

A wide range of adhesive performance properties can be obtained with aqueous systems which allows for their use in a wide variety of applications. Table

Table 21-6. Formulations of Pressure Sensitive Emulsions

HIGH PEEL ADHESION AND TACK	
Dispersion of 83°C softening point glycerol ester of partially hydrogenated rosin	50 parts
35% styrene carboxylated SBR latex, low to medium molecular weight	50 parts
90-degree quickstick (oz/in.)	67
180-degree peel adhesion (oz/in.)	118 CF[a]
Polyken Tack (g/cm^2)	1768
178-degree shear adhesion (min)	
(1 in. × 1 in. × 1000 g)	4800
(1 in. × ½ in. × 1000 g)	476
Dispersion of 83°C softening point glycerol ester of partially hydrogenated rosin	50 parts
25% styrene carboxylated SBR latex, medium molecular weight	50 parts
90-degree quickstick (oz/in.)	41
180-degree peel adhesion (oz/in.)	93 CF[a]
Polyken Tack (g/cm^2) (1 in. × 1 in. × 1000 g)	1218
178-degree shear adhesion (min)	1826 CF[a]
GOOD ELEVATED TEMPERATURE SHEAR ADHESION (TO CORRUGATED)	
Dispersion of 83°C softening point glycerol ester of partially hydrogenated rosin	30 parts
50% Styrene carboxylated SBR latex high molecular weight	70 parts
90-degree quickstick (oz/in.)	37
180-degree peel adhesion (oz/in.)	75
Polyken Tack (g/cm^2)	660
178-degree shear adhesion to corrugated (min) (1 in. × 1 in. × 1000 g)	
At room temperature	>7000
At 50°C	> 60

4. Bennett, H., Bishop, J. L. Jr., and Wulfinghoff, M. L. *Practical Emulsions,* Vol. I, *Materials and Equipment.* Chemical Publishing Company, New York, 1968.
5. Linsay, E. C. Aqueous Resin Dispersions. Presented at the ASTM D-14 Water Based System Seminar, Boston, Mass., October 11, 1977.
6. *AQUATAC Water Based Adhesive Tackifiers: High Solids Resins, Water Dispersions.* Sylvachem Product Data Sheet.
7. Hercules, Inc. Development Data Sheet.
8. *Products for Water-Based Adhesive Systems, A Reference Guide for the Adhesives Formulator.* Hercules, Inc.
9. Tenneco Paper Chemicals Canada. Product Data Sheets.
10. Jagisch, F. C. Emulsion Tackifiers for Pressure Sensitive Adhesives. Presented at the Pressure Sensitive Tape Council Technical Seminar, Itasca, Ill., May 8, 1987.
11. Zawilinski, A. Water-Based, Pressure Sensitive Adhesives—Formulation and Performance. Presented at the Pressure Sensitive Tape Council Technical Seminar, Chicago, Ill., May 1984.

to acrylic polymers. Proper modification of these two parameters by use of resin can give adhesives with properties ranging from low-tack-removable to substrate-destroying-permanent over a range of temperatures.

Use of tackifying resin can also alter the surface energy of the adhesive system and can be used to develop specific adhesion. Inclusion of certain functional groups in the resin or use of different resin types can develop adhesion to low energy substrates, such as polyethylene and polypropylene, as well as polar substrates like paper, cellophane, and metals.

Acrylic is a broad classification term used to describe a class of polymers derived from chemicals of the general structure:

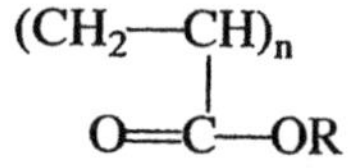

This ester exhibits very different properties dependent on alcohol chain length (—OR), for instance, material properties range from poly(methyl acrylate), (R = CH_3), (T_g = 10°C), to poly(butyl acrylate) (R = C_4H_9), a soft rubbery elastomer (T_g = −54°C), to poly(hexadecyl acrylate) (T_g = 35°C), (R = $C_{16}H_{33}$), a hard, waxy solid.

Acrylic polymers used for pressure sensitives are somewhat narrower in their overall physical properties, but the term ''acrylic'' can be almost as vague, telling little about the T_g and polarity of the polymer. This can cause quite a dilemma when choosing a tackifying resin for use in an acrylic polymer. This situation can be further complicated by inclusion of other vinyl-containing monomers such as vinyl acetate and styrene. These monomers are often used to lower the cost of the acrylic polymer.

RESIN-POLYMER SOLUBILITY

In a recent study, three different polymer latexes were used to model the diversity of acrylic polymers available in the marketplace, as shown in the following list:

Type	Composition (%)
Acrylic A	75 BA 25 STYRENE
Acrylic B	60 2-EHA 40 MA
Acrylic C	83 2-EHA 10 MA 7 VA

where performance is frequency related. Tan δ for a polymer modified with a soluble resin is shown in Figure 22-1.

Other parameters can also be measured in dynamic mechanical experiments. For viscoelastic materials, dynamic experiments allow separation of properties into elastic and viscous components. For example, complex shear modulus, G^*, can be decomposed into shear elastic or storage modulus, G', and shear viscous or loss modulus, G''. The ratio G''/G' is also tan δ.

$$G^* = G' + iG''. \tag{1}$$

In fact, in miscible polymer-resin systems, the T_g of the adhesive is linearly related to the respective T_g of the individual components by Equation (2).[1]

$$\frac{1}{T_{g_a}} = \frac{W_r}{T_{g_r}} + \frac{W_p}{T_{g_p}}, \tag{2}$$

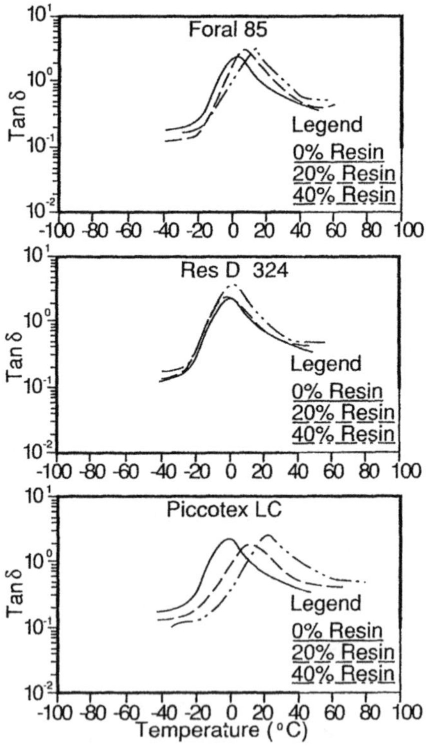

Fig. 22-1. Tangent of delta for the polymer A modified with soluble resins.

Table 22-1. Apparent vs. Calorimetric T_g

		ACRYLIC		
RESIN	DSC	A	B	C
Piccotex LC	+40	+75	+80	+107
Foral 85	+35	+32	+34	+34
RES D-324	−14	+4	−1	+3

Plots of $1/T_{g_a}$ versus W_p are linear for all three resins studied, but give appreciably different slopes due to their differing T_g(app). Values for T_g(app) calculated from the intercept compare to DSC values for resin as shown in Table 22-1.

There are two possible reasons to account for the difference between DSC measured T_g and T_g(app). First, the dynamic mechanical method used to measure tan δ, in which the sample of adhesive is subjected to a sinusoidally oscillating strain, is frequency dependent. This applied frequency corresponds to a higher measured T_g, since molecular movement must be within the frequency of applied strain. A T_g shift attributable to frequency effects alone can be seen by comparing T_g obtained by dynamic mechanical analysis (DMA) versus DSC. The T_g obtained for acrylics A through C by both methods are shown in Table 22-2.

Second, if both components are not completely miscible, the value of T_g (app) will be changed. For example, if there is an appreciable gel fraction in the polymer, T_g (app) for the resin will be raised, since the resin will only be modifying the ungelled portion of the polymer. In this way, T_g (app) can be used to qualitatively judge polymer-resin solubility.

In the case of Acrylic A, the butyl acrylate/styrene copolymer, resin solubility follows the order Foral 85 resin > RES D-324 > Piccotex LC resin.

Polymer-resin blends in which either Piccolyte S90 resin or Piccotac 95 resin is blended with Acrylic A show gross incompatibility. This is indicated by no movement in the polymer T_g peak along with the appearance of a secondary peak attributed to resin (Figs. 22-3 and 22-4). Therefore, based on DMA evaluation, appropriate resin chemistries for this polymer are Foral 85 resin, RES D-324, or Piccotex LC resin.

Table 22-2. DSC vs. DMA T_g Adhesives with 40% RES D-324

RESIN	DSC	DMA
Acrylic A	−11	1
Acrylic B	−29	−12
Acrylic C	−34	−13

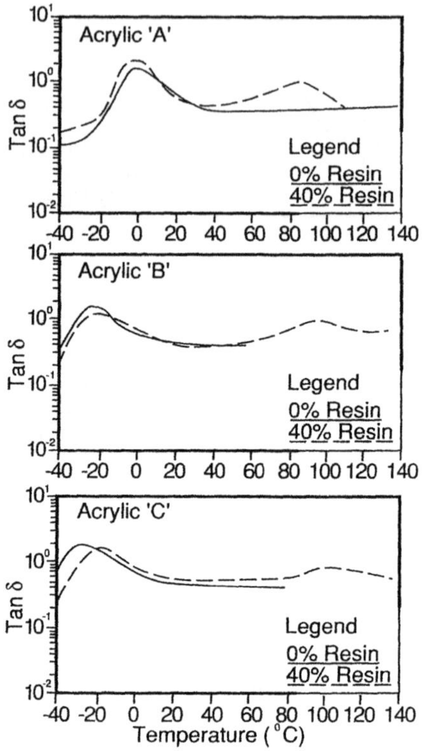

Fig. 22-4. Piccotac 95/acrylic polymer blends.

polymer solubility from Acrylic A to Acrylic C. In Acrylic A, the more aromatic polymer, these resins appear to be totally insoluble as seen for the tan δ spectrum of adhesive blends containing this resin. The low-temperature tan δ peak associated with polymer is unchanged, while a second peak appears due to resin insolubility. When Acrylic C is used, however, there is movement in the T_g of the first peak to higher temperature, which is indicative of resin solubility (Figs. 22-3 and 22-4). The second peak is still present, indicating that the resin is only partially soluble in the polymer, partitioning between the polymer and a separate resin phase.

RESIN-POLYMER VISCOELASTICITY

Equally important as resin-polymer solubility in ensuring adequate performance in a pressure sensitive adhesive based on an acrylic polymer is a polymer-resin combination with the proper viscoelasticity. Various viscoelastic criteria for

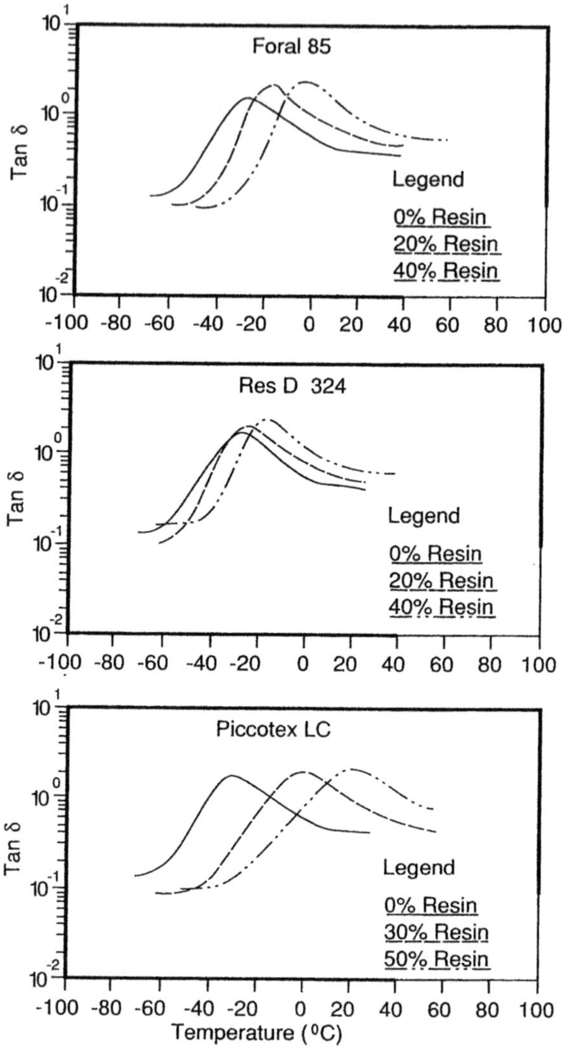

Fig. 22-6. Tangent of delta for the polymer C modified with soluble resins.

This can be illustrated by examining the viscoelastic properties of a permanent and a removable adhesives with temperature. It has been shown previously that the effect of temperature and frequency on viscoelastic properties are inversely related. For this reason, high-frequency response of an adhesive can be determined by measurement at low temperature. Comparing the shear storage modulus, G', for a removable and a permanent adhesive, we see that although shear storage modulus, G', at 25°C is identical for both adhesives, the slope

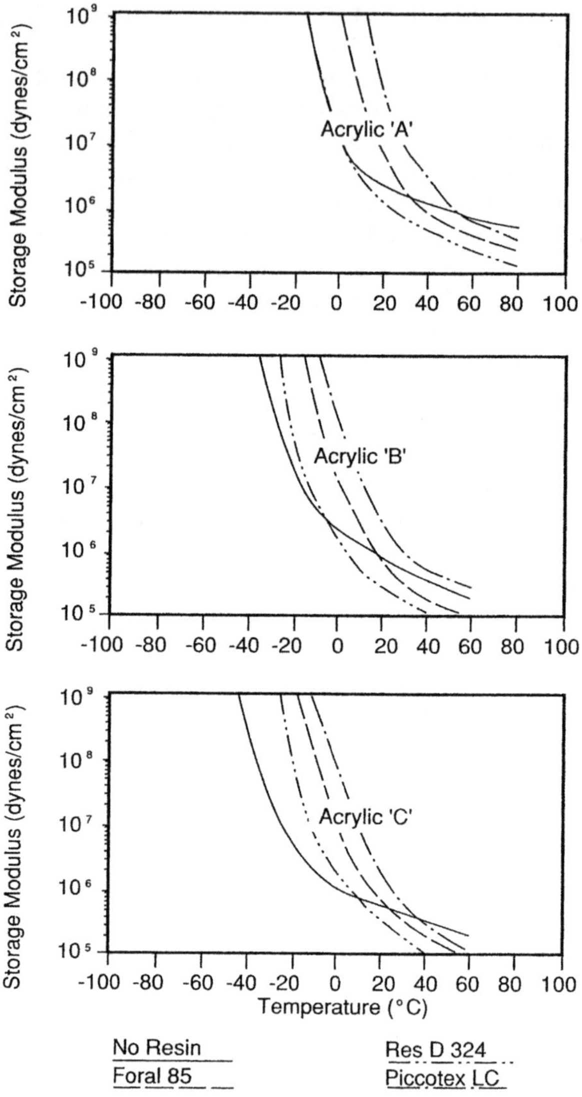

Fig. 22-8. Effect of soluble resins on G'.

causing G' to be moved further out of the pressure sensitive range of $5.0 \times 10^5 < G' < 1.0 \times 10^6$ dyne/cm^2 and $-10 < T_g < 10$°C, as defined by Foley and Chu.[4]

Acrylic B, with its lower T_g of -22°C, is more effectively tackified as compared to Acrylic A. Both RES D-324 and Foral 85-55WKX depress G'

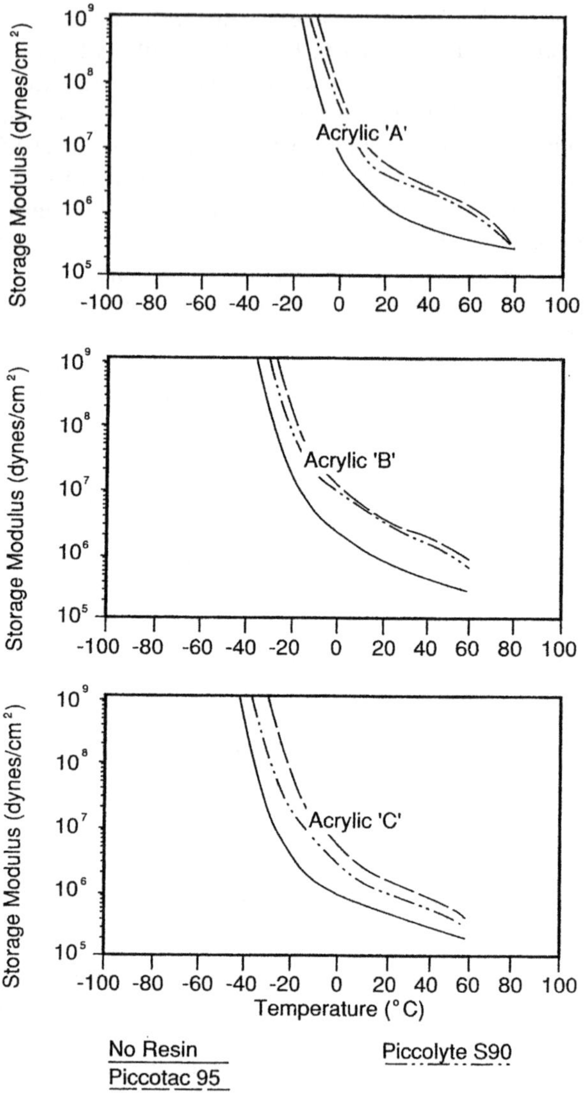

Fig. 22-9. Effect of insoluble resins on G'.

and SBR[6] has been measured by contact angle experiments with application of theory by Fowkes,[7] Owens and Wendt,[8] and others, this cannot be done for typical pressure sensitive adhesives. Data generated by contact angle for pressure sensitive adhesives are unreliable due to the solubilization of low polarity contact angle liquids into the substrate. Use of high polarity, hydrogen-bonding

however. Zosel[5] showed that pressure sensitive adhesion to a substrate obeyed the following conditions: if $\gamma_{adh} \geq \gamma_{sub}$, then measured adhesion would be dependent on the surface energy of the substrate. If $\gamma_{adh} < \gamma_{sub}$, then adhesion would be independent of substrate surface energy. The value at which adhesion would be independent of substrate surface energy would be an estimate of the adhesive surface energy, $\gamma_{adh} \approx \gamma_{sub}$.

Rosin-derived resins have long been used as sizing agents in the paper industry. These materials change the polarity and surface energy at the cellulose interface, making the paper lower in surface energy and more repellent to water. Rosin and hydrocarbon tackifying resins also have the effect of modifying adhesive surface energy while affecting adhesive viscoelasticity to enhance adhesion to low-energy substrates.

Changes in the surface energy of Acrylic B and Acrylic C with tackifying resin modification can be seen by measuring peel strength as a function of substrate surface energy. Five substrates were used: Teflon (γ_s = 18 dyne/cm), polyethylene (γ_s = 30 dyne/cm), polypropylene (γ_s = 33 dyne/cm), poly(ethylene terephthlate) (γ_s = 43 dyne/cm), and stainless steel (γ_s > 50 dyne/cm). For both Acrylics B and C, RES D-324 (52°C softening point) and RES D-2070 (42°C softening point) showed superior adhesion to polyethylene versus the unmodified acrylic polymer or polymer modified with rosin ester chemistry (Fig. 22-11). This is presumably due to a change in the surface energy of the adhesive of between 30 and 33 dyne/cm. Above 33 dyne/cm, the peel strength of the adhesive is primarily viscoelastically controlled. This is evidenced by peel force that is essentially independent of surface energy in this region.

CONCLUSION

There are three important criteria to consider in selecting a tackifying resin for use in acrylic polymer modification for pressure sensitive adhesive application. Of primary importance is resin-polymer solubility, which is a function of polymer and resin functionality and molecular weight. Once resins that are soluble in the specific acrylic polymer are chosen, resin must have the proper T_g(app) to give viscoelastic performance that allows flow at low rates of deformation (bonding) while giving adequate resistance force at high rates of deformation (debonding).

The two previously mentioned criteria are needed to establish good pressure sensitive adhesion to high energy substrates such as stainless steel. Additionally, resins also affect the surface energy of the adhesive system. To enhance adhesion to low-energy substrates, such as polyethylene, proper resin chemistry must be chosen. RES D-324 and RES D-2070 resins enhance adhesion to low-energy solids when used with acrylic polymers.

4. Foley, K. F., and Chu, S. G. *Adhesives Age* **9:** 24 (1986).
5. Zosel, A. *Colloid Polym. Sci.* **263:** 541 (1985).
6. Gent, A. N. and Schultz, J. *J. Adhesion* **3:** 281 (1972).
7. Fowkes, F. M. *Ind. Eng. Chem.* **56**(12): 40 (1964).
8. Owens, D. K., and Wendt, R. C. *J. Appl. Polym. Sci.* **13:** 1741 (1969).
9. Fowkes, F. M., Tischler, D. O., *et al. J. Polym. Sci., Polym. Chem. Ed.* **22:** 547 (1984).

coatings exhibit the same critical surface tension, but their release level varies depending on the degree of coating cross-linking: highly cross-linked coatings exhibit an easier release than uncross-linked ones. Furthermore, coatings, such as some fluoropolymers, exhibit a lower critical surface tension, yet the release from such surfaces is not as easy as that from polydimethylsiloxane coated surfaces. This suggests an additional mechanism perhaps related to the rheological properties of the coating. The presence of a more complex and unexplained mechanism in the release from silicone coated liners has also been discussed by Owen.[1] The experimental evidence appears to eliminate the splitting of silicone release coatings as a possible release mechanism. A weak boundary layer in the case of polyethylene film, however, is suspected to contribute to low adhesion, lower than that to polypropylene, which has a lower critical surface tension.[2]

The surface to which the release coating is applied should be reasonably nonporous to hold the release coating on the surface and to prevent the flow of soft pressure sensitive adhesive into the pores and irregularities, causing an increase in the unwind force. Backsize coating, which might or might not have the release characteristics, is often used on fabric and paper surfaces to provide a smooth and impermeable surface for the release.

A release coating is expected to meet the following requirements.

1. It should provide an adequate release. The release level should be sufficiently low to provide an easy unwind without unduly stretching the tape backing, but it should be high enough to prevent flagging when the tape is wrapped over its own backing. The release should be slow at a low peel angle and rapid and easy at the 90-degree and higher angles, normally used for the tape unwind.
2. The level of release should be reproducible. This requires that the release level is not very sensitive to the amount of release agent used.
3. The release agent should be firmly anchored and should not transfer to the adhesive surface, causing a decrease in tack.
4. The release coating should be resistant to aging and the unwind should not change much with prolonged aging.

There might be additional requirements for specific applications. Masking tape requires that the release coating have a good solvent resistance and that the paint adhere reasonably well to the release coating. Otherwise, the paint chips may fall off and remain adhered to the freshly painted surface.

Release coatings may be classified according to their approximate release level. Table 23-1 lists the release coatings according to this classification.

Low and normal release levels are mainly found in silicone coated liners. For tapes wound without an interliner, the release level is higher, in the range

are difficult to wet may serve as release coatings. Thus, good release denotes only inadequate adhesion, usually due to poor wetting and not some other phenomenon (the term abhesion is sometimes used suggesting the presence of some other mechanism different from that affecting adhesion). Easy release, of course, also depends on the adhesive, and some adhesives will separate from surfaces that are not considered suitable for release. Such adhesives either do not wet the surface well, or they exhibit a low peel because of their rheological properties. Surface wetting is discussed in more detail in Chapter 35.

Zosel[3] has shown that adhesion increases with increasing surface tension of the adherend, until the surface tension of the adhesive is reached. Any further increase of the adherend surface tension does not increase the adhesion, indicating that a complete surface wetting has been achieved. Such data employing polyisobutene adhesive (γ_c = 32.5 mN/m) is shown in Figure 23-1.[4] It has not been shown whether polar systems, such as acrylic pressure sensitive adhesives, perform the same way. Quickstick measurements using an acrylic emulsion adhesive have shown different results (Fig. 23-2).[4] The critical surface tension of the adhesive film was lower than 30 mN/m, and the maximum adhesion lies above the surface tension of that adhesive. Other authors have found a maximum tack for critical surface tension values that were equal to the surface tension of the adhesive.[5,6] Still other authors found continuously increasing tack with increasing surface tension of the adhesive.[7,8]

POLYMER COATINGS

The first release coatings for pressure sensitive tapes were natural and synthetic polymers, such as shellac, starch, zein, casein, and nitrocellulose. Used on

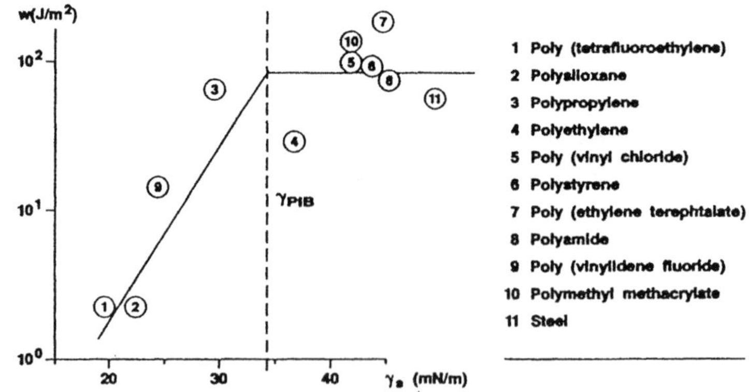

Fig. 23-1. Adhesive failure energy w (according to Zosel) as a function of the surface tension γ_s of various substrates. (*Courtesy of BASF AG, Ludwigshafen.*)[4]

fluoroethylene film and coated fabric tapes have been constructed without release coatings.[12]

WAXES

Various waxes are used as additives to film-forming polymer coatings to improve the slip, blocking resistance, and release. The effectiveness of these waxes is based on their incompatibility with the polymer and their ability to migrate to the surface of the coating. Such release coatings should be carefully evaluated for application with pressure sensitive adhesives. Waxes may transfer to other surfaces and can easily affect the tack and adhesion properties. For this reason, waxes are not widely used for pressure sensitive applications. Polyamide waxes with melting points above 90°C have been added to film formers, such as nitrocellulose, vinyl chloride-vinyl acetate copolymers, and to polystyrene for applications as release coatings for pressure sensitive adhesives.[13] Polyamide waxes are generally useful as slip, antitack, antiblocking, and release compounds. When added to the polymer coating, these compounds exude to the surface, providing a surface more characteristic of the amide than of the base polymer. Various amide waxes are used for such applications: oleamide, *N,N′*-ethylenebis stearamide, bisoleorylamide, and others.

Hydrocarbon wax blended with polyvinyl butyral is suggested as a release coating for tacky release surfaces.[14] Polyethylene waxes and modified polyethylenes can be used as release coatings for polar adhesives, especially acrylates.

SILICONES

Silicones are the most widely used materials for pressure sensitive release applications. They yield coatings of a low release level that exhibit easy release at low peel rates. Silicone release coatings are discussed separately in Chapter 24. The easy release of silicone coatings is not needed, and indeed not desirable, for tape applications. Therefore, pure silicone coatings are rarely used for tapes, but are employed as additives to a film-forming coating. Such blends give coatings with a much tighter release.

Silicone release agents have been used with nitrocellulose.[15] Polysiloxanes have been used with alkyl ether-maleic anhydride copolymer as the carrier for the release agent. The higher the amount of polysiloxane in the blend, the easier the release.[16,17] A mixture of cured aqueous organosiloxane emulsion blended with water-soluble polyvinyl methyl ether homopolymer or with vinyl methyl ether-maleic anhydride copolymer has been used as a release coating.[17] Table 23-2 shows the variation of release level with the varying amount of

Table 23-4. Release Coating Formulation

INGREDIENT	PARTS ADDED	PERCENT SOLIDS
Organopolysiloxane emulsion	30	40
Organotin salt catalyst	6	20
Acrylic emulsion polymer	63	45
Colloidal silica	0.25	40
Glacial acetic acid	0.75	100

the data. Phenolic cross-linking resin was used with some of the adhesives.[20] Diorganopolysiloxane use as release coatings has been discussed.[21]

Blends of polymethyldihydroxysiloxane with dihydroxypolybutadiene gave release values of 125–600 g/25 mm.[22] A release agent consisting of a mixture of cross-linkable acrylic emulsion, silicone release agent, catalyst, and a slip agent was used for vinyl floor tiles coated with pressure sensitive adhesive. Such tiles could be stacked without using release paper. The release coating was applied at a level of 5 g/m^2. A typical formulation is given in Table 23-4.[23]

A series of aqueous release agents has been prepared consisting of a cross-linkable vinyl acetate copolymer [vinyl acetate/*N*-methylol acrylamide (94:6)] to which release agents are added. Various release agents can be used; organosilicone compounds (dimethyl polysiloxaanes), chrome complexes (Quilon by Du Pont), and fluorocarbons (FC-805 and FC-808 by 3M Co.).[23]

An aqueous release coating for masking tape has been prepared from an acrylic coating containing a release agent that migrates during drying to the surface and becomes immobilized by the reaction between the release coating and the acrylic polymer.[24]

The release level of silicone coatings can be controlled by corona discharge or by flame treatment.[25] Oxidative treatment disturbs the molecular structure of the silicone surface and decreases its easy release properties. The silicone polymer molecules have a tendency to realign again to the original structure. If the surface is in contact with a pressure sensitive adhesive, the realignment does not take place, and the higher release level is maintained.

LONG CHAIN BRANCHED POLYMERS

Polymers with long side chains are waxy compounds exhibiting sharper melting points, quite unlike corresponding polymers with short side chains. Apparently long chains crystallize. Such compounds are suitable as release coatings of medium release value, which are especially desirable for pressure sensitive tapes. The polymers are used alone or as a part of the backsize. Many release coating patents describe the use of such long chain compounds.

Release coatings consisting of polyethylene imines acylated with higher fatty acids, such as palmitic, stearic, behenic, and myristic, are described.[31] These coatings are especially useful for cellulosic backings. Poly-*N*-acyl imine was prepared for release coating applications.[32] Solution polymers of vinyl stearate, allyl stearate, or vinyl octadecyl ether with maleic anhydride have been prepared and found useful as release coatings.[33] *N*-Stearyl polyacrylamide release coating applied from a dilute solution in toluene over nitrocellulose coated masking tape exhibited good anchorage and adequate release.[34] The unwind force at a speed of 50 m/min was 1250 g/25 mm width after aging for 24 hr at room temperature, and practically did not change after aging at 50° and 65°C for 24 hr. This release agent can be compounded with phenol-, urea-, or melamine-formaldehyde to give a cured coating with improved solvent and heat resistance.

Solution polymerized stearyl itaconate, polymonacetyl itaconate, and polymonobehenyl itaconate were useful as release agents over nitrocellulose coated kraft paper backing.[35] Copolymers of *N*-substituted long straight chain alkyl maleamic acids and vinyl comonomers formed release agents for products such as creped masking tapes with rubber-resin adhesive.[36] The release agents were applied from a 2% solution in toluene-isopropanol blend over the backsize coating, or added to the backsize coating in the amount of five parts per hundred on a solids basis. The backsize consisted of short oil alkyd (80 parts), butylated melamine formaldehyde (10 parts), and carboxylated polyvinyl chloride (10 parts). The performance of this release agent is shown in Table 23-6.

A release agent composition specific for silicone pressure sensitive adhesives has been described.[37] A solution of stearic acid is treated with vanadium oxytrichloride, molybdenum pentachloride, or antimony pentachloride in an alcohol solution of morpholine or another secondary amine.

A most unusual way has been suggested for obtaining release properties in a tape by the addition of a small amount of glyceryl monostearate, or a

Table 23-6. The Release Properties of Octadecyl Maleamic Acid and Ethyl Acrylate Copolymer

	RELEASE FORCE (g/cm)			
AGING CONDITIONS	RELEASE OVER THE BACKSIZE	RELEASE ADDED TO THE BACKSIZE	NO RELEASE	PEEL ADHESION TO GLASS (g/cm width)
No aging	80	125	240	510
66°C, 16 hr	87	138	315	510
66°C, 7 days	100	150	475	490
66°C, high humidity, 7 days	110	160	490	475

Table 23-7. Release Properties of Stearato Chromic Chloride

		INITIAL TEST		PRESSURE TEST[a]	
	COATING WEIGHT (g/cm^2)	PEEL ADHESION (g/cm)	PERCENT RETAINED ADHESION	PEEL ADHESION (g/cm)	PERCENT RETAINED ADHESION
Control	None	125	83	161	80
Chrome complex	0.6	36	91	72	105
	1.2	18	97	18	63

[a] In the pressure test, the release coated paper and the adhesive tape were compressed for 60 sec at a pressure of 175 kg/cm^2.

The cure is fast at 60–93°C. Light coatings of 0.5–1.5 g/m^2 are sufficient to impart the release properties. Properly cross-linked coatings are well anchored and do not transfer to the adhesive side.

The compound is manufactured by Du Pont and is available in several grades: Quilon M is the myristic acid (C_{14}) complex and is more stable than Quilon S, which is a complex of stearic acid (C_{16}). Quilon C is also a stearic acid complex, but of improved stability. Quilon H and L are higher strength grades. Quilon L and S are generally used with papers intended for pressure sensitive products, and Quilon C is applied over film surfaces. On surfaces that are not wetted by water, such as polyester, polyethylene films, or aluminum foil, Quilon C from 1–10% isopropanol solution may be used. The treated substrate should be dried at 120°C, if possible. The release properties of Quilon C applied over a glassine surface are shown in Table 23-7.

Quilon blends with a film-forming polymer can be used as release coatings over porous substrates. Quilon or S or C blends with polyvinyl alcohol have been applied over porous surfaces, yielding surfaces of improved release.[42]

A reaction product of hydroxyethyl cellulose and chromium stearate was used as a release coating for kraft paper tape.[43] Stearate chromic chloride with phenolic aldehyde binder gave an improved release coating for polyester film tapes. This curable coating was also suitable for masking tapes.[44]

FLUOROCARBON COPOLYMERS

Fluorocarbon polymers have a low critical surface tension and adhesives do not wet such surfaces easily. Several release coatings based on these polymers have been described in the patent literature.

N-Ethyl-perfluorooctyl sulfanamidoethyl methacrylate was copolymerized with acrylic acid and *n*-octyl acrylate in solution. Such polymers are useful as release coatings; the release property is provided by fluoroalkyl groups, and

Directly printable release coating for oriented polypropylene film tapes is suggested. Such coatings may be compounded from cellulose acetobutyrate, nitrocellulose, polyamide resins, acrylonitrile acrylate copolymer, chlorinated polyolefin, vinyl copolymer, polyester resin, and ethylene-vinyl acetate copolymer.[52]

A release coating consisting of water-dispersible aldehyde or imide resin, emulsifying agent, and an acrylic film-forming polymer, such as ethyl acrylate, acrylonitrile, and methacrylic acid terpolymer, is proposed for direct printing.[53] This release coating migrates through the ink, providing a continuous release surface, as if the coating were applied over the print. Another release coating printable with flexographic inks has been prepared from cellulose acetate butyrate and copolyester or ethyl acrylate-methyl methacrylate copolymer.[54] The coating is suitable for polyester and unplasticized vinyl films.

A directly printable release coating has been compounded from higher alkyl acrylate or methacrylate copolymer with acrylic or nethacrylic acid, fluorine-containing release agent, and a cellulose ester.[55] Such a backside coating was successfully used over various film tapes, such as oriented polypropylene, polyester, and polyvinyl chloride. Similarly, a release coating based on a copolymer of stearyl methacrylate and acrylonitrile (80:20) is effective as a release coating for pressure sensitive tapes and is also printable with standard flexographic inks.

A release coating has been prepared from a vinyl film former, a hard thermoplastic acrylic resin, and a polyamide plasticizing agent, which acts as an adhesive promoter for inks.[56] It is suggested, however, that the release coating is best applied over the print.

Additives to inks to make them transferproof are suggested. Fluorinated surfactants are added to flexographic inks. These surfactants are soluble in solvents used for flexographic inks, but insoluble in the adhesive.[57] Similar fluorine-containing surface active agents are suggested in another patent.[58]

REFERENCES

1. Owen, M. J. Silicone Release Coatings—Relationship Between Surface Properties and Release Behavior. *1981 Hot Melt Adhesives and Coatings.* Short Course Notes of the Technical Association Pulp and Paper Industries, pp. 13–19. San Diego, Calif., June 1–3, 1981.
2. Wu, S. *Polymer Interface and Adhesion,* p. 449. Dekker, New York, 1982.
3. Zosel, A. *Colloid Polym. Sci.* **263**(7): 541–553 (1985).
4. Druschke, W. Adhesion and Tack of Pressure-Sensitive Adhesives. Presented at the AFERA Meeting, Edinburgh, Scotland, October 1–4, 1986.
5. Toyama, M. *et al. J. Appl. Polym. Sci.* **14:** 2039 (1970).
6. Toyama, M., and Ito, T. *Poly. Plast. Technol. Eng.* **2:** 161 (1973).
7. Sherriff, M., Knibbs, R. W., and Langley, P. G. *J. Appl. Polym. Sci.* **17:** 3423 (1973).

40. Iler, R. K. U.S. Patent 2,273,040 (1942) (assigned to E. I. DuPont de Nemours & Co., Inc.).
41. Trebilcock, J. W. Chromium Complex Treatments for Papers. In *Proc. 1983 Paper Synthetics Conference,* Tappi Press, pp. 299–302. Lake Buena Vista, Fla., September 26–28, 1983.
42. *Quilon Chrome Complexes.* Du Pont Company Publication E-68203, 1986.
43. Martin, J. B., and Funk, C. S. U.S. Patent 2,803,557 (1957) (assigned to Crown Zellerbach Corp.).
44. Bradstreet, J. A. U.S. Patent 3,236,667 (1966) (assigned to Johnson & Johnson Co.).
45. Dixon, G. M. U.S. Patent 3,318,852 (1967) (assigned to Minnesota Mining & Manufacturing Co.).
46. Dabroski, W. C. U.S. Patent 4,513,059 (1985) (assigned to Permacel).
47. Estes, P. W. U.S. Patent 3,690,924 (1972) (assigned to W. R. Grace and Co.).
48. Ritson, D. D., and Reynolds, W. F. U.S. Patent 3,394,799 (1968) (assigned to American Cyanamid Co.).
49. Cristmas, H. F. U.S. Patent 3,240,330 (1966) (assigned to Adhesive Tapes, Ltd.).
50. Crone, J. W. Jr., and Pike, C. O. U.S. Patent 3,282,727 (1966) (assigned to Shuford Mills, Inc.).
51. Webber, C. S. U.S. Patent 2,822,290 (1958) (assigned to Norton Co.).
52. Galli, G., Penzo, L., and Pina, F. U.S. Patent 4,617,199 (1986) (assigned to Manuli Autoadesivi SpA).
53. Dabroski, W. C. U.S. Patent 3,067,057 (1962) (assigned to Johnson & Johnson Co.).
54. Lambert, R. J. U.S. Patent 3,967,031 (1976) (assigned to Minnesota Mining & Manufacturing Co.).
55. Wu, M. S. U.S. Patent 4,587,156 (1986) (assigned to Minnesota Mining & Manufacturing Co.).
56. Blum, A. U.S. Patent 3,978,274 (1976) (assigned to Borden, Inc.).
57. Milutin, I. C., Bartell, C., and Blum, A. U.S. Patent 3,681,105 (1972) (assigned to Borden, Inc.).
58. Milutin, I. C., Bartell, C., and Blum, A. U.S. Patent 3,773,538 (1973) (assigned to Borden, Inc.).

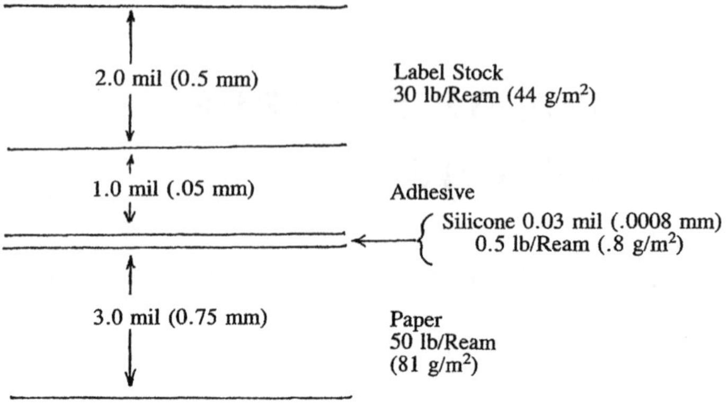

Fig. 24-1. Schematic cross section of a pressure sensitive laminate.

used for labels. The release liner, coated with a very thin film of cured silicone coating, permits the laminate to be rolled up or stacked flat. Without the liner, the adhesive would stick to anything it touched and be impossible to store, ship, or handle. Sufficient adhesion exists between the adhesive and the silicone coating to hold the laminate together during handling. The label and its adhesive are easily peeled from the protective, silicone coated release sheet when needed.

The major component of a cured silicone release coating is polydimethylsiloxane. Curing the silicone coating with a cross-linker and a catalyst produces a cross-linked film that resists penetration by the organic adhesive with which it is laminated. Without cross-linking, the adhesive molecules would readily penetrate the thin layer of silicone because of the flexibility of the siloxane polymer chain. Cross-linking the silicone also provides an elastic film that can pull away from the adhesive.

PHYSICAL PROPERTIES

One of several important physical properties contributing to the release characteristics of the polydimethylsiloxane coating is its low surface energy—only 22 to 24 dyne/cm compared to a range of 30 to 50 dyne/cm for most organic adhesives.[2] This differential helps prevent the two from bonding tightly by preventing the adhesive from intimately wetting the silicone surface. The result is enough adhesion to keep the adhesive in place, yet not enough to prevent easy parting of the silicone adhesive interface.

Being nonpolar, silicone coatings exhibit low attractive forces for other molecules. Also, the polydimethylsiloxane is incompatible with organic polymers, including those on which most pressure sensitive adhesives are based. The polymer requires little energy to activate viscous flow, making it relatively

Pressure sensitive adhesive release is a very complex phenomenon. Table 24-1 lists the many factors that affect release characteristics.

One of the most important factors is cross-link density, or the number of polymer cross-links per unit volume in the cured film. Cross-link density is a function of the molecular weight, or the amount and the structure of the polymer and cross-linker. The relationships provide the ability to vary the release characteristics of a silicone coating from very low, or easy release, to very high, or tight release.

Figure 24-2 illustrates how the molecular weight of the polymer determines cross-link density and release performance. All other parameters being equal, the higher the molecular weight, the higher the force required to release a given adhesive, and vice versa.[3] Higher molecular weight polymers mean longer polymer chains between cross-links, which means greater flexibility of the polymer chains. This greater flexibility allows more penetration of the adhesive polymer chains into the silicone than a more highly cross-linked silicone coating. The difference in release force resulting from the difference in the cross-link density can be demonstrated by comparing the release force obtained when measuring this value for two different coatings under the same conditions, as shown at the bottom of Figure 24-2. Syl-off 23 has a solution viscosity of 30,000 cps at 30% solids, indicating that it contains a high molecular weight polymer. Syl-off 291, which is 100% solids, has a viscosity of 3500 cps, indicating that its molecular weight is significantly less than that of Syl-off 23. Syl-off 23, which is higher in molecular weight and thus lower in cross-link density, has a typical release value of 125 g/in., whereas Syl-off 291, which has a lower molecular weight and thus a higher cross-link density, has a typical release value of 50 g/in.

When discussing a specific release system, it is imperative that the adhesives

Table 24-1. Factors Affecting Release Level

Nature of the adhesive	*Substrate variations*
Chemical type	Roughness
Thickness	Porosity
Modulus	Sizing or plasticizers used
Diluents	Surface energy
Nature of the silicone coating	*Laminate characteristics*
Chemical composition	Paper age
Coating weight	Laminate age
Film continuity	Thickness and modulus
Degree of cure	Mode of adhesive application
Cross-link density	*Stripping operation*
Modulus	Speed
	Angle
	Physical dimensions

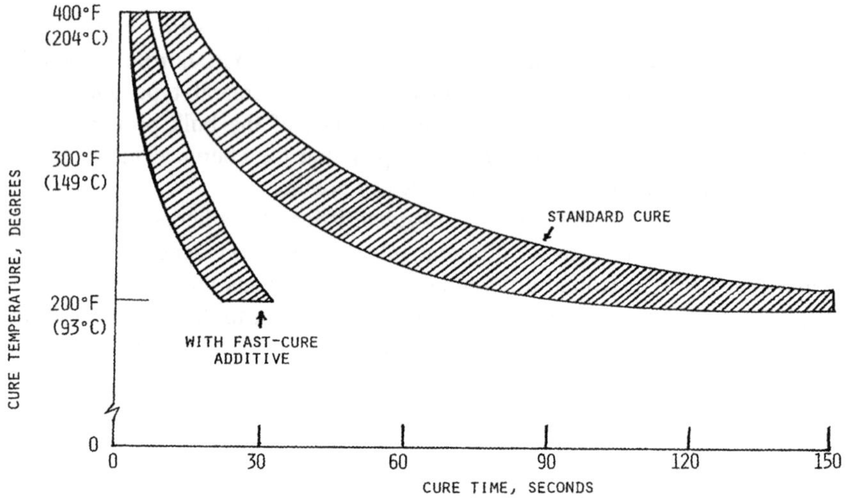

Fig. 24-3. Typical cure time of solventborne (Syl-off 23, 30%) silicone release coatings as a function of temperature with and without the use of a fast cure additive.

accelerates the reaction, and the silicone typically cures to a nonmigrating state within 20–30 sec at 120–150°C (Fig. 24-3, standard curve).

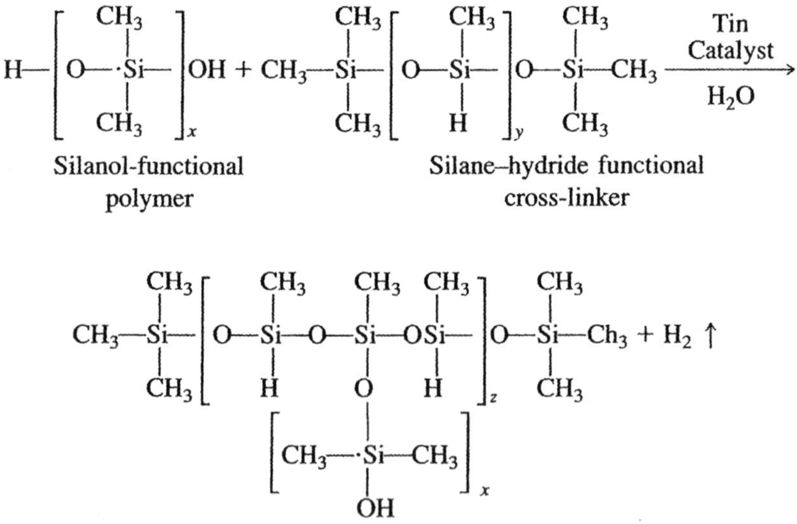

(1)

The most commonly used catalyst is dibutyltin diacetate, although other dialkyltin esters may be used. They are not true catalysts, since they are present

control over release characteristics of the silicone coating. Differential release can be achieved using different polymers and additives. Release equalization is not a serious problem.

In the addition reaction, a vinyl-functional polydimethylsiloxane reacts with a silane–hydride functional cross-linker in the presence of a noble metal catalyst [reaction (4)].

$$\equiv Si-CH=CH_2 + HSi\equiv \xrightarrow{\text{Catalyst}} \equiv Si-CH_2-CH_2-Si\equiv \qquad (4)$$

Addition reaction systems cure in 30 sec at 75–175°C, forming a three-dimensional polymer cross-linked by silicon–carbon–carbon bonds, in contrast to the silicon–oxygen–silicon bonds of those formed by the condensation reaction. Vinyl-functional polydimethylsiloxane polymer as shown below

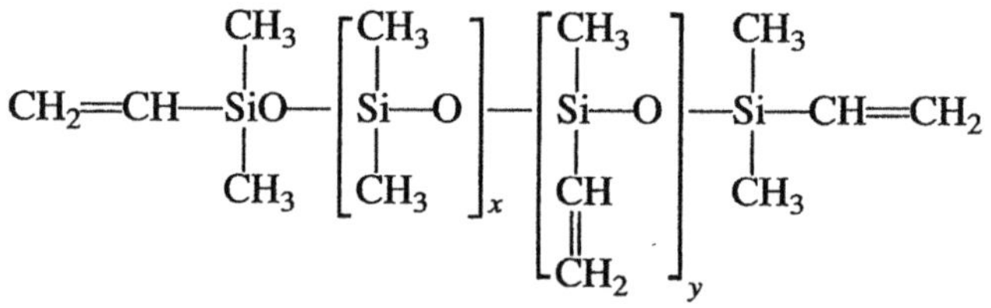

reacts with a silane–hydride functional cross-linker in the presence of a Group VIII noble metal catalyst:

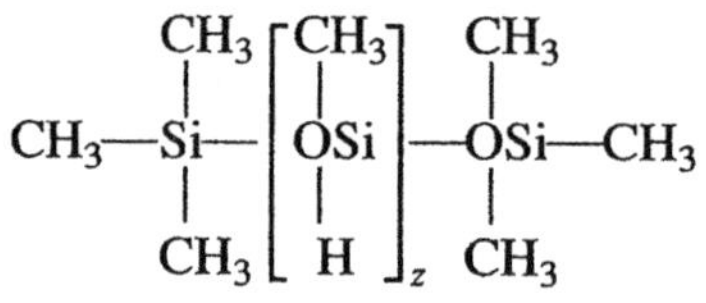

The catalyst is a metal complex of either platinum or rhodium. Used in small quantities ranging from 50 to 150 ppm of metal per silicone, it promotes the addition reaction and is a true catalyst.

The cross-linking reaction will occur at room temperature and the coating systems are formulated with inhibitors to extend the bath life and prevent premature gelation. These systems are designed to deactivate at an elevated temperature. As in the case of the tin catalyzed systems, it is best to make small batches to minimize bath age. The addition-cured systems are more prone to inhibition and poisoning than the condensation-cured systems. Inhibition is defined as the slowing of cure, and poisoning is defined as preventing the cure. Among the types of materials that will slow or stop the addition reaction from occurring are the tin catalyst used for the condensation reactions, chromium complexes

Table 24-2. Comparison of the Properties of the Three Release Coating Systems

TYPE OF COATING	CURE RATE	RELEASE RATE	COMMENTS
Solventborne coatings	Ranges from fast to slow	Ranges from easy to tight	Dull to glossy coatings Good anchorage
Emulsion coatings	Ranges from medium to slow	Ranges from easy to medium	Some coatings require post-cure Limited range of release
Solventless coatings (100% solids)	Ranges from fast to medium	Ranges from easy to tight	Glossy coatings Premium release

35% of the global pressure sensitive markets. As might be expected, this market share is increasing globally for reasons of ecology and economy.

Table 24-2 summarizes the basic properties and characteristics of the three types of silicone release coatings. Table 24-3 summarizes the applications and performance properties.

Table 24-3. Summary of Silicone Pressure Sensitive Release Coating Application and Performance Properties

TYPE OF COATING	SOLVENTBORNE	AQUEOUS	SOLVENTLESS
Primary substrates	S2S kraft	Machine-finished (MF) kraft	Two-side size Super calendered S2S kraft
	Clay coated kraft PE coated kraft		
Primary methods of applications	Direct gravure Wire-wound rod	Wire-wound rod	Offset gravure Transfer coater
Application solids (%)	2–7	5–10	Up to 100
Typical cure cycle			
temperature (°C)	80–150	120–150	80–175
time (sec)	5–30	30–40	1–60
Additives	Bath-life extender Fast cure Adhesion aid High release	Holdout agent Wetting agent Defoamer	Bath-life extender Fast cure Adhesion aid High release
Processing latitude	Wide	Narrow	Medium
Typical release values (g/cm)		10–18	8–63
SBR adhesive	10–63	10–20	12–69
acrylic adhesive	10–47		

A wide range of release is possible with these systems. Release of a typical commercial styrene-butadiene rubber (SBR) adhesive may vary from 10 to 65 g/cm. Values from 10 to 50 g/cm may be achieved with a typical commercial acrylic adhesive.

Additives. The most widely used additives are bath-life extenders to prevent the moisture-sensitive coating from gelling in the bath. Typical bath-life extenders include methyl ethyl ketone, which also reduces static buildup at the coating head during coating, isopropyl alcohol, and acetic acid. The use levels are from approximately 1–5% depending on which extender is used. Since the bath-life extenders tend to slow the cure, they are best used above 150°C.

Although not absolutely necessary, most users employ fast cure additives. These additives permit faster coating speeds at a given temperature, or, conversely, lower cure temperatures at a given speed. This improves the ability to coat polyethylene coated paper and film. Some fast cure additives improve adhesion of the coating to the substrate. Figure 24-3 illustrates the effect of the addition of a fast cure additive to Syl-off 23. The use level of these additives is from 3% to 6%. Care should be taken in choosing a fast cure additive because some are designed for in-line use whereas others are designed for off-line use. (In-line refers to the siliconizing of the substrate followed immediately by the lamination step.) If the wrong fast cure additive is chosen, release problems may develop.

Anchorage additives may be used when adhesion of the silicone coating to the substrate is a problem. The additive is multifunctional. One functional component reacts with the substrate and a second with the silicone. The use level usually is 2–4%. They will often extend the bath life as well.[10]

High release additives can be used to increase the release of the coating. These materials are resinous and act as tackifiers. They work best in-line, while off-line coatings tend to experience decreased release forces with time. These additives can be used up to about 70%, based on silicone solids.

Catalysts. Catalyst selection also affects bath life and cure time. In general, catalysts that increase cure rates decrease bath life, because of their greater activity. Dibutyltindiacetate is the usual catalyst, since it is the most active catalyst. The first zone of the curing oven should be below 150°C to prevent volatilization, since this catalyst's boiling point is 142–145°C. For situations where this is not possible, dibutyltin-di-2-ethylhexoate can be used. In situations where the liner requires FDA food contact status, dibutyltindilaurate is used. The catalysts are used at 3–10%, and when the siliconizing is done in-line, the coating must be well cured. For these situations, a fast cure additive is often used.

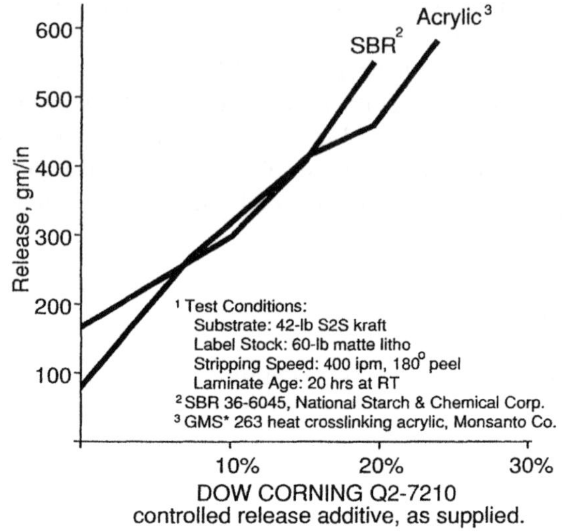

Fig. 24-4. Release performance of Syl-off Q2-7362 coating with Dow Corning Q2-7210 controlled release additive.

Catalyst. The common platinum catalyzed solvent systems in the United States supply the catalyst in the coating. This is in contrast to the tin system where the catalyst is provided separately. Additional platinum is available, as mentioned in the preceding section. The platinum catalyst is more easily inhibited or poisoned, and care must be taken to keep these coatings segregated from tin catalyzed silicone systems. Also rubber parts should be sulfur-free, and amine- and chromium-containing materials should not come in contact with these systems while they are in the uncured state.

Solvents. The coatings are usually used at about 5% solids and typically coated at 0.8 g/m^2 (0.5 lb/ream) on paper. The solvent used is usually heptane or a mixture of toluene and heptane. Toluene tends to slightly slow the cure rate.

WATERBORNE SYSTEMS

Historically, aqueous emulsion-type silicone release coatings have been used primarily for non-pressure sensitive applications. Typical applications include releasing sticky products, such as rubber or asphalt, from containers, and as release sheets for processing sticky foods, such as baked goods and candy.

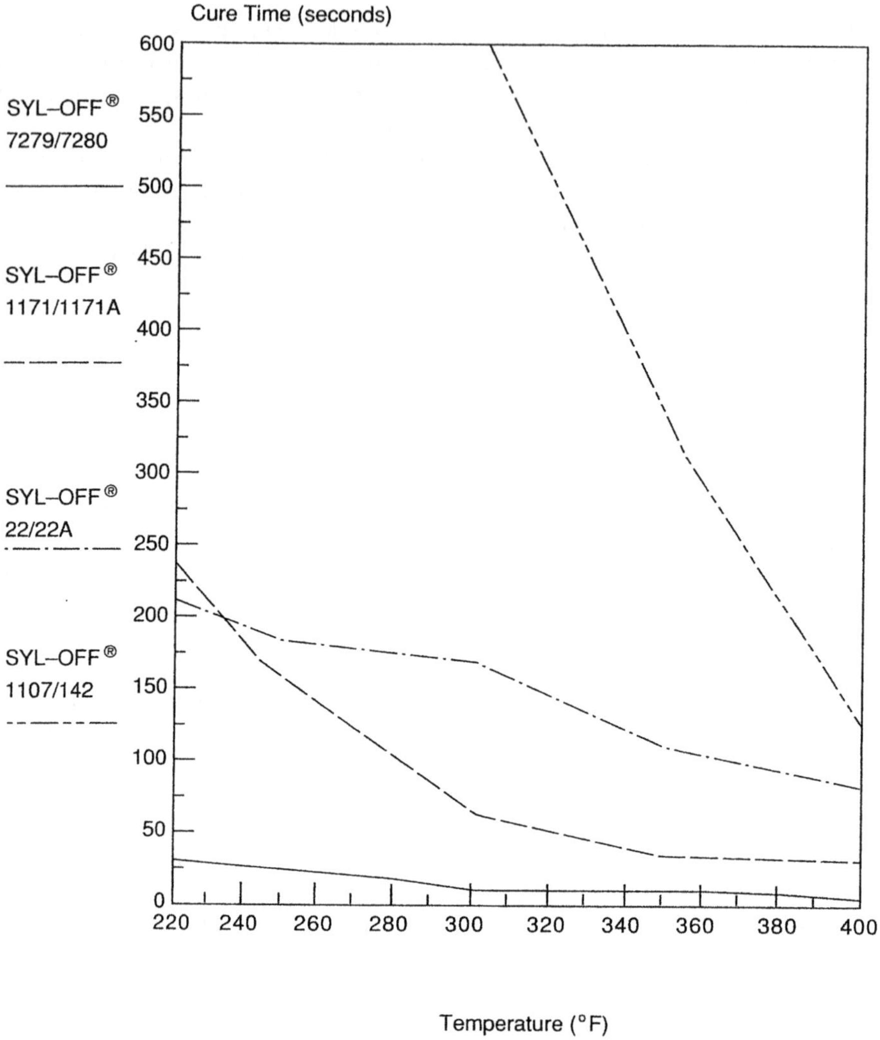

Fig. 24-5. Comparison of cure time for several silicone release coating emulsions.

The tin catalyzed emulsions are supplied in the same configuration as the tin catalyzed solvent coatings. The coating contains the silicone polymers, and the hardener contains the tin catalyst. Release values for these coatings are typically 75 g/in. when measured using a rubber-based adhesive, and 50 to 110 g/in. when an acrylic-based adhesive is used. The condensation cure chemistry of the tin systems described earlier also applies to these systems.

Table 24-4. Shear Stability of Emulsion Release Coatings

	TIME (MIN)					
SYSTEM	0	15	30	45	60	120
A. Syl-off 1171 paper coating Dow Corning 1171A catalyst		Gel				
B. Syl-off 22 paper coating Dow Corning 22A catalyst			Gel			
C. Syl-off 7007 paper coating Dow Corning 142 catalyst				Gel		
D. Syl-off Q2-7279 emulsion coating Dow Corning Q2-7280 catalyst emulsion						Oil

Test: Prepare coating bath and initiate run on two-roll coater. Roll speed is 175 rpm. Measure time to reach gel formation or oiling, which indicates failure.

Catalyst. Since the catalyst is platinum, the precautions noted earlier should be followed. That is, the system can be inhibited or poisoned by tin compounds and sulfur.

Emulsions require special handling procedures to ensure that they do not break prematurely. Conditions that will lead to premature breaking of the emulsion are: temperature extremes; shear; hard water; emulsion shock (adding concentrated emulsions together); foaming; high or low pH or sudden pH changes.

The technology of emulsion-type silicone release coatings allows the user to use less expensive grades of paper to coat, which is a great cost savings. Environmental and energy concerns also play a role in the selection of this type of coating.

SOLVENTLESS SILICONE COATINGS

Toward the end of the 1960s, work was begun to produce solvent-free 100% solids release coatings. Environmental pollution problems, rising oil prices, and increasing concern over worker health hazards associated with large-scale use of solvents contributed to the desirability of such coatings.

Unfortunately, the traditional condensation chemistry used for solventborne and emulsion systems proved totally unsuitable for solvent-free systems because of severe limitations on the working life of the system at 100% solids. Fortunately, addition curing systems were already known in many areas of curable silicone compositions, and these systems proved readily adaptable to the new requirements.

Today solventless coating is the major segment of the silicone release

Most commercial systems will package the catalyst with the base polymer and inhibitor, while keeping the cross-linker separate, resulting in a two-pack system, in which the real catalyst is already dispersed in the major component and the second, smaller, pack is actually a polymer and not a catalyst.

The low molecular weight polymer used results in solventless silicone release coatings, giving very easy release—too easy for a great many applications; consequently it gives rise to the need for a high release additive. These additives are usually based on a silicone resin, itself a solid, dispersed in a silicone polymer. This latter polymer is normally the same polymer as the base material.

Catalyst and inhibitor are incorporated into the resin/polymer blend. This constitutes the high release polymer and it can be blended in all proportions with the base material to give a wide range of release values.

The effectiveness of the high release polymer is very much a function of the adhesive. With many adhesives release values can be achieved that are many times the unmodified levels, but some adhesives make it difficult to even double the unmodified level. The typical system has been a three-part system: base polymer, high release polymer, and cross-linker. Today some multicomponent systems are available that consist of polymer, catalyst, inhibitor, cross-linker, and high release polymer, thus allowing the user greater formulating latitude.

Solventless systems have evolved substantially since their introduction with major improvements in bath life combined with faster curing at lower temperatures. Solventless systems offer fast curing at high speeds, freedom from blocking in double-sided coatings, explosion-risk-free coating, and much improved reliability in the process. There are some potential problem areas, however. The chemistry of the base paper can affect cure rate. This effect can range from mild inhibition to total prevention of cure. The holdout of the base paper is also more critical than for solvent-based coatings. The coating equipment frequently involves a very fine cell gravure, and these are prone to both physical damage and to build-up of gelled silicone in the cells.

One subject that has become much more urgent with the widespread use of solventless systems is the determination of state of cure. In the past a coating was said to be cured when it did not smear, did not migrate, and did not rub off. With solventless systems, these criteria still have some validity, but at that state of cure, there are still a large number of unreacted vinyl and silyl hydride groups. These groups can, and do, react with certain adhesives, causing poor release characteristics. Using such adhesives—mainly, but not only, solvent-borne curable acrylics—calls for either prolonged curing cycles or several days of postaging.

For use against other adhesives, such as hot melts, the traditional criteria are still a useful guideline. Waterborne adhesives do show some sensitivity to

by smell problems associated with mercapto materials. Nevertheless, they do offer even faster curing speeds than acrylate systems.

Ultraviolet-Cured Coatings

Ultraviolet radiation penetration is much less than EB, and as such has less tendency to affect the substrate properties. This factor alone makes UV a viable alternative to EB. The coatings themselves are of similar composition to the EB systems, but require the use of photoinitiators and, in some cases, promoters and side reaction suppressors.

All radiation curable coatings present the same coating problems as thermally cured 100% solids materials. Because of the low temperature associated with radiation curing, several heat-sensitive substrates can be silicone coated. These include polystyrene, polyethylene, polypropylene, and cellulosic films.

Coating Application

Silicone release coatings are applied by a converter. The converter may be a paper company that also manufactures the base substrate, or it may be a company that specializes in converting paper or film manufactured by others. The converter may or may not then go on to produce finished labels or tape.

There are two basic types of converting operations—off-line and in-line. In an off-line operation, only the silicone coating is applied. Adhesive and label stock are mated to the coated release roll in a subsequent operation. In-line converting equipment has two or three coating heads and a laminating station. The entire pressure sensitive laminate is made in one continuous operation. First, the silicone release coating is applied and cured. Then the adhesive is applied to the silicone coated stock and dried. Finally, the label stock is laminated to the adhesive.

Line speeds for off-line converting are usually faster than in-line. Speeds of in-line operations are usually limited by the adhesive drying rate. The increased use of hot melt adhesives may change this equation, with the silicone cure rate becoming the limiting factor. Line speeds may vary from 31.5 to 450 m/min. Most are in the range from 95 to 200 m/min.

The adhesive is usually transfer coated. That is, the adhesive is applied to the silicone coated sheet and transferred to the label stock in the final laminated step. In this way, a more efficient use of the adhesive can be realized.

Die-cutting labels and rewinding tape is a separate operation. It should be pointed out that die-cut labels require a dimensionally stable release sheet,

they may prohibit the use of volatile, low molecular weight catalysts, such as dibutyltin diacetate, at oven temperatures above 150°C.

Infrared heating may be used to supplement oven curing. Heaters are usually located after the oven tunnel.

COATING EVALUATION

The methods of evaluating silicone release coatings can be classified into two broad categories.

1. Quick tests are used on the coating line to check coating quality during the run, to determine the end of the useful life of the coating bath, and to check the degree of cure.
2. Control tests quantify release values, subsequent adhesion, and coating weight.

Quick Tests

As silicone coatings are almost invisible on many substrates, it is difficult to judge coating uniformity and state of cure by casual observation. The following quick tests serve these purposes.

Coating adequacy and uniformity can be judged by applying lengths of cellophane tape to the coated surface, rubbing it with the finger, and then pulling it off. The amount of force required to lift the tape is a quick indication of coating quality.

When the tape has been removed, it is doubled back on itself so the sticky sides just touch. The tape is then pulled apart. Any loss of tack, as compared to that of a fresh piece of tape, indicates that silicone has transferred to the tape, and may be evidence of incomplete cure. With experience, an operator can grade tack loss as "none," "very slight," "slight," "some," and "gross."

Dye tests are used to check the state of cure and film uniformity of solvent and solventless release coatings. They also form a permanent record of a particular run. Coated test strips are smeared with a dye solution and wiped dry with absorbent paper. One frequently used dye solution is 1% Rhodamine 6 GDN extra (DuPont) in diacetone alcohol. The quality of the coating is judged by the amount of color retained on the test strip and the amount striking through to the substrate. Very little color is retained by a good silicone coating. The dye test is less effective for testing coatings containing water-soluble extenders and coatings applied to nonstaining substrates, such as polyethylene coated paper.

force. The same instruments used for measuring release force are used to test subsequent adhesion, and the values are reported in ounces or grams per inch or centimeter.

Release tests for high-speed label stock are usually designed to duplicate the actual stripping conditions as closely as possible.

There are several methods for determining the weight of silicone coating solids deposited on the substrate.

Material Balance: One of the most commonly used methods is by material balance. The amount of silicone used is divided by the amount (area) of paper coated. Errors are introduced by uncoated paper, waste paper, and unused bath, unless accounted for in the material balance.

Actual Pickup Weight Method: An untreated paper is subjected to the same cure conditions as a coated paper, and its weight per unit area is determined. The difference in weight from that of the same unit area of coated paper is the weight of the applied coating. The error in this method can be quite large (±100%) due to variations in the basis weight.

Aluminum Foil Pickup Weight Method: Pass a piece of aluminum foil of known weight through the coating system. Determine its unit weight after processing. Any increase is the weight of the coating applied. There can be ±50% error in this method due to differences in transfer to aluminum foil versus paper.

X-Ray Fluorescence: There are several X-ray fluorescence machines available and two or three laboratory-sized portable ones. These can determine silicone coat weight very accurately, but do need calibrating from known standards.

REFERENCES

1. Dennett, F. L. U.S. Patent 2,588,367 (1952) (assigned to Dow Corning Corporation).
2. Gordon, D. J., and Colquhoun, J. A. *Adhesive Age* **19**(6): 21–30 (June 1976).
3. Bey, A. E. *Adhesive Age,* **15**(10): 19–32 (October 1972).
4. Colquhoun, J. A., *et al.* British Patent 1,374,792 (1974) (assigned to Dow Corning Corporation).
5. Chandra and Rowland. U.S. Patent 3,960,810 (1976) (assigned to Dow Corning Limited).
6. Grenoble, M. E. U.S. Patent 4,162,356 (1979) (assigned to General Electric Company).
7. Hockemeyer Ger. 2,736,499 (1982) (assigned to Wacker Chemie).
8. Koerner *et al.* U.S. Patent 4,306,050 (1981) (assigned to Th. Goldschmidt AG).
9. Pressure Sensitive Products, Skeist Laboratories, Inc., July 1984.
10. Ault and Salisbury. U.S. Patent 3,076,726 (1963) (assigned to Dow Corning Corporation).
11. Kookootsedes, *et al.* U.S. Patent 3,445,420 (1969) (assigned to Dow Corning Corporation).
12. Eckberg. U.S. Patent 4,256,870 (1981) (assigned to General Electric Company).

ADHESIVE TAPES

Hospital uses of pressure sensitive adhesives involve many general, as well as specific, -purpose tapes. Coated or uncoated cotton fabric was used for early hospital tapes. These backings are still used for strapping tapes that require higher tensile strength. Synthetic fiber fabrics, nonwoven fabrics, and films were later introduced as backings for hospital tapes.

The proliferation of different backings and different adhesives for hospital tapes was the result of the need to provide the best-suited products for various applications, rather than offering a single multipurpose tape. Many improvements in the hospital tape field were achieved because of this trend, the most important being the development of tapes that caused a lower degree of skin irritation.

Tapes are used for many different applications in the hospital and health areas, but basically perform one of two functions. Tapes are used either to pull something into place and restrict movement, as in strapping applications, or are used to hold something in place, such as wound dressings.

In the first category, tapes are used for therapeutic strapping—semirigid or elastic support of soft tissues. Adhesive tape is used for the treatment of sprains, ruptured muscles or tendons, and separated joints. Tape support of soft tissues is analogous to the splint support of fractured bones. The tape provides compression and restricts movement. Higher tensile strength fabrics and elastic fabric-backed tapes are used for these applications. Similarly, tapes are used for compressing venous circulation in the legs and strapping fractured ribs, in conjunction with plaster casts.

Athletic trainer tapes are heavy fabric tapes that are used to prevent injury or to protect injured joints by restricting movement. The use of athletic tapes has spread from professional to college and high school sports, especially for contact sports, such as football and boxing.

Adhesive tape strips may be used in place of sutures. The strips are used to bring the wound edges together and hold them in that position (see Fig. 25-1). Such wound closures are thin adhesive coated nonwoven fabric or film strips mounted on a release coated card. As well as being used as primary sutures, adhesive strips are used to protect the wound after suture removal. ''Butterfly'' bandages are a familiar example that have been used for a long time for minor skin injuries in place of sutures or clamps.

Holding a wound dressing in place is the earliest and most common application of tapes. Lighter weight, lower strength, nonocclusive, less skin-irritating tapes were developed primarily for this application. Nonwoven fabric, lightweight synthetic fabric, and polymeric film backings are used.

Other hospital uses of pressure sensitive adhesive tapes include the use of transparent film tapes for cosmetic purposes to secure facial dressings, such as eye patches. Since occlusive dressings increase the effectiveness of applied

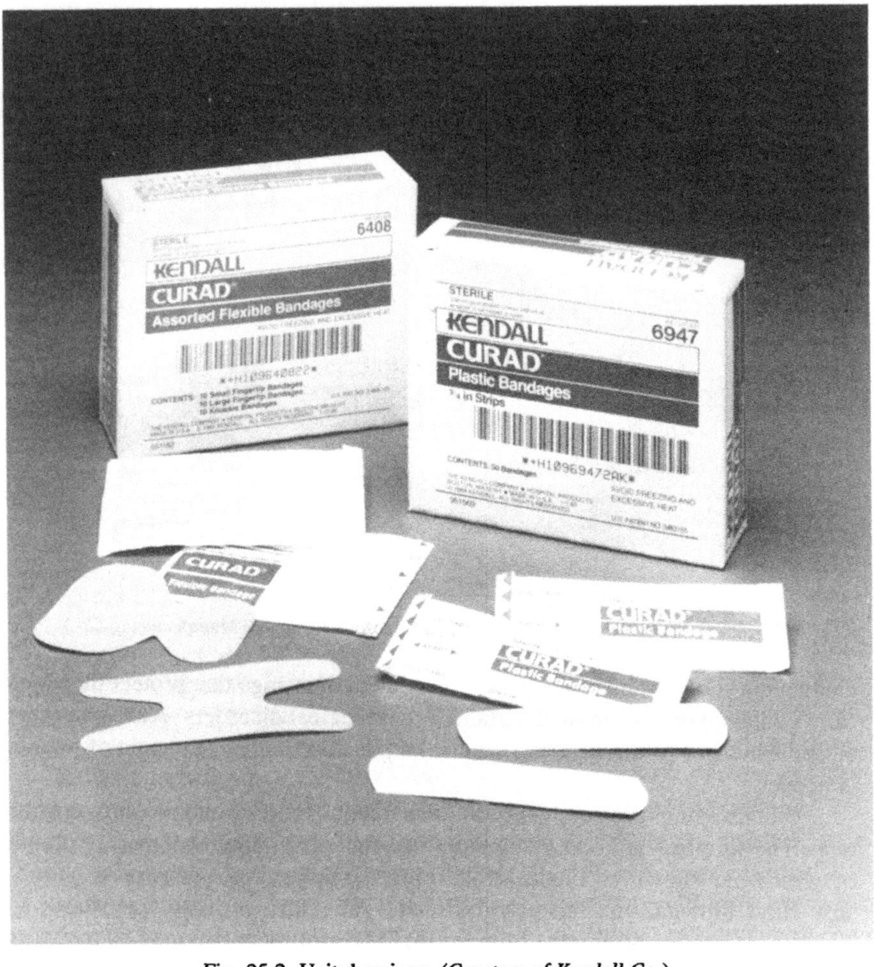

Fig. 25-2. Unit dressings. (*Courtesy of Kendall Co.*)

Adhesive tapes are used to bundle wrap drapes and other surgical products for sterilization. Tapes are used as indicators to determine whether a particular instrument or bundle has been sterilized. Tapes that change color when sterilized by steam or ethylene oxide are available. Such tapes may be multipurpose: they hold the bundle together, provide a surface for notes identifying the contents, and serve as sterilization indicators.

Intravenous needles are secured by IV tube holders that are held in place by two adhesive coated strips. Pressure sensitive adhesives secure liquid-crystal thermometers to the skin, and are used for securing blood bag temperature gauges and for labeling other blood products.[5]

Fig. 25-4. Surgical drape (*Courtesy of Minnesota Mining & Manufacturing Co.*)

"windows"—that accommodate various surgical procedures. Adhesive tapes, especially those used in the operating room, must be lint-free in order to reduce the number of particles in the air.

TRANSDERMAL DRUG DELIVERY

Transdermal drug delivery is a relatively new area where pressure sensitive adhesives have found an application.

Pressure sensitive adhesives can be used in various ways in transdermal drug delivery systems. The adhesive is used to secure the drug-containing pad to the skin. The requirements for this application are the same as those for other medical pressure sensitive adhesives. Patent literature contains several instances of pressure sensitive adhesives used in this manner.[6–9]

The active drug may be dispersed in the adhesive mass. In such cases, the adhesive must be compatible with the drug and must allow its diffusion to the adhesive–skin interface at the required rate. A schematic diagram of a transdermal drug delivery pad is shown in Figure 25-5.[10] Most dressings of this type are occlusive, that is, they are constructed using an impermeable backing in order to ensure drug diffusion in only one direction. Acrylic adhesives have been used as carriers for acne-treatment drugs,[11] and for arthritis analgesics.

use of pressure sensitive adhesives. Various types of skin irritation have been reviewed by Seymour.[14]

Chemical irritation is caused by the diffusion of chemical irritants, present in either the adhesive or the backing, into skin tissues.

Allergic irritation is caused by an adhesive or backing component to which a particular individual is allergic. It has been shown that accelerators, antioxidants, and certain tackifying resins in natural rubber formulations can cause allergic contact dermatitis. An allergic irritation is easily distinguishable from a mechanical one. The mechanical irritation is generally confined to the skin that was covered by the tape, while an allergic irritation will spread beyond the area over which the tape was applied. Polyacrylates and other single-component pressure sensitive adhesives that do not require compounding, are less likely to cause allergic skin reactions. These polymers do not require antioxidants and do not have proteinaceous impurities, such as are found in natural rubber, but consist of a single compound, thus decreasing the chances of skin irritation.

Tapes interfere with the normal physiological activity of the skin. Occlusive tapes can cause accumulation of fluid, resulting in a softening of the skin or maceration. Skin, weakened by maceration, may split upon tape removal. Overhydrated skin is also more easily penetrated by chemical irritants, making it more susceptible to irritations of this type.

Sometimes the area beneath an occlusive tape may become completely anhydrous, due to the blockage of sweat duct openings. The dehydrated tape–skin interface is then suitable for the growth and multiplication of bacterial flora, resulting in infection. In studies investigating bacterial growth under adhesive tape, it was found that cloth tape backings were especially ideal environments for the spread of microorganisms. Microporous nonwoven tape backings, however, yielded a bacterial population equivalent to or, in some cases, even less than that found on uncovered skin.[15]

NONOCCLUSIVE TAPES

Nonocclusive or breathable tapes show better adhesion to skin surfaces since moisture does not accumulate at the skin–adhesive interface, causing a decrease or loss in adhesion.[16] Nonocclusive tapes decrease the level of skin irritation and minimize the incidence of bacterial infection. The approach in producing a nonocclusive product consists of providing a porous backing and a porous, or at least a lighter adhesive, coating.

The early work in the area of nonocclusive dressings was done by T. J. Smith & Nephew, Ltd. They prepared a porous adhesive mass by blowing air[17] and used porous polyvinyl chloride films.[13]

A porous tape is prepared by applying the adhesive mass over a release

sive tape has been prepared from a fibrillated polymer film consisting basically of monofilament strands held together by interconnected fibrils.[34] Porous films can be prepared by dispersing finely divided particles of calcium carbonate or some other nonhygroscopic salt into the polymer. This is achieved through milling, forming, and stretching the film. Leaching the salt particles is not necessary to obtain porosity.[35] Since porous nonwoven fabric tapes are easily soiled, a porous film may be used for overlamination.[36]

ADHESIVES

Natural-rubber-based adhesives were the most commonly used for hospital tapes. Polyisoprene, polyisobutylene, and other synthetic elastomers have been used to replace natural rubber. Some typical formulations are given in Table 25-1. Vulcanized oil (factice) may be used to reduce the cold flow of the adhesive mass. It is especially useful in synthetic polymer formulations.

These adhesives have been largely replaced by acrylic copolymers such as a copolymer of 2-ethyhexyl acrylate/ethyl acrylate/acrylic acid or adhesives of similar compositions. To a lesser degree, silicone- and polyvinyl-ether-based adhesives are also used for medical applications. The main characteristic of these polymers is their high level of water vapor permeability.

Adhesives that adhere well to wet surfaces can be obtained by compounding a suitable elastomer with a water-soluble hydrocolloid, such as pectin, gelatin, or carboxymethycellulose. A pressure sensitive adhesive capable of absorbing 7% of its own weight in water adheres well to wet as well as dry skin, and is especially suitable as a stoma seal adhesive for ostomy mounts. Examples of some typical formulations are shown in Table 25-2.[37]

Table 25-1. Natural Rubber and Synthetic Elastomer-Based Pressure Sensitive Adhesives

SOLVENT-BASED ADHESIVE		CALENDERABLE ADHESIVE		SYNTHETIC POLYMER-BASED ADHESIVE	
Smoked sheet	100	Smoked sheet	100	Polyisobutylene (80,000)	50
Stabilized rosin	75	Zinc oxide	100	Liquid polyisobutylene	35
Zinc oxide	50	Stabilized rosin	100	Hydrated alumina	62
Lanolin	10	Starch	40	Factice	29
Antioxidant	2	Factice	15	Glyceryl ester of hydrogenated rosin	24
		Lanolin	10		
		Antioxidant	3	Lanolin	20
				Titania	14
				Beeswax	9
				Antioxidant	1.5

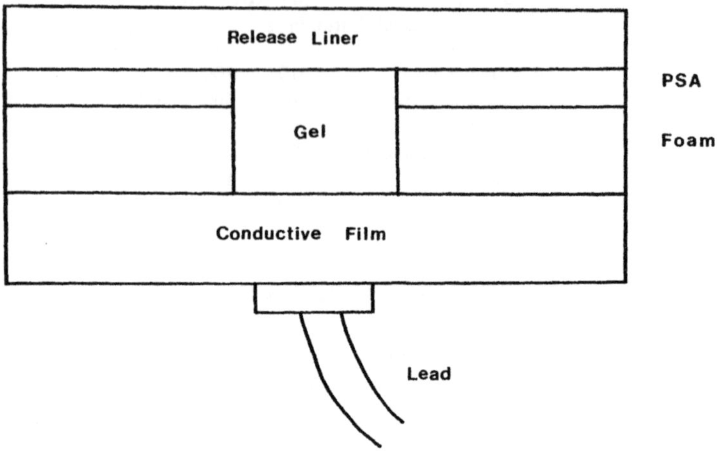

Fig. 25-6. Electrosurgical grounding pad.

is very important for skin adhesives, and testing on a hygroscopic surface, such as collagen, is of interest in predicting the adhesive's capability to retain a bond under moist conditions. For determining product performance other than interaction at the skin–adhesive interface, the rheological properties of the adhesive are the most important. The characteristics of a good hospital adhesive are:

1. Absence of skin-irritating ingredients.
2. Good water resistance, so that the bond is maintained when the skin perspires.
3. Sufficiently high cohesive strength to allow clean removal of adhesive from the skin.
4. Preferably a high water vapor transmission rate to decrease tape occlusivity.
5. Rheological properties that accommodate skin movement without losing the bond and without excessive mechanical skin irritation.

The standard properties of pressure sensitive adhesives used for medical applications are listed in Table 25-3.[5] However, meeting these properties does not assure that the adhesive will be suitable for application to skin and that it will remain attached under the moist conditions and under the continuous flexing that is experienced in applications to skin.

Wettability of the adhesive mass may be measured by the contact angle

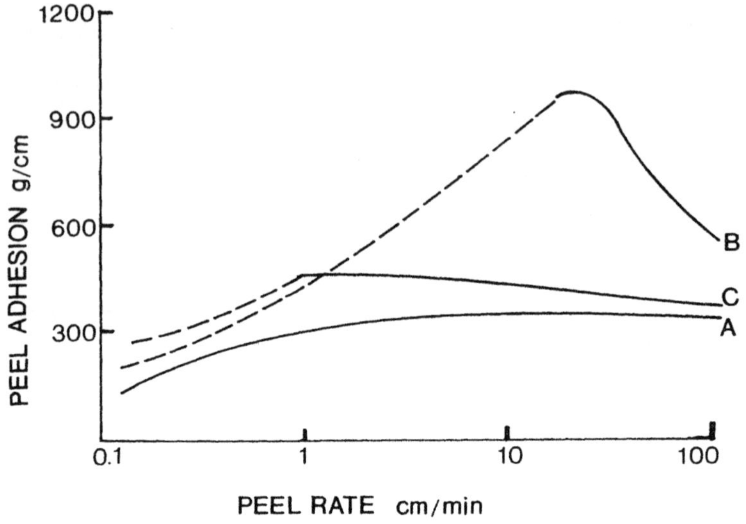

Fig. 25-7. Dependence of peel adhesion on peel rate for some polyacrylic adhesives. A = High cohesive strength industrial adhesive. B = Low cohesive strength adhesive. C = Hospital tape adhesive.

GOVERNMENT REGULATIONS

Pressure sensitive adhesives used for medical devices or in conjunction with drugs fall under Food and Drug Administration regulations. Unlike food packaging adhesives, there are no set rules for clearing medical adhesives on a compositional basis. Instead, adhesives require testing and approval on a product-to-product basis.[39]

Adhesives are tested on animals and on humans in order to develop data, indicating their suitability or lack of suitability, for their use on human skin. Adhesive patches are applied to shaved skin areas on rabbits. The level of irritation is commonly rated on the Draize scale, ranging from zero for no irritation to eight for severe irritation.[40] The Draize test evaluates primary irritation after a single application. In addition to primary irritation, the potential for sensitization to a given adhesive is also important.

The possibility of an allergenic reaction or some other type of sensitization is determined by repeated tests over the same skin area. Such repeat insult patch tests are carried out on a panel of randomly selected human subjects. The adhesive tape is placed on the skin for 24 hours, and then removed. The skin is then allowed to rest for 24 hours before the tape is again applied. Such tests are often carried out 10 to 15 times in succession.[5]

36. Dunshee, W. K., and Vier, C. E. U.S. Patent 3,908,650 (1975) (assigned to Minnesota Mining & Manufacturing Co.).
37. Doehnert, D. F., and Hill, A. S. U.S. Patent 4,505,976 (1985) (assigned to Johnson & Johnson).
38. Komerska, J. F., and Moffett, N. Collagen Films as Test Surfaces for Skin-Contact Pressure Sensitive Adhesives. In *Proc. Technical Seminar*, pp. 108–114. Pressure-Sensitive Tape Council, Itasca, Ill., May 8–10, 1985.
39. Rosenberg, L. Pressure Sensitive Adhesives: FDA Regulations. In *Proc. Technical Seminar*, pp. 177–83. Pressure-Sensitive Tape Council, Itasca, Ill., May 8–10, 1985.
40. Draize, J., Woodard, G., and Calvery, H. *J. Pharmacol.* **82:** 377–90 (1944).

of adhesive tapes not to be greatly affected by aging at 155°C or even at 180°C. Another application requirement is the tolerance to the resin and lacquer impregnants.

The tapes must undergo rigid quality control in order to ensure that products of constant properties are delivered to the user. The test methods must be closely related to the actual application. Naturally, these tests must give reproducible results to all testers. These tests have been nationally and internationally standardized after a careful selection of testing procedures. Only in exceptional cases are so-called house methods used after an agreement is made between the manufacturer and its customer.

The most important test methods will be discussed. These are related to tests designed for other electrical insulation materials.

DIELECTRIC STRENGTH

The value of dielectric strength gives the user an indication of the stability of the pressure sensitive adhesive tape in strong electrical fields. The dielectric strength is defined as a quotient of breakdown voltage and the thickness of the tape being tested. Breakdown voltage is defined as the effective value of a sinusoidal alternating voltage at which the insulation capability of the adhesive tape is suddenly lost because of its structural destruction.

The test apparatus for breakdown voltage determination consists of transformers with a high rated voltage, equipped with controls that allow an increase in voltage at a desired and constant rate from zero to the breakdown value. The voltage is imposed on the test specimen with metallic electrodes in the form of plates, spheres, or rods of different dimensions. For accuracy of measurement, it is important to locate the electrode centers in a homogeneous electric field that decreases in strength near the electrode edges.

The dielectric strength is not a material property, but depends on many outside factors. For example, a spherical pair of electrodes will give a different breakdown voltage than the one obtained by employing plate electrodes. It has also been observed that a thicker tape will give a lower dielectric strength value as compared to a thinner tape of identical chemical composition. Other factors that affect the dielectric strength are the frequency and curve form of the test voltage, as well as the rate of voltage increase.

Generally, the short time breakdown voltages are determined. This means that the voltage in the test apparatus is increased at a constant rate over a short time period until breakdown occurs. In practical applications, it is often more important to know the stability of the tape against a constant voltage of a longer duration. This is obtained by measuring the so-called withstand voltage and *n* minute test voltage. The withstand voltage is the voltage that can be

Tab

TEST
Shor
ar
Shor
te
15-
st

withstood
conditions
voltage th
lower tha
In m
23 ± 2°
might be
strength
An
to volta
and for
were tes
voltage
hr and
exposur
strength
tape a
T
nation
of a r
are st
detern

ELEC

Press
elect

Such corrosion can increase the resistance of thin wires by decreasing their cross-sectional area and even cause a break in the current flow. This phenomenon is the result of the ionic conductivity of the adhesive tape. It is especially accelerated by direct current.

The corrosive action of the adhesive tapes cannot be definitely predicted by a chemical analysis. Dependable results can be obtained only through physical measurement methods. The most common test methods are the indirect measurement of electrolytic corrosion, the wire tensile strength method, and the visual inspection method.

The most often conducted test is the indirect measurement of electrolytic corrosion. By this method, the conductive constituents of the adhesive tape are determined by a quantitative measurement of the electrical resistance. A layer of an adhesive tape is placed between two electrode clamps. The electrodes are made from brass. Their size and the distance between electrodes is fixed. The sample is aged for 18 hr in a chamber at 23°C and 96% R.H. The resistance at the electrodes to a direct voltage of 100 volts is determined immediately after aging. The results are reported in megohms (MΩ).

A direct method of determining the corrosive action of an adhesive tape is the so-called wire tensile strength method. Thin, bare copper wire are arranged as anode and cathode on the adhesive tape by being placed paralled with fixed distances between them. A potential of 250 volts is imposed on the wires for 24 hr in an atmosphere at 50°C and 100% RH. The effect of electrolytic corrosion is determined by the measurement of the tensile strength of the copper wires. If the adhesive tape has ionic ingredients, the cross-sectional area is changed, causing a decrease in tensile strength. The ratio of the tensile strength after to before the effect of the adhesive tape is called the corrosion factor. It has a value of 1.0 for noncorrosive adhesive tapes and a value of < 1.0 for corrosive tapes.

The visual inspection method allows a direct observation of the effect of electrolytic constituents in the tape. The tape to be tested is pressed against two juxtaposed brass foil electrodes, which are connected to a direct current at 100 volts. The test is conducted for four days in a chamber at 40°C and 93% RH. If the adhesive tape contains ionic materials, color changes are observed on the brass foils. At the anode, zinc is removed and the excess of copper causes red coloration. At the cathode, oxidation causes a black coloration. The appearance of the brass foils is compared to the reference color tables and a value is assigned. Unobjectionable adhesive tape might have a value of $A = 1.0$, while very bad tape might have a value of $B = 3.0$.

Which of the preceding methods for corrosion determination is chosen depends on the technical requirements of the application. All methods have their advantages and disadvantages. It should be mentioned that experience

temperature, and voltage. They give the user a useful indication of the suitability of the insulator, in our case the pressure sensitive adhesive tape. For example, the use of pressure sensitive adhesive tapes in insulation technology should give low dielectric losses. Their maximum should be shifted to the highest possible temperature range in order to avoid heat penetration. If the pressure sensitive adhesive tape is used for condenser construction, a high dielectric constant would be advantageous.

The dielectric constant is the factor by which the capacity of a condenser increases, when another insulator such as a dielectric is placed between the condenser plates instead of a vacuum. This factor is brought forward by polarization effects that are related to the structure of the insulator.

Energy losses in the electric alternating field are caused by the rhythmic polarization changes. A measure of these losses is the dielectric dissipation factor. It is expressed as a tangent of the angle of effective and reactive currents. It should be remembered that the effective current, which causes the energy loss in the insulator, always has the same direction as the corresponding voltage. The capacitative reactive current precedes the voltage by a one-quarter period.

In practice, the use of a loss index is preferred. It is a product of the relative dielectric constant and the dielectric dissipation factor.

A general qualitative rule is important that, in case of nonpolar or slightly polar materials, the dielectric values are nearly constant in a broad frequency range. In the case of polar substances, the dielectric constant varies depending on the frequency and temperature. In the case of the dissipation factor, several maxima are observed as the result of dependence on frequency and temperature. The dielectric properties are determined by bridge, resonance, or other methods. The values obtained by different methods are comparable.[13–16]

TEMPERATURE-TIME LIMITS

The size of electrical equipment is decreasing for weight and cost reasons, such as in the case of electric motors, and the materials used must be capable of meeting stricter thermal requirements. Determination of the temperature-time limit is an increasingly important method for stating the thermal expectations of electrical pressure sensitive adhesive tapes.

The temperature-time limit is the highest temperature at which the adhesive tape can be used continuously for a period of 20,000 hr. During this time, the electrical and mechanical properties must not fall below a minimum value. For testing purposes, the dielectric breakdown voltage is considered as representative of all electrical properties and the weight loss representative of all mechanical properties. The dielectric breakdown voltage was chosen as a representative of all electrical properties, because during thermal aging tests, informative curves

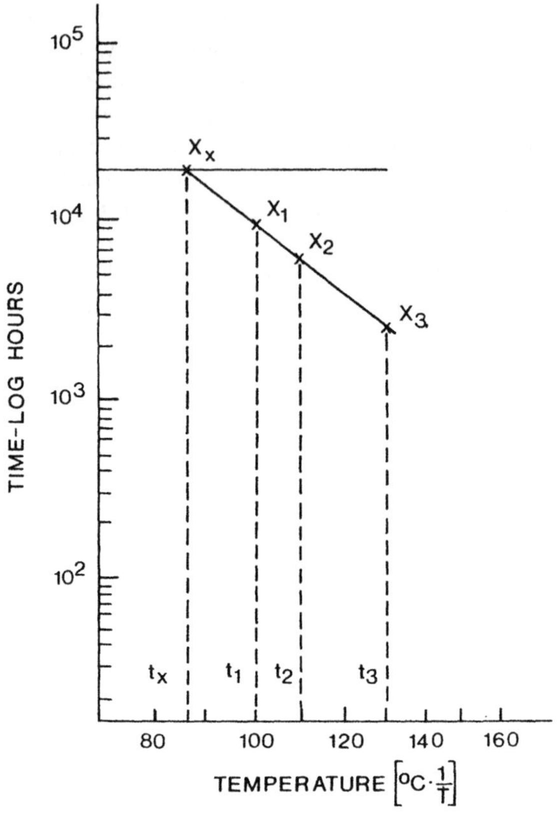

Fig. 26-2. The thermal aging curve for determination of the temperature limit.

temperatures, a second diagram is constructed (Fig. 26-2) by plotting the time logarithm versus reciprocal temperature. The resulting points X_1, X_2, and X_3 lie approximately on a straight line that is extended to the time of 20,000 hr. The temperature at the intersection X_x is the sought temperature limit.

The determination of the temperature limit based on the weight loss data is carried out in an analogous manner.[17–21]

OTHER IMPORTANT TEST METHODS

Several test methods have been discussed in some detail because they are considered important for electrical pressure sensitive tapes. For the sake of completeness, other methods that are equally important for some segments of the electrical industry should be mentioned. In many specialty tapes, the temperature curing

Sheet 12: Requirements for polyethylene film tapes with nonthermosetting adhesive.

Sheet 13: Requirements for combined cellulose–viscose woven fabric tapes, one side covered with plastic coat, the other side with thermosetting adhesive

The national branches of the commission work actively not only within IEC, but are also prompted to include the results in their own specifications. In the near future, the specifications for electrical pressure sensitive adhesive tapes will be uniform worldwide, only written in different languages.

PROBLEMS IN SPECIFICATIONS FOR TAPE DEVELOPMENT

Tapes, such as PVC film tapes, are consumed in large quantities for electrical applications, so their specifications were therefore written first (IEC 454). The development of tapes to meet these requirements is discussed in the following paragraphs.

High adhesion, good tensile strength and elongation values, and high water resistance are required from plasticized PVC film tapes. The tapes must have a high dielectric strength, and must not cause electrolytic corrosion. They must also be difficult to ignite and easily handled at low temperatures.

Like other tapes, plasticized vinyl film electrical tapes consist of the adhesive and the backing. The adhesive may be applied as an aqueous dispersion or as a solution. The backing consists of several ingredients: PVC powder, plasticizer, fillers, colorants, stabilizers, and lubricants.

Lubricants may create difficulties in tape development. If the lubricant is present in a large quantity on the film surface, wetting problems may be experienced when coating with aqueous dispersions. The adhesive coating recedes and breaks up into separate drops. However, adhesive coating from a solution is not problem-free either. Although the adhesive will wet the film surface well, it will contact the backing and the lubricant present on its surface after drying and winding. The transfer of the lubricant to the adhesive surface will cause a reduction in tack.

Other difficulties arise from the plasticizer. Plasticizers that are effective in softening vinyl are also prone to migration. In adhesive tape, the plasticizer will migrate from both the coated side and the backside, until an equilibrium concentration between the backing and the adhesive is established. This is accompanied by a decrease in adhesion and sometimes by a decrease of backing flexibility. Migration may be avoided by the employment of proper plasticizers, by the incorporation of the required amount of plasticizer into the adhesive, or

prove to be too temperature-sensitive and might stick to the dryer idlers. The solution is to lower the drying temperature and thereby, the tape manufacturing rate, resulting in increased tape manufacturing costs.

This short example serves to illustrate the plasticized PVC tape problems that the developer of specification PVC tapes must resolve. The problems in the case of electrical tapes of other compositions are not more easily solvable; they are only different.

ELECTRICAL TAPES FOR SPECIAL APPLICATIONS

Flame-Resistant Tapes

It is difficult to fulfill the need for completely nonflammable tapes. Flame-retarding, difficult-to-ignite, combustion-resistant, or self-extinguishing tapes are easier to produce. Such tapes are needed for automotive and aviation industry uses. This need is further enhanced by the additional requirements of low smoke density, low toxicity of combustion products, and a combustion gas temperature that is not too high.

The tapes are tested by various procedures: some of the most important tests for tapes are enumerated below.

For the building trade a burning chamber test that follows DIN 4102 is used. The specimens are arranged vertically in the chamber, and subjected to an intense flame for a long period of time. A strong air stream flows through the air chamber. The burn-damaged area and the combustion gas temperature are determined.

According to UL specification UL 510, a metal wire is wrapped in several layers with the tape, then hung vertically. The sample is exposed to a Bunsen burner flame of prescribed intensity for several consecutive time periods. The attached marks should not be burned.

IEC 454 and AFERA 4009 tests consist of igniting a freely hanging electrical tape with a hydrated cellulose leaflet. The duration and the extent of the burning are observed.

Tapes used in the aviation industry are tested according to other specifications, namely ATS 1000 in FAR-Test Part 25,25853. The tapes are placed vertically, horizontally, and at angles of 45 and 60 degrees in a special combustion chamber and then exposed to an exactly defined Bunsen burner flame for a specified time period. After the flame is removed, the burning time and burn extent must not exceed prescribed values. In addition, the smoke density and the toxicity of combustion products are checked.

In the development of flame-retardant pressure sensitive adhesive tapes, the behavior of the backing and of the adhesive are regarded separately.

are often produced, and at least one important flame retardant is suspected of being a carcinogen; its continued use depends on the development of more specific information. Flame retardants are discussed in various publications.[26–29]

Adhesive Tapes that Are Stable at High Temperatures

High-temperature-resistant tapes are increasingly needed for specialized uses in the electrical industry. Such tapes must remain functional when exposed to high temperatures for a longer time period, that is, their mechanical and electrical properties must not show excessive change at such conditions. One example of such use is the construction of electrical motors. Generally, while long-term exposure to temperatures of up to 130°C are encountered, the size of the motors has been decreasing for space and price reasons. For the same power motors, insulators capable of withstanding 155°C and higher temperatures are needed.

Most polymers decompose at temperatures ranging from 90° to 120°C, for example, PVC, polyethylene, polycarbonate, and polyester. Only a small number of select polymers are stable at higher temperatures, and are therefore used for high-temperature-resistant tapes. From a chemical structure point of view, polymers with a high fluorine or aromatic ring content, or with a high cross-linking density, are the most temperature resistant. The next group is the fluorinated plastics with varying fluorination levels. In addition to a high chemical resistance, good electrical properties, and flame resistance, these plastics also exhibit a broad temperature performance range. Polytetrafluoroethylene is useful over a temperature range of −200°C to 260°C.

Polysulfone and polyether sulfone films can be utilized at temperatures up to 165°C. These films also have outstanding electrical properties. They are self-extinguishing and resistant to acids, alkalis, and most common chemicals. Unfortunately, they are not resistant to many important saturants and encapsulating resins, thus reducing their utilization in some electrical applications.

Very good high-temperature-resistant tape backings are papers, textiles, and nonwoven fabrics made from aromatic polyamide fibers with a high-temperature limit of 190°C. Chemical resistance and electrical properties are also good. This material does not burn or melt, but simply decomposes at high temperatures.

Aromatic polyimide films have especially good properties. They are stable up to 400°C. Films with similar structure, such as polyetherimide polyamidimide, are also in this class. Their disadvantage is their high price due to the production technology used.

As tape substrates, glass fabrics and the combinations of various materials already discussed should not be forgotten. A frequently used backing material ester film and polyamide woven or

or temperatures above 180°C. Adhe-
cohesive strength above 150°C, so
ss-linked for higher temperature use,
tion. Cross-linking must be carefully
xcessively.
tural rubber are also suitable for use
hese adhesives must be cross-linked.
ally through the employment of alkyl
oxides or di- or tri-isocyanates, and
liation. High-temperature-stable adhe-
ications.[30–33]

used in motors, transformers, and
encapsulated in resins. The leads
resistant to mechanical abuse. The
nt conditions, acids, alkalis, etc., is
g the operation is conducted through
through a good insulator,
lly, repairs are carried out
ion when saturant-covered

lacquers and recently with
urant application processes
ckle process. It is important
adhesive tape, used only to
resistant to the saturant,
the saturant.
facturers as supplied
ensively compounded
al-rubber-based adhe-
alkyl phenolic resins
ing lacquer solvents.
sin cure. As a result,
esulting in diminished
ked with peroxides,

may also present problems. Peroxides affect the cure of the saturating resins, and these resins are usually not compatible with the adhesives. Silicone adhesives are best suited, but they are expensive and some branches of the electrical industry have traditionally avoided their use. Silicones are suspected of being capable of covering electrical contacts in relays and rendering them useless. Acrylic adhesives are very polar and may cause stress corrosion of the coil wire lacquer insulation, which may lead to insulation problems. Therefore, the best universal adhesive for saturation-resistant tapes is a natural-rubber-based product without oil-soluble resins. The adhesive is cross-linked by some other method, such as beta or gamma radiation.

The application of saturants is much more complicated than discussed in these short paragraphs. Compatibility of the saturant with the adhesive tape is mentioned only briefly. It is important that the compatibility be checked out by testing, since incompatibility may lead to the formation of voids between the tape and lacquer. Similar problems can appear due to the air bubbles that accumulate during the saturation of impervious tape. The bubbles interfere with the electrical field and may cause premature dielectric failure due to sparks.

These problems may be avoided by using saturant-permeable tapes. These tapes are constructed using open-weave fabric or a nonwoven fabric, and a discontinuous adhesive applied in a line or dot pattern. Additional information may be found in the literature.[34–36]

Conductive Adhesives

Pressure sensitive tapes, because of their composition, are generally good electrical insulators. The backing is made from a high molecular weight polymer, and the adhesive consists mainly of polymeric materials. Polymers and plastics are nonconductors. However, there are some uses that require electrically conductive tapes.

For some applications, it is required that the tape act antistatically, as in the manufacture of microchips and in the application of MOS construction blocks over conductive plates. These tapes remove the electrostatic charges that may destroy the sensitive MOS blocks. Other applications for conductive tapes are flexible electrical conductors and heating bands. Such tapes are also useful for high-frequency interference shielding.

A long time ago, plastics were made conductive by the addition of carbon black or metallic powders. The use of graphite, iron oxide, metallized glass spheres and fibers, copper and aluminum particles, and carbon fibers is described in the patent literature. Tape backings can be made conductive by employing these additives. Fabrics, nonwoven fabrics, and papers can be impregnated with conductive carbon black. Films may be vapor coated with conductive

18. IEC Publication 216.
19. DIN 40 633, Blatt 1.
20. VDE 0340, Teil 1.
21. VDE 0304, Teil 2.
22. IEC Publication 254–2.
23. VDE 0340, Teil 3.
24. SEMKO 26–1964.
25. NEMKO 1963.
26. *Federal Aviation Regulation,* Part 25, 25–853.
27. *Airbus Industry Material Test Specification* 1000.001.
28. Troitzsch, J. *Kunststoffe* **69:** 557–562 (1979).
29. Troitzsch, J. *Brandverhalten von Kunststoffen, Grundlagen Verschriften-Pruefverfahren.* Carl Hauser Verlag, 1982.
30. Schroeder, K. F. *Adhaesion* **1971:** 290–296.
31. Fitz, H. *Kunststoffe* **70:** 27–33 (1980).
32. *Plastverarbeiter* **33:** 1099–1101 (1982).
33. *Plastverarbeiter* **34:** 784–786 (1983).
34. Stender, A. *Elektrotechnik* **65:** 74–75 (1983).
35. Dr. Beck. *ABC der Lack- und Kunstharzisolierung fuer die Elektrotechnik.* Second edition. BASF, 1974.
36. Braeunling, E., Kuhlmann, K. D., and Lutz, P. D. B. (Ger.) Patent 2,409,277 (1974) (assigned to Beiersdorf AG).
37. Mair, H. J. *Gummi, Asbest, Kunststoffe* **38:** 122–124 (1985).
38. Wessling, B. *Kunststoffe* **75:** 375–378 (1985).
39. Menke, K., and Roth, S. *Chemie in unserer Zeit* **20:** 1–10 (1986).

Table 27-1. Tapes Used for Packaging Applications

TAPE BACKING	TENSILE STRENGTH (N/mm)	(lb/in.)	TOTAL THICKNESS (μm)
Film tapes			
biaxially oriented polypropylene	4.4–6.3	25–36	50–57
polypropylene strapping tapes	16–39	90–225	110–125
polyester, label protection	3.9–4.4	22–25	50
polyester, shrink packaging	3.5–4.9	20–28	62
polyester, heavy duty	6.8–8.8	39–50	75–82
vinyl, rigid, general packaging	3.5–3.9	20–22	60–67
vinyl, rigid, carton sealing	5.3	30	62
vinyl, rigid, label protection	4.9	28	62
vinyl, rigid, heavy duty	7.9	45	98
vinyl, plasticized, colored	5.3–5.6	30–32	57–68
cellophane	5.3–5.6	30–32	57–68
acetate	3.9–4.4	22–25	59–62
polyethylene	3.5–12	20–67	60–200
Filament reinforced tapes			
polypropylene film, glass filament	16–83	90–475	140–240
polyester film, polyester filament	32–56	185–320	150–240
saturated paper, glass filament	53–56	300–320	275–310
reinforced paper	21–60	120–340	250–300
Paper tapes			
kraft paper	5.6–8.8	32–50	120–210
saturated paper, crepe, general purpose	3.7–4.0	21–23	150–210
saturated paper, crepe, bandoleering	3–3.9	17–22	185–200
saturated paper, flatback, general purpose	3.2–3.9	18–22	175
saturated paper, flatback, heavy duty	7	40	300
saturated paper, flatback, box sealing	9	50	175–200
saturated paper, rope stock, medium strength	3.7	21	125
saturated paper, rope stock, high strength	7–8	40–45	175–200
Fabric tapes			
vinyl coated cotton sheeting	7–8	40–45	275–300
polyethylene coated cloth	7–8	40–45	275–300
coated cloth, extra strong	9.6	55	350
polyethylene coated nylon	2.6	15	250
Foil tapes			
aluminum foil, sealing	3.5–5.3	20–30	90–125

low priced packaging tapes. Rigid vinyl film tapes in Europe, kraft paper tapes in Japan, and saturated paper tapes in the United States still enjoy a fairly wide use. BOPP tape, however, is not only replacing other pressure sensitive tapes, but also the gummed paper tape. Although the latter is less expensive, BOPP tape is easier to apply and forms the bond quicker.

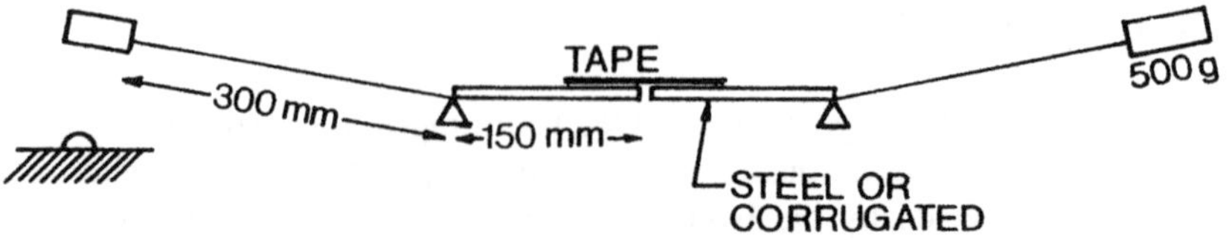

Fig. 27-1. A tape tester simulating the forces exerted on a carton closure tape.

shear and to low angle peel forces. The adhesive must wet the corrugated board surface well to establish a good bond quickly and it also must have a sufficient creep resistance to sustain a long-term load. The important part of ASTM test method D2860, Adhesion of Pressure Sensitive Tapes to Fiberboard at 90-degree Angles and Constant Stress, was to specify a standard fiberboard surface for testing, instead of conducting the tests to steel.[3] The test has been adopted by Pressure Sensitive Tape Council as PSTC-14.[4] These tapes are also tested by hanging a loaded corrugated container over a bar so that the force is exerted over the taped cover. A special testing device has been used in which the tape is exposed to shear and peel forces simultaneously, thus approximating the actual use conditions. A schematic diagram of such a tester is shown in Figure 27-1.

Narrowly slit pressure sensitive film tapes are used as tear strips for various packaging applications, including tearing open the corrugated board boxes. The tear strips are applied at high speeds on corrugated machines and the tape must therefore be supplied in large rolls. This led to the development of spooling of tapes, which enable the winding of as much as 15,000 meters per spool. Figure 27-2 shows a spool winder.

Fig. 27-2. Spooling system with shear slitting head.

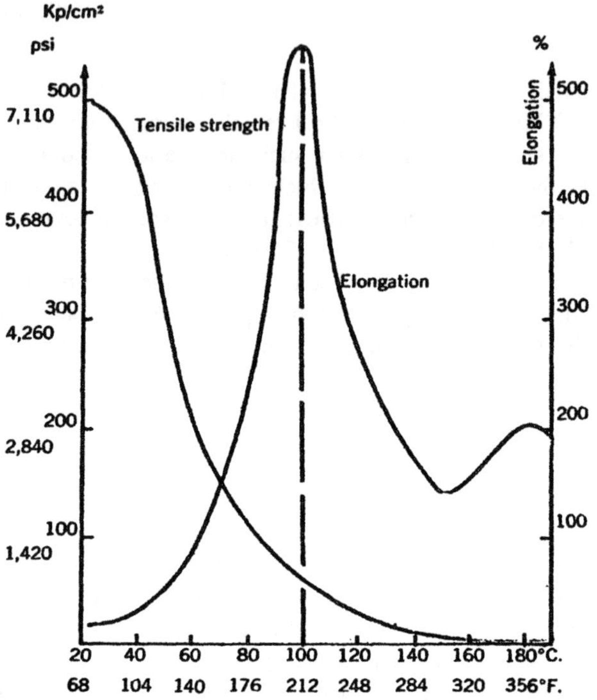

Fig. 27-3. Tensile strength and elongation of vinyl film backing as a function of temperature.

Table 27-4. Solvent Resistance at 20°C[a] (MD-Oriented Rigid Emulsion PVC Film, 40 μm Thick)

SOLVENT	TIME TO RESIST DIMENSIONAL CHANGE (sec)	BOILING POINT (°C)	RELATIVE EVAPORATION RATE
Tetrahydrofuran	20	66	2.6
Acetone	30	56	2.1
Methyl ethyl ketone	35	80	3.3
MEK/Toluol, 1:1	60	—	4.2
Ethyl acetate	150	76	2.8
Methyl isobutyl ketone	600	116	7.5
Butyl acetate	1700	121	12.5
Toluol	2300	110	6.1

[a] Data from *Adhesives Age* **15**(3): 17–23 (1972).

for strapping beef and other slaughterhouse products. Paper backing accepts the inspection stamp better.

Some tapes may be constructed so that the filaments are also attached to the film backing,[11] including placing the filaments between two film layers,[12] and by employing a laminate of two different films.[13] It is claimed that the firm bonding of filaments directly to the backing prevents their dilamination, which results in puckering. Most of such improvements have not come on the market yet.

Hot melt pressure sensitive adhesives based on block copolymers, and solvent adhesives based on natural rubber-resin (and sometimes block copolymers), are most often used for reinforced tapes. Some tapes are also made by extrusion of the adhesive.

Specialty filament-reinforced tapes with nonstaining adhesive are used for securing the doors of appliances. Tapes are offered with a printable surface, and special solvent- and oil-resistant adhesives are used for electrical applications.

SATURATED PAPER TAPES

Saturated paper tapes developed for masking applications have found many uses in packaging. However, their packaging uses have been steadily declining, because of the increased use of film tapes, but they are still used for freezer tapes, holding produce, carton sealing, labeling, packaging refrigerators, and as bandoleering tape to hold electronic components, staples, nails, etc. Such bandoleering tapes might be needed temporarily during manufacturing; some might become a permanent part of the packaging, such as supplying nails for a gun application.

KRAFT PAPER TAPES

Kraft paper tapes are constructed by extrusion coating the paper with a 18–20 g/m^2 polyethylene coating, corona discharge treating the surface, and applying 0.6–0.8 g/m^2 silicone release coating. Clupak-type kraft paper is used with an elongation of 7–10% instead of 4% for regular kraft paper. The higher elongation at break results in a much tougher tape. The properties of Clupak kraft paper used for tape backings are shown in Table 27-5.

A pressure sensitive adhesive coating of 35–42 g/m^2 is applied over the paper. The adhesive is usually natural rubber-resin type coated from solution but hot melt adhesives are also used for this product. The adhesive is applied directly to the paper, and it releases from the silicone release coated side. The

tions. Natural-rubber-based adhesives are most often used. The reverse side of the backing is overcoated with shellac or synthetic resin and a release coating. It is a fairly popular tape in Japan.

FABRIC TAPES

Historically, the industrial fabric tapes have developed from the hospital tapes, which were produced by calendering. Calendering is still an important method in the manufacture of these tapes. Polyethylene coated fabric tapes can be manufactured by calendering polyethylene and adhesive coatings in the same pass on a three-roll calender. A schematic diagram of this process is shown in Figure 34-44.

Polyethylene coated fabric tapes can also be manufactured by applying a polyethylene coating by extrusion over a low cost primed rayon or cotton fabric. In some cases, a release coating is used over polyethylene. Polyvinyl chloride and acrylic resin coated fabric tapes also have packaging applications.

Various adhesives are used for fabric tapes: the most common ones are natural rubber, SBR, reclaimed natural rubber, and block copolymer-based adhesives.

Packaging applications of fabric tapes are mainly sealing and waterproofing. A smaller volume of high-strength glass fiber fabric tape is used for bundling and strapping applications. Woven fabric tapes have an unequaled flexibility and conformability for the given tensile strength and bulk. Fabric tapes can also be made to tear across easily, thus simplifying their dispensing and application.

SURFACE PROTECTION TAPES

Surface protection is a packaging-related use. Pressure sensitive saturated paper, kraft paper, polyethylene film, vinyl film, and other materials are used to protect large polished metal or plastic sheeting surfaces from damage during shipping and subsequent handling, including cutting and stamping operations. Low tack pressure sensitive adhesive, sometimes only lightly compounded natural rubber latex, is used. The adhesive should not leave any residue that might require later cleaning. Some of the adhesive requirements might be quite demanding, requiring not only a slow initial peel adhesion, but also a low adhesion build up during adverse aging conditions.

The easily removable adhesives have been formulated in various ways.

TAPE DISPENSING

Tape dispensing for various packaging applications is very important for the successful use of tapes. There are over 150 types of tape dispensers, starting with hand applicators on up to powered and automated systems. In manual

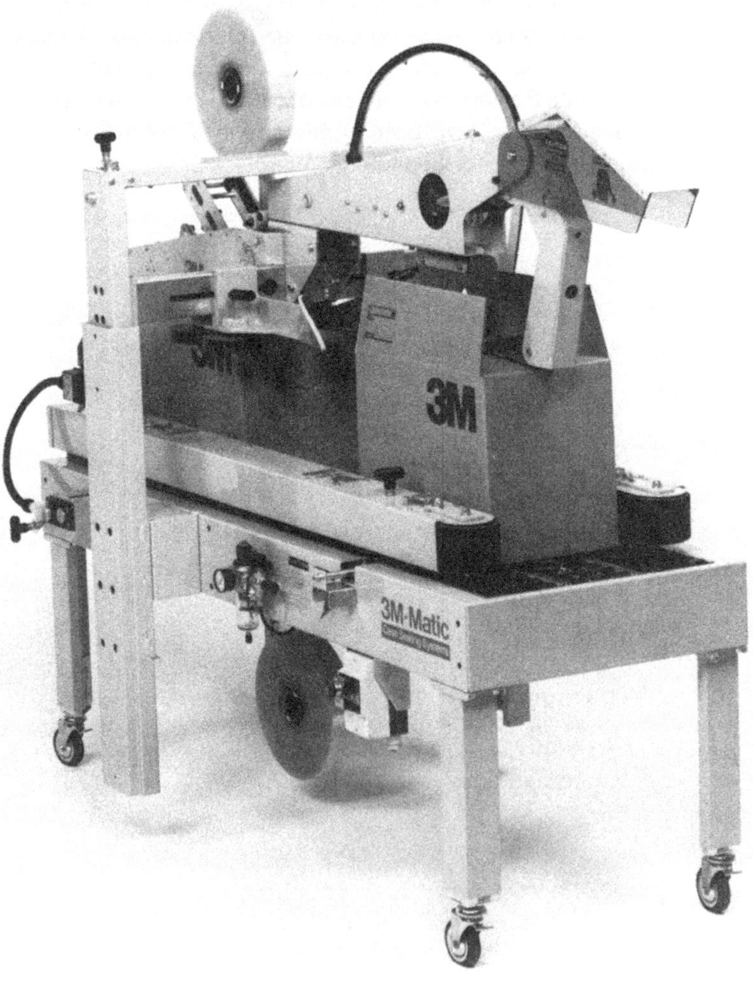

Fig. 27-5. Automatic adjustable case sealer. (*Courtesy of Packaging Systems Division/3M*)

16. Davis, J., and Sirota, J. U.S. Patent 3,535,295 (1970) (assigned to National Starch & Chemical Co.).
17. Merrill, R. F., Jr., and Courtney, H. R. U.S. Patent 3,857,731 (1974) (assigned to Minnesota Mining & Manufacturing Co.).

Tensile Strength

Tensile strength is usually defined as the strength of a 2.5-cm (1-in.)-wide sample about 10 cm long pulled apart on a pendulum tensile tester at 30 cm/min (12 in./min). The tensile strength is also tested in strain-gauge-type tensile testers at speeds of 30 cm/min. Actually, the test speed of a pendulum-type tester is lower, perhaps about 5 cm/min. The pendulum-type tester is actually slower than the set speed, because of variable jaw separation during testing, whereas the strain-gauge tensile tester operates at a constant speed.

For creped masking tapes, a tensile strength range of 3 to 4.5 kg/cm width (18 to 25 lb/in.) is used, with a more common range of 3.5 to 4 kg/cm (20 to 22 lb/in.). The tensile strength values vary with the supplier, depending on the raw paper source and saturation formula. Generally, the tensile strength is not critical, if it is within the range stated.

Wet Tensile Strength

Wet tensile strength is measured the same way as the dry tensile strength. It is not important for most applications, but it might be important for some packaging uses where the item might get wet. In general, 50% retention of the dry tensile strength is considered to be good, and 75% is excellent. The wet tensile strength is also indirectly important in that it can reflect the degree of cure of impregnant and it can be useful as a control test.

Elongation

Elongation is measured at the same time one measures the tensile strength. It depends on the amount of crepe in the raw paper, type of raw paper, and type of saturant. Elongation is especially important in those paint masking applications where the tape is applied around curved surfaces. A tape of low elongation may tear at the edge and yield a jagged paint edge, whereas one of higher elongation will yield a smooth edge. Elongation below 6% is considered too low, 8% is acceptable, and 10–12% very good.

Tear

The tear resistance can be measured by a number of methods: Elmendorf, edge tear, trapezoid, Finch, etc. All of these indicate the resistance the tape offers to tearing. The most critical operation requiring a good tear strength is

tear more easily. There are a number of ways of measuring stiffness. One often used in the industry is the Gurley stiffness test.

Porosity

The most important value of the porosity test is in measuring the resistance the paper offers to subsequent release, or to coating penetration into the paper. It indicates the ability of the adhesive coating to penetrate into or through impregnated paper. If the porosity is too high, the adhesive and release coatings can penetrate through the backing. If the adhesive penetrates the backing during the processing, the opposite side of the tape becomes sticky, can stick to the rollers, and cause web breaks. This property can be measured by Gurley's Densometer.

Sensitivity to Solvents

This property is not usually measured, but could be of concern in masking applications. During the painting of masked surfaces, the solvent in paint may solubilize the impregnant by soaking in at the edges. Usually, a good release coating will help to prevent such solvent attack if the impregnant is of the type affected by the solvent.

Color

The color of creped masking tapes is usually natural. It can vary from a very light, almost white, for bleached papers, to a dark tan. The color depends largely on the color of raw paper: bleached paper being white, semibleached being light tan, and unbleached being darker tan. The impregnant can also influence the shade. If a darker raw paper is used and a lighter shade of the tape is desired, some TiO_2 may be incorporated into the impregnant formula.

Interaction Between Adhesive and Saturant

After coating the impregnated paper with pressure sensitive adhesive, sometimes a lowering of tack and adhesion on aging is observed. In most cases, this is due to the resin migrating from adhesive into the impregnant. There are several ways to avoid or minimize this phenomenon. One simple way is to use an impregnant into which the resins will not migrate, such as a butadiene acrylonitrile

leather may use this system and have certain advantages. But the process is rarely used for thin pressure sensitive tape backings.

RAW PAPER

The physical properties of impregnated sheet are largely determined by the raw paper. The impregnant will improve upon the properties of the paper, but the degree of improvement is determined by the properties of raw paper. Creped raw paper is made by a small number of paper companies. Under a microscope, the paper looks like a bundle of flattened straws with a relatively high amount of void space. It is believed that adhesion between fibers is primarily due to hydrogen bonds. Raw paper is a weak structure somewhere in the area of 1.3–1.8 kg/cm in tensile strength. On saturation, the rubberlike material fills the voids and much of it tends to clump around fiber-to-fiber contact points, gluing them together. It also forms a network of rubber between fibers. This increases the tensile strength to 3.2–4.4 kg/cm as well as improving other physical properties, especially the delamination resistance. To improve the properties at a minimum cost is a difficult task, and the properties of the raw paper become important.

On continued refining of the pulp, the strawlike fibers making up the paper are split and hairlike fibrils extend from them. These fibrils increase the number of contact points between fibers and increase the tensile strength. Simultaneously, the sheet becomes denser with less void space; it becomes stiffer and more difficult to saturate. The paper also becomes less stretchy and tears easier. Thus, there is an optimum amount of refining necessary to give a sheet of optimum properties that are well-balanced. The effect of different pulps and treatment is discussed by Page[7] and by Dunlap.[3] The correct pulp or pulp blend must be chosen. Relatively long, softwood pulp fiber is preferred in order to obtain the properties discussed: sufficient void space for saturation and resulting good elongation and tear properties. The fiber length of hardwood pulp is too short and does not give the best properties.

The paper just discussed is primarily used for masking tapes. Usually it is a semibleached color—light tan. Besides creped papers, flat and special papers are also used. These can be bleached white for printable tapes and other products, semibleached, or unbleached, depending on the end use. The paper may be considerably refined to give a relatively dense stiff sheet, or it may be lightly refined to give a less dense, soft, and stretchy sheet.

Usually, the paper is oriented so that the strength ratio in machine/cross direction is approximately 2/1–2/1.5. There is a type of paper highly oriented in the machine direction to give a higher tensile strength. A 30-lb/ream paper normally has 3.2–4.4 kg/cm. If made with special fibers (hemp or high tensile

SATURATED PAPERS

Two main types of saturants are used: solvent-based and water-based latexes. The solvent-based saturants were the earliest developed, as illustrated in a series of patents.[8–10] Water-based latexes were developed later and are by far the most extensively used today.

Solvent-based Impregnants

Solvent-based saturants coat all fibers uniformly. This, together with the fact that there are no water-sensitive materials in the saturant, as there might be in water-based latexes, tends to give a sheet of better water resistance. However, the sheet tends to be weaker, as compared to one saturated by water-based latex. Also the number of impregnating polymers that can be used is limited. Natural rubber and SBR are the preferred polymers. Pollution problems and the high price of solvents are the other reasons favoring water-based impregnants. However, the solvent-impregnated papers are necessary for electrical paper tapes.

Water-based Impregnants

Latexes are by far the most commonly used impregnants by tape manufacturers as well as by independent saturators for tape backings. There is a wide variety of latexes available that are used alone or blended. Latexes are much simpler to use as compared to solvent-based impregnants, which require breaking down, compounding, and then the rubber must be dissolved.

Some of the synthetic latexes used are enumerated below.

SBR. SBR improves the overall properties of paper and is most widely used. A significant advantage is its low price. A disadvantage is that the adhesive used on such a backing may lose tack and adhesion on aging due to migration of resins from adhesive into SBR impregnant.

NBR. NBR was widely used from 1940 to 1960 because it produced excellent backing properties and stable adhesive properties, and suffered no loss of adhesion and tack due to migration of resins into NBR on aging. Its disadvantage is its high price.

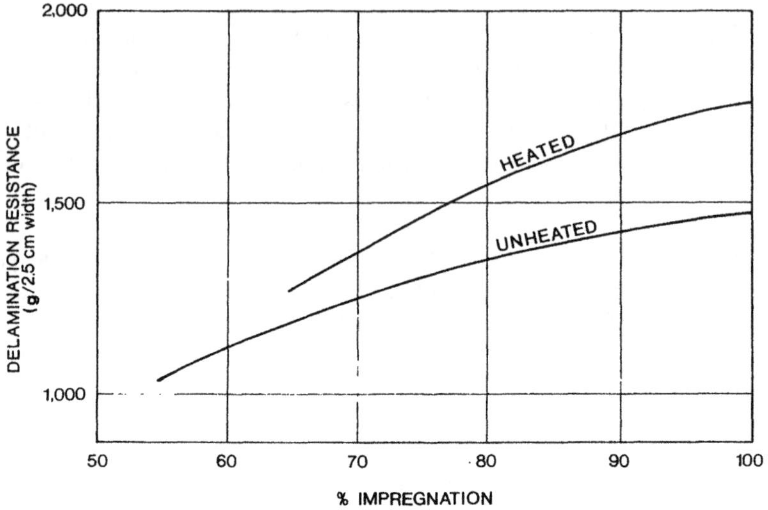

Fig. 28-2. Delamination resistance as a function of impregnation.

affecting the tape performance are delamination resistance and edge tear. The effect on delamination resistance by the impregnation level is illustrated in Figure 28-2 by Weschler and Bartell.[18] In this case, which uses a conjugated diene-acrylic acid copolymer emulsion as saturant, an additional increase in delamination resistance is obtained on heating at 163–193°C. For a creped masking tape of 500 g/cm width peel adhesion, 80–100% saturation is recommended.

In general, the firmer the saturating polymer, the higher the delamination resistance and tensile strength, but this gives a stiffer backing and lower edge tear. Butadiene/styrene 75/25 copolymer is soft and gives products of lower delamination resistance, lower tensile strength, and higher edge tear than butadiene/styrene copolymer in the range of 35/65. The same considerations are valid for butadiene/methacrylic ester copolymers, as described by Weschler and Bartell[18] and by Finnegan and Miller.[20] In the latter reference, incorporation of polycarboxylic-acid-containing copolymers were shown to develop higher delamination values.

Early work to increase the delamination resistance at the lowest possible levels of impregnation led to curable impregnants, as illustrated in Bartell's patents.[15,16] Curing effect increased with increasing concentration of curing agent and increased heat treatment. The higher the degree of cure, the higher the delamination resistance, but the lower the edge tear and elongation, so that a compromise between these properties is needed. This is illustrated in Table 28-1, which is based on Bartell's patent.[16] Note that although an excellent delamination resistance was obtained at a curing resin concentration of 13%,

holding, bundling, packaging, and other applications. A relatively low cost paper is sufficient with a minimum delamination resistance of 330 g/cm, though higher delamination resistance might be desired depending on the adhesive-release coat system and aging conditions. Tensile strength of 3.4–4 kg/cm and elongation of 8–12% are adequate. The impregnant and the adhesive must be compatible to avoid adhesion drop-off. This is true for all impregnated papers.

In just about all commercial products a release coat is used, which in general is considered confidential. One such release coating is described.[19] The backing is backsized with a mixture of 100 parts of a urea formaldehyde resin, 13.6 parts of copolymer of vinyl acetate and stearyl maleate, and 0.45 part of phosphotic acid. The backsized sheet was dried at 93°C and cured at 177°C. A water-based release coating is described by Hechtman and Greenman.[32] Various other release coatings are covered in a separate chapter in this book. A number of companies are offering polymer emulsions for use as release coatings (Rohm and Haas Co., National Starch & Chemical Co., Reichhold Chemicals, Inc.).

This tape may or may not have primer, depending partly on the paper impregnant. A primer can be made up of a solution of 320 g of a mixture composed of a 25% solution of pale creep rubber, 40 g of a 50% solution of *p,p'*-diphenylmethane diisocyanate, and 210 g of toluene. Various blends of SBR and NBR, with and without resins, are also used.

Upon the primed side of the backing an adhesive coating of 40–50 g/m^2 is applied. Such an adhesive based on natural rubber is described by Collins[22] (see Table 28-2).

Adhesive formulas based on block copolymers have been proposed. These are suitable for hot melt and high solids adhesives. A possible hot melt formula for a general-purpose masking tape is given below.

Kraton 1107	100
Escorez 5280	125
Shellflex 371	25
Butyl Zimate	2

Table 28-2. Formula for an Adhesive Based on Natural Rubber

INGREDIENT	PARTS
Crude rubber	100
Dehydrogenated rosin	90
Aluminum hydrate	90
Mineral oil	30
Antioxidant	2

Elongations are low, in the range of 4% to 8%. Both natural (light tan) and dark tan (kraft brown) colored backings are used.

Release coating, primer, and adhesive systems similar to general-purpose masking tape may be used. In general, good adhesion, tack, and hold to kraft boxboard are desired.

High Tensile Strength Flatback Paper Tapes

Regular kraft paper tape meeting the Military Specification PPP-T-76C (Carton Sealing) is a special packaging tape with stringent government specification requirements. The tensile strength of at least 8 kg/cm and usually in the range of 9 to 10 kg/cm is offered. A minimum wet cross-sectional tensile strength of 2.1 kg/cm and a minimum Elmendorf tear resistance of 75 g after ultraviolet (UV) and accelerated aging are required. Higher weight, properly designed papers are necessary to meet these specifications. The adhesive also has to meet certain strict military specification requirements. Minimum peel adhesion as received and after aging must be at least 390 g/cm. It must have at least 110 g/cm adhesion to its own backing before and after aging, as well as holding power of at least 96 hr.

Another variety of these tapes is high tensile strength rope paper tapes. These tapes are also used for packaging and require tensile strengths of 8–10 kg/cm in the machine direction. Tensile strength in the cross-machine direction is low and unimportant. These requirements are met by oriented rope (hemp) paper. This paper also has the advantage of being thinner, about 0.1 mm versus 0.15 mm for standard high tensile flatback paper, and also less stiff. The adhesive must have a high holding power. In addition, the white goods market requires it to be nonstaining. A noncured adhesive with very high Mooney viscosity rubber, or a phenolic resin cured with a lower Mooney viscosity, are suitable. Release coats and primers are similar to the ones used in other constructions.

Since hemp used in the rope paper is in short supply and is expensive, a considerable amount of work has been done by paper manufacturers in developing all-kraft or kraft-hemp blend substitutes to provide a similar backing material. These have in some measure replaced all-hemp papers. The kraft paper tends to be somewhat thicker and stiffer than all-rope paper.

Protective Flatback Paper

These papers are used to cover and protect stainless-steel, chrome-plated steel, and lacquered brass. These metal stocks are sold in roll form to companies that cut, drill, and bend them into various articles, such as handles for furniture,

combinations in solvents, and are cured with phenolic resins. Curable solvent-type acrylic adhesives are also used.

REFERENCES

1. Hechtman, J. F., and Jayne, J. E. *Industrial and Specialty Papers,* Vol. II (Mosher and Davis, eds.). Chemical Publishing Co., New York, 1968.
2. Hoch, R. C. Impregnating Paper for Use in Pressure Sensitive Adhesive Tape. In *Proc. Industry Technical Conference,* pp. 24–29. Pressure-Sensitive Tape Council, Itasca, April 5–7, 1983.
3. Dunlap, I. R. *TAPPI* **40**(8) (August 1957).
4. Latex Manual HL-2. B. F. Goodrich Chemical Co.
5. Finnegan, L. P., and Miller, V. A. *TAPPI* **41**(12) (December 1958).
6. Hechtman, J. F., and Bonnslaeger, S. R. U.S. Patent 3,464,848.
7. Page, D. H. *TAPPI* **62**(9) (September 1979).
8. Drew, R. G. U.S. Patent 2,236,527 (1941) (assigned to Minnesota Mining & Manufacturing Co.).
9. Kellgren, W. U.S. Patent 2,410,078 (1946) (assigned to Minnesota Mining & Manufacturing Co.).
10. Tierney, H. J. U.S. Patent 2,438,195 (1948) (assigned to Minnesota Mining & Manufacturing Co.).
11. Thommen, E. K., and Stannett, V. *TAPPI* **41**(11) (November 1958).
12. Heyse, W. T., Sarkanen, K., and Stannett, V. *J. Appl. Polym. Sci.* **3**(9) (1959).
13. Engel, E. W., Grolito, C. J., McGarry, J. A., Morris, V., and Werdner, C. L. U.S. Patent 2,592,550 (1952).
14. Eger, L. W., and Engel, E. W. U.S. Patent 2,726,967 (1955).
15. Bartell, C., and Weschler, J. R. U.S. Patent 2,848,105 (1958).
16. Bartell, C. U.S. Patent 2,848,355 (1958).
17. Bartell, C., Weidner, C., and Smith, R. M. U.S. Patent 2,937,109 (1960).
18. Weschler, J. R., and Bartell, C. U.S. Patent 2,963,386 (1960).
19. Bartell, C. U.S. Patent 3,197,330 (1965).
20. Finnegan, L. P., and Miller, V. A. U.S. Patent 3,156,581 (1964).
21. Hechtman, J. F., and Greenman, E. G. U.S. Patent 3,026,217 (1962).
22. Collins, W. C. U.S. Patent 3,012,913 (1961).
23. Bailin, M. M. U.S. Patent 3,127,284 (1964).
24. Weschler, J. R. U.S. Patent 3,068,121 (1962).
25. Hechtman, J. F., and Greenman, E. G. U.S. Patent 3,066,043 (1962).
26. Engel, E. W., and Dunlap, I. R. U.S. Patent 2,899,353 (1959).
27. Kellgren, W., Birchwood, and Marschall. U.S. Patent 2,633,430 (1953).
28. Gustafson, K. H., Excelsior, and Picard, L. E. U.S. Patent 3,503,495 (1968).
29. Cluett, S. L. U.S. Patent 2,624,245 (1953).
30. Dunlap, I. R. U.S. Patent 3,055,496 (1962).
31. Sonnichsen, H. M., and Bartell, C. U.S. Patent 3,245,863 (1964).
32. Hechtman, J. F., and Greenman, E. G. U.S. Patent 3,066,043 (1962).

be lightly release coated, depending on the adhesive type. Such films cannot be used with very soft adhesives, which conform to the liner surface irregularities, thus increasing the adhesion, since such adhesives may split during liner removal.

Adhesives

The pressure sensitive adhesives used are based on natural rubber, synthetic rubber, a mixture of both, acrylic solution or emulsion polymers, hot melts, and electron-beam curable adhesives. The future for the last two adhesive types for double-face tapes is uncertain. Hot melt adhesive application has the advantage of lower equipment costs, and neither hot melt nor radiation curable adhesives emit solvent vapor. Their performance, however, has not yet reached that of acrylic adhesives.

The optimum adhesive coating weight is important for the manufacture of adhesive tapes. Even minor irregularities in the adhesive thickness can strongly affect the adhesion values. Coating and drying conditions, including the air humidity, must be constant. The equilibrium humidity concentration in the dried adhesive film must be watched: too high a moisture content in the adhesive may affect its properties because of plasticization.

Carrier

The carrier is the material by which both sides are adhesive coated. It may be paper, fabric, or foam, as well as various films: cellophane, cellulose acetate, polyvinylidene chloride, plasticized and rigid polyvinyl chloride, and blends with other polymers, such as nitrile rubber, polypropylene homo- and copolymer, polyester, polyamide, polyurethane, polyethylene, and polytetrafluoroethylene. The choice of carrier material depends on application. For transparent films polyester, polypropylene, or polyvinylchloride are most often used. Polyolefin, polyurethane, or fluorocarbon polymer films may be used for their solvent resistance.

For adhesion to rough surfaces, various foams, such as polyethylene, polyurethane, polyvinylchloride, or polychloroprene rubber, may be used. Radiation or chemically cross-linked closed-cell polyethylene or polyethylene copolymer foams have occupied the dominant position as a carrier for double-face foam tapes.

The carrier material must be suitable for tension- and wrinkle-free processing on coating equipment.

9. Slitting of jumbo rolls
10. Packaging

Compounding

The adhesive mass is prepared according to standard mixing and dissolving processes. Natural rubber, for example, is supplied in bales, cut into small pieces, and then degraded in a mixer in order to decrease its molecular weight and to make it more easily soluble. The rubber is dissolved in an appropriate solvent in a masticator or intensive mixer, and then blended with other components. It is filtered and stored in special storage tanks or transportable containers. Synthetic rubber and acrylic polymer solutions, as well as emulsion adhesives, are also mixed with various additives in similar mixers, filtered, and stored.

Intermediate storage in big reservoirs is necessary, in order to allow the mass to age (defoam) and to reach homogeneity in a production charge as large as possible. In most cases, several mixing batches are combined for better control of the uniformity of a production charge.

The use of in-house polymerized acrylics is increasing. This allows the tailoring of adhesive properties by varying the polymerization conditions.

Coating and Drying

While the coating process might appear simple at first glance, complex and expensive equipment is required in order to produce double-face tapes economically. The times are over when 1200 meters per day of finished product were produced on a simple coating unit, requiring three workers, and the product was wound in 100 meter rolls.

Knife and reverse roll coaters are most often used for adhesive solution coating; reverse roll, rod, and air-knife coaters are preferred for emulsions. The transfer coating process is often used for double-face tapes.

Care must be taken that the adhesive film is bubble-free after drying. This requires gentle drying at the early drying stages in order to delay the formation of a film over the adhesive surface. Most of the solvent (about 80%) goes off in the first third of the dryer, and the rest is evaporated at slightly harder conditions in the remaining two-thirds of the drying oven. The drying process is controlled by employing appropriate solvents, air flow and temperature, and the processing speed. Drying speeds can vary from 3 to 60 m/min.

The coating process is completed by laminating the carrier material to the adhesive coated release liner (when the transfer process is used), or by laminating

norms in Europe; the usual roll length is 25, 50, 100, and 200 m. The maximum reasonable roll diameter is about 400 mm.

If a longer length is needed, as for continuous processes, the tape may be wound in spools. Thin tapes up to 19 mm in width with film or paper carriers can be spooled with overlapping paper (fingerlift); thicker foam tapes can be spooled only up to 12 mm in width, because these tapes cannot be wound with overlapping release liner. It is important for proper spool winding that the tape does not stretch, that is, that sufficient winding tension is developed.

APPLICATION CONDITIONS

Application Pressure

The most important and most often neglected factor in applying double-face tape is the application pressure. Only if a sufficiently high application pressure is employed, can a good bond be formed (Fig. 29-1). Application of pressure by a 10-kg roll results in a larger adhesive contact surface than by a 1-kg roll. Adhesive compliance, carrier type, and other properties affect the dependency of bond development upon the application pressure.

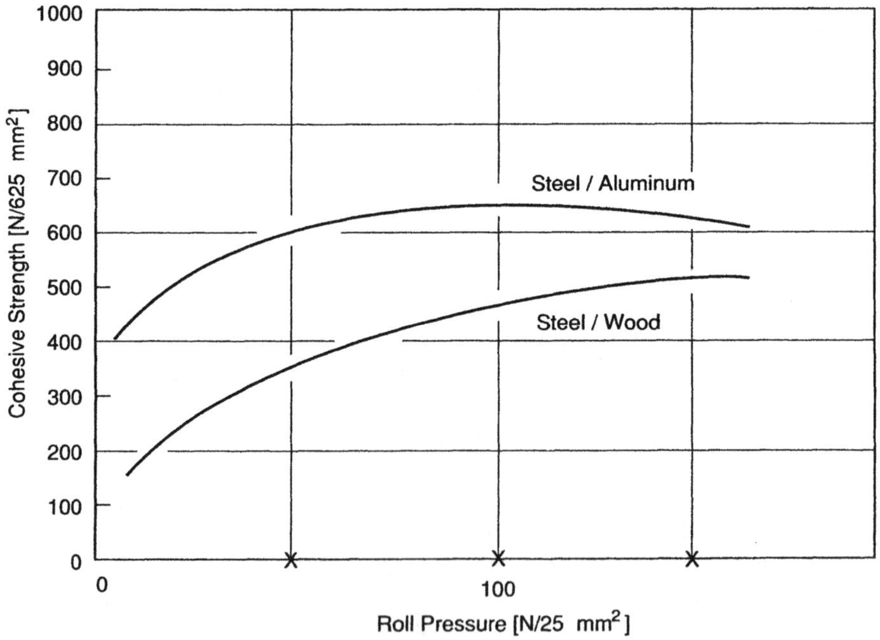

Fig. 29-1. Cohesive strength as a function of the roll pressure applied during mounting.

Bonding that is carried out below 18°C usually results in a lower initial adhesion. This is more noticeable in the case of highly cohesive adhesives. In the case of pure acrylics, this adhesion reduction is more noticeable than in the case of soft rubber formulations. In order to compensate for the low initial adhesion of pure acrylic adhesives, especially for applications in the automotive industry, such as fender protective strips, the adhesive tape and the adherend surface, in this case the fender surface, are brought to a temperature of about 40°C before bonding. Products have been developed (Duplomont 910/911 and the product family Duplomont 930/931) that do not require heating, as experience in the European automotive industry indicates, and bonding at room temperature results in a sufficient adhesion (Fig. 29-3). If a part of the laminate requires preheating, because of inadequate adhesive weight/area ratio or other reasons, it should be ascertained that the adherend surface and the tape are at the same temperature in the bonding zone. A temperature differential will unavoidably cause a retention of stresses after bonding, which will reduce the bond strength. An often used practice, heating up the adhesive edges, improves the bond strength through stress relaxation.

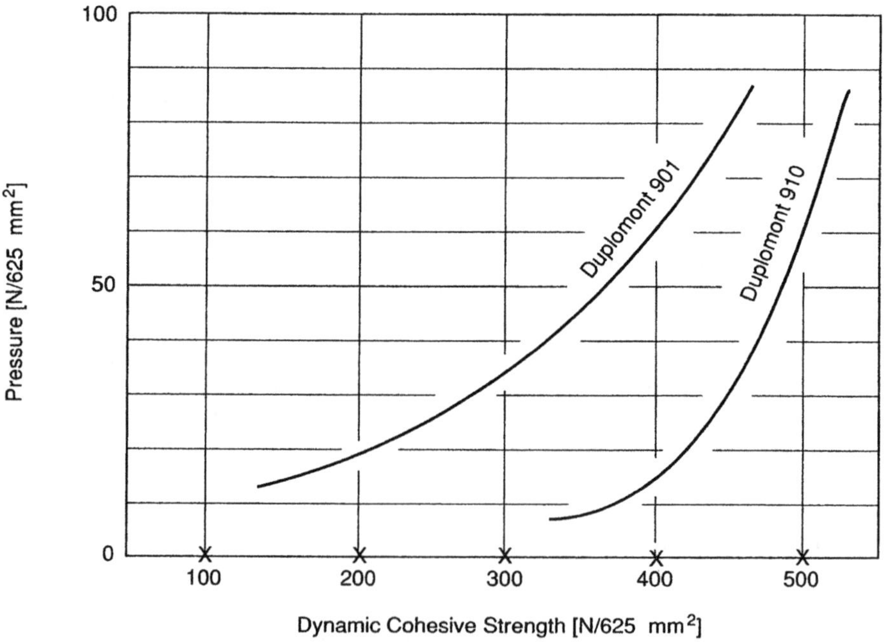

Fig. 29-3. Cohesive strength of Duplomont 910 compared to another product.

range, exhibits an aggressive tack, and is suitable for difficult-to-adhere-to surfaces. It is suitable for universal application and exhibits a higher immediate adhesion.

Double-face tapes are put up in a minimum width of 6 mm and a maximum width of 1000 mm; tape length ranges from 300 to 1000 m. The standard core size diameter is 76 mm. Special tapes, normally and reverse wound, may also be wound on 1-in.-diameter dispenser cores. Transfer tapes are available in many different put-ups: as jumbo rolls on 76-mm- or 1-in.-diameter cores, normally and reverse wound. These tapes are important products used in various industries, for example, the foam industry, the automotive industry, the advertising industry, paper and paper converting industry, and the label industry.

When bonding to curved surfaces, transfer tape assures that the laminated substrate conforms to surface irregularities. A permanent bond to deformable, elastic materials, such as rubber, foam, and foils, is assured.

A maximum-use temperature is often determined by the weakest element—the adhesive carrier, which is eliminated in transfer tapes. Exposure to a maximum temperature of 150°C is possible with transfer tapes. Transfer tapes allow the placement of a thin (< 50 μm) adhesive layer of constant thickness on flat surfaces. The adhesive also provides a layer capable of dimensional adjustment and stress relaxation, in the case of laminated substrates of different temperature expansion coefficients.

Application of the transfer film is possible in a circular or wavelike pattern, without formation of creases.

Acrylic transfer films are resistant to weathering, solvents, and ultraviolet (UV) radiation. This makes such tapes useful in many applications. For example, transfer tapes are used as splicing tapes to join paper and film, for making labels, and paper, foil, plastic, and metal nameplates. They are used for mounting trim, emblems, fascias; for laminating glass, wood and foams; for making moldable plastic foils, photos, display cards, and self-adhesive office forms.

Automotive Industry

The automotive industry offers a sizable market for double-face tapes. This market is growing through the innovative and increased use of these tapes.

The securing of decorative and protective fender moldings is the best known application of double-face tapes in the automotive industry (Figs. 29-4 and 29-5). Acrylic and other adhesive tapes may be used for these products. The carrier may be polyethylene vinyl acetate copolymer closed-cell foam, based on either LDPE (temperature range of −40° to 95°C), or LLDPE (temperature

range −40° to 120°C). The most common tape thickness is either 0.9 or 1.2 mm.

Such tapes may be used to secure extruded PVC protective strips with or without a plasticizer barrier foil. In the case of PVC strips without barrier foil, an adhesive of sufficient resistance to plasticizers (up to 30% active plasticizer) is required. Injection-molded polyurethane strips are also used for these products. Cleaning and priming of the back surface is required, and special primers are available. Polypropylene and EPDM strips must be either primed or corona discharge treated for sufficient adhesion. In-line methods and equipment for surface preparation are available.

Polyethylene/vinyl acetate (PE/VA) copolymer foam systems have some advantages over neoprene, polyurethane, and pure polyethylene foam carriers. PE/VA foam is more elastic and compressible, and therefore assures a larger contact surface immediately after bonding. The contact area approaches 100%, barring construction errors. These foams have a lower thickness tolerance (± 0.05 mm) as compared to the thickness tolerance for neoprene foam (±0.15–0.2 mm). PE/VA foams have a smoother surface, and therefore double-face tapes exhibit a higher tack. The adhesive surface of these tapes is glossy as compared to a matte surface for neoprene foam tapes. Because of these deficiencies and the low level of initial adhesion of hard foam carrier tapes, double-face tape and the adhered surface must be heated to a temperature of 40°C before assembly. This is not required with PE/EVA foam tapes, and the application of such double-face tapes is easier and less expensive.

Polyurethane foam carriers have the same deficiencies as neoprene foam. In addition, the foam is not very stable in hydrolysis, and its structure consists of only partially closed cells, resulting in higher water absorption.

Polyethylene foam has a poorer temperature resistance than PE/VA copolymer foam. It has a lower density and a lower initial tear strength, as well as a lower force required to propagate the tear.

Double-face tapes are used by every automobile manufacturer to attach identification letters, numbers, emblems, and labels (Figs. 29-6 and 29-7). These are the same tapes as are used for the attachment of protective strips. Tapes are supplied in a roll or as die-cut pieces that are protected on both sides. The material used for emblems is usually injection-molded ABS, or in the Scandinavian market, stainless steel or aluminum.

Double-face tapes are used for mounting spoilers, door sills, and hubcaps. A combination of curable elastomeric adhesives and double-face tapes is preferred for the installation of injection-molded polyurethane door sills and hubcaps. The tape is employed for temporary holding until the elastomeric adhesive is cured. Cleaning and priming of the bonding surfaces is required. Spoilers made from an integral polyurethane–foam composite are attached mechanically. The adhesive foam tape is used as a vibration dampener and as a barrier against

moisture penetration. It is customary to apply tapes with an adhesive thickness of 100–250 μm.

Self-adhesive extruded rubber profiles usually are made from EPDM elastomer, which is a nonpolar difficult-to-bond material. In-line laminating methods have been developed for such profiles, which produce a high bond strength with polar acrylic adhesives. In a specially constructed laminating facility, it is possible to prepare the surface of the extruded profile by activation through heat, primer application, or corona discharge treatment. Such treatments can be applied separately or in combination.

Such rubber profiles have the following advantages:

1. There is no damage to the rubber during installation.
2. There is no adhesive squeeze-out, causing a visual distraction.
3. Additional C_w value (a measure for the air resistance of the car body) reduction is possible without any prior arrangements.
4. Elimination of profile reinforcement.
5. Seal to moisture penetration underneath the profile.
6. Simplified installation through preassembled parts.
7. Delivery process suitable for robot handling. Stiffening foil based on curable compounds is used to dampen vibration and to stiffen the metal skin. Such foil is applied with a pressure sensitive adhesive before painting the metal skin, and it is cured at a temperature as high as 150°C. The adhesive of the double-face tape cures together with the foil during the heating cycle, leaving no soluble compounds.

Foam, felt, or foil double-face tapes are widely used for the installation of self-adhering vibration dampening sheeting, especially for the bonding of polyester or polyether polyurethane acoustic foams on the inside of the motor hood. An acrylic transfer tape about 50 μm thick is used.

An acrylic adhesive transfer tape is used for ABS thermoformable sheeting. These products may be subjected to vacuum forming after the adhesive tape is laminated. Heavy transfer tapes (250 μm thick) protected by a polyethylene film release liner may be stretched as much as 500%. The process of providing this self-adhesive feature begins with lamination of the tape to the exterior-type ABS sheet. The sheets may be smooth or grained. After freeing the laminated transfer tape of creases and air bubbles, the sheeting is plasticized by subjecting it to temperatures as high as 200–210°C. By applying a pressure of 1 bar from the adhesive side, the laminate is drawn over a vacuum form. Such thermoformed self-adhesive ABS covers are used for the steering column and for covering the instrument panel.

Self-adhesive wheel axle covers and license plate frames made from metal or various plastics are secured by double-face tapes.

30. Automotive Applications

Whiteford D. Grimes
Avery Graphic Systems Division—Automotive
Farmington Hills, Michigan

Pressure sensitive usage in the automotive industry surely began with masking tape in its traditional role as an aid to manufacturing, protecting trim areas during painting. This logical use, which made it easier to offer assembled units of painted metal, glass, and bright trim, is typical of the thousands of applications for adhesive coated products that save time and labor, while providing improved utility to a product. From this basic use, pressure sensitive materials have grown with the industry to become essential engineering materials in a wide range of applications.

In general, uses of pressure sensitive adhesives in the automotive industry encompass every known characteristic of such materials. The reasons for their use, likewise, run the gamut of justifications described in industry trade literature and sales promotions. In short, pressure sensitive adhesives are used for temporary benefits and to add permanence; they are used to save money, to increase functionality, and to achieve a higher profit per unit of sale.

Other chapters of this book deal with the mechanics of adhesion and adhesive chemistry. In this section, the focus will be on applications, materials, performance, and benefits.

The function of pressure sensitive adhesives in all automotive applications is to fasten one material to another. In the paint masking example, the adhesive fastens a protective paper to trim and glass areas, but retains its characteristic of easy removal so that the paper can be discarded when the paint has dried. In a more sophisticated use, such as the application of body side moldings, an adhesive is used to attach a shaped vinyl extrusion to a plastic foam, which is then affixed permanently to the painted side of the vehicle with yet another layer of adhesive. These two examples illustrate the complexity of pressure sensitive uses for the automobile; they bring to mind questions of adhesion, durability, chemistry, mechanics, and even utilization of work force.

the car may not be affected. The data illustrated were determined under laboratory conditions. That is, room temperature was controlled at 21°C; paints were formulated under close supervision; test panels were properly prepared before painting, including priming, curing, light sanding, and cleaning; application was made by controlled standard practices. Actual application on a vehicle assembly line would not be so controlled. Many normal variances of a manufacturing operation would exaggerate the differences of the data. For example, a typical application in a cold weather climate, where cars are stored outside immediately after finishing, could result in adhesion dramatically different from that pictured, even to the point where a bond may not exist.

BRIGHT METALS

Many surfaces, both interior and exterior, on a car are either aluminum, chromium, or stainless steel. Pressure sensitive application to each again varies with the type of surface and the consistency within each type. Metal treatment is an inexact science, and where such operations as anodizing and plating are involved, wide differences in surface character are common between production lots. These differences result in differences in adhesion levels and aging characteristics. Even laboratory applications on anodized and plated surfaces often vary widely. Generally, these differences can be traced to the quality of the metal surface, but the frequency of "inexplicable" data on such surfaces raises doubts as to their suitability for laboratory evaluation.

Stainless steel, on the other hand, is a standard surface for adhesive characterization. This material is readily available, easily cleaned, and makes no apparent chemical contribution of its own to the total adhesion structure in a specific adhesive evaluation. In fact, the accepted method of characterizing a pressure sensitive adhesive (not a particular application but only the adhesive itself), is to coat the adhesive on a "nonreactive material" such as a polyester film and then to apply it to a stainless-steel test specimen. Thus, no chemical interaction between film and adhesive, or stainless steel and adhesive will be a factor in the characterization, and a true reading of adhesion, heat aging, humidity resistance, etc., will be obtained.

PLASTICS

Plastic materials are increasing rapidly in use as structural and surface materials for automotive parts. ABS (acrylonitrile-butadiene-styrene), vinyl (polyvinyl chloride), glass-reinforced polyesters, polycarbonates, and polyolefins are a few of the plastic resin materials in wide use on today's automobiles. There are others and, surely, more to come. These materials are growing in popularity because of their ease of fabrication, durability, low cost, low density (potential for vehicle weight reduction), chemical resistance, and strength characteristics. Each plastic material has its own identity in relation to pressure sensitive adhesives. Some have a strong affinity for adhesive bonding, while others totally reject conventional adhesive mechanisms. Fortunately, most plastic structural parts are painted and, although their chemistry may have made it difficult for the painting process, the paint itself may be a more consistent surface for pressure sensitive adhesion. For direct application to plastics, specific materials must be matched for optimum performance and, in some cases, fabrication with adhesives may either be impractical or require materials and techniques not easily obtained.

Considering all surfaces, it is important to understand that the character of the material to which an adhesive product is to be applied may have an overriding impact on the suitability of that product to the application. Performance on one surface by no means implies equal performance on another. Additionally, one should keep in mind that any automotive surface must be properly prepared for adhesive application, for example, cleaned, lightly sanded, and conditioned to a temperature within the application range for the particular adhesive being used.

APPLICATIONS

In the following pages, specific pressure sensitive applications for the automotive market are reviewed. This is not an all-inclusive list, but only illustrations of uses, provided for reader interest in hopes of furthering understanding of pressure sensitive adhesives and the versatility of such materials in fulfilling market needs. Commodity uses, such as masking for painting and tape bindings for electrical harnesses, are not discussed here.

The most visible pressure sensitive applications on automobiles are those that fulfill aesthetic objects, such as striping, decals, exterior tops, and woodgrains. Less recognized applications hold down wiring, seal joints, and seams from moisture, provide instructions, affix trim, eliminate rattles, and compensate for loose tolerances. The most complex of these items is probably an exterior woodgrain film because of its stringent requirements for both design and performance. An in-depth explanation of this product is presented in the following section to facilitate understanding of all pressure sensitive materials and their use on automotive products.

EXTERIOR WOODGRAIN FILMS

The use of wood as a decorative trim on automobiles is almost as old as the industry itself. Early in their history of infatuation with the automobile, Americans realized the practicality of combining a utility vehicle with an active family life-style and came up with the station wagon. The wooden body was a natural reminder of country living and became the standard for luxury and social recreation. For more than 40 years, real wooden bodies were the top of the line, but eventually the cost of fabrication and problems with durability sounded the death knell of this classic reminder of another day.

Today "wood body" station wagons are simulated by applications of woodgrain decals. Through gravure engraving and subsurface printing, beautiful woodgrains can be reproduced in exact detail, providing durable, realistic woodgrains at minimal cost.

There are two types of films used to produce exterior-quality woodgrain decals, cast vinyl and calendered vinyl. The film type is derived from the process used for production. Casting is the process of forming a film from a liquid compound by spreading it to a desired thickness over a support web or carrier that has suitable release characterisitcs. An external energy source is used to fuse, cure, or dry the film while still on its carrier. Calendering is the process of forming a film from a melted compound by passing it successively through narrow passages formed between a series of hot milling rolls until it reaches a desired thickness and width, whereupon it is cooled sufficiently to set film properties.

Cast vinyl films are characterized by exceptional dimensional stability that results from the carrier nature of the process. Since the film is supported through all of its processing, there are no residual stresses to cause shrinkage in the web. This characteristic is quite important to woodgrain decals because they are supplied to the automotive assembly line in precut shapes that must fit within carefully defined tolerances. Cast vinyl films are produced only from dispersion vinyl resins whose high initial costs are compounded by excessive materials-handling requirements and thus entail a substantial cost penalty.

Calendered films, on the other hand, are produced from bulk-handling suspension resins, the commodity class of PVC resins in the United States. Fortunately, calendered films can be made dimensionally stable by additional processing and are quite suitable for woodgrain products. A further advantage of the calendered product is the versatility that evolves when a product is a commodity. This versatility in use has resulted in a mature compounding technology, not available to cast products. Proven compound additives and select suspension resins provide the wherewithal to achieve exceptional resistance to the degrading effects of ultraviolet light and heat.

With either film, the ultimate performance against the effects of sunlight

of varying depth, several thousand per square inch, that serve as inkwells in the printing process. The engraved cylinder is plated with chromium to provide a durable wear surface and thus consistent design reproduction.

In printing, the design is applied to the clear vinyl film so that the woodlike effect is visible only through the film. In this way, the protective properties of the vinyl formulation are provided for the aesthetic effects of the inks and design. The print may be made direct to the film in a reverse order, that is, grain, tone, and ground coat, or in the more natural order, like ground, tone, and grain, on a release carrier for subsequent transfer to the film under heat and pressure.

In a manner similar to the transfer of printing, the woodgrain film is made pressure sensitive by first coating an adhesive on a release paper and laminating this combination to the ink side of the printed film. The finished laminate (film, print, adhesive, and release carrier) has only to be die-cut to specification dimensions to become a decal, ready for application on a vehicle assembly line.

Adhesive

The adhesives used for exterior decals are selected to provide a balance of properties suitable for the film/ink combination, application techniques, and performance as an exterior automotive component. The use establishes the controlling criteria. Specifications issued by the original equipment manufacturer's engineering departments spell out requirements for adhesion levels and time of achievement, resistance to environmental conditions as represented by laboratory testing, actual exposure to hostile climates, and application techniques. Although the adhesive is a critical part of such evaluations, it is the total package of film, ink, and adhesive, and the interaction of all three with the vehicle surface, that is under test. Experience has shown that only highly cohesive acrylic adhesives thus far have been able to satisfy the demanding range of properties necessary for automotive use.

Performance Criteria

Each of the automobile manufacturers has its own requirements for qualifying a woodgrain product for use on its vehicles. Although these specifications vary somewhat in detail, the principal concern of all is the same: application on the assembly line and durability in use.

Adhesion: Two lb/in. of width, 1 hr after application and 3½ lb or more after 72 hr.

Environmental Cycle: Exposure to alternate conditions of heat and humidity over a two-week period. A typical cycle is 17 hr at −30°C, 72 hr. at 80°C, and 24 hr. at 38°C and 100% RH, then repeat for the second week.

Gravelometer: Subfreezing exposure and impingement of a specific gravel mixture at road velocities.

Weather Resistance: Exposure in standard Weather-Ometer cabinets (controlled temperature and humidity cycles) with prescribed ultraviolet arc lamps. 1000–2000 hrs. of exposure are generally required.

Hostile Environment: Exposure at independent test sites in Florida and Arizona for two years at a prescribed angle to the sun.

Other typical controlled tests include heat aging, abrasion, humidity, exposure, gasoline resistance, and cleanability.

In most of the tests, the principal evaluation is that of appearance. An acceptable product is not expected to exhibit significant color change, crazing, blistering, or other indications of film degradation.

STRIPING FILMS

The use of exterior striping products has developed rapidly in recent years. Several major developments can claim responsibility for this popular practice of decorating automobiles and other vehicles with striping and decals of every description: (1) the trend to sport models that developed in the early 1970s reflecting a youth-oriented society; (2) the high cost of retooling automotive assembly lines that has extended the life of body styles to five years or more and necessitated other means of providing product differentiation; (3) the energy situation that is forcing weight reduction and resultant similarity in all car models as manufacturers struggle to meet government-imposed limitations on miles per gallon of gasoline; and (4) the development of durable vinyl films, and high-performance adhesives that make vehicle striping easy, distinctive, and lasting.

Whereas the history of striping differs greatly from that of woodgrains, the technologies of the two are quite similar. Cast films are predominantly used; printing is an important part of the striping process; and design is the key to achieving desired appearance results. Solid-color stripes may be made by film pigmentation, surface printing by screen process techniques, and subsurface printing by the gravure process. Multi-colored products, such as the large hood decals of birds and insects, are generally printed by screen processing

such as the polyolefins, or where destructibility of the decal is required (as with an EPA conformance label) to demonstrate that a label was provided, rubber-based adhesives may also be used.

Automotive manufacturers have complete specifications for every label and decal used on their vehicles. Each specification also provides test method information and application criteria.

EXTERIOR AUTOMOTIVE TOPS

With the growing popularity of air-conditioning for cars that developed in the early 1960s came the near obsolescence of the once-popular convertible. Rather than let the appeal of a convertible become a thing of the past, the automotive designers came up with a new body style, popularly known as the "hardtop convertible." This new design was merely a two-door coupe without door posts, and a roof line designed to preserve the convertible look. It was not too long before the vinyl industry capitalized on the popularity of this new body style by developing an exterior coated fabric for application to the roof that made the hardtop convertible all the more realistic. By the mid-1970s, more than 50% of all American-made automobiles came off the assembly line with vinyl tops.

Application of vinyl tops was achieved by the spraying of contact cements on both the painted roof of the car and the prefabricated vinyl coated fabric. This was a complicated means of application that promised uncertain results at best. In effect, the quality of the vinyl top became the quality of the spray application, with a poor spray resulting in delamination, rusting, and a short life. In 1977, Stauffer Chemical Co. conceived the idea of an exterior top fabric with a factory-applied pressure sensitive adhesive. The result was an exterior vinyl top that provided improved adhesion, added corrosion resistance, and application security, while eliminating spraying operations and attendant environmental health concerns. This application of pressure sensitive adhesives is a classic example of the way these products can save labor, improve product performance, and solve problems of manufacturing.

ADHESIVE SYSTEM TAPES

The advent of adhesive systems in the form of double coated and transfer tapes has been of considerable value to the automotive industry. With a variety of adhesives and product structures, these materials successfully save labor, reduce weight, solve manufacturing problems, and improve product performance. The body side molding tape mentioned in the beginning of this chapter is a

31. Plastic Tape Pipeline Coatings

G. M. Harris
The Tapecoat Company
Evanston, Illinois

HISTORY

Plastic tapes were introduced to the pipeline industry as pipeline coatings more than 30 years ago. Since then, the worldwide consumption of plastic tape pipeline coating has grown to the point that plastic tapes are used yearly on more than 50% of all new underground pipeline projects. The rate of growth in the use of plastic tape coatings was slow but steady through the first 10 to 15 years, but more recently the rate of growth of plastic tapes as pipeline coating has accelerated rapidly.

In 1976, the annual use of plastic tape pipeline coatings reached 100 million m^2. The usage continued to accelerate for the next decade. This was brought on by the sudden drop in crude oil prices that occurred in the early 1980s. Since then the annual use of plastic tapes as pipeline coatings has remained constant, largely due to the increased use of multiple-layer tape specifications and the increase in very large diameter steel water pipelines coated with plastic tape systems.

While plastic tapes continue as the major coating for projects where the coating is applied over-the-ditch, the amount of plastic tape pipeline coatings applied to pipe at coating plants has increased.

BASIC COMPOSITION

Although there is a wide range of plastic pipeline coating tapes commercially available made by several different manufacturers, all of the products have several common properties:

1. Th struction i
2. Th a backing
an based on
3. Th the use
qu
4. Th pplied in
ho

The th polymers u
nyl chlori ene copoly
far the mo used polym
The e ed as the
rubber, eth ne rubber,
elastomers ded with
pene, co alt in ord
some case the adhes
must be easy-rele
unwound In other
The i be a pla
nterliner s to manu
ers are m . There
where ta ners wer

acceptance of plastic tape coatings was on small sections of pipe, such as welded field joints, where the tape was applied by hand. It could be said that the early success of joint wrap plastic tapes was the forerunner to the extensive use and broad acceptance of plastic tapes as pipeline coatings that exist today.

LONG LINE TAPE COATINGS

Plastic tapes used as the only coating on pipeline projects, not just as joint wraps, are supplied in long lengths (up to 800 ft) and wide widths (up to 18 in.). They are applied by power-driven equipment both over-the-ditch and at coating plants. They can be used on any size pipe. Many thousands of miles of 1400-mm pipe have been coated with plastic tapes outside of the United States. In 1980, a 2250-mm water pipe project in Colorado was coated with a plastic tape plant coating system.

Long line plastic tape coatings are tougher and have much greater tensile strength than joint wrap tapes and can only be applied smoothly, tightly, and with high conformability by machines designed for tape application.

Tapes used as the corrosion protective coatings on pipelines usually have backings and adhesives of close to the same thickness—but most often the backing is thicker. Typically, the backing will make up 55% to 60% of the total tape thickness. Until recently, a tape with 0.50-mm total thickness was the most commonly used, but now the tendency is toward use of a tape with 0.625-mm overall thickness, or greater in some special cases.

Plastic tapes for corrosion protection have always been used with an outerwrap to afford mechanical protection to the plastic tape. Outerwrap materials such as kraft paper, saturated asbestos felts, and plastic films have been widely used for this purpose in the past, but now almost 100% of the time, the outerwrap is a second tape consisting of a plastic backing and a suitable adhesive. These double-layer plastic tape systems have all the properties required to meet the mechanical and corrosion mitigation needs of a pipeline coating. Plastic tape pipeline coatings are most often applied over-the-ditch, but the consumption of plastic tapes as plant coatings has increased significantly since 1978.

The application of plastic tapes over-the-ditch is well established in terms of specifications, procedures, and equipment. The first over-the-ditch experience using power-driven equipment with plastic tape coatings occurred in 1953.

A typical over-the-ditch tape coating specification includes the following elements:

- Cleaning by use of wire brushes
- Priming
- Application of the corrosion protective wrap (inner wrap)

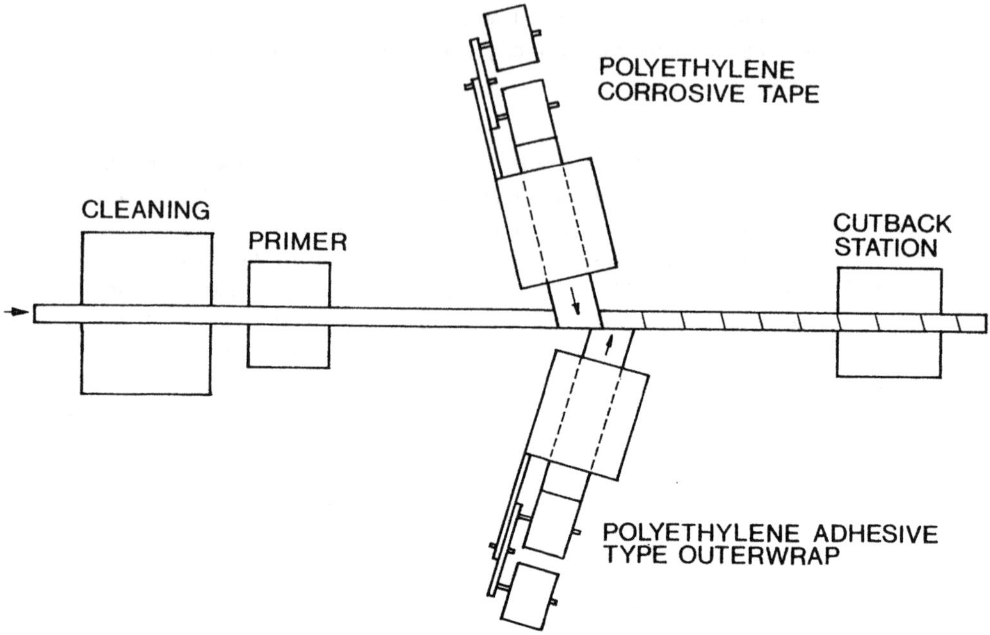

Fig. 31-1. A schematic diagram of multilayer polyethylene tape plant coating process.

each separate joint of pipe as the coated joints of pipe leave the coating machine and proceed to the outgoing racks.

POLYETHYLENE TAPE PLANT COATING SYSTEMS

The multilayer polyethylene processes and systems offer important advantages in the area of product performance. These product advantages are:

1. The corrosion protection layer is completely protected during outdoor storage. The outerwrap used as part of the system completely covers and shields the corrosion protective layer from ultraviolet rays, cold, heat, and moisture. With all other types of plant coatings, including epoxy and extruded or sintered polyethylene, the corrosion protective layer during outdoor storage is exposed directly to deterioration from ultraviolet rays, cold, heat, and moisture. Any deterioration of the corrosion protective layer will reduce the level of corrosion protection after the coated pipe is installed underground.
2. It remains flexible in cold weather. The polyethylene-based tape and

outerwrap remain ductile and flexible down to −40°C. Other plant coatings become brittle at temperatures considerably above −40°C.

3. It resists bending damage because of good ductility and flexibility throughout a wide temperature range.
4. It can be concrete coated. Pipe coated with the multilayer polyethylene tape coating systems can be coated with concrete by all of the commonly used concrete coating techniques.
5. Coating is tough and resists mechanical damage during handling and transportation. The polyethylene outerwraps absorb the mechanical forces that cause damage and prevent damage to the primary corrosion protection layers. With all other plant coating systems, all of the mechanical forces that cause coating damage act directly on the primary corrosion protective coating, resulting in damage to the primary corrosion coating.
6. Field joints can be coated and repairs can be made easily. Polyethylene tape is used to make field joints and repairs when required.

SPECIFICATION

The specification for the application of long line plastic tape and joint wrap tape are identical in concept and principle but vary in that the long line tapes involve power-driven mechanical means (Fig. 31-2), while joint wrap application involves mostly manual steps (Fig. 31-3). A typical application specification includes the following:

1. Removal of moisture and foreign material
2. Removal of corrosion products and mil scale
3. Application of primer
4. Application of tape
5. Application of outerwrap

With joint wrap tapes, the outerwrap may be omitted, but it is usually substituted for by two layers of the joint wrap tape.

The specifications of the various plastic tapes vary, as would be expected from manufacturers, but they all have a strong similarity. The American Water Works Association (AWWA) Standard C209 includes a specification for joint wrap tapes.[3] Table 31-1 will serve as a good example of a specification for joint wrap tapes.

The AWWA Standard C214-83, "Tape Coatings for the Exterior of Steel Water Pipelines,"[4] was published in 1983. This standard for plant-applied tape systems accelerated the already established use of multiple layer polyethylene tape systems for very large-diameter steel water pipeline projects (Fig. 31-4).

Fig. 31-2. Application of polyethylene plastic tape to large-diameter pipe by line travel equipment under winter conditions in Canada. (*Courtesy of The Kendall Co.*)

Fig. 31-3. Application of hand wrap plastic tape coating to 16-in. gas pipeline. (*Courtesy of The Tapecoat Co.*)

Table 31-1. Joint Wrap Specifications

PROPERTY	REQUIREMENT	TEST METHOD
Thickness	20 mils ± 5%, min. Type I and II 10 mils ± 5%, min. Type III	ASTM D1000
Water vapor transmission (perms)	0.25, max.	ASTM E96, method BW
Dielectric breakdown (V/mil)	400	ASTM D1000 ASTM D149
Insulation resistance (megohms)	500,000, min.	ASTM D1000 ASTM D257
Adhesion (oz/in. width, with primer)	20, min.	ASTM D1000, method A
Cathodic disbondment	2 in. diameter (average of 5 test samples)	ASTM G8 method A, 30 days

Fig. 31-4. This 168-in. (4267-mm) -diameter steel penstock for a hydroelectric project is coated with 3 layers of polyethylene tape, which meets the AWWA C214-83 Standard. (*Courtesy of The Tapecoat Co.*)

The physical property tables in the AWWA C214-83 Standard serve well as specifications for the inner-layer tape, the outer-layer tape, and the total minimum two-layer system.

Tables 31-2, 31-3, and 31-4 show the important properties and their standard values.

Table 31-2. Physical Properties of Inner-layer Tape for Long Line Pipe

PROPERTY	REQUIREMENT
Width deviation	± 5% of width or ¼ in. (6 mm), whichever is smaller
Thickness	20 mils (508 μm) ± 10%
Adhesion (with primer)	200 oz/in. width (2190 N/m width) minimum
Water absorption (24 hr.)	0.2% maximum
Water vapor transmission	0.2 perm (1.1×10^{-11} kg/(Pa·s·m^2) maximum
Dielectric strength (V)	6000 minimum/single thickness
Insulation resistance	500,000 μΩ minimum
Tensile strength	20 lb/in. width (3500 N/m width) minimum
Elongation (%)	100 minimum
Temperature resistance	80% of original

Table 31-3. Physical Properties of Outer-layer Tape for Long Line Pipe

PROPERTY	REQUIREMENT
Width Deviation	± 5% of width or ¼ in. (6 mm), whichever is smaller
Thickness	30 mils (760 μm) ± 10%
Adhesion to Inner-Layer Tape	20 oz/in. width (200 N/m width) minimum
Tensile Strength	40 lb/in. width (7000 N/m width) minimum
Elongation, %	100 minimum
Temperature Resistance	80% of original

Table 31-4. Physical Properties of Two-layer Tape System for Long Line Pipe

PROPERTY	REQUIREMENT
Thickness	50 mils (1270 μm) ± 10%
Dielectric strength (V)	12,000 minimum
Cathodic disbondment	2 in. (50 mm) diameter (average of 5 samples)
Impact resistance	25 lb/in. (2.8 N·m) minimum
Penetration resistance	20%—maximum penetration at 22°C (72°F) 30%—maximum penetration at 60°C (140°F)

IN-SERVICE PERFORMANCE

A short review of the in-service record that plastic tapes have compiled since they were introduced as a means of mitigating corrosion on pipelines will establish their credentials.

The performance of plastic tape coatings has been monitored continuously by the "Methods for Evaluating In-Service Field Performance" given in NACE Standard RP-01-69, Section 5, table 6.[5] Two of the methods given in this table are for measuring the leakage conductance of pipeline coatings and the rate of change in the current required for cathodic protection. These methods have been employed as recommended by making a comparison of initial measurements with subsequent periodic measurements.

The continuous record of in-service performance of a polyethylene tape pipeline coating over a 27-year period was reported in Boston at the 1986 Conference of The National Association of Corrosion Engineers.[6] The very good performance of the polyethylene tape used on over 5000 miles of large-diameter crude oil pipelines (20 in.–26 in. diameter), as reported in this technical paper, is summarized in the following list.

	Coating Conductance	Current Requirements
first 14 years =	4.6 micromhos/ft.^{2} (49 micromhos/m^2)	2.0 to 2.5 microamperes ft.^{2}
second 13 years =	6.7 micromhos/ft.^{2} (72 micromhos/m^2)	(24 microamperes m^2)

Table 31-5 compares the year-to-year leakage conductance of a polyethylene pipeline coating that was applied to 48 km of 150 mm pipe in 1957. The latest leakage conductance measurements were made by a corrosion consultant in July 1978.[7]

The leakage conductance of the polyethylene tape coating was determined to be 78 micromhos/ft^2 (845 micromhos/m^2). This is a low level of leakage conductance after 23 years and confirms that plastic tape coatings retain the integrity of their physical, chemical, dielectric, and moisture resistance properties.

It is generally accepted in the pipeline coating industry that plastic tape coatings exhibit a high level of corrosion protection in conjunction with cathodic protection.

The worldwide use of plastic tape coatings since the mid-1950s has proved that these coatings provide a high level of corrosion protection as measured against cathodic protection current requirements. In general, plastic tapes have shown better corrosion protection qualities than other types of coatings. Table

Table 31-5. Coating Conductance Performance of a Polyethylene Tape and Coal Tar Coatings After 16 and 12 Years Underground in Comparable Environments

	COATING CONDUCTANCE	
AGED UNDERGROUND (years)	COAL TAR ENAMEL (micromhos/m^2)	POLYETHYLENE TAPE (micromhos/m^2)
1	80.2	—
2	178.2	45.9
3	467.0	70
4	258.2	76.5
5	903.8	78.3
6	1248.2	70
7	1635.0	106.4
8	1108.3	98.3
9	1291.2	108.9
10	1527.9	188.1
11	2388.7	239.0
12	2238.1	212.4
16	—	520
19	—	700
23	—	845

31-6 illustrates this point by making the comparison of a tape coated section and a coal tar coated section of a 500-mm products pipeline built in 1971. In this case, the coal tar enamel coating requires greater than four times the amount of cathodic protective current to achieve acceptable protective pipe to soil potential levels than the plastic tape coating. Many other similar case histories of the performance of plastic tape coatings are cited in the literature.[8]

During 1981 and 1982 in the United States a two-layer polyethylene tape pipeline coating system was applied to a 42-in. natural gas pipeline of greater than 800 miles in length. The choice of plastic tapes as the coating for this large pipeline project testifies that plastic tape pipeline coatings fulfill the engineering criteria of good performance and good economics.

Table 31-6. Current Requirements on a Product Pipeline Coated with a Polyethylene Tape Coating System and Coal Tar Enamel (Construction: 1971)

Pipe diameter	500 mm (20 in.)
Total length of line	234 km
Length of polyethylene tape section	159 km
Length of coal tar enamel section	76 km
Current Requirements: 1974	
Polyethylene Tape	Coal Tar Enamel
31.2 μA/m^2	132.3 μA/m^2

REFERENCES

1. Smith, T. E. Arctic Construction Experience on the Trans-Alaska Pipeline (Paper 15). Presented at the Interpipe Convention, 1976.
2. Benson, T. D. PG & E's Polyethylene Wrap. *Gas Ind.* **16**(11) (September 1972).
3. *AWWA C209–84* "Cold-Applied Tape Coatings for the Exterior of Special Sections, Connections, and Fittings for Steel Water Pipelines."
4. *AWWA C214–83:* "Tape Coating Systems for the Exterior of Steel Water Pipelines."
5. NACE Standard RP-01-69, Recommended Practice—Control of External Corrosion on Underground or Submerged Metallic Piping Systems (NACE August 1969).
6. Seager, W. H., and Robinson, H. C. A Look at Polyethylene Pipeline Tape Coating Performance Today After 27 Years in Ground Services, Paper 284. In *Proc. NACE Corrosion Engineers,* Boston, Mass., 1985.
7. May 1976 Coating Conductance Test, The Kendall Company (1976), unpublished.
8. Harris, G. M. A Review of Recent Developments and the Performance of Tape Coatings for Underground Pipelines. *Mater. Prot.* (September 1979).

32. Graphic Art Applications

Norman DeBastiani
CPG International Company
Leeds, Massachusetts

Pressure sensitive products are currently used extensively in many segments of the diverse graphics industry. These products feature repetitive uniformity, ease of application, versatility, and reproducibility. They are now constant elements used in the preparation and composition of commercial art, advertising, charts, graphs, visual aids, etc., minimizing tedious manual drawing and typesetting. The utilization of graphic art pressure sensitive products has become a significant cost saving factor.

Four general groups of products will be discussed:

1. Tapes
   a. Stripping tape
   b. Projectable tape
   c. Colored matte and gloss tape
   d. Crepe tape
   e. Fluorescent tape
   f. Printed graphic tapes
2. Films
   a. Color films
   b. Shading films
   c. Pattern films
   d. Masking films
   e. Blockout films
   f. Drafting applique film
3. Die-cut Products
   a. Die-cut lettering

   b. Die-cut symbols
   c. Printed circuit electronic die-cut symbols
4. Dry Transfer Lettering

The fabrication and manufacture of these products require the implementation of numerous specialized and technical skills, processes, and disciplines.

This chapter describes the general physical properties, construction, and specialized end uses of these products.

GRAPHIC TAPES

The many different styles, kinds, and types of pressure sensitive tapes used in the graphic arts industry are designed to eliminate long hours of repetitive hand work. They are excellent aids for printers, publishers, designers, industrial engineers, architects, commercial artists, and chart, graph, and map makers.

Litho Blockout Tape

Litho blockout tape, otherwise known as stripping tape, is designed for blocking out work when setting up negatives for printing forms. it is also used for opaquing and edging negatives in photographic labs. Stripping involves the application of a light-blocking pressure sensitive tape to a negative for the purpose of blocking out unwanted graphics. The light-blocking tape must photograph clear when producing a positive.

The first produced blockout tape was an opaque black paper that was applied over the unwanted graphics. The development of a transparent red tape was a significant improvement because the user could see through the tape and see the graphics he wanted blocked out.

The most common backing materials used in its construction are cellophane, acetate, polyester, and polyvinyl chloride films. The light-blocking compounds are added into the resin matrix and extruded into a solid red film. This involves the selection of the correct light-blocking ingredients (pigments, dyes, absorbers, etc.) that will block out the radiation frequencies to which orthochromatic photo film is sensitive.

The red wavelength radiation emitted by the red blockout film does not affect the orthochromatic photo film, and after processing, the blocked out area will result in a clear window in the positive produced. The clear window can now be used by the artist to insert new graphics without remaking the entire artwork. This positive is now converted to a second-generation positive, which will then be used to generate a printing plate.

Table 32-1. Physical Properties of Litho Blockout Tapes

PSTC# 1	PSTC# 1	PSTC# 6	PSTC# 7	PSTC# 31	PSTC# 33
Peel Adhesion Initial 24 hr g/2 5 mm	24 hr. g/2 5 mm	Quickstick cm	Shear hr	Elongation %	Thickness mm
600	800	2.5	22	40	0.063

The pressure sensitive adhesive applied to the red film is usually a clear, medium tack, high shear system demonstrating a good balance of adhesive and cohesive properties (Table 32-1).

The adhesive is applied smoothly to the red film. Figure 32-1 illustrates a typical litho blockout tape construction.

The most common widths used range from 1⁄32 to 3 in.

Projectable Tapes

Projectable pressure sensitive tapes are constructed similarly to stripping tapes. The adhesive properties are usually very similar to those used for blockout tapes as illustrated in Table 32-1. Diacetate or triacetate backing is preferred, because of its superior clarity, cuttability, and heat resistance; PVC, cellophane, and polyester films are also sometimes used.

The backing is colored or tinted by coating the color uniformly on one or both sizes of the backing film. Another method is to laminate two clear films together with a colored adhesive. The film is then release coated on one side and pressure sensitive adhesive is applied on the other side, again similarly to stripping tape.

These projectable transparent colored tapes are used for preparing overhead transparencies and overlays. There are 11 or more colors available, ranging in width from 1⁄32 to 1 in.

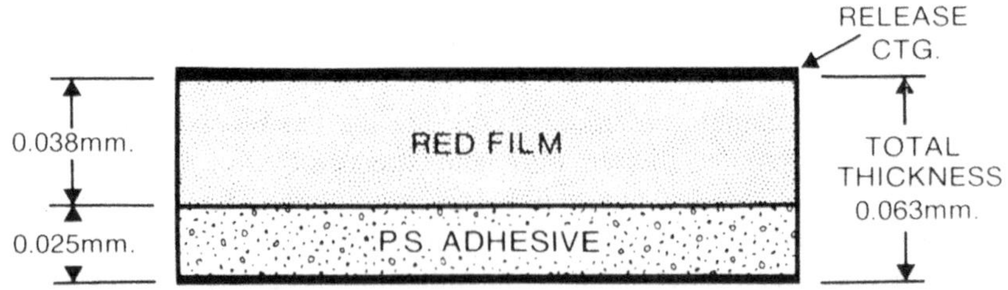

Fig. 32-1. Construction of litho blockout tapes.

Colored Matte and Gloss Tape

A very popular family of graphic tapes are colored matte and gloss tapes. These tapes are designed to be used in preparing visual presentations such as bar charts, graphs, posters, signs, and maps. Their brilliant colors add a professional appearance to all artwork.

Gloss and matte tapes are constructed exactly the same way, except for their surface finish. Matte tapes are used where nonreflective colors are needed, as in photographic reproduction or in presentations where glare from surrounding light might make viewing difficult.

The physical requirements for colored matte and gloss tapes are brilliant opaque colors, easy tearability, good tensile strength, and permanent adhesion with good cohesive strength. Typical physical properties are illustrated in Table 32-2.

The backing is usually a laminate construction consisting of paper and film. The laminating adhesive is pigmented to produce the desired colored backing. The laminating adhesive must be permanent to prevent the film and paper from delaminating. The surface of the film is either matte or gloss to produce the desired finish, while the color is common to both matte and gloss tapes. The film is generally diacetate or polyester. The paper is usually hemp fiber paper. Hemp fiber paper possesses a high tensile strength in the grain direction combined with easy tearability in the cross-grain direction. This is an important property to the artist end user who insists on easy tearability when preparing charts.

A typical example of chart tape construction is illustrated in Figure 32-2. The pressure sensitive adhesive must be permanent, exhibiting a good balance between adhesive and cohesive properties that will yield good creep resistance and adhesion to a wide variety of receiving surfaces.

Colored Crepe Tapes

Colored crepe tapes are used by artists and draftsmen where tape is required to make curves down to one inch or less of arc. These tapes are extremely

Table 32-2. Physical Properties of Colored Matte and Gloss Tape

PSTC# 1	PSTC# 6	PSTC# 7	PSTC# 31	PSTC# 33
Peel Adhesion Initial	Quickstick	Shear	Elongation	Thickness
g/2 5 mm	cm	hr	%	mm
1200	2.5	24+	1–4	0.135

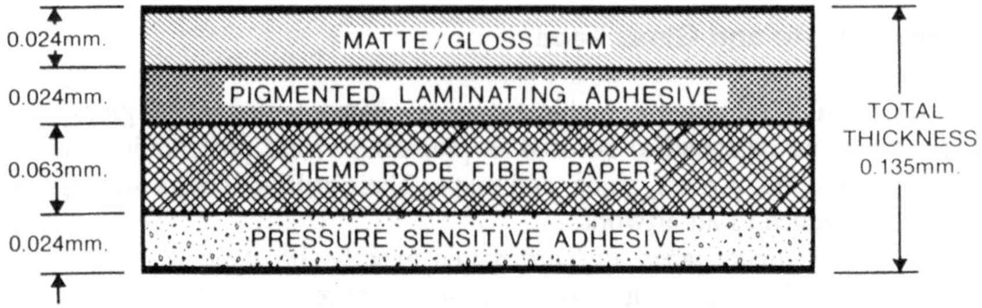

Fig. 32-2. Construction of colored matte/gloss tapes.

flexible and curve easily with little or no distortion. Uses include graphs, topographical maps and cross-sectional drawings.

The backing is colored prior to the creping process. The crepe in the backing will allow the tape to curve without stress. When combined with a permanent pressure sensitive adhesive, the crepe tape will stay firmly in place without any stress buildup.

Fluorescent Tapes

These tapes are constructed similarly to the colored crepe tapes. The backing is usually a stretched-out crepe paper. These tapes are used where there is a need for eye-catching emphasis to a chart or drawing. They can also be used with black light for extra emphasis.

Printed Graphic Tapes

These tapes are printed with various graphics and are used as:

- Decorative border tapes
- Pattern tapes
- Plant and office layout tapes
- Statistical and symbol tapes
- Mechanical production tapes
- Register and crop mark tapes

These tapes are of the same construction; only the graphics vary. The most common construction is a two-ply lamination of the same film. The first film is printed with the desired graphics and the second film is laminated over the printing to protect the graphics. Figure 32-3 illustrates a typical two-ply-construction.

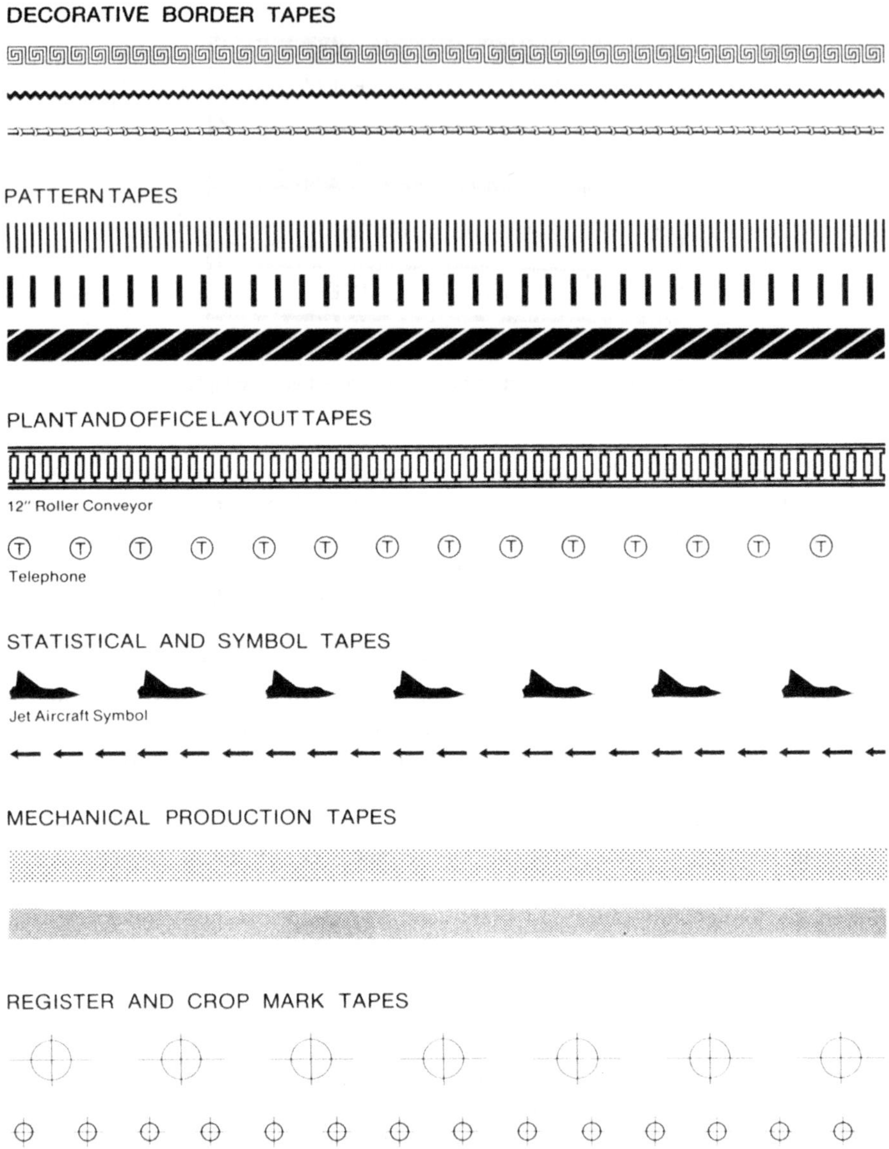

Fig. 32-4. Examples of printed pattern tapes.

Table 32-3. Physical Properties of Drafting Tape

PSTC# 1	PSTC# 1	PSTC# 33	PSTC# 31	PSTC# 31
Peel Adhesion				
Initial 24 hr	24 hr	Thickness	Elongation	Tensile
g/2 5 mm	g/2 5 mm	mm	%	g/2 5 mm
400	500	0.16	9	500

Black Photo Tape

Black photo tape is an opaque dense black crepe tape designed for optimum photographic masking and sealing of light-sensitive materials such as films and plates. This tape has an excellent long-term holding strength. It is resistant to moisture, will not creep, and is easy to use.

GRAPHIC FILMS

In the graphic arts industry, there is as much demand for assorted pressure sensitive graphic films as there is for tape. These films are also designed for the purpose of eliminating many hours of repetitive hand drawing for artists, architects, engineers, or draftsmen.

Color Film

Color films are produced by applying a carefully controlled solid color to one side of the film. Cellulose diacetate or triacetate film is preferred because it can be cut easier than polyester film with an art knife. Both mattc and gloss finish are available.

The most popular films use a repositionable adhesive in its construction. The films can be applied, cut out, and removed from an art drawing without damage to the rest of the work. Ordinary pressure sensitive adhesive would tear the surface or lift the paper fibers and damage the art drawing. The repositionable adhesive must perform on a variety of receiving surfaces such as vellum and drafting film, gloss polyester and acetate film, art board, and assorted inexpensive art papers.

The adhesive system must be clear so as not to interfere with the color of the film. For example, an adhesive with a tendency to yellow on aging causes a blue color to appear green.

The final step in the construction of color film is the lamination of film

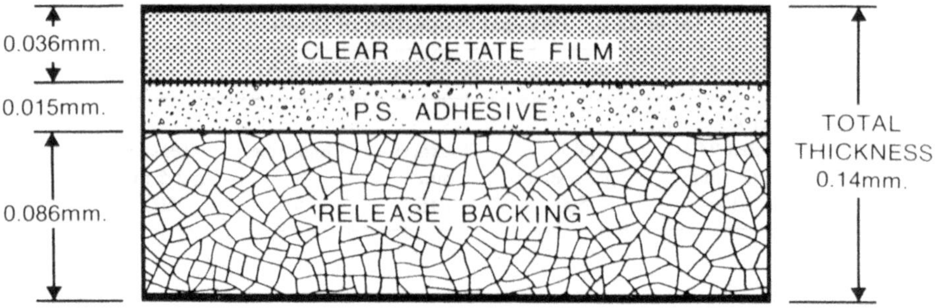

Fig. 32-5. Typical construction of color films.

and adhesive to a suitable paper release backing. The release backing will protect the film from damage and at the same time aid the artist in its use. The release properties of backing paper have to be matched with the adhesive properties for optimum performance. Should the adhesion between the adhesive and backing be too light, the sandwich construction will have a tendency to tunnel. Should the adhesion be too strong, the film will have a tendency to stretch and tear when it is removed from the release backing, making it very difficult to use.

Figure 32-5 shows a typical color film construction, and Table 32-4 identifies typical adhesive properties on various receiving surfaces.

Color films are provided in over 200 opaque and transparent colors. There are also many special color effects that can be provided.

Projectable Color Film

Projectable color films are constructed basically the same as opaque color films. The difference is that the colors must be clear and transparent. These films

Table 32-4. Physical Properties of Color Film

RECEIVING SURFACE	PSTC# 1 INITIAL (g/2 5 mm)	PSTC# 1 PEEL ADHESION 24 HR (g/2 5 mm)
Stainless steel	200	250
Vellum	35	50
Drafting film	175	200
Polyester film	220	275
Art board	75	125
Bond paper	35	100

are usually used in the preparation of overhead projector slide mounts and transparencies. The adhesive system must also be clear and must adhere to polyester or acetate slide mounts without air pockets or lumps that distort the colors or interfere with projection.

Shading and Pattern Film

Shading and pattern films are basically constructed the same as color films, that is, on gloss or matte acetate and also on polyester films. An assortment of topographic, architectural, thematic, grids, textures, line tints, screens, graduated tones, and patterns are printed on them. They are used by artists to tone and shade their line artwork. Just like the graphic tapes, they function as timesavers for the end user. A few examples may be seen in Figure 32-6.

The black printing should be of sufficient density to yield good photo reproduction. Sometimes the image is reverse printed and laminated to the adhesive side of the film. This technique will protect the printing from scratches and abrasion while being used. However, with the advent of new ink technology, these patterns can now be surface printed to produce films with excellent scratch and abrasion resistance.

Masking Film

Masking film consists of a red or amber coating applied to a clear, dimensionally stable polyester film. The red or amber coating is usually applied from a cellulose nitrate resin solution that has been pigmented and compounded to block out orthochromatic light. The amber film provides better "see-through" when used over dark and complex artwork, but it will not block out as well as the red film.

The artist can use masking film to create color separations and overlays rapidly, to cut intricate lines, and to add dropout marks.

Masking films have a unique adhesive system that permits readhesion to itself, to its own polyester backing, or to other receiving surfaces, while demonstrating very easy removability.

The adhesive is the first coating applied to the polyester backing. It is applied very thin. The second coating is the red or amber coating that is applied to a thickness of 1.5 mil. It usually requires two coating passes to achieve that coating thickness.

After the cellulose nitrate coating has dried, the adhesive will transfer and become attached to it. Some manufacturers offer products with several

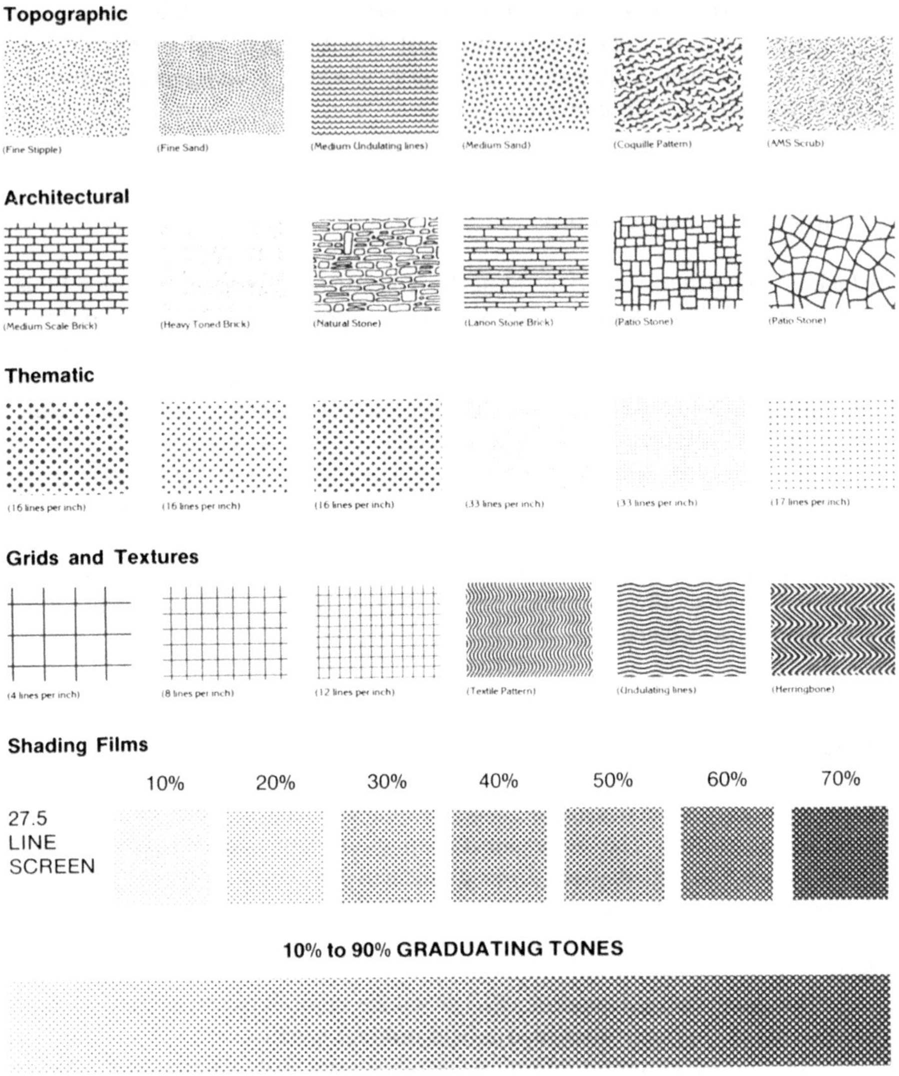

Fig. 32-6. Examples of pattern and shading films.

adhesive levels. The most popular level average is 8–15 g/cm. Figure 32-7 illustrates a typical masking film construction.

The film is offered on a 3- or 5-mil polyester base. The 3-mil base is the most popular.

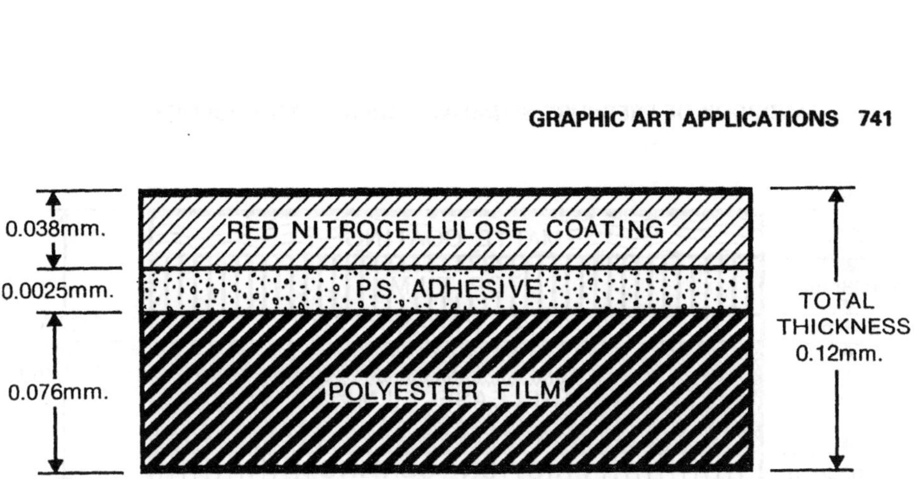

Fig. 32-7. Construction of a masking film.

Blockout Film

Blockout films are constructed differently than masking film. They consist of a red acetate film coated with a low tack adhesive and laminated to either an opaque or translucent silicone coated release backing. The construction is very similar to color films, except that the red acetate film is pigmented like stripping tape but laminated to a release liner.

Use of blockout films also differs from masking film in that it is mostly used to block out negatives and positives rather than to separate colors. It is used mostly in conjunction with litho blockout tapes where large areas of negative need to be blocked out.

The adhesive used for blockout films must be repositionable and must apply free from air bubble entrapment.

Drafting Applique Film

Drafting applique film is primarily designed for "scissor drafting" applications. These films have a specially formulated matte coating that will accept many common marking and printing systems, such as india ink, drafting pencil, typewriter, ball-point pen, felt marker, xerography, and offset printing. Images or lettering that are repetitive on drawings can be imaged in multiples on drafting applique film via xerography or standard printing methods and cut out and applied to the subject drawing. It is a major time-saver for the end user.

Drafting applique film is a laminate construction consisting of a matte coated polyester top film and either a polyester or paper silicone coated release liner. The adhesive is a permanent, nonremovable, nonyellowing system that cannot be separated from the drawing once it is applied. For applications where removability or repositionability are desired, a removable adhesive system is available. The back side of the release film is coated with a suitable friction

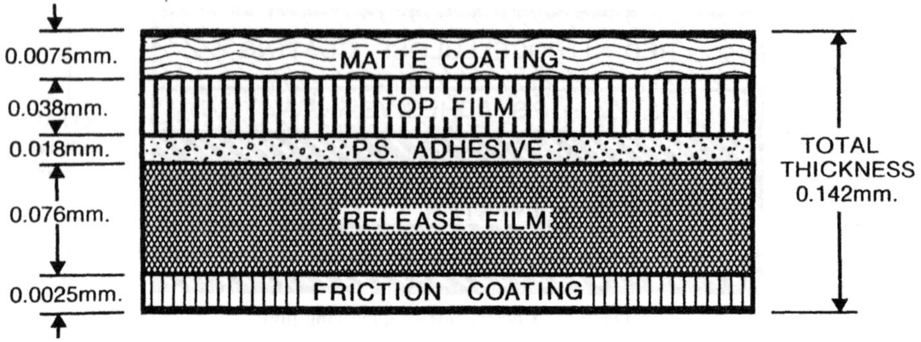

Fig. 32-8. Construction of a drafting applique film.

coat to assist drafting applique film through the feed mechanism of the various plain paper copiers available today. Figure 32-8 illustrates a typical drafting applique film construction.

DIE-CUT PRESSURE SENSITIVE PRODUCTS

Die-cut Lettering

Die-cut lettering is produced on a 4-mil vinyl stock and is backed with an adhesive that is suitable for both indoor or outdoor applications. These items are designed to assist the architect, building contractor, maintenance engineer, autobody painter, hobbyist, or homemaker who has specific lettering or sign requirements. They can be applied to painted or sealed wood, plastic, and metal receiving surfaces. The adhesive is specially formulated to resist heat and cold, and it is weatherproof.

The adhesive coated vinyl is laminated to a suitable paper release liner. The lettering is die-cut on the release liner, which also serves as a spacing system.

Die-cut Symbols

Die-cut symbols are produced similarly to the die-cut lettering. They are used for logos, insignias, etc.

There is a second family of die-cut symbols produced for the visual communication industry. These products are die-cut from a variety of materials such as colored tapes, crepe tapes, and films. They are used indoors for visual communication on such things as charts, maps, and drafting films.

Electronic die-cut symbols, cut from black tape, are used by electronic engineers, designers, and draftsmen in the design of printed circuits. They require precision tolerances and an aggressive adhesive system. They can be applied quickly and accurately to a drawing, thus eliminating the need of repetitive drawings of subject symbols.

DRY TRANSFER LETTERING

Dry transfer lettering is a unique and unusual product. It consists of a plastic carrier film, usually styrene, polyethylene, or polyester that has been coated with a pigmented coating formed in the shape of letters or indicia and overcoated with a unique adhesive.

A silk screen printing process is used to apply the pigmented coating and form the indicia. The adhesive is formulated so it can be flood coated evenly over the indicia and dry to a nontacky surface.

The adhesive will become activated when the indicia is burnished or rubbed from the opposite side through the carrier film. This will cause the indicia to adhere to the receiving surface before the carrier film is lifted away and positioned for the next transfer.

The original product was invented in England some 30 to 35 years ago. The inventor capitalized on the fact that a cellulose nitrate lacquer will not adhere to a styrene or polyethylene film.

The pigmented coating or ink must silk-screen print indicia with good edge definition and density. The indicia must transfer cleanly and easily without distortion and cracking. The indicia must adhere to a wide variety of receiving surfaces, such as art board, bond paper, vellum, drafting film, polyester film, acetate film, and glass. Transfer errors should be correctable by rubbing the indicia off with a rubber eraser or by lifting them off with a piece of a pressure sensitive tape.

Dry transfer lettering is widely used in place of typesetting, offering a real savings when creating compositions, layouts, and other artwork prepared for generating printing plates. There are hundreds of letter styles, sizes, and colors available, making this product very convenient to use. Custom sheets of any design, color, or multiple of colors are also manufactured. These custom sheets represent a fast way to apply frequently used logos, technical symbols, trademarks, etc., eliminating the need for the artist to draw each one. Dry transfer lettering is also used in drafting to take advantage of the convenience it offers.

Figure 32-9 illustrates a typical dry transfer construction. The carrier film

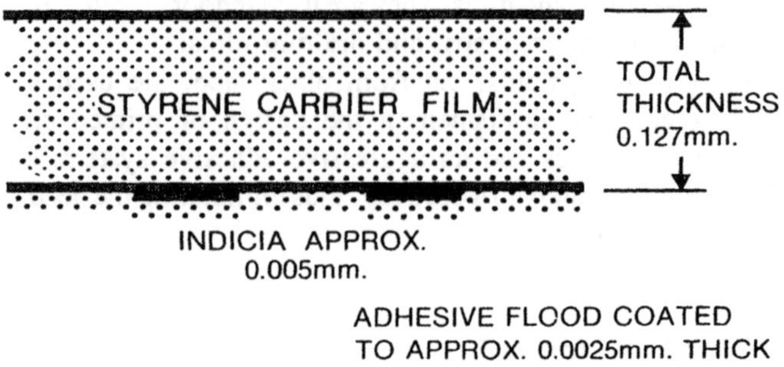

Fig. 32-9. Construction of dry transfer lettering.

thickness will vary from 3 to 6 mils, depending on the manufacturer's preference. The thickness of the indicia is normally about 0.2 mil and the adhesive about 0.1 mil.

Most manufacturers flood coat the adhesive over the indicia, covering the entire carrier film. This method requires that the adhesive be formulated so that there will be no legging or stringiness at the indicia and adhesive interface when transfer is taking place. A second requirement is that the adhesive be dry enough to prevent the sheets from blocking or pretransferring when stored in piles for a long period of time. The adhesive must also display five or more years longevity to ensure consistent indicia transfer.

33. Labels

Donatas Satas
and
A. Maria Satas
Satas & Associates
Warwick, Rhode Island

The development of pressure sensitive labels took a different route (and was carried out by different companies) than that taken by pressure sensitive tapes. Only relatively recently have the two applications merged. Pressure sensitive label use has grown, and still does, at the expense of water-remoistenable labels and decals. Therefore, pressure sensitive labels had rather narrow constraints on their acceptable price, as well as on materials that could be used in their manufacture.

Labels are made from printed, precut flexible materials such as paper, film, or foil and are affixed to a product for purposes of identification, information, instruction, or decoration. There are over 2000 commercial printers in continental North America that are involved in the label printing business; about one third of these companies specialize in labels.

The manufacturing of labels involves three separate operations: stock coating, printing and diecutting, and label application.

Over 60% of labels are applied with glue at the point of labeling. Pressure sensitive labels are supplied with the adhesive precoated on the label stock. The label is mounted on a release coated sheet from which it is removed prior to application.

Pressure sensitive labels are more expensive than glue applied labels; they require a silicone coated release liner, and the adhesives used are generally more expensive than those employed in direct glue application. Nevertheless, pressure sensitive adhesive labeling has occupied a considerable segment of the market for several reasons. Despite the higher label cost, the overall costs of pressure sensitive adhesive labeling are not necessarily higher. Label application equipment costs less for pressure sensitive labels than for glue applied labels, the setup and cleanup times are shorter, and therefore pressure sensitive labels are often preferred for short product runs. This condition arises most frequently in the case of secondary labels. Secondary labels are labels that carry additional information, such as that indicating a special product feature or a sale price, in addition to the product information already shown on the primary label. Since such products are usually run in smaller volumes, pressure sensitive labels are most often chosen.

Labels that are constructed from moisture-impervious materials, such as films or foils, are usually applied employing pressure sensitive adhesives. Water glues do not dry well through such substrates, because of the water trapped in between two impervious surfaces. Labels of intricate form are also best applied when carried on a release liner rather than from a label stack. Pressure sensitive adhesives adhere better to plastic surfaces, especially to polyolefins, than water glues, and are therefore preferred for such applications. Since plastic packaging has been growing at a fast rate, the use of pressure sensitive labels has also increased. In-mold labeling, however, is a competing process for large-volume plastic bottle label applications.

There have been some attempts to construct linerless labels. Pressure sensitive adhesive tapes have been used as labels by printing on the tape face. Patents are claimed for release agents that can be overprinted with ink and still function as release agents.[1,2] Tapes with such release agents can be directly printed. Otherwise, a release coating must be applied over the preprinted tape.[3] Various other methods, such as spacing between the adhesive and the label face, are employed in order to avoid the need for a release liner. Such products are of little commercial importance, however, and most of the labels are supplied on a release liner.

Future growth of pressure sensitive labels is expected in the primary labeling area, which is currently dominated by glues.

LABEL STOCK MANUFACTURING

The manufacture of pressure sensitive label stock is a large-volume business, mainly carried out by converters. The adhesive coating may be applied as a solution, latex, or hot melt. Pressure sensitive adhesive coatings are applied by the transfer technique: the adhesive is first applied to the release paper, dried in the oven on the release paper, and then the label face stock is laminated to the adhesive. A schematic diagram of the transfer coating process is shown in Figure 33-1.[4] In the transfer coating process, the label stock is not exposed to the higher temperature of the drying oven, and therefore any possible damage to the more sensitive face stock is minimized. Most coating operations are carried out on wide roll machines. The rolls are then slit or sheeted to the required size for printing equipment. Narrow-width hot melt coaters are also available for the production of label stock and are used by some label manufactur-

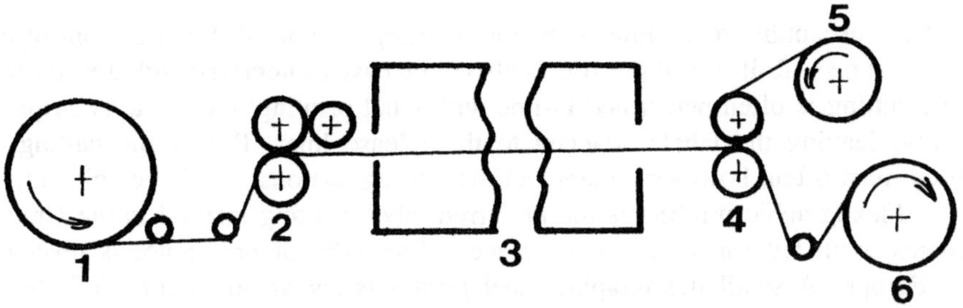

Fig. 33-1. Transfer coating process: (1) release paper unwind, (2) coater, (3) drier, (4) laminating rolls, (5) label face supply roll, (6) rewind.

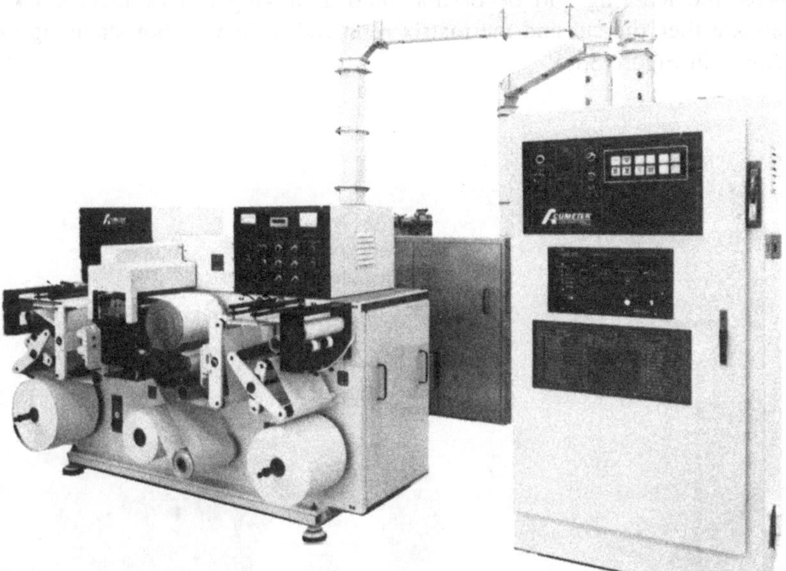

Fig. 33-2. Hot melt coating machine. (*Courtesy of Acumeter Laboratories, Inc.*)

ers to produce label stock in-house. Such a narrow hot melt machine is shown in Figure 33-2.

LABEL PRINTING

Label stock is printed in either roll form or in sheets, with the roll printing method dominating in the manufacture of pressure sensitive labels. Roll printing is less susceptible to curling problems. Curling is more difficult to control on a sheet press. Roll-to-roll printing is also more easily integrated with die cutting. Die cutting is often performed in-line with label printing; the waste is stripped away, leaving the labels attached to the release liner. Rotary die cutting is used most often; however, reciprocal die-cutting stations are also employed.

Flexographic printing is the preferred label printing method in the United States, with letterpress in second place.[5] The order of preference is reversed in Europe. A small flexographic label printer is shown in Figure 33-3. Such machines may consist of many stations, allowing printing in several colors as well as the application of a clear, protective coating (Fig. 33-4). Ultraviolet (UV) curable inks and coatings may also be used on such label printers. Hot stamping is a less often used printing technique that employs a heated die to transfer the lettering and decoration onto a moving roll of label stock. The labels are then die-cut and the matrix removed in-line. A hot stamping printer is shown in Figure 33-5.

Fig. 33-3. Small flexographic label printer. (*Courtesy of Mark Andy, Inc.*)

Fig. 33-4. Multistation flexographic label printer. (*Courtesy of Mark Andy, Inc.*)

LABEL FACE STOCK

The least expensive and most often used material for labels is one-side coated bleached kraft paper 50–60 lb (81–97 g/m^2) weight. White clay coatings dominate, with yellow coatings also being used occasionally. Other label colors are used in smaller volumes for specialty identification and marking purposes. Glossy clay coated paper holds an important place for label stock. Such glossy coatings are applied by cast coating and similar techniques. Foil/paper laminate labels are popular for various soft drink, beer, and beauty aid products containers. Metallized paper is also increasingly used as new metallizing technologies are developed. For applications where the label might be exposed to excessive moisture, such as bumper stickers, or where a higher backing strength is required, saturated paper is used.

Plastic film labels are preferred for plastic containers. Because they are flexible, film labels conform to the flexing of the container and are less likely to form wrinkles and creases. A metallic appearance can be achieved by the

Fig. 33-5. Hot stamp printer. (*Courtesy of Franklin Co.*)

use of metallized polyester film. Because of the easy printability, polyvinylchloride films are used most often for labels. Thick gauge plasticized vinyl film is offered to screen printers for manufacturing signs and decals that are widely used for more permanent applications than are possible to obtain from paper substrates. Thinner gauge unplasticized vinyl film is useful as a label substrate for labels that are subjected to frequent flexing.

RELEASE LINERS

The development of silicone coated release liners made possible pressure sensitive labels, as we know them today. A successful release coating for pressure sensitive labels must show a sufficiently easy release at low, and not only at high peel rates. The labels are removed from the release liner during their application by bending the release coated carrier sharply over an edge. This principle of label removal requires that the release be easy at low peel rates. Regardless of the label application rate, each individual label starts at a zero peel rate and then quickly accelerates to a rate corresponding to that of the label application rate. While there are many release liners that show low peel forces at high peel rates, only silicone release coatings exhibit low peel force at low peel rate. This is illustrated in Figure 5-3. Some fluorocarbon polymer coated release liners approach the properties of the silicone coatings, but they are expensive.

Release level is measured by a peel test and the value obtained is referred to as the Keil release. Such release-level measurements are usually carried out by peeling a 1-inch-wide sample at an angle of 180 degrees and a rate of 30 cm/min. There is a higher peel rate Keil test that employs a 2-inch-wide strip peeled at a rate of 1500 cm/min.[6]

Ideally, the liner should have a sufficiently low release, even if a highly tacky adhesive is used, and should allow easy removal of the matrix after die-cutting. On the other hand, the release should be sufficiently tight so as to hold labels during die-cutting, printing, and other converting operations prior to label application. Needless to say, such a universal release liner does not exist, since the adhesives and processing conditions used vary. Therefore, various levels of release are required depending on the particular product. Table 33-1 shows an approximate subdivision of release liners according to release value.

Many different substrates are used for release liners. Paper dominates: it is the least expensive sheet material with the bulk and stiffness properties required. Supercalendered and machine-finished kraft papers dominate the release liner field. However, other papers are also important for pressure sensitive labels. Polyethylene coated kraft paper is widely used, because of its smoother surface and decreased susceptibility to moisture and therefore to curling. Curling of label stock can be a real problem, especially for sheet printed labels. Double-side polyethylene coated paper is sometimes used to minimize moisture uptake after drying. Clay coated kraft paper has some use for label stock, because of its lay-flat property. Glassine paper had a much more important market share as a release liner, but it has lost ground to polyethylene coated papers.

Release coated films are becoming more important as release liners. The

Table 33-1. Release Levels of Silicone Coated Liners

LINER CATEGORY	KEIL RELEASE (g/in.)
Super low release	4–8
Low release	8–25
Normal release	25–50
Moderate release	50–150
Tight release	150–250

growth of transparent film labels requires a very smooth adhesive surface in order to eliminate the inclusion of air after label application. This is achieved by employing plastic film liners, and the use of polyester film for this application dominates. There is also a growing tendency to apply pressure sensitive labels at higher application rates. Paper liners tear more easily so that film liners are preferred. Various films are used for such liners: polyester, polyethylene, and polypropylene. A foamed polypropylene film liner has been offered on the market. It has a good balance of strength and stiffness. Foaming is an effective way to increase web thickness and therefore web stiffness, without a corresponding increase in weight.

The smoothness of the release surface affects the smoothness of the adhesive

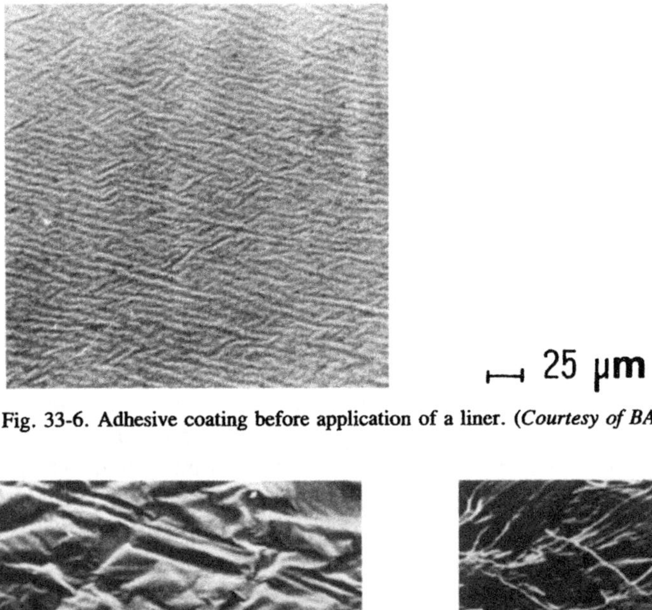

Fig. 33-6. Adhesive coating before application of a liner. (*Courtesy of BASF, Ludwigshafen.*)

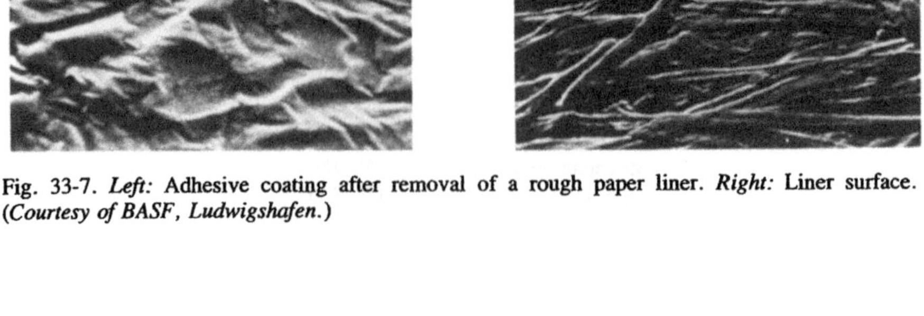

Fig. 33-7. *Left:* Adhesive coating after removal of a rough paper liner. *Right:* Liner surface. (*Courtesy of BASF, Ludwigshafen.*)

and will therefore have an effect on initial tack and on the clarity of a transparent label. The following photographs illustrate the effect of the release liner on the adhesive surface. Figure 33-6 shows the adhesive surface before contacting the release surface. Figure 33-7 shows the surface of a rough release liner and the adhesive surface after its removal from the release surface. Figure 33-8 shows the effect on adhesive by a smoother paper surface. Figure 33-9 shows a coated release liner surface and the corresponding adhesive, and Figure 33-10 shows the surface of a film liner and the corresponding adhesive surface. Inspection of these surfaces clearly shows the reason for using film or other smooth surface release liners for clear film labels, where a smooth adhesive surface without air entrapment is required.

Silicone coating is applied to the liner and cured. Many larger label stock

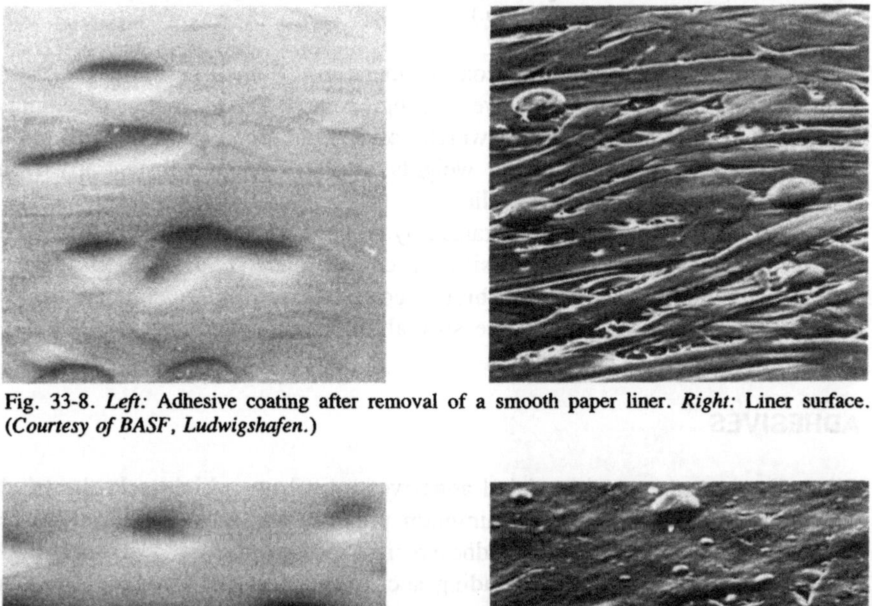

Fig. 33-8. *Left:* Adhesive coating after removal of a smooth paper liner. *Right:* Liner surface. (*Courtesy of BASF, Ludwigshafen.*)

Fig. 33-9. *Left:* Adhesive coating after removal of a coated paper liner. *Right:* Liner surface. (*Courtesy of BASF, Ludwigshafen.*)

Fig. 33-10. *Left:* Adhesive coating after removal of a film liner. *Right:* Polyethylene film liner surface. (*Courtesy of BASF, Ludwigshafen.*)

producers have integrated release coating operations. Sometimes, release coating is even done in-line with adhesive coating. A very small amount of silicone coating is required: the average weight of silicone coating is about 0.6–0.8 g/m^2 (0.4–0.5 lb/ream). Coating weights have decreased substantially from the approximate 12 g/m^2 used earlier.

A special release problem is raised by silicone pressure sensitive adhesives that do not release from regular silicone coated liners. Fluorocarbon polymer coated liners are available that exhibit a good release from silicone adhesives. Claims have been made that some special silicone coatings are also suitable.

ADHESIVES

Generally, pressure sensitive label adhesives must have a high tack: the labels are often applied quickly with minimum pressure and must bond well under those conditions. A good label adhesive is expected to bond well to various surfaces used in packaging, including such plastics as polyethylene. A good label adhesive suitable for use on various packaging surfaces should show peel adhesion to low surface energy substrates, such as polyethylene, not substantially lower than that to steel. Many polar adhesives do not have this property. Labels are rarely expected to carry a load or otherwise be exposed to shear forces during use; therefore, shear resistance is less important than in most tape adhesives. The adhesive coating thickness is lower than that used for tapes, that is, 15–35 g/m^2 of adhesive is usually sufficient.

Label adhesives, as well as tape adhesives, might have many special requirements because of their specific application, for example, solvent resistance is required for labels used in automotive and other applications with exposure to

Table 33-2. Solvent Solution Pressure Sensitive Adhesive

INGREDIENT	PARTS BY WEIGHT
Pale crepe rubber	75
Styrene-butadiene polymer	25
Polyterpene resin (S.P.115°)	50
Pentaerythritol ester of hydrogenated resin (S.P.104°)	25
Calcium carbonate pigment	25
Antioxidant	2
n-heptane	600

solvents. Resistance to water is required in some applications, and easy removability by water in others. Some labels are applied at low temperature (frozen food) to moist surfaces and require specially compounded adhesives. Label adhesives may be applied directly to produce or to fruits and are required to meet FDA regulations for adhesives in contact with food.

Pressure sensitive labeling adhesives can be subdivided into three categories on the basis of the vehicle used; this also determines the method of adhesive application. The three categories are solventborne, waterborne, and hot melt.

Solvent systems are composed of an adhesive dissolved in a suitable solvent, with the solids content generally not exceeding 45% on a weight basis. Aqueous systems consist of emulsions with a solids content of up to 65%. Hot melts are adhesives applied as 100% solids in molten form.

Solution adhesives usually consist of natural rubber or SBR tackified with a resin and dissolved in a petroleum solvent, such as heptane and, sometimes, toluene. Higher solids label adhesives may be produced by employing a block copolymer as the elastomeric ingredient. A typical adhesive formulation is given in Table 33-2.[6]

Aqueous emulsion adhesives are usually acrylic, although styrene-butadiene (S/B) emulsions are also used. Table 33-3 shows an example of such an adhesive.[7]

Hot melt pressure sensitive adhesives used for labels are usually based on block copolymers. However, ethylene-vinyl acetate copolymer adhesives are used occasionally. Table 33-4 shows a typical block copolymer-based pressure

Table 33-3. S/B-Natural Rubber Latex Pressure Sensitive Adhesive for Freezer Grade Labels

INGREDIENT	PARTS BY WEIGHT ON DRY BASIS
S/B latex (Dow PSA 578A)	40
Natural rubber latex	20
Tackifier (Hercolyn D)	40

Table 33-4. Hot Melt Pressure Sensitive Adhesive

INGREDIENT	PARTS BY WEIGHT
Block copolymer (Kraton 1102)	100
Hydrocarbon resin (Super Sta-Tac 80)[a]	90
Hydrocarbon resin (Super Sta-Tac 100)[a]	40
Polyterpene resin (Nirez 1010)[a]	20
Antioxidant	1

[a] Reichhold International

sensitive adhesive useful for label stock and Table 33-5 gives an adhesive based on EVA copolymer.

Pressure sensitive labels are divided into three categories based on the intended application. Each of the applications requires a special adhesive. The categories are permanent, removable, and low temperature.

The majority of pressure sensitive labels are coated with permanent adhesives. Once applied, a label with a permanent adhesive cannot be removed without destruction of the label stock.

Removable adhesives allow the applied label to be removed and repositioned. Removability is a property dependent on many extraneous conditions: the surface type to which the label is applied, time elapsed since application, temperature to which the label has been exposed, etc. The ease of removal might depend on conditions to which the label has been exposed.

There are several approaches in producing removable adhesives. One way is to compound an adhesive that does not wet the surface very well and does not produce a good bond. Such adhesives are highly elastic, exhibit a low level of viscous flow, are not very tacky, and might have a high shear resistance. Lightly tackified natural rubber and other elastomers (SBR, polyurethane, etc.) produce such adhesives. Such adhesives exhibit two main deficiencies: a low tack level, and some may show an increase in adhesion on prolonged aging, especially at higher temperature.

Table 33-5 EVA-Based Hot Melt Pressure Sensitive Adhesive

INGREDIENT	PARTS BY WEIGHT
Ethylene/vinyl acetate copolymer (Elvax 460)[a]	20
Paraffin wax (155°F AMP)	35
Hydrocarbon resin (Poly-Tac 100)[b]	45
Antioxidant	0.2

[a] Du Pont Co.
[b] Reichhold International

Another type of removable adhesive is a highly tacky, low shear resistance product, which exhibits a relatively low peel force. Highly plasticized pressure sensitive adhesives behave in this manner.

It is also possible to incorporate various particles into the adhesive that prevent complete contact formation. Glass spheres, large polymeric spheres, and other such materials have been used to produce removable labels and other products. A formulation of a removable pressure sensitive adhesive based on natural rubber is given in Table 33-6.

There is a need for label adhesives that can be removed by water or by alkali. Such adhesives may be used for reusable bottles. Alkali-removable pressure sensitive adhesives have been prepared from hydroxyl group containing acrylic polymers.[8] Adhesives suitable for wash-off labels consisting of maleic acid/methyl vinyl ether are described.[9] Many other adhesive compositions that can be easily removed in bottle reprocessing are known.

Some pressure sensitive labels are required to have adhesives that are dispersible in the paper manufacturing process to allow reprocessing of packaging. This demand for recyclable paper packaging is expected to grow.

Low-temperature adhesives are employed for labels that are applied at freezer temperatures to a surface that might be moist. This requires an adhesive that has a low glass transition temperature, and therefore remains tacky at low application temperatures.

In addition to these three main categories of label pressure sensitive adhesives, many other specialty adhesives are used: delayed tack adhesives, water-soluble, dispersible, suitable for direct application to fruit and other food, etc.

Delayed tack adhesives are nontacky prior to their activation, and therefore such label stock does not require a release liner. Such adhesives are activated by heating and they remain tacky for a long time: days or even months. The main additive that allows the compounding of such adhesives is a solid plasticizer—sulfonamide or diphenyl phthalate—that is not compatible with the adhesive at room temperature and does not plasticize and tackify the adhesive until heated. Raising the temperature causes the plasticizer to melt and increases its

Table 33-6. Removable Label Adhesive

INGREDIENT	PARTS BY WEIGHT ON DRY BASIS
Natural rubber latex[a]	100
Tackifier (Piccopale 85)[b]	40
Tackifier (Piccovar AP-10)[b]	14
Butyl Zimate	3
Santowhite crystals[c]	3

[a] Hartex 103 by Firestone Synthetic Rubber & Latex Co.
[b] Hercules, Inc.
[c] Monsanto Industrial Chemicals Co.

Table 33-7. Delayed Action Label Adhesive

INGREDIENT	PARTS BY WEIGHT
Vinyl pyrrolidone/ethyl acrylate copolymer emulsion (40% solids, VP/EA ratio 60/40)	50
Dicyclohexyl phthalate	58
Glycerol tribenzoate	43
Pentaerythritol tetrabenzoate	7

compatibility with the adhesive, rendering the coating pressure sensitive. In time the plasticizer may start separating and blooming to the surface, causing bond loss. Such labels have found use for pharmaceutical, cosmetic, and food products packaging. Table 33-7 shows a delayed action adhesive formulation that employs two solid plasticizers.

When heated, these plasticizers form a eutectic mixture that has a lower freezing point than either of the plasticizers.[6] The molten eutectic tackifies the polymer, giving a single-phase adhesive instead of the multiphase one that is present prior to heating.

LABEL APPLICATION

Pressure sensitive labels are easy and convenient to apply. The principle of label removal from the release carrier is simple and is well illustrated in Figure 33-11. Labels are removed from the release liner by making a sharp bend over the edge of a peeling blade. While the release liner bends, the label does not follow the bend and comes off the release liner at the point of bending. The ease of label removal depends on the relative stiffness of the two materials: a very soft label on a rigid liner might not come off easily.[4] After its removal from the release liner, the label is applied to a package.

The simplest example of pressure sensitive labeling is retail pricing with a hand-held gun, an example of which is shown in Figure 33-12. Labels are also applied on automatic or semiautomatic machines. Semiautomatic machines

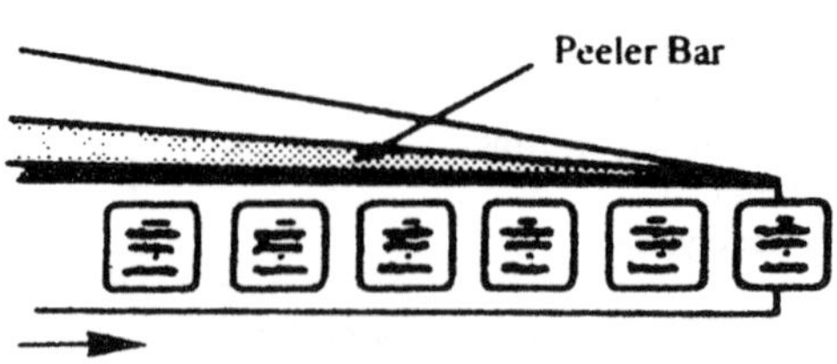

Fig. 33-11. The principle of pressure sensitive label removal.

Fig. 33-12. Hand dispenser of pressure sensitive labels. (*Courtesy of Monarch Marking Systems.*)

are equipped to apply the labels automatically, but the parts to be labeled are hand-fed to the conveyor. Automatic machines include automatic handling of the packages as well. Labeling machines for pressure sensitive labels are less expensive than glue application equipment and are simpler to operate. Table 33-8 compares label application speeds for several methods.

A pressure sensitive labeling line consists of the following components: label supply, label sensor, peeling blade, drive system, label applicator, rewind station, product-handling system, product-sensing device.

Pressure sensitive labels are supplied die-cut to the desired shape in roll form, as shown on the labeler in Figure 33-13. Sometimes fan-folded labels are used, since continuous label feeding is more easily accomplished than with the roll form; it is not necessary to stop the operation to change rolls. To overcome this problem, many newer pressure sensitive labelers are equipped

Table 33-8. Labeling and Marking Speeds for Various Methods

OPERATION	SPEED (pieces/min)	EQUIPMENT COSTS
Glue labeling	350–400	High
Pressure sensitive	300	Low
Heat seal labeling	200–300	High
Direct screen printing	100–125	Moderate
Heat transfer	60–80	Moderate

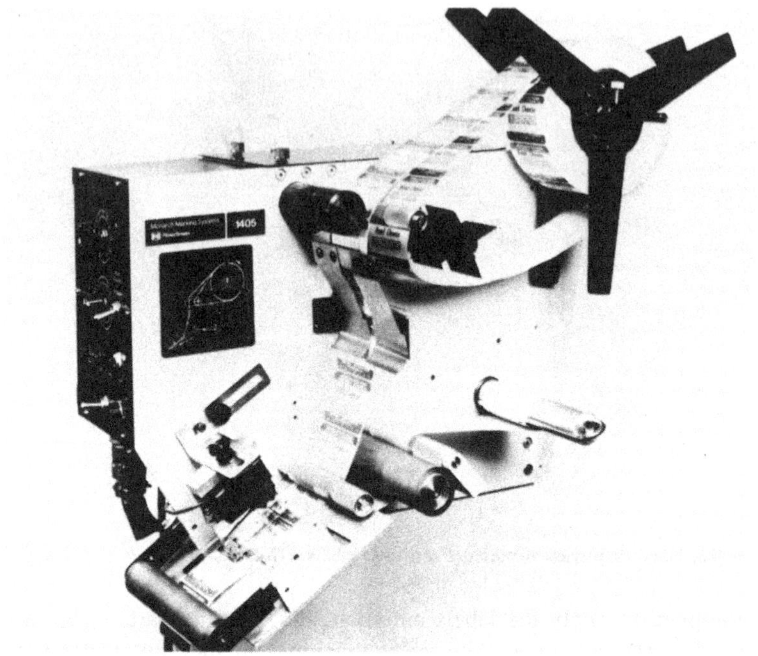

Fig. 33-13. Pressure sensitive label applicator to apply labels to products moving on a conveyor. (*Courtesy of Monarch Marking Systems.*)

with two label supply stations: when one station runs out of labels, the other automatically takes over.

Photoelectric sensors are commonly used to monitor label supply. These are of two types: direct scanner, which senses the difference in opacity between the label and the carrier, and the reflective scanner, which senses printed register marks. Transparent labels that are difficult to detect by photoelectric sensor are often monitored using air sensors. This method works by sensing holes present in the matrix between labels.

A peeling blade removes the label from the release liner and the label is applied to the surface of the container by one of several methods. It may be rolled and adhered onto the surface by a roller. A vacuum grid may be used to position the label. The vacuum is then reversed, and the label is blown onto the surface. A device employing a high pressure air blast for label application has been described.[10] An impressor cylinder with a vacuum port may be used to position the label, which is then pressed onto the surface. These three methods of pressure sensitive label application are shown schematically in Figure 33-14. Examples of automatic pressure sensitive labelers are shown in Figures 33-15 and 33-16.

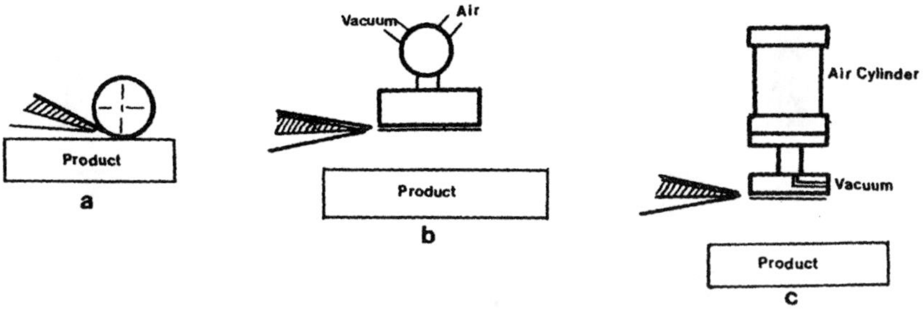

Fig. 33-14. Pressure sensitive label application methods: (a) roll-on, (b) blow-on, (c) impressor method.

Fig. 33-15. Automatic labeler. (*Courtesy of Labeling Systems, Inc.*)

USES OF PRESSURE SENSITIVE LABELS

Labels are divided into two categories according to their use: primary and secondary labels.

Primary labels are the main labels that are applied to retail goods. They

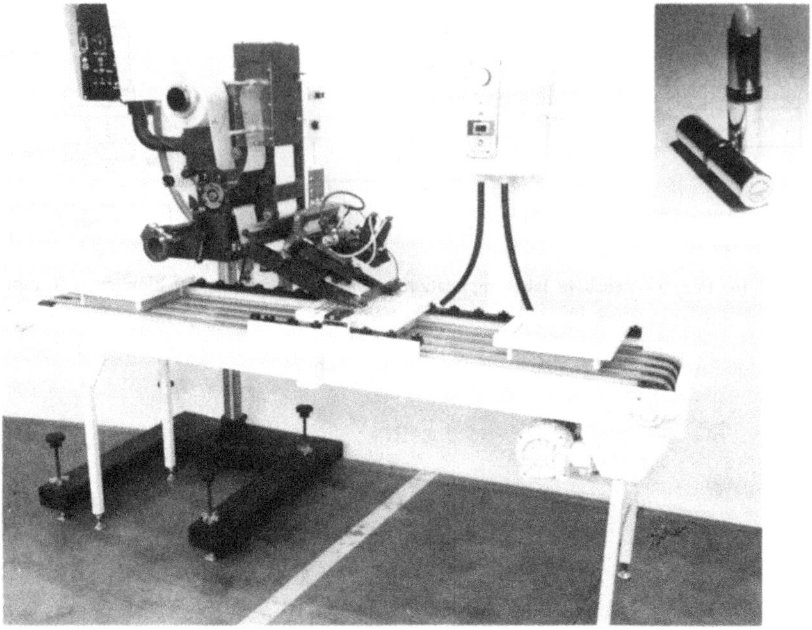

Fig. 33-16. Pressure sensitive labeler for application of small cosmetics labels (*Courtesy of NJM, Inc.*)

are intended to identify the product, content, manufacturer, and in the case of some food items, nutritional information.

Secondary labels are intended for pricing, promotional uses, special discounts, etc. Pressure sensitive labels dominate the secondary label market. Secondary labels are applied in smaller volume, and pressure sensitive labeling is often more economical for small production runs, because of less expensive application equipment and less downtime for cleaning.

Primary Labels

Primary labels are applied at the time of packaging. The penetration of pressure sensitive labels into primary label applications has been small, but it is increasing. Pressure sensitive labels are the main label type used in certain retail sectors, such as pharmaceuticals, toiletries, and cosmetics. The reason for the successful penetration of these product areas is the acceptable overall costs of pressure sensitive labeling as compared to glue application in the case of small runs.

Pressure sensitive labels are also replacing delayed action adhesives for pharmaceutical products where such labels have a well-established acceptance.

Primary label consumption is very large and it is expected that the main growth of pressure sensitive labels will be in this area.

Pricing Labels

Pressure sensitive adhesive labels have replaced other means of price marking various products in retail stores. Label stock is usually 60 lb (97 g/m^2) or lighter paper one-side coated for litho printing (C1S) or similar paper with permanent pressure sensitive adhesive. The paper may be prescored so that the label, once affixed, cannot be removed without being destroyed, thus preventing price tampering.[11]

For special applications, the label stock may be coated with water-removable adhesive, making the label easy to remove by simply wetting it, for use on dishes and other houseware items.[12]

Delayed tack adhesive labels are used to price food items, such as meats, poultry, and deli products that are individually wrapped and weighed. Weighing of such products and application of labels may be automated. Bakery products and other high turnover and/or freshness date-coded products are also priced with delayed tack adhesives.

Office and Inventory Labels

These labels are supplied in both roll and sheet form, depending on volume of use. The labels are coated so as to accept inks.

Electronic data processing (EDP) labels are used for addressing, shipping, content identification, and inventory control. With the advent of the computer, especially the small business computer, EDP labels are used extensively. Address labels are supplied in fan-folded form, so that they stack neatly prior to printing, unfold as they go through the printer, and then refold in a neat stack.

Duplicate-copy labels are available. The release liner might have encapsulated ink that is activated on printer impact.

Single-part business forms are manufactured with removable pressure sensitive adhesives, so that the form can be detached and repositioned as needed. Duplicate label forms can be simultaneously printed in one step, and the labels are removed in consecutive steps—for quality control, for recordkeeping and inventory control, and for shipping—with a final label remaining for product identification. Incorporating pull tabs into the label construction allows for a combination shipping label with unit pricing labels that are used when the package has reached its destination.

Piggyback label constructions are based on a double release liner system. The label can be used, and repositioned and reused at a later time. Magazine publishers and other end users employing mass mailing often use piggyback address labels.

Transparent labels are used on microfiche and engineering drawings for identification and titling purposes. The adhesive coated film is specially treated so as to accept ink or toner. Transparent film envelopes for packing lists are attached with pressure sensitive adhesives to the sides of cartons, pails, and boxes. Once the envelope is in place, the contents can be removed only by destroying the envelope.

Embossable and Imageable Labels

Embossable labels are constructed from a heavy vinyl film, a pressure sensitive adhesive coating, and a release liner. The film is formulated with a specific metallic hydride that blushes when deformed. Word messages are embossed on such tapes by a special embossing tool, yielding a permanent and attractive label with white letters on a colored background.

Ultraviolet light imageable tapes compete for the same applications. Such tapes are imaged one digit at a time with a special tool like the embossable tapes.

Other imageable tapes are constructed from a coated film that can be imaged by applying pressure with a stylus or a typewriter key. The coating may be designed to accept ball-point or typewriter inks, including computer printer inks.

Wire Markers

Wire marker labels are widely used in the electrical and electronics industries for identification purposes. A wire marker label is a small strip that is coded by either number or color, identifying an electrical terminal and/or the wire leading to that terminal. The marker is generally film or coated fabric with a pressure sensitive adhesive coating mounted on a release coated card for easy application. Wire markers are used extensively in computers, aircraft, ships, and factory machine controls. A solvent-resistant adhesive is required. Products similar to wire markers are used for marking piping.

GRAPHIC ARTS

Pressure sensitive numbers, letters, and symbols are used in the graphic arts and architectural fields to provide a neat, professional appearance for drawings, displays, and signs. For a more detailed discussion, see Chapter 32.

AUTOMOTIVE LABELS

Pressure sensitive labels are used in passenger cars, vans, and trucks to carry operating instructions, specifications, and warrants on batteries, jacks, and engine parts. The labels are constructed from various materials, such as paper, film, foil, and various laminates, and may require solvent and high-temperature resistant adhesives.

Pressure sensitive labels are applied at service centers to indicate date and nature of automobile servicing.

Special film tire labels are attached to tire treads by the manufacturer, giving specific information, such as allowable air pressure.

NAMEPLATES AND DECALS

This is a broad category, encompassing a variety of decorative, identifying, and instructional stickers and labels.

Nameplates are generally used for information and identification on motors, appliances, and various mechanical parts. Coated and metallized films and foils, including aluminum sheeting, are often used as facing materials, while the back of the nameplate is coated with a pressure sensitive adhesive designed for a long service life. Most nameplates are furnished as individual pieces and are usually placed into position by hand. The adhesive may be conveniently applied by using transfer tape.

Decals are used for product decoration and promotion. The difference between a label and a decal is not always clear; decals are generally more elaborate labels that are designed for long use, and imitate decalcomanias applied employing water soluble glue. Clear films are usually used for decals so as to suggest that only the printed image has been transferred. The variety of decals is almost endless, ranging from children's stickers to football helmet insignia to manufacturers' logos that identify a particular product. Polyester film 13–40 μm (0.5–1.5 mil) thick is most often used for decals.

SIGNS

Signs are basically large labels designed for application with pressure sensitive adhesives to various surfaces. The facing material is usually plastic film. The face material is printed by any printing method that is convenient to handle the material in sheet form. It may be printed by screen, offset flexography, or letterpress. The printing may be overcoated or overlaminated, or the printing may be on the second surface in order to protect the print and increase the lifetime of the sign. Many elaborate printing techniques are used to obtain special effects.

The adhesive is usually applied after printing, by laminating with a double-face tape or a transfer tape. If the sign is to be applied to a rough surface, a heavy adhesive coating is required, or a double-face foam tape that helps the adhesive to conform to the uneven surface is often used.

PROTECTIVE SHEETS

Clear pressure sensitive adhesive coated films are used as protective surfaces in conjunction with labels. Oriented polypropylene, polyester, and vinyl films are used with a clear adhesive.

REFERENCES

1. Blum, A. U.S. Patent 3,978,274 (1976) (assigned to Borden Inc.).
2. Dabroski, W. C. U.S. Patent 3,067,057 (1962) (assigned to Johnson & Johnson Co.).
3. Funk, M. C., and Rogalska, E. A., Jr. U.S. Patent 3,575,788 (1971) (assigned to American Can Co.).
4. Satas, D. Labels and Decals. *Plastics Finishing and Decoration,* pp. 372–385 (D. Satas, ed.). Van Nostrand Reinhold, New York, 1986.
5. Fairley, M. C. Labels and Labeling Machinery. *Wiley Encyclopedia of Packaging Technology,* pp. 424–429 (M. Bakker, ed.). Wiley, New York, 1986.
6. Komerska, J. Labels and Decals. *Handbook of Pressure-Sensitive Adhesive Technology,* pp. 478–497 (D. Satas, ed.). Van Nostrand Reinhold, New York, 1982.
7. Hickman, A. D. The Design of S/B Latex Pressure-Sensitive Adhesives. In *Proc. Technical Association of the Pulp and Paper Industry,* pp. 61–68. Lake Buena Vista, Fla., Sept. 26–28, 1983.
8. McKenna, L. M., and Gardner, D. M. U.S. Patent 3,763,117 (1973). (assigned to Monsanto Co.).
9. Mullen, J. N. U.S. Patent 3,758,433 (1973) (assigned to Imperial Chemical Industries Ltd.).
10. Reich, R. L. U.S. Patent 3,093,528 (1963) (assigned to Kleen Stik Products, Inc.).
11. Huber, E. W. U.S. Patent 2,845,728 (1958) (assigned to Topflight Corporation).
12. Mullen, J. N. U.S. Patent 3,758,433 (1973) (assigned to Imperial Chemical Industries Ltd.).

34. Coating Equipment

Donatas Satas
Satas & Associates
Warwick, Rhode Island

Pressure sensitive adhesives are applied as solutions, emulsions, or hot melts over various flexible substrates—films, papers, and fabrics—in a relatively heavy layer. Most of the adhesive coatings are in the thickness range of 20 to 70 μm. A coating of 25 μm (1 mil or about 25 g/m^2) is considered to be thin, although it is sufficient for label stock and thin film tapes. Coatings 50 μm thick are quite common for tapes, however, while the adhesive coatings may be as heavy as 150 μm for some products. If the adhesive coating is applied as a solution (25–50% solids), or as an aqueous emulsion (50% solids), the liquid coating must be correspondingly heavier before drying. Therefore the coating techniques used must be suitable for the application of heavy adhesive layers. The choice of the coating technique largely depends on the physical state of the adhesive—emulsion, solution, hot melt, calenderable solid, or 100% solid reactive material.

The manufacturing of pressure sensitive products may also require application of release and prime coatings, which are deposited as lightweight coatings and require different coating methods than the heavier adhesive coatings. Silicone release coatings are applied in the coating weight range of 0.8 to 1.5 g/m^2. If the coating is applied from a solution, it can be diluted to keep the solids low and to apply a coating sufficiently heavy for the application technique used. Thus dilution only increases the accuracy of the deposit. But introduction of 100% solids reactive coatings placed a heavy demand on the equipment to deposit such coatings in thin layers within the acceptable accuracy limits. Therefore, dilution cannot be used any longer as a technique to improve deposit accuracy. This demand prompted the development of new coating techniques.

Most of the pressure sensitive adhesive products are manufactured by direct coating, that is, the adhesive is applied directly to the backing material. Some products, notably labels, are manufactured by transfer coating: the adhesive is applied to the silicone coated release paper, dried, and then laminated to the label face stock at the end of the coating line. In addition to label stock, transfer coating is also used to apply the adhesive to heat-sensitive or bulky backing materials, which are difficult to handle in the drying oven and which also require release liner to protect the adhesive. Thus foam tapes, adhesive-backed carpeting, double-face tapes, and other products are usually made by the transfer process.

Transfer coating requires a laminating station at the end of the line. This station consists of a steel- and a rubber-covered roll, at least one of which is driven, and a separate unwind stand for the web to be laminated.

Two different functions of the coating process should be distinguished: coating application and metering of the deposit thickness. In some coating equipment these two functions are combined; in some they may be carried out by different pieces of equipment. Some coating equipment applies a coating of predetermined thickness independently of any variations in the backing caliper. Other coating methods coat to a constant total thickness, resulting in a coating of nonuniform thickness if the backing gauge varies (Fig. 34-1).

Many different coating methods are used for pressure sensitive adhesive coating. Reverse roll, knife-over-roll, and roll-over-roll are the most commonly used methods for solution coatings; a wire-wound rod is often used for emulsion adhesives and lighter weight coatings; a slot orifice is used for hot melt adhesives. Wire-wound rod, gravure, and transfer roll coaters are used to apply lightweight coatings such as are used for release or priming. Coating equipment has been discussed in numerous publications, including several books.[1-3]

REVERSE ROLL COATERS

Reverse roll coater is a versatile and accurate coating machine. The normal viscosity range is from 1 to 6 Pa·s (1000 to 6000 cps), but coatings of much higher viscosity (20 Pa·s) have been applied successfully. Lower viscosity coatings require pan-fed coaters, because the edge dams cannot contain fluid coatings. Coating speeds of 100 m/min are common, and speeds of 300 m/min are claimed.

The operation of the reverse roll coater consists of two steps: metering of the coating onto the applicator roll and transfering the coating to the substrate

Fig. 34-1. Difference in the coating processes. (a) Coating to constant total thickness. (b) Coating to constant coating thickness.

(Fig. 34-2). The adhesive coating is metered onto the surface of the applicator roll by the gap set between the metering and applicator rolls. The metering roll usually rotates in the opposite direction of the applicator roll at a speed 2–20 times slower than that of the applicator roll. Higher viscosity requires slower rotational speed of the metering roll in order to keep the film surface smooth. In some cases, the metering roll may be kept stationary and rotated only for cleaning purposes. The adhesive is transferred to the substrate at the nip between the applicator and rubber-covered backup rolls. The backup roll rotates in the direction opposite to that of the applicator roll (hence reverse roll coater) and the adhesive is wiped off cleanly by the substrate. The roll speed ratio between the applicator and backup roll can be varied between 0.6 and 4 (most often between 1 and 2). The higher the ratio, the faster the rotation of the applicator roll in relation to the web and the thicker the adhesive coating applied to the substrate. This roll speed ratio, called wipe ratio, as well as the preset gap between the metering and applicator rolls, is used to control the coating thickness. The wipe ratio remains constant when the web speed is changed. The gap between the applicator roll and the rubber-covered backup roll is set a little below the web thickness so that the web is slightly compressed in the nip.

Several designs of reverse roll coaters are used. The main variations are the type of feed—either nip or pan (or the more recently proposed fountain feed)—the location of the feed—top or bottom—and the number of rolls used—three or four. Figure 34-2 shows the most popular top-nip-fed three-roll reverse roll coater. Similar coater with bottom feed is shown in Figure 34-3. This

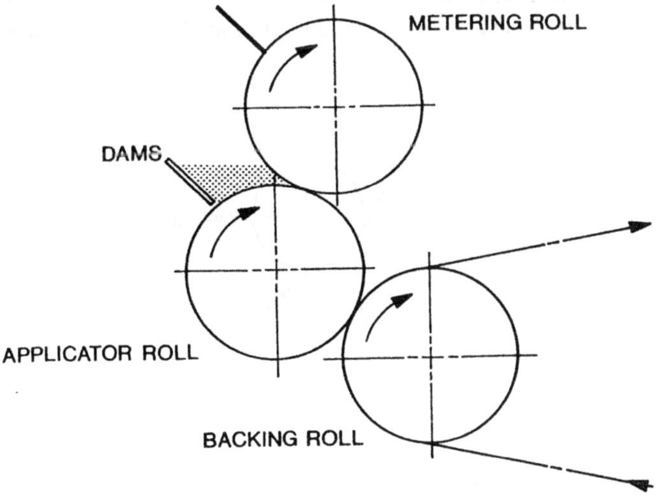

Fig. 34-2. Top-nip-fed three-roll reverse roll coater.

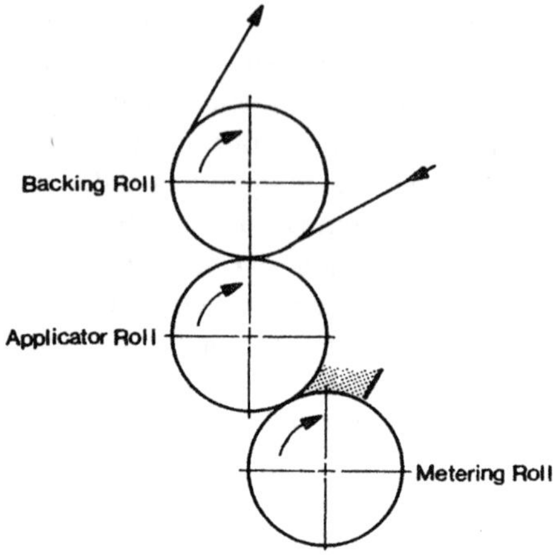

Fig. 34-3. Bottom-nip-fed three-roll reverse roll coater.

arrangement allows the operator to clearly see the coated surface that is in front of him. The pan-fed three-roll coater, suitable for lower viscosity coatings, is shown in Figure 34-4. Pan feed coaters require edge wipers in order to remove the coating from roll edges and to prevent it from flowing under the web edges. The pan-fed four-roll coater is shown in Figure 34-5. Reverse roll coaters have been made as wide as 480 cm, but the most popular widths are 137 cm (54 in.) and 183 cm (72 in.).

Reverse roll coaters are versatile but expensive machines. The rolls are constructed from chilled cast iron and are chrome plated for corrosion resistance.

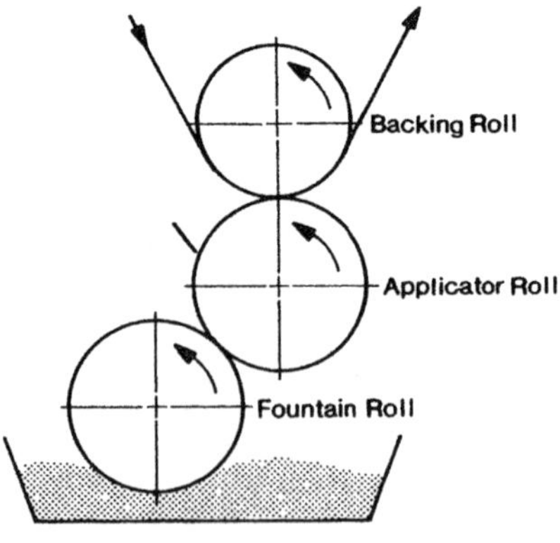

Fig. 34-4. Pan-fed three-roll reverse roll coater.

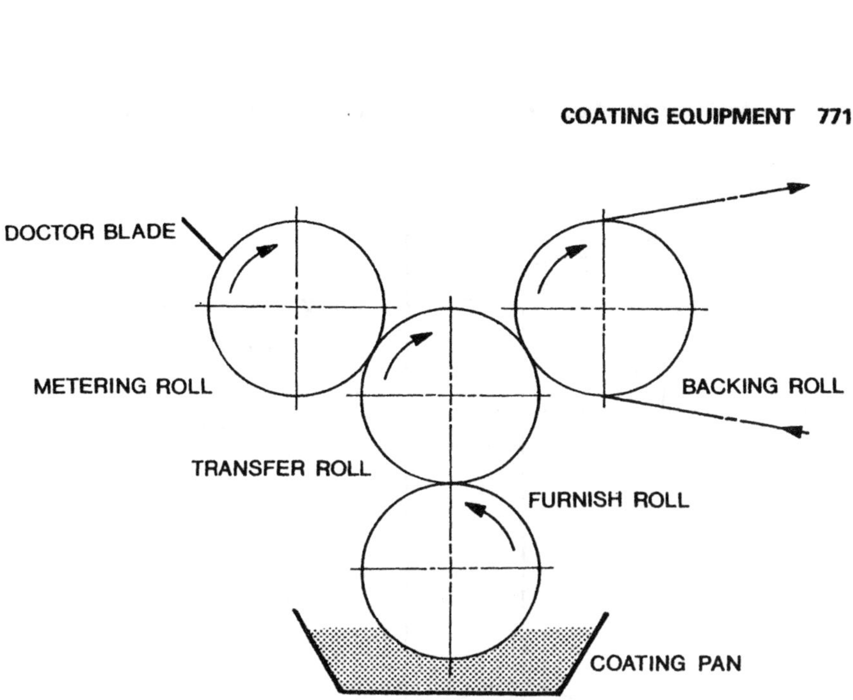

Fig. 34-5. Four-roll pan-fed reverse roll coater.

Although individual rolls may be machined to the TIR* of 1 μm, the total error of a reverse roll coater is 6–12 μm.[4,5] If we are applying a 50 μm dry coating from a 25% adhesive solution, the wet coating will be 200 μm thick and the error about 10 μm (about 5%). A coating of emulsions containing 50% solids increases the coating error to about 10%.

Because of the film splitting between the applicator and metering rolls, longitudinal striations are formed in the adhesive, and these may be carried over to the finished product, unless the adhesive coating levels sufficiently well before drying. This phenomenon is observed in all adhesive types, but it can be especially serious with aqueous emulsions. The severity of these surface disturbances depends on two processes: flow of the adhesive during coating, which takes place at high shear rates (10^4 sec^{-1}), to which the adhesive is exposed on the reverse roll coater, and leveling, which takes place at low shear rates (0.01–0.1 sec^{-1}). Leveling depends on such coating properties as viscosity and surface tension and is discussed in Chapter 35. Striation formation is related to coating equipment geometry and coating conditions.

A schematic illustration of the adhesive behavior in the nip between applicator and pickup (metering) rolls and the pressure profile is shown in Figure 34-6.[6] Pressure maximum at the nip entrance helps to keep the nip fed with the adhesive, while pressure minimum on the exit side creates cavities that develop

* Total indicator runout (TIR) is the deviation of a roll as read from the total reading of a dial indicator. It includes deviation because of roll eccentricity, as well as deviation from a straight surface.

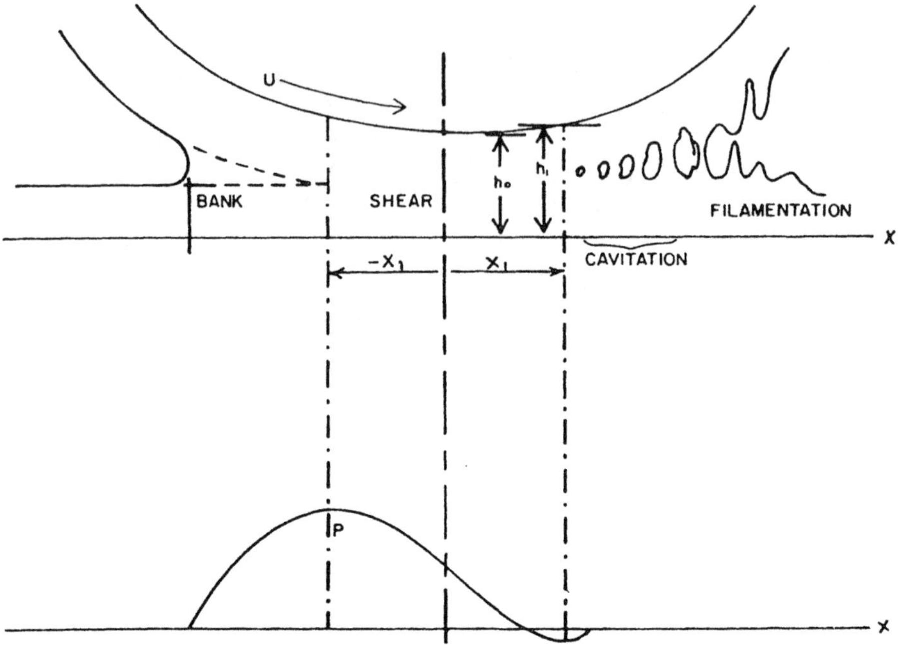

Fig. 34-6. Regions in a roll nip and the pressure profile created, according to Meyers. (Reprinted by permission of John Wiley.)

into strings. The strings break and snap back, creating the ribs-striation marks. The ribs start leveling even before they are transferred to the web, which is wiping the applicator roll surface. The wiping action helps to reduce the size of the ribs, and further leveling takes place before the coating becomes immobile as it dries.

The closer together the striation marks (frequency) and the smaller the size (amplitude) of the individual ribs, the smoother the coating. Less flow is required of the coating during the leveling process to smooth out a finely ribbed coating than a coarse one. The spacing (frequency) of the striation marks depends on the nip clearance. The narrower the nip, the higher the frequency of the striation marks and the smoother the resulting coating. At a given wet coating thickness, the required nip opening depends upon the transfer efficiency of the coating. Viscosity at the high shear rates determines the transfer efficiency: the higher the viscosity, the thicker the coating on the applicator roll after splitting. Viscosity has no effect on the rib frequency, only on the transfer efficiency. The leveling process depends on the viscosity at low shear rates and on the thickness of the wet coating. Higher film thickness favors better leveling; thin films $<10\mu m$ have a highly striated appearance. Thus an adhesive of a high viscosity at high shear rates and a low viscosity at low shear rates is the best for obtaining a smooth coating.

An empirical equation was developed relating the film thickness to material properties and coating process variables.[7]

$$L = k\alpha \frac{\eta_H}{\gamma} \cdot l \frac{R_p^{1/2} R_a^{3/2}}{R_l} \tag{1}$$

where

L = dry film thickness
k = dimensionless coefficient
η_H = viscosity at high shear rate
γ = surface tension
α = percent volume solids
l = nip clearance
R_p = pickup (metering) roll speed
R_a = applicator roll speed
R_l = backup roll (line) speed

The first member $\alpha(\eta_H/\gamma)$ depends on the material properties of the adhesive, and the second member $l(R_p^{1/2} R_a^{3/2})/R_l$ depends on the operating variables of the coater.

Under fixed operating and product conditions $\eta_H \cdot l$ is constant and

$$\frac{l}{l_0} = \frac{(\eta_H)_0}{(\eta_H)}$$

Twofold increase in the viscosity will result in decreasing the nip clearance by half, which will substantially increase the frequency of striation marks and improve the coating smoothness. Since it is not possible to run a reverse roll coater at conditions where no rib formation will take place, nip clearance is the main factor to control the rib frequency. Nip clearance is determined by roll speeds and material constants as shown below.[7]

$$\frac{l}{l_0} = \left(\frac{R_{p0}}{R_p}\right)^{1/2} \left(\frac{R_{a0}}{R_a}\right)^{3/2} \left(\frac{R_l}{R_{l0}}\right) \left(\frac{\eta_{H0}}{\eta_H}\right) \left(\frac{\gamma}{\gamma_0}\right) \left(\frac{\alpha_0}{\alpha}\right) \tag{2}$$

KNIFE-OVER-ROLL COATERS

Knife-over-roll coaters are simple and direct machines for applying adhesives in relatively heavy coatings (over 50 μm wet) over various backings. Adhesives

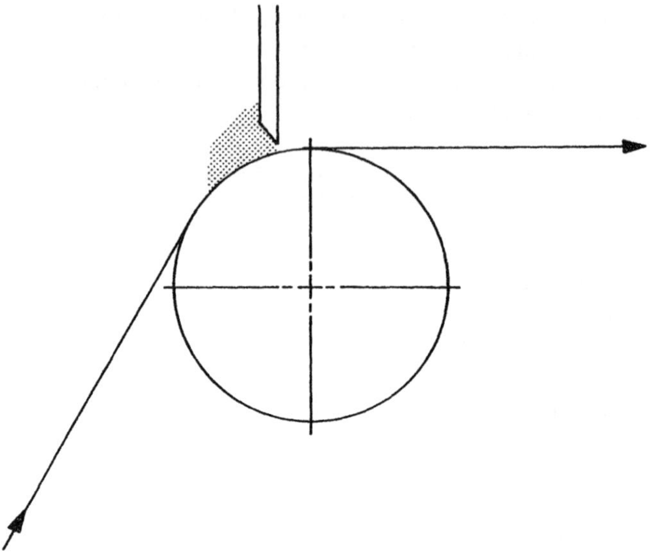

Fig. 34-7. Knife-over-roll coater.

over a wide viscosity range can be used. The minimum viscosity is limited to about 4 Pa·s by the capability of the dams to hold the adhesive, while the maximum is about 50 Pa·s and is determined by the formation of scratches and poor leveling inherent to high viscosity coatings. Figure 34-7 shows the principle of this coating technique. The web-supporting roll is machined to a high precision and is chrome plated. The heavy steel knife is also accurately machined and mounted slightly off center. The coating thickness is determined by the gap between the knife and the roll. The gap is adjusted by an elevating mechanism at both ends of the knife. Some of the knives may be equipped with contour adjustment, which is accomplished by bending the knife by pressure bolts.

Knife coater develops high shear rates: a high-speed knife-over-roll coater develops a shear rate in the range of 1000 to 10,000 sec^{-1}. Problems may be experienced when coating dilatant coatings on knife-over-roll machines.

Knives of various designs are used in knife-over-roll coaters. Some of these designs are shown in Figure 34-8. The sharper the knife, the smaller the radius on the leading edge, and the higher the shear force exerted on the coating. For pressure sensitive adhesives large-radius knives are most often used, such as shown in Figures 34-8d and 34-8e. A knife-over-roll coating head is shown in Figure 34-9.

The knife-over-roll coating method is not suitable for coating emulsion adhesives because of the scratches and lines left by the knife. If the knife is

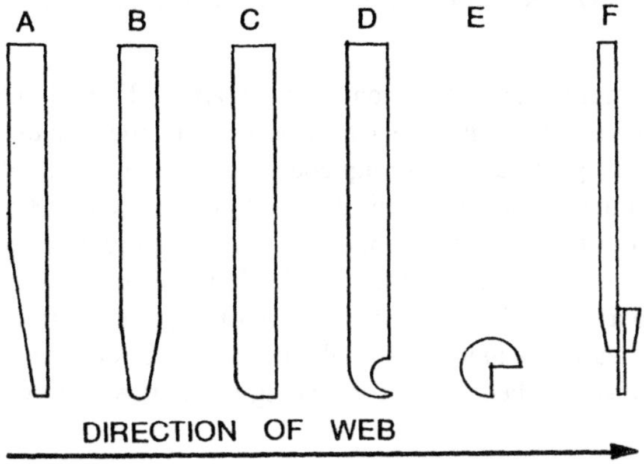

Fig. 34-8. Blade designs. (A) Sheeting knife. (B,C) Radius knives. (D) Hook knife. (E) Bull nose. (F) Spanishing knife.

Fig. 34-9. Knife-over-roll coating head. (*Courtesy of Liberty Machine Co., Inc.*)

inclined at an angle, a smoother coating is obtained. Flexible-blade coater with the blade at a very low angle, such as is used in the manufacture of coated paper, reportedly gives a smooth aqueous pressure sensitive adhesive coating.

Knife coating is used in many different arrangements. Knife-over-blanket coating is similar to knife-over-rubber-roll, and floating-knife coating is shown in Figure 34-10. In the latter method the knife is pressed against the web, which is supported on both sides of the knife by either rolls or nonrotating supports, such as a channel. The method is suitable for backcoating of textile materials and also has been used for coating pressure sensitive latex adhesives.

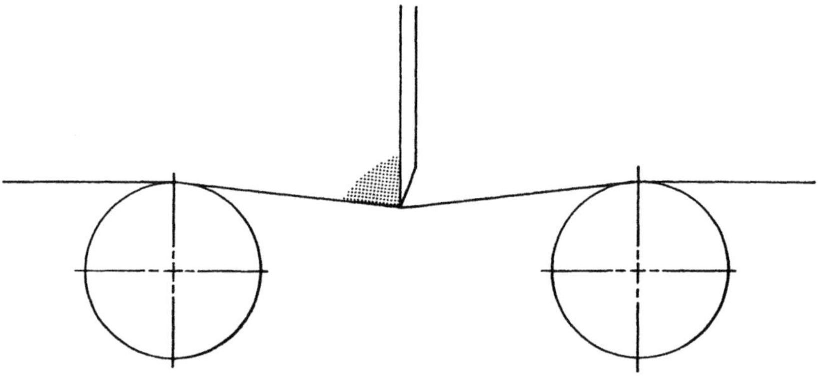

Fig. 34-10. Floating-knife coater.

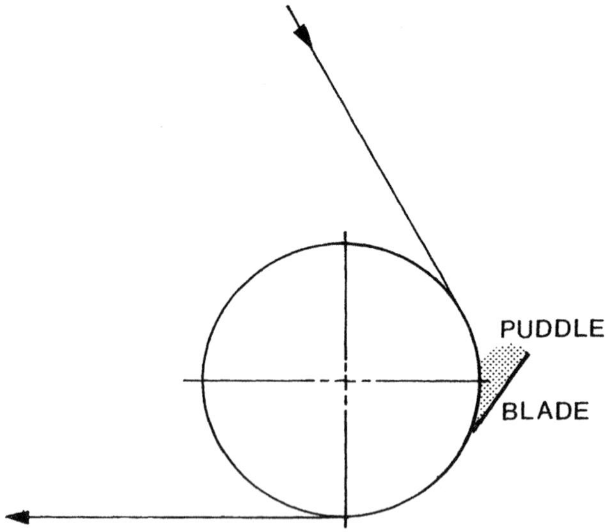

Fig. 34-11. Trailing blade coater.

Blade coaters are similar to the knife coaters, except that a flexible blade is used instead of a rigid one. The blade wipes the surface of a web against a backup roll. The coating can be a puddle in front of the blade, as in the case of the trailing blade coater shown in Figure 34-11. The coating can also be preapplied by a roll and the blade used for metering purposes only. The blade can be used in the inverted position and in several other combinations. The trailing blade coating is important for high-speed paper coating. In pressure sensitive product applications, the method is applicable for release coating and other lightweight applications. Blade coating has been tried with some success for the application of aqueous pressure sensitive adhesives. The development of a blade coater with an inflatable pressure tube near the blade tip provides accurate control of the blade pressure and allows the deposition of heavier coatings.

Knife coating is a high shear process. Its mechanism is complex and has not been studied that much. Mathematical models proposed required many simplifying assumptions. Freeston[8] has published a theoretical analysis of knife coating over porous substrates. A knife coating model as shown in Figure 34-12 has been analyzed. The analysis is based on a simplified geometric model of the knife/coating/substrate interface and on several other simplifying assumptions. Freeston derived the equation

$$p_m = \frac{4\mu U}{l\theta^2} \tag{3}$$

where

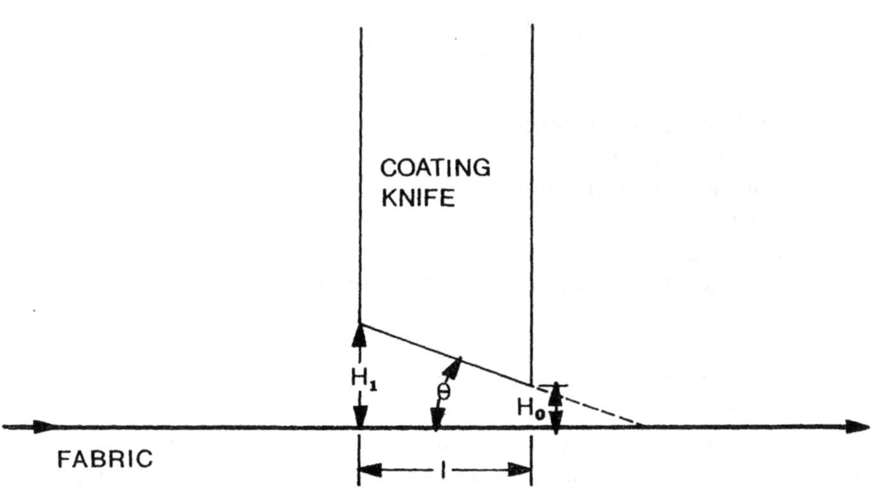

Fig. 34-12. The geometry of a coating knife.

p_m = the difference between the total pressure and the hydrostatic pressure
U = web velocity
μ = viscosity
l = knife width
θ = knife angle

The expression shows that the higher the web velocity, the thinner the knife, and the smaller the knife angle, the greater the dynamic pressure generated under the knife. Increasing the coating viscosity increases the pressure, but the viscosity is a complex variable and the relation is not as simple as shown in Equation (3).

The penetration of the coating under the knife into the porous substrate is also discussed by Freeston.[8] He concluded that the depth of penetration is mainly determined by the pressure developed under the knife. The depth of penetration is independent of web velocity, but increases with an increasing degree of coating pseudoplasticity. The penetration is increased by decreasing the angle between the blade and the substrate and by increasing the pore size of the substrate.

Hwang[9] has presented a detailed analysis of the hydrodynamic forces involved in a coating operation employing a rigid blade and assuming a Newtonian fluid. Two different analyses are presented: a simplified one employing five variables and a complete one employing fifteen variables. In the simplified version, the coating thickness is related to gap width, liquid viscosity, density, surface tension, and the web speed. The relationship is shown in Equation (4). A schematic diagram of the coating geometry is shown in Figure 34-13.

$$T = \frac{h}{2} + \frac{1}{12\mu}\left(\frac{\sigma}{2h^2} + \rho g_x\right)\frac{h^3}{u_0} \tag{4}$$

where

T = coated liquid film thickness
h = coater gap setting
σ = surface tension of the liquid
μ = viscosity
ρ = liquid density
g_x = gravity constant in the direction of x coordinate
u_0 = web speed

The results calculated by Equation (4) are compared to experimental data in Figure 34-14.

Particles in the adhesive can clog the gap between the knife and the substrate, causing streaks. Therefore, it is desirable to coat from the largest possible gap, so that these particles would be able to pass under the knife. Usually,

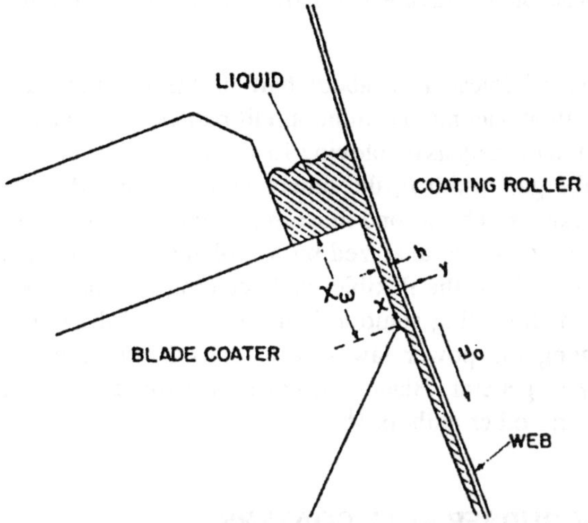

Fig. 34-13. The geometry of a blade coater.

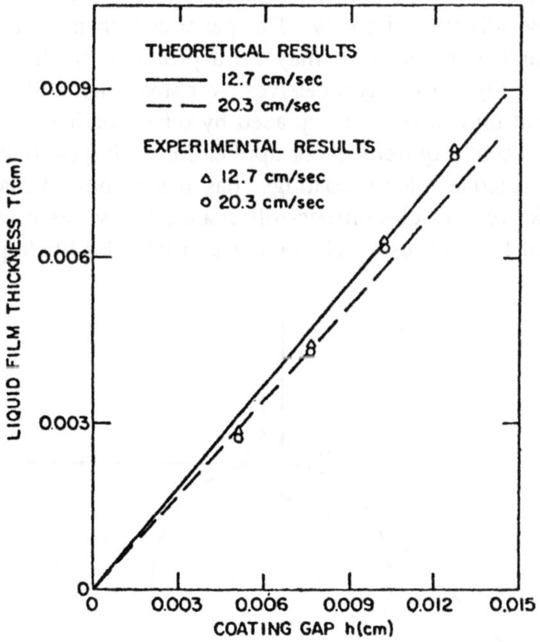

Fig. 34-14. Comparison of theoretical calculations based on Equation (4) with experimental data.

the liquid film thickness T is about half of the coating gap h. An accurate coater optimization leading to a minimum film thickness for a given gap thickness requires a complete analysis employing 15 variables. The most important variable, however, is the gap width h; the other variables can effect only 10% change in the film thickness. The second most important variable is the blade thickness.

Blade coating is also analyzed by Middleman[10] for both rigid and flexible blades. The analysis of the flexible blade coating becomes more complex. The case of Newtonian coating solution is discussed, and the nonlinear behavior of a fluid following the power law is also considered as well as the effect of viscoelasticity in general. Blade and knife-over-roll coating has also been discussed by several other authors.[11–13]

KNIFE-OVER-RUBBER ROLL COATERS

A knife-over-roll coater of similar appearance, but based on a different operating principle, is shown in Figure 34-15. The roll is covered with a resilient rubber cover and the coating thickness is determined by the blade pressure applied against the rubber covering, causing its deformation. The coating thickness is also affected by adhesive viscosity. The generated shear forces are high and the strike through a porous web may be a problem. Such coaters had been used for applying high viscosity adhesives over strong textile backings to manufacture tapes, but they have been replaced by other coating techniques.

The more recent requirement for application of lightweight coatings, such as 100% solid silicone release coatings, has placed new demands on coating equipment. A knife-over-elastomeric-roll coater, known as a Dahlgren coater, has been offered for such uses. The coater consists of a blade pressing against

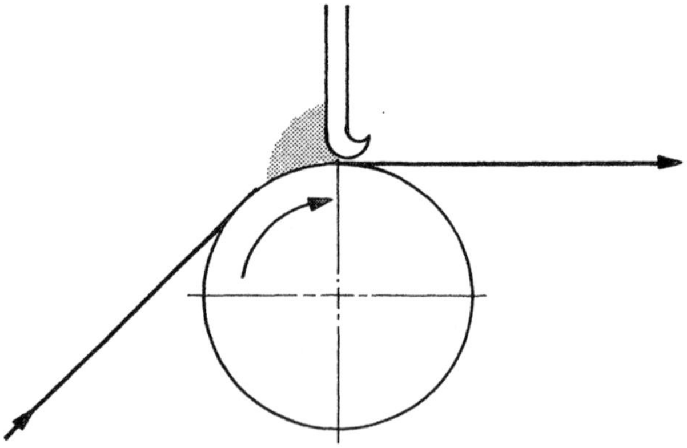

Fig. 34-15. Knife-over-rubber-roll coater.

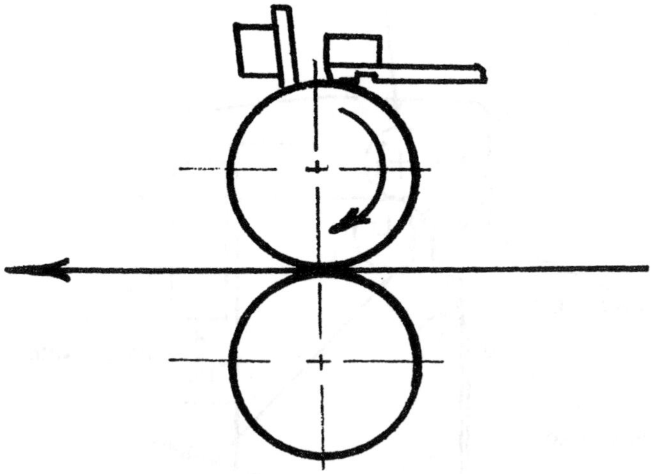

Fig. 34-16. Dahlgren coater.

an elastomeric roll covering (Fig. 34-16). The amount of the coating deposited onto the roll surface depends on the blade pressure and angle.[14] The coating gets past the blade and is carried by the roll to the nip where it is transferred to the web. Coating deposits as low as 0.7 g/m^2 are accurately applied.

ROLL-OVER-ROLL-COATING

The roll-over-roll coating method (also known as Levelon) is similar to knife-over-roll coating, except a roll is used in place of the knife. Thus the coater consists of two accurately machined steel rolls, or a steel and a rubber roll (Fig. 34-17). Such coaters generate less shear in the coating nip than knife-over-roll machines and produce a smoother surface with less streaks. They are used for pressure sensitive adhesive coating, and have been found useful for the coating of aqueous adhesives. Roll-over-roll coaters are available in several modifications, including a pan-fed coater as shown in Figure 34-18.

AIR-KNIFE COATING

The air-knife coating has been developed for clay coating of paper and has been used for coating of aqueous pressure sensitive adhesives. Another of its uses is for label stock where the adhesive is applied to the label face directly rather than to the release paper. It is most suitable for low coating weight applications of, say, up to 20 g/m^2 per pass.

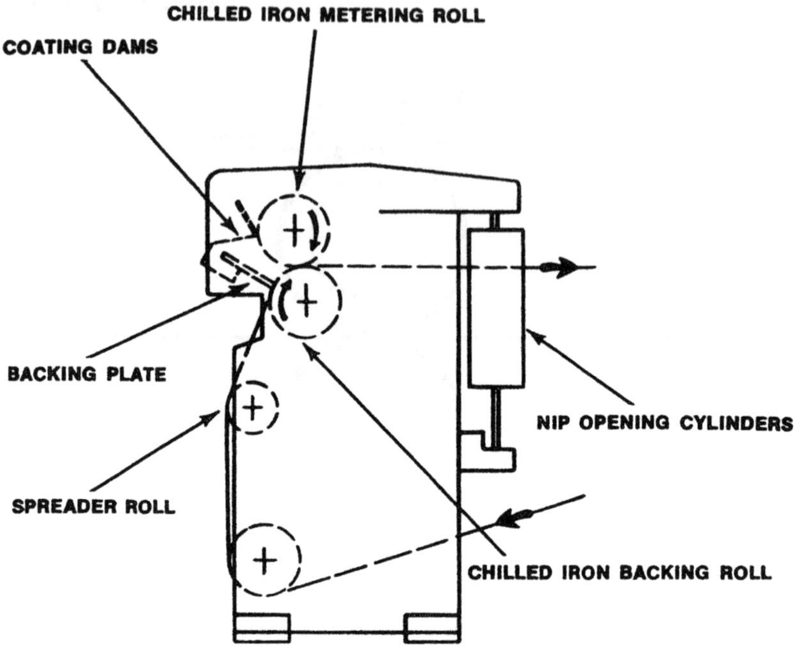

Fig. 34-17. Levelon coater vertical reservoir type. (*Courtesy of Midland-Ross Corp.*)

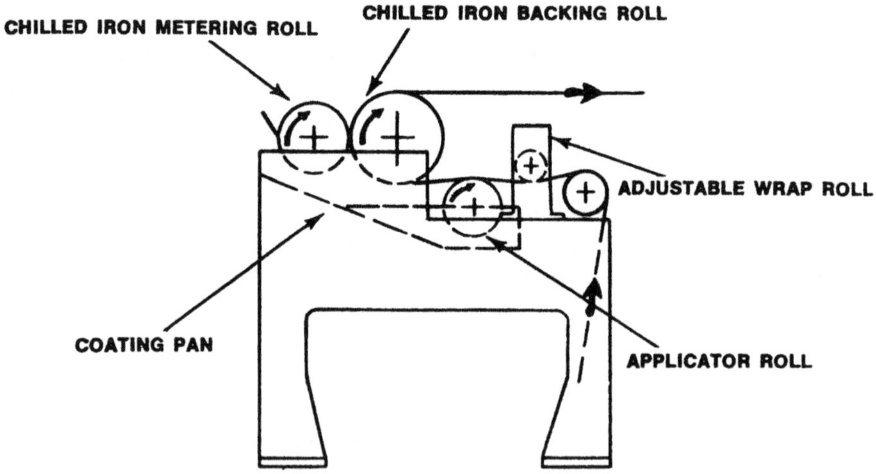

Fig. 34-18. Levelon coater pan-fed type. (*Courtesy of Midland-Ross Corp.*)

Air-knife coating is a metering system that removes excess coating applied to paper. Unlike knife-over-roll or blade coating methods, which coat to a constant total thickness, air-knife coating follows the contours of the paper: it coats to a constant coating thickness, similarly to the roll coaters.

The air knife can be used in three different ways. In metering a highly filled latex coating on absorbent backing, the air knife removes the outer, less viscous layer. The absorbent material removes the water from the coating and the layers closer to the web are higher in solids and viscosity and are more difficult to remove by shearing. The shearing is done by the air stream forced through a gap between the blades, as shown in Figure 34-19. This coating mechanism is similar to the formation of a filter cake on a filter screen. In the case of nonabsorbent webs or less filled lower viscosity coatings, the air knife removes a portion of the coating. The air velocities used are lower in this case. Air knife and brush air doctor can be used for finishing purposes only. A light stream of air helps to level the coating that has been applied by some method.

A schematic diagram of an air-knife coater is shown in Figure 34-20. The excess coating is removed by the air knife and is recirculated. The aqueous coatings are inclined to foam, and various designs of the vessel receiving the excess coating are available, facilitating the deaeration of the coating. Judicious use of defoaming agents can also help. The coating can be applied by various methods: single roll as shown in Figure 34-20, multiple roll system, fountain, or some other method.

The air knife as shown in Figure 34-19 consists of a plenum chamber and a set of blades that can be adjusted to regulate the gap A. The main

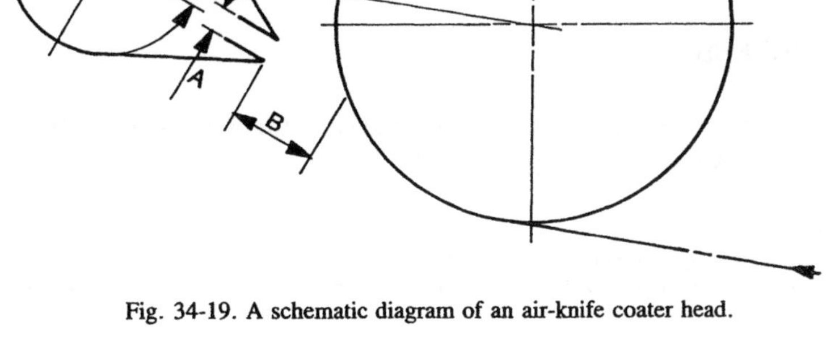

Fig. 34-19. A schematic diagram of an air-knife coater head.

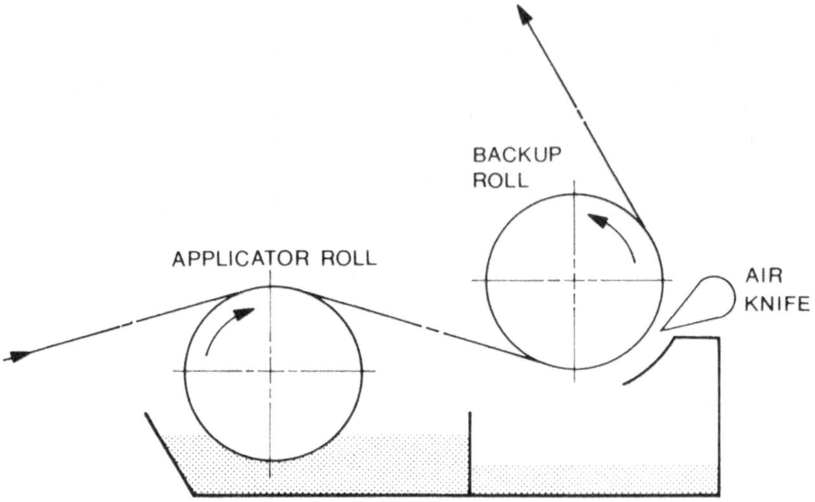

Fig. 34-20. Air-knife coater.

variables are the air velocity, distance between the lip and the drum B, lip gap A, angle of attack (θ). The range of these variables is given below.

Lip gap, 0.25–0.75 mm
Lip distance, 3.5–5.5 mm
Angle of attack, 5–12 degrees
Air velocity, 30–140 m/min
Air pressure,* 0.1–0.6 atm
Air volume, 2–10 m^3/hr/cm
Coating weight, 3–30 g/m^2
Coating speed, 10–750 m/min

The effect of various process variables is discussed by Taylor.[15]

BAR COATERS

Bar coaters are widely used for metering and smoothing operations. Wire-wound rod coaters, also known as Mayer rods, are the most popular version of bar

* The air pressure range is given for coating of absorbent materials, such as paper. The air pressure for nonabsorbent materials is 5–10% of that, and the air velocity and air volume are correspondingly lower.

coaters and they are often employed for applying lightweight pressure sensitive adhesive as well as release and prime coatings, although coatings as heavy as 100 μm have been applied. The viscosity should be below 1 Pa·s to be suitable for applying both solvent and aqueous coatings at speeds as high as 200 m/min. Bar coaters are metering devices, and the coating is applied by some other method, usually a kiss roll (Fig. 34-20).

The wire-wound rod is constructed by winding a wire around a small-diameter (6–8-mm) rod. The rod wipes off the excess coating, except what passes between wires. The amount of coating remaining depends on the wire diameter: the larger the diameter, the more coating remains on the substrate. Wire of diameters 0.07–0.8 mm and heavier are used for wrapping the rod. To minimize clogging, the thickest possible wire should be used.

Figure 34-21 shows schematically the wire-wound rod and the spaces between the windings that control the coating deposit. The area of one gap is equal to $2r^2 - r^2/2 = 0.49r^2$. Ideally this would yield a coating thickness of $0.2145r$, if the gap area were the only factor that determined the coating weight. However, the amount of coating deposited depends on variables such as web tension, wrap angle, rheological properties of the coating, web speed, bar rotating speed, and the absorbency of the substrate. The amount of the coating must therefore be determined experimentally, and the wire diameter used gives only a general idea of the expected coating deposit.

The rod is usually rotated in either direction by an independent drive to equalize the wire wear and to dislodge the particles that might be jammed between the wires. The rod should be installed in such a way that the rotation tightens the wire to the rod. Sometimes two or more rods are used in series in order to obtain a smoother coating. The first rod is wound with a slightly heavier wire than the second rod, although identical rods are also used in a two-rod coating assembly.

Various rod holder designs, including magnetic holders, are used. The holder keeps the rod in place and allows it to rotate without excessive friction. The holder may consist of simple clamps holding the ends of the rod sufficiently

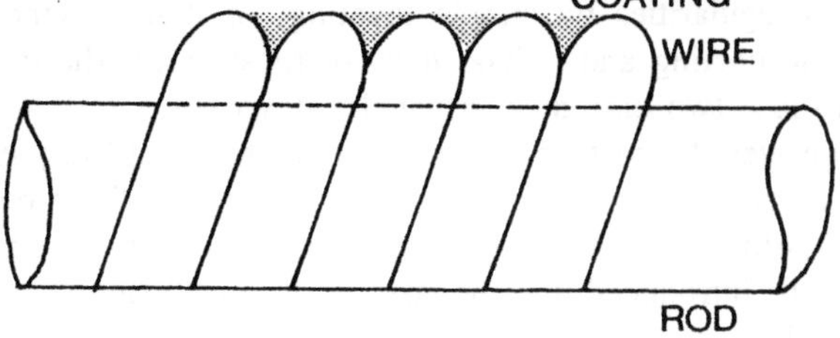

Fig. 34-21. A schematic diagram of a wire-wound rod.

Fig. 34-22. Wire-wound rod coating head. (*Courtesy of American Tool and Machine, a Wolverine Company.*)

firmly, but exerting minimum resistance to rotation. Figure 34-22 shows a photograph of a typical rod coating assembly.

Bars are also used to smooth out the adhesive coatings. A common coating imperfection is ridges in the machine direction formed by the splitting of the coating in the roll coating machines. Such polishing bars consist of a smooth rod or tube that is brought in contact with the adhesive surface. It may be stationary or it may be equipped with a drive in order to rotate the bar, usually in the direction opposite to the web movement. The polishing bar smoothes the surface irregularities by simple leveling, or it may remove a substantial amount of the coating and redeposit it. In most cases, the mode of operation is between these two extremes. Mushel[16] describes a fast rotating smoothing bar that improves the surface smoothness not by completely eliminating these striae, but by decreasing their amplitude. The flow of the coating completes the leveling before it dries. Polishing bars can also help to remove air bubbles, which might be introduced into the coating during mixing or during the coating operation itself.

DIRECT ROLL COATERS

The mechanism of roll coating involves compression of a thin wedge of a liquid and splitting it at the other side of the nip. The porosity of the backing, coating thickness, coating speed, and the rheological properties of the adhesive influence the coating process. Miller and Meyers[17] have studied the film-splitting mechanism and divide the flow in the nip into four regions: laminar flow, cavitation, cavity expansion, and filamentation. Surface tension and viscosity resist the expansion of the cavity after its formation. Air present in the cavities provides PV work to the system during expansion. Surface tension is the most important factor in the cavity expansion. Various regions in a roll nip and the pressure profile created are shown in Figure 34-6. Minimum clearance h_0 and clearance at the point of cavitation h_1 together provide an important geometrical parameter.

Roll coating was also analyzed by Middleman.[10] Coating thickness is related to geometric, rheological, and operating variables. The analysis is carried out, assuming Newtonian behavior of the coating fluid and also a case in which the coating fluid follows a power law.

The dependence of wet-film thickness was investigated experimentally by Schneider[18] in a two-roll setup. Roll speed had no effect on the wet-film thickness; the wet film was directly proportional to the roll clearance. The increase in viscosity caused a slight increase in the film thickness, but the surface tension had no effect on the film thickness. A simple mathematical expression

$$t = K(c/2) \tag{5}$$

is proposed, where

t = film thickness on either of the two equal-diameter rolls rotating at equal speeds
c = clearance between rolls
K = factor including the effect of all the other variables. The maximum variation of K is $1 \leq K \leq 4/3$. For conditions in which the inlet bank wets the top roll surface at about 10 degrees, $1.192 \leq K \leq 4/3$.

The equipment employing the direct roll coating principle is of interest for the manufacture of pressure sensitive adhesive products. Kiss roll coater (Fig. 34-20) may be used as a direct roll coater and several other coating heads utilize this coating method.

Transfer roll coater is an important machine for the application of low weight coatings and it utilizes the direct roll coating principle. Transfer roll

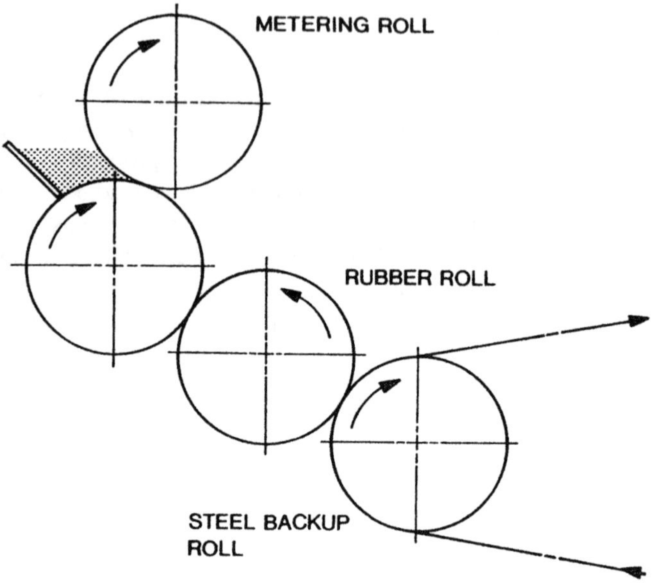

Fig. 34-23. A schematic diagram of a microtransfer coater.

coaters were originally developed for applying clay coatings on publication paper.[1] These coaters have since been replaced by air-knife and more recently by blade coaters. Transfer roll coating was revived again for the application of 100% solid silicone release coatings. Such second-generation transfer coaters are also known as multifilm split or differential draw coaters.[19] Another variation is known as microtransfer coater[20] (Fig. 34-23). Such a coater is suitable for an accurate application of very light coatings, as low as 0.4 g/m^2. As in reverse roll coater, premetering is accomplished in a conventional way. The coating is contained in a bank between two accurately machined-chilled iron rolls and is transferred to the rubber-covered roll that rotates 10–20 times faster than the chilled rolls in the premetering section. The coating is transferred onto the web in the next nip between the rubber applicator roll and the backing roll.

A similar machine is shown schematically in Figure 34-24. Each roller is driven individually and their speeds can be adjusted independently.[21] Every second roller is covered with an elastomeric cover. The gaps between the rolls can be adjusted with an accuracy of 0.001 mm. The coating is introduced into the nip between the first two rolls, and each subsequent roll is run at a higher speed carrying a thinner coating layer. The total surface speed differential in such a coater can be 1:30. The applicator roll is run at the web speed. Such a coater can apply the coating at speeds as high as 400 m/min.

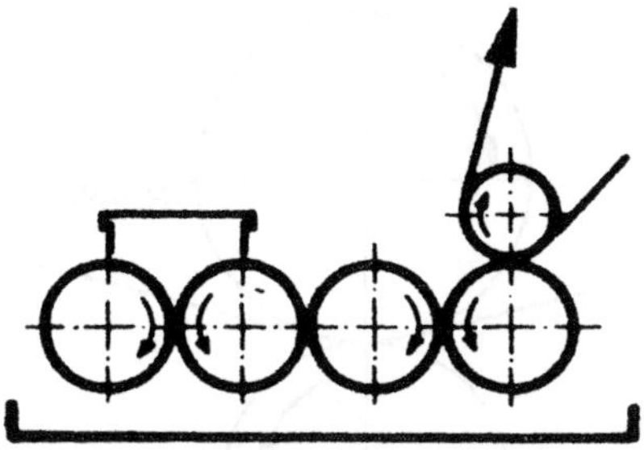

Fig. 34-24. A schematic diagram of a transfer roll coater.

GRAVURE COATERS

Gravure coating is a highly reproducible method suitable for application of low weight, 6–50-μm-thick wet coatings. This excludes gravure coating from applications that require heavier adhesive coatings. The gravure method is often used for application of the release and prime coatings required for pressure sensitive adhesive products.

A gravure coater consists of an engraved cylinder, a pan from which the coating is applied, a doctor blade, and a rubber-covered impression roll that carries the web. Smoothing bar may be employed if the coating does not level sufficiently. The basic machine is shown in Figure 34-25. Instead of employing a pan, the coating may be supplied by a fountain or fed off a blade, as shown in Figure 34-26.

Several types of engravings are used for coating applications. Pyramidal

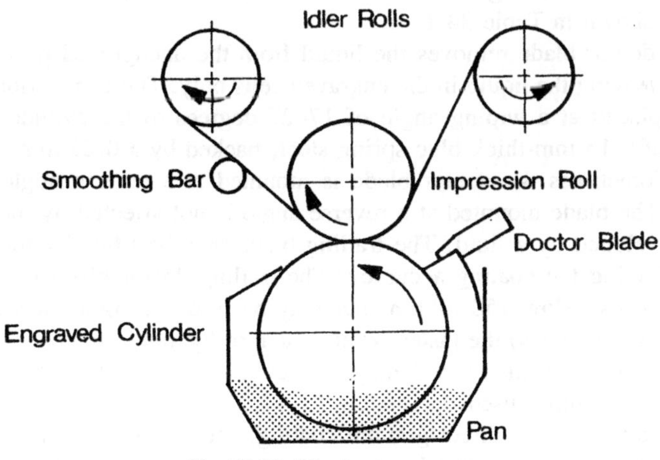

Fig. 34-25. Direct gravure coater.

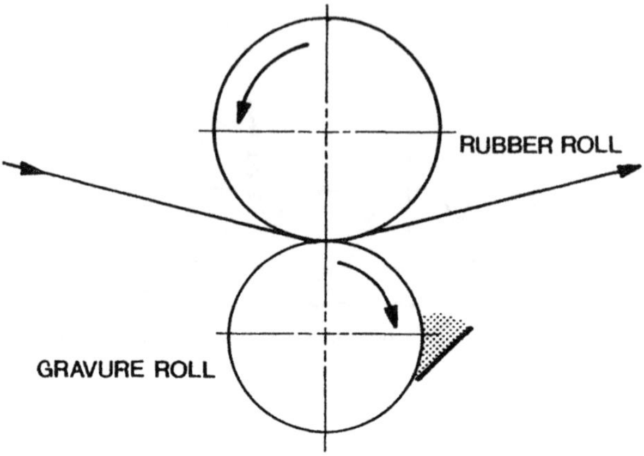

Fig. 34-26. Direct gravure coater with coating reservoir between the roll and the blade.

engraving has cells in the shape of a four-sided pyramid. This engraving type is most suitable for low weight deposits. Quadrangular engraving has cells in the shape of a truncated pyramid and is mainly used for direct gravure applications. Hexagonal gravure cell cylinders serve purposes similar to those of the quadrangular pattern. The former engraving pattern is more popular in Europe. Cylinders engraved with a trihelical pattern are suitable for heavier deposits, but the coating is rougher and may require smoothing. Quadrangular channel gravure cylinders have a quadrangular-type engraving in which the cells are interconnected. The channel increases the volume of the engraved area and reduces the force developed between the fluid and the doctor blade. These engraving patterns are shown in Figure 34-27. The characteristics of some of the engraving types are shown in Table 34-1.

The doctor blade removes the liquid from the unengraved portions of the cylinder, leaving the liquid in the engraved cells only. The conventional trailing blade is placed at a wiping angle of 17–27 degrees to the cylinder axis.[22] It consists of 0.15-mm-thick blue spring steel, backed by a 0.25-mm blade (Fig. 34-28). Sometimes the doctor blade is mounted at a reverse angle of 30–35 degrees. The blade mounted at a reverse angle is not affected by the hydraulic forces developed by a fluid. The trailing blade may be lifted by the hydraulic forces affecting the coating accuracy. The trailing doctor blade works well at coating speeds below 150 m/min, but may be lifted at higher speeds. Higher viscosity contributes to the faster development of hydraulic forces. The reverse-angle blade is more likely to damage the cylinder, so plastic rather than steel blades are sometimes used.

In addition to the direct gravure coating, offset gravure is used (Fig. 34-29). The coating is first applied to the rubber-covered offset roll by the engraved

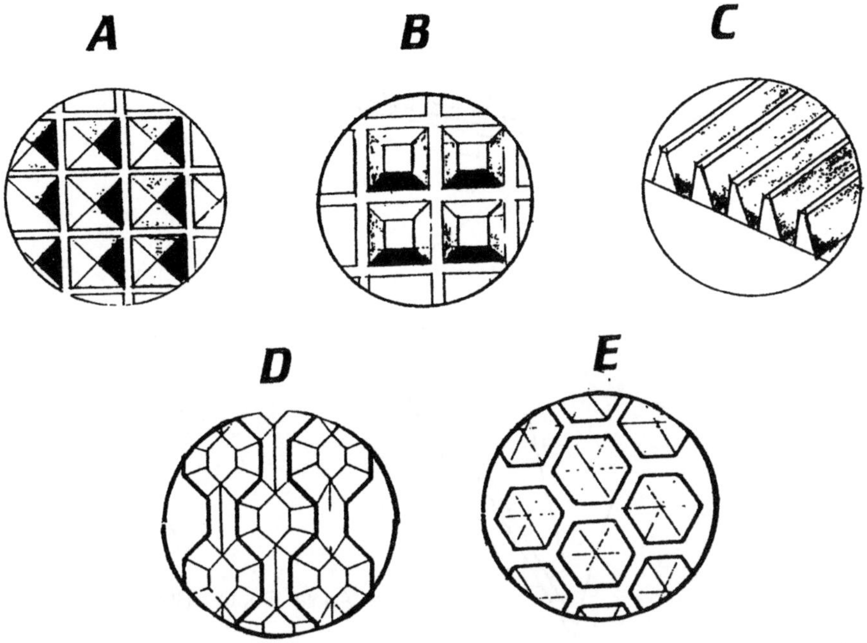

Fig. 34-27. Engraving types. (A) Pyramidal; (B) quadrangular; (C) trihelical; (D) quadrangular channel; (E) hexagonal.

cylinder. The offset coating method is better suited to coatings that do not level easily and for rough substrates that require higher impression pressures. Higher pressure can be applied on the offset gravure coater without exposing the engraved cylinder to higher forces and potential damage.

The reverse offset gravure coater has been adopted for the deposition of low coating weights, such as those required in 100% solids silicone coatings.

Table 34-1. Characteristics of Mechanically Engraved Gravure Cylinders

TYPE OF ENGRAVING	SCREEN (lines/in.)	(lines/cm)	CELL DEPTH (mm)	DEPOSIT[a] (g/m^2)
Pyramidal	200	79	0.003	2.4
	40	16	0.037	27.6
Quadrangular	200	79	0.005	3.6
	16	6	0.065	49.5
Trihelical	120	47	0.010	11.7
	24	9	0.055	63.1

Note: For more information, see reference 3.
[a] Wet coating.

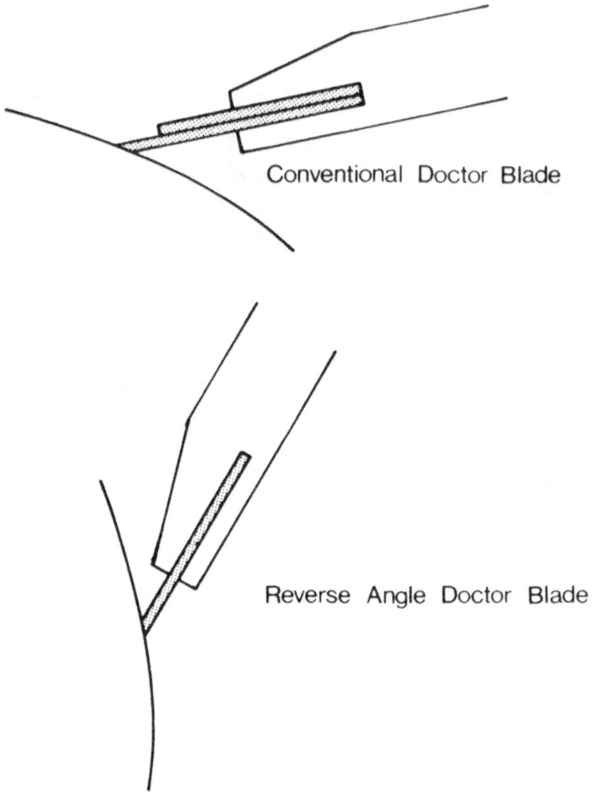

Fig. 34-28. Doctor blade types.

In this coater, the impression roll rotates in the direction opposite to that of the direct gravure coater. In one arrangement, the gravure roll is run at 20–25% of the line speed; the rubber roll is run at the line speed. This allows the reduction of the coating deposited on the rubber offset roll.[23] The doctor knife is usually set at the reverse angle in order to better control the amount of coating picked up by the gravure roll. The reverse roll offset gravure method is in competition with the five roll transfer coater for the application of 100% solids silicone release coatings. The former is the preferred coating method in the United States and the latter is preferred in Europe and the Far East.

HOT MELT COATING

Hot melt pressure sensitive adhesives have become important, especially with the advance of block copolymer/tackifying resin adhesives. The early hot melt

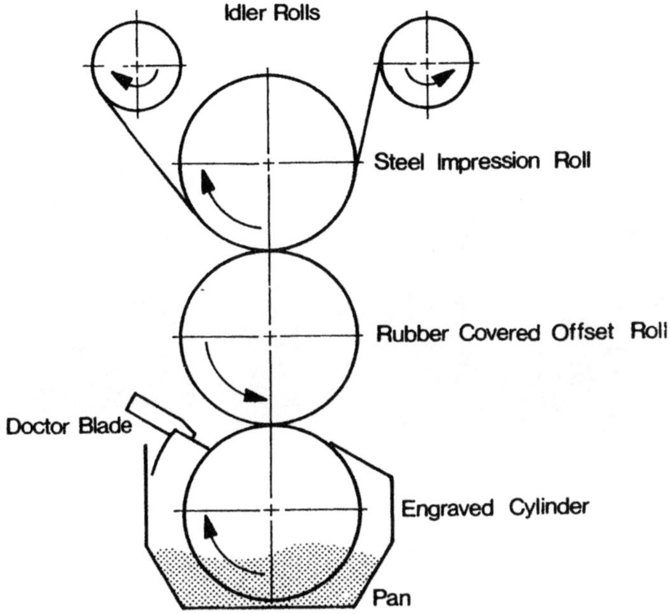

Fig. 34-29. Offset gravure coater.

adhesives were based on ethylene-vinyl acetate (EVA) copolymers, and their acceptance was rather limited, primarily because of the low cohesive strength.

The advantages of hot melt adhesives are obvious. Drying is not required, thus eliminating space occupied by expensive ovens needed for processing of solvent or aqueous adhesive coatings. Also, the energy requirements are much lower. The main disadvantage is somewhat limited adhesive formulation capabilities and the inherent low shear resistance at elevated temperatures. Some of these disadvantages might be eliminated in the future by new materials, or by posttreatment, such as irradiation. The development work on new polymers and resins for hot melt applications is more active than the development of other pressure sensitive adhesives.

Hot melt coating technology has penetrated several product areas: biaxially oriented polypropylene (BOPP) film tapes, label stock, floor tiles, some paper masking tapes, and few other products. Hot melt equipment is quite suitable for small-scale operations, and this has created the opportunity for many label manufacturers to produce their own stock. Small machines resembling laboratory equipment are used for production purposes. Because of high coating rates, coaters can supply sufficient stock for a small converter. Such a coater for processing of a 16-in. web is shown in Figure 34-30.

Typical coating weight for label stock is 18–27 g/m^2 (12–18 lb/1000 ft^2).

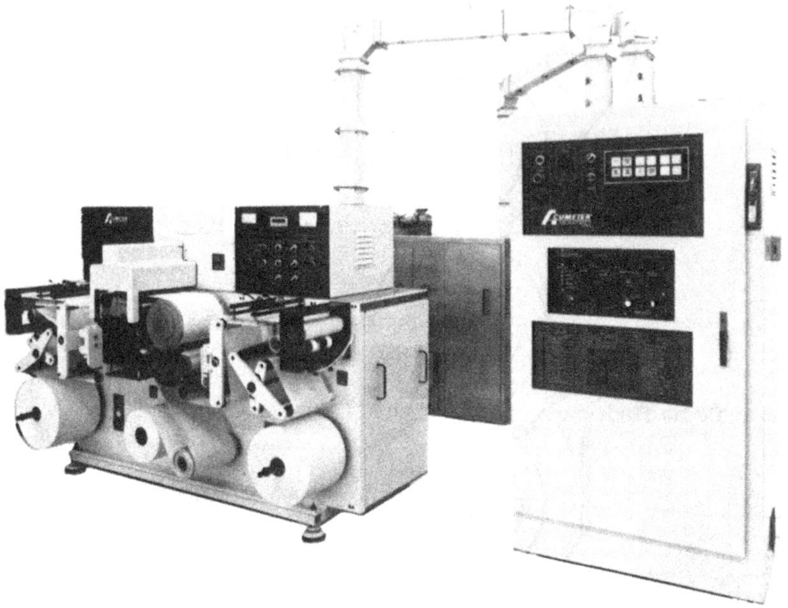

Fig. 34-30. Narrow width hot melt coater. (*Courtesy of Acumeter Laboratories, Inc.*)

There are some applications where a higher coating weight is needed, such as applications to difficult-to-adhere-to surfaces: corrugated packaging or automobile tires. These high coat weights accentuate the differences of hot melt adhesives in die cutting. Hot melts are softer materials and do not fracture under the high shear of a knife edge, like some of the adhesives based on different formulations. Because of this it is important that the depth of the cut be controlled, so that the knife edge cuts through at least 80% of the adhesive mass. Otherwise label lifting may occur or the waste matrix could break.

Hot melt coating has developed from wax coating technology. Waxes are applied at low viscosity and many coating methods are suitable for their application, although direct roll coating predominates. The difference between waxes, hot melts, and extrudable or calenderable coatings is mainly the viscosity. Typical application temperatures and viscosities for these materials are: waxes, 0.02 Pa·s at 100°C; hot melts, 10 Pa·s at 160°C; extrudable and calenderable adhesives, 200 Pa·s at 120°C.

Direct roll application of hot melt pressure sensitive adhesives is possible, but slot orifice application, discussed next, has become the most often used method.

Slot Orifice Coaters

Slot orifice coaters have been developed from the extrusion coaters: in both the polymeric coating is delivered through a slot in a die. Since the hot melt viscosity is much lower than that of extrudable coatings, less massive dies are used for slot orifice coaters. More recently slot orifice coaters have found application in emulsion and solution adhesive coating.[24] It yields a smooth coating, free of striation marks, and has become an important method in applying aqueous pressure sensitive adhesive coatings.

The adhesive in a slot orifice is formed into a thin sheet by the orifice and deposited onto the web, which is usually supported by a roll (see Fig. 34-31). The amount of the coating deposited is controlled by the pump that forces the coating through the orifice. There is another, less frequently used version of slot orifice coating, where the amount of the coating is metered by a scraping device. The lip of the die set at a gap may serve as the scraper. This technique is used more often for precision coating of thin films than for pressure sensitive adhesives.[19]

The most often used hot melt coating mode, where the amount of material deposited is equal to that delivered through the slot, has several variations. These are discussed in more detail by Booth.[3] One type of slot coater is the curtain coater, which is an important machine for depositing heavyweight coatings, but is not used for pressure sensitive adhesive application. The first type

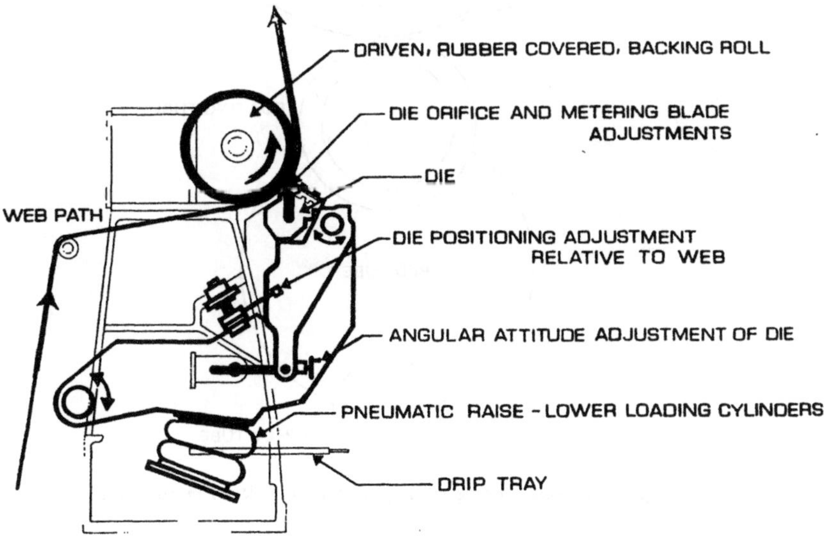

Fig. 34-31. Hot melt slot orifice coater. (*Courtesy of Black Clawson Co.*)

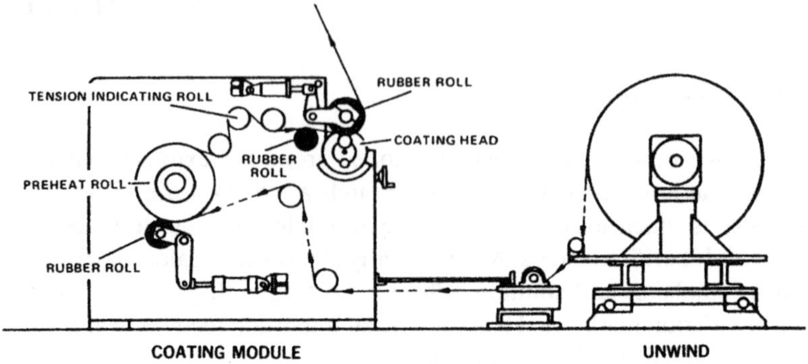

Fig. 34-32. Schematic diagram of a Park coater. (*Courtesy of Bolton Emerson, Inc.*)

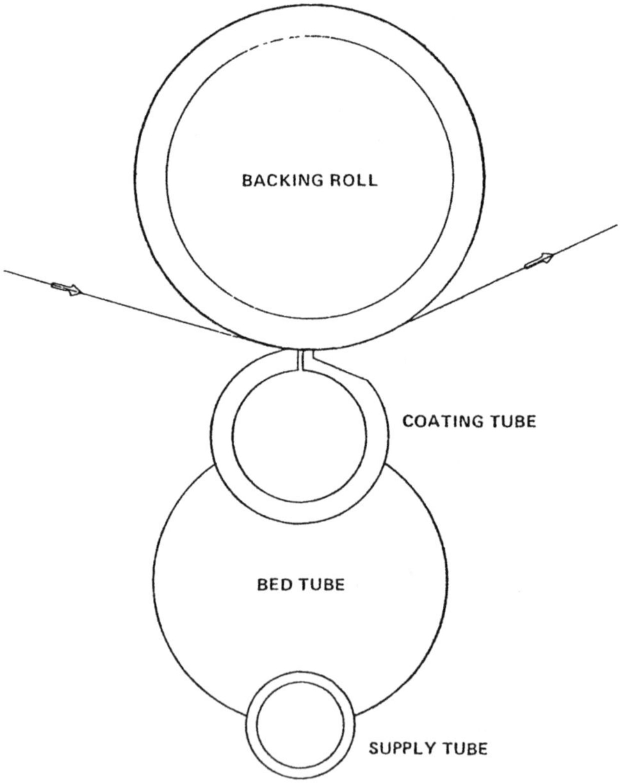

Fig. 34-33. Park coating head. (*Courtesy of Bolton Emerson, Inc.*)

of slot orifice coater was the Genpac coater.[25] The first and the most successful machine for the application of pressure sensitive adhesives was the Park coater.[26] A schematic diagram of this coater is shown in Figure 34-32. The main difference between this machine and previous equipment was the special design of the coating head (see Fig. 34-33). Similar machines have been made by several other manufacturers, some including various improvements.[27] A rotating smoothing bar has been added to some machines to smooth out the coating. A new die design that claims to produce a more uniform flow across the web width has been offered more recently. It has the liquid pumping station built into the die[28] (Fig. 34-34). The flow direction of the liquid after the pump is in the web direction, thus overcoming the flow problems common to the coat hanger die design. This new die design has found applications in the coating of aqueous and solvent adhesives.

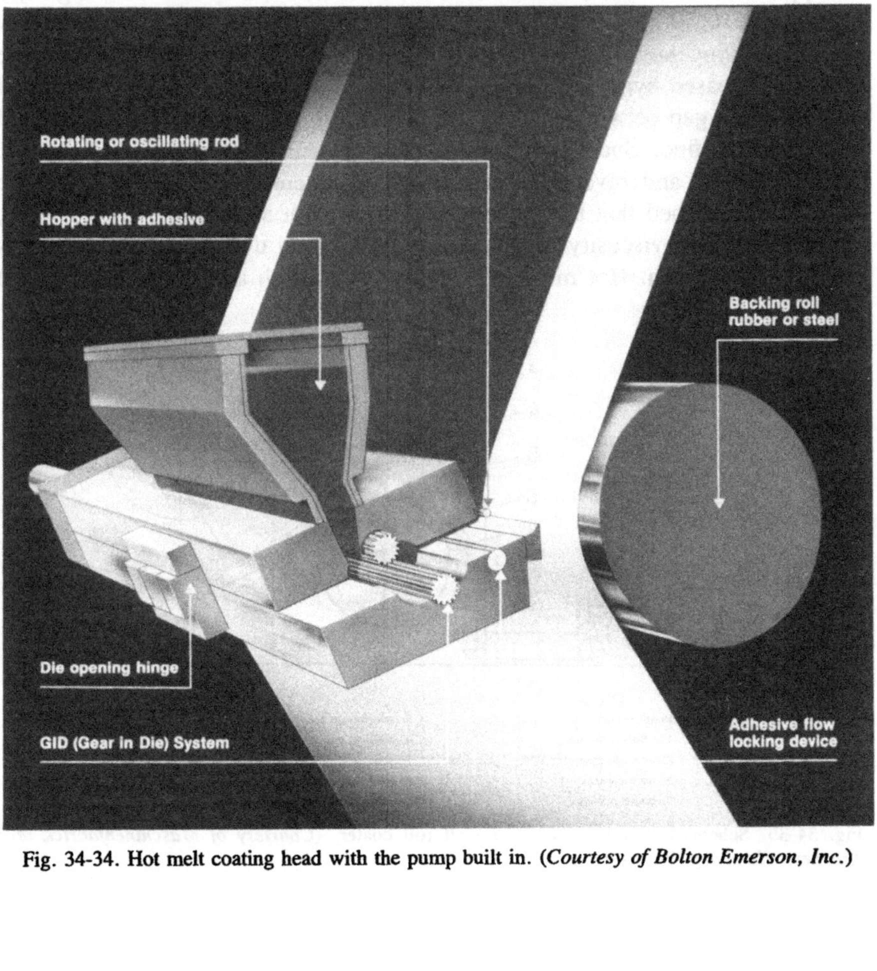

Fig. 34-34. Hot melt coating head with the pump built in. (*Courtesy of Bolton Emerson, Inc.*)

Slot orifice coaters can apply coatings over a wide range of coating weight—0.5–600 g/m^2—and a coating weight tolerance of ±5% is claimed.[29] Hot melt viscosities up to 200 Pa·s can be handled, and the coating may be applied at speeds of up to 300 m/min. Pattern coating is easily carried out with a slot orifice coater by simply blocking parts of the slot.

Roll Coaters

Roll coaters are also used for hot melt pressure sensitive adhesives. Hot wax coatings have been traditionally applied by roll coaters, and the technique was quite naturally adapted for higher viscosity hot melt application. A schematic diagram of a hot melt roll coater is shown in Figure 34-35. Hot melt is pumped between two heated rolls: the metering roll, which usually is not rotating, and the applicator roll, which runs at about 10% of the web speed. The originally heavy (0.25–0.50-mm) adhesive coating on the applicator roll is wiped off by the fast-moving web. The accuracy of such machines is high, because the error is decreased by the roll speed difference. The coating weight is adjusted by setting the gap between rolls for coarse adjustment and by setting the speed differential to fine. Such machines can also be run as a reverse roll coater. Both the direct and reverse coating mode arrangement are shown in Figure 34-36. It is claimed that the machine can work with an accuracy of ±1 g/m^2, that the optimum viscosity range is 10–30 Pa·s, and that the speed can be as great as 100 m/min. Hot melts of viscosity of as high as 200 Pa·s have been

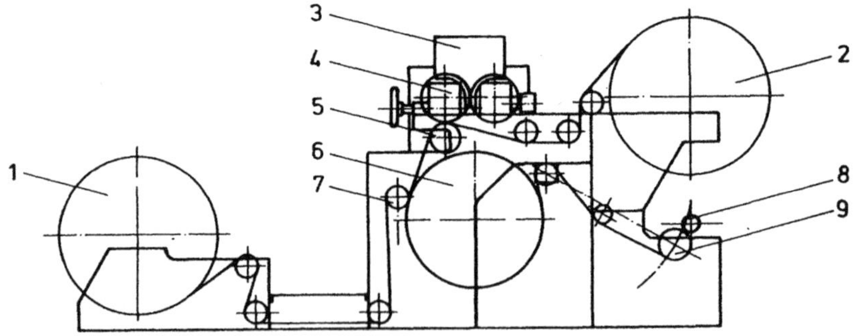

Positioning
1. unwind U 100
2. unwind diameter 800
3. hotmelt tray
4. coating roller
5. counter pressure roller
6. cooling roller
7. pressure roller
8. rewind diameter 800
9. adhesive tape rewinding system

Fig. 34-35. Schematic diagram of a hot melt roll coater. (*Courtesy of Maschinenfabrick Max Kroenert, Hamburg, West Germany.*)

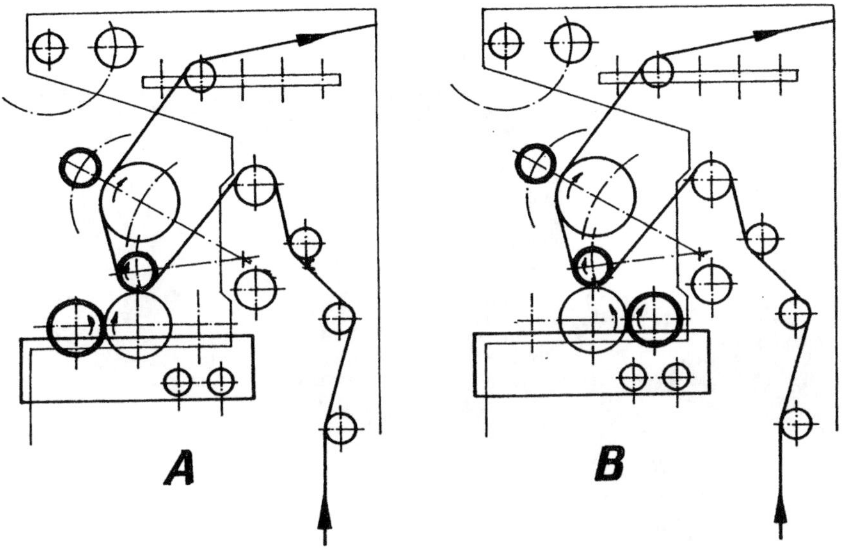

Fig. 34-36. Various arrangements of a hot melt coating machine. (PAK 600, *Courtesy of Maschinenfabrik Max Kroenert, Hamburg, West Germany.*) (A) Reverse roll coating. (B) Direct roll coating.

coated on such machines. The lower viscosity limit is 5 Pa·s.[19] A photograph of this machine is shown in Figure 34-37.

The roll coaters are less suitable for the coating of porous substrates—strike through a porous web is possible—and for the coating of heat-sensitive thermoplastic films. The adhesive is also exposed to the atmosphere for a longer time, which might be undesirable for adhesives of poor heat stability. Roll coaters are also known to cause the formation of striation marks running in the machine direction, which might be difficult to level. High viscosity hot melts (>100 Pa·s) produce high pressures in the roll nip and may cause roll bending which, unless compensated, would result in a nonuniform coating thickness across the web width. This may be corrected by roll crowning or roll crossing.[30]

Remelters

Hot melt coating requires the delivery of molten adhesive to the coating head. If the adhesive is compounded in-house, it may be delivered directly from an intermediate storage tank. For large coating operations the adhesive may be compounded continuously in an extrusion mixer. For smaller operations a batch high shear mixer, such as a double-arm mixer, may be used, and in some cases a simple low shear mixing tank may be employed. The block copolymers

Fig. 34-37. Pressure sensitive adhesive hot melt coater, PAK 400. (*Courtesy of Maschinenfabrik Max Kroenert, Hamburg, West Germany.*)

and resins used may degrade if exposed to an elevated temperature for a longer time period. An inert gas blanket is often used to minimize the degradation, but the most important factor is the length of exposure to high temperatures. Figure 13-22 illustrates the effect of the thermal history on the degradation of block-copolymer-based hot melt adhesives. Continuous mixing requires the shortest exposure time to high temperature, and therefore results in minimum degradation of the adhesive. The molten adhesive is usually filtered: a 100 mesh screen is sufficient.

If the adhesive is purchased, it requires remelting. Remelters of several constructions are available.

Bulk Tank Remelters. These are of simple construction, consisting of an agitated jacketed tank. Hot melt is maintained in its fluid state in the tank, and cold material is added periodically as required. Molten hot melt is kept in the tank for a long time, and it may be subjected to severe thermal degradation.

High Shear Mixing. This method is a faster way to melt a batch of adhesive. Double-arm mixers are used most often, and these machines are also useful for hot melt compounding. The adhesive temperature is raised gradually under continuous mixing. It is important to make sure that the batch temperature rises uniformly and that two phases—molten liquid and solid pieces—are not formed. In the latter case, the mixing time is prolonged excessively.

Grid Melters. Grid melting is a convenient and often used method. It allows the adhesive to melt at a higher temperature on the grid. The adhesive is then stored at a lower temperature below the grid in the main tank. A simple grid melter is shown in Figure 34-38. The melting rate is adjusted to the process requirement by feeding the solid adhesive as required. Several designs of grid melters are available.

Drum Melters. Drum melters are constructed on the same principle as grid melters. The plate of a drum melter is placed over the drum with the cover removed. The hot plate causes the adhesive to melt and flow along the grooves to the intake of a gear pump, which delivers the molten adhesive to the coater. Only the top 5–15 cm are molten at any given time, thus avoiding long exposure of the adhesive to the elevated temperature.[31] A photograph of a drum melter is shown in Figure 34-39.

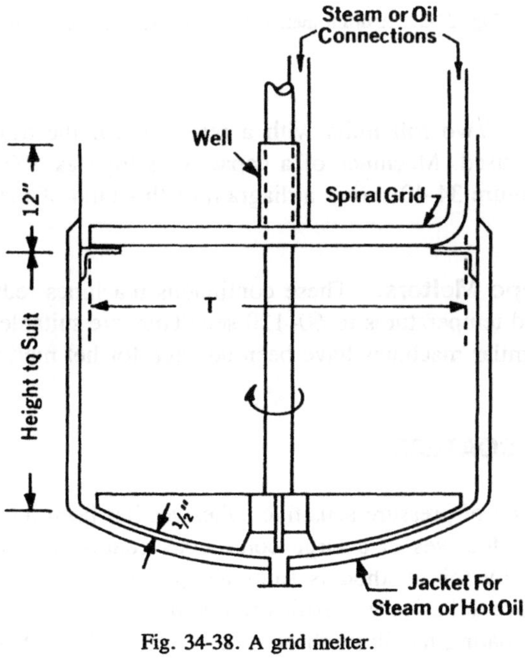

Fig. 34-38. A grid melter.

Fig. 34-39. A drum melter. (*Courtesy of Nordson Corp.*)

Roll Melters. Two roll mills with a reservoir for the molten adhesive on the bottom are used. Machines of a capacity as high as 400 kg/hr have been constructed. Figure 34-40 shows a diagram of this kind of melter.

Extrusion-Type Melters. These continuous machines reduce the exposure time at elevated temperatures to 60–120 sec. They are suitable for high volume processing. Similar machines have been adopted for hot melt compounding.

EXTRUSION COATERS

Extrusion coating of pressure sensitive adhesives has been introduced to handle high viscosity adhesives in a large continuous coating operation. One of the products made by this method is saturated paper tape. The normal hot melt pressure sensitive adhesive viscosities range in the area of 5–50 Pa·s at 177°C. The extrusion coating handles adhesives of viscosity 0.25–1.5 kPa·s at 177°C,

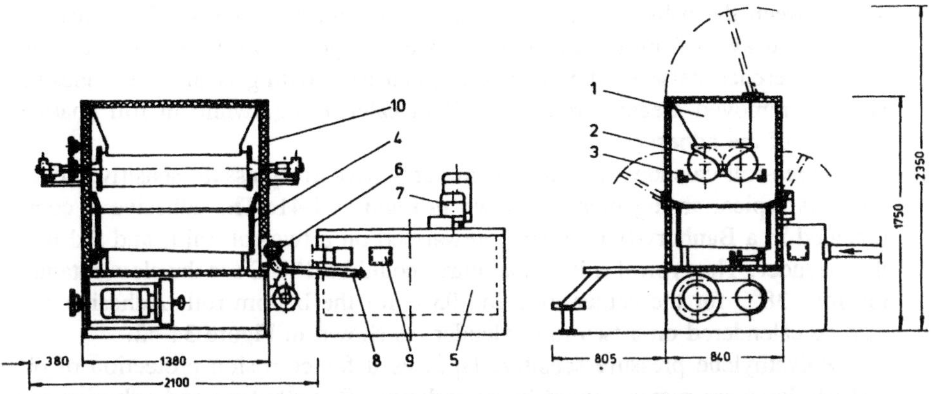

Fig. 34-40. A roll melter. (*Courtesy of Maschinenfabrik Max Kroenert, Hamburg, West Germany.*)

and reportedly could process the adhesives in the range of 0.02–2 kPa·s at 177°C.[32]

The method allows mixing and coating pressure sensitive adhesives in one pass with a total residence time of 2–3 min. The adhesive is exposed to high temperature for only a short period of time, and the problem of thermal degradation is minimized. The substrate used does not need to have thermal stability; poly(vinyl chloride) and polypropylene films have been coated by this method.

Standard plastic extrusion equipment is not suitable for the handling of pressure sensitive adhesives. The adhesive sticks to the flight of the screw. A special single-screw extruder has been developed.[33] It is equipped with an auxiliary wiper screw in the feed section that continuously clears the flights of the feed screw of any material that might be sticking to it. In order to obtain proper mixing of all ingredients, a large length-to-diameter ratio is required (40:1). The adhesive, after passing through the extruder, is properly mixed and it is then extruded through a die onto the substrate. This method is suitable for a large-volume operation: a 88.9-mm extruder mixes 270 kg/hr, and a 152-mm one mixes 900 kg/hr.

CALENDERING

Calendering has been used for a long time in coating of pressure sensitive adhesives. The early hospital and electrical friction tapes were coated by calendering and many cloth and polyethylene tapes are still made by the calendering method.

In calendering, the material softened by heat is dragged through a narrow gap between the rolls, producing a sheet of polymeric material. This sheet is then pulled away from the roll either alone or together with the backing. The main difference between the calendering and roll coating is that the adhesive sheet is removed cleanly from the roll in calendering, while in roll coating splitting of the adhesive takes place.

Three-roll calenders are most commonly used for pressure sensitive adhesives. A typical arrangement is shown in Figure 34-41. The adhesive is compounded in a Banbury-type mixer, prewarmed on a two-roll mill, and fed into the calender. The typical roll temperatures could be 127°C for the slow rotating top roll, 38°C for the center roll, and 93°C for the bottom roll. Adhesive can also be calendered on a two-roll calender, as shown in Figure 34-42.

Polyethylene pressure sensitive tapes used for corrosion protection of oil and gas lines are manufactured by calendering. Polyethylene and adhesive are

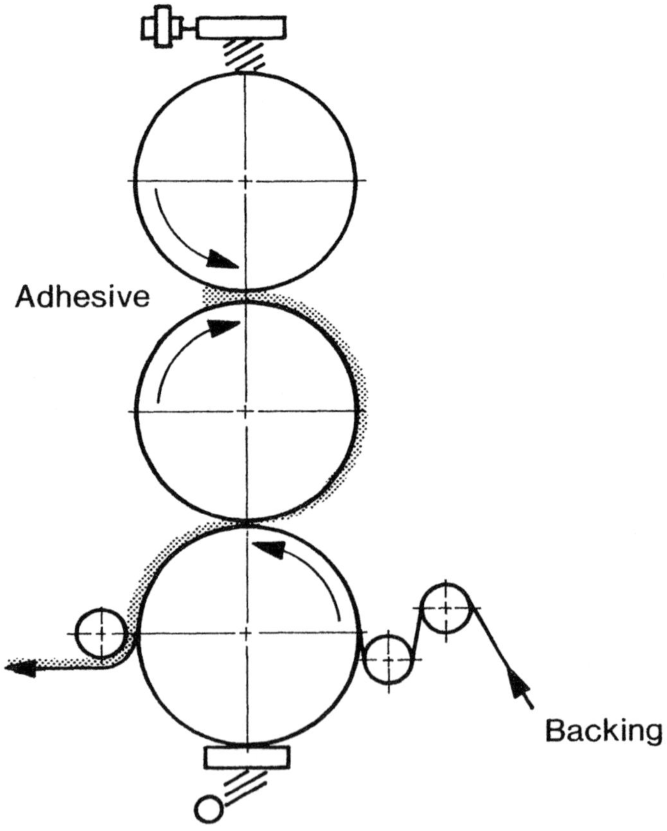

Fig. 34-41. Three-roll pressure sensitive tape calender.

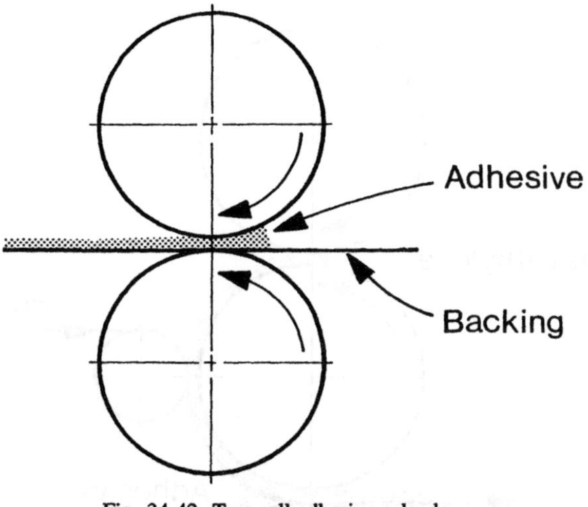

Fig. 34-42. Two-roll adhesive calender.

calendered simultaneously, as shown in Figure 34-43.[34] Similarly, a tape with a supporting scrim in the center can be made, as shown in Figure 34-44.

The adhesive is stripped off the calender roll at a fairly high force, despite a high roll polish. Therefore, the backing must have a sufficiently high tensile strength. The use of low tensile strength backings requires special processing.

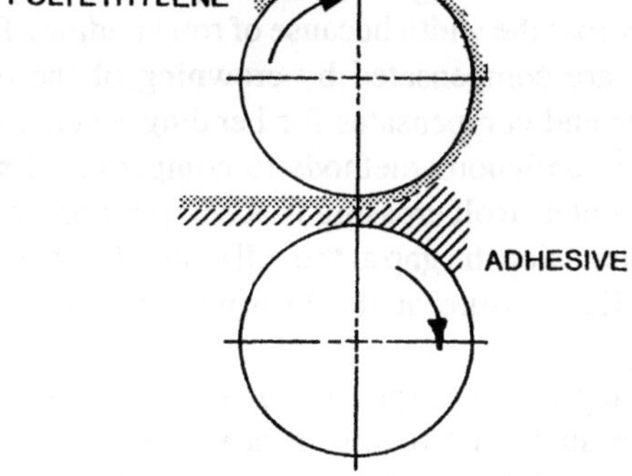

Fig. 34-43. Simultaneous polyethylene and adhesive calendering to produce polyethylene tape.

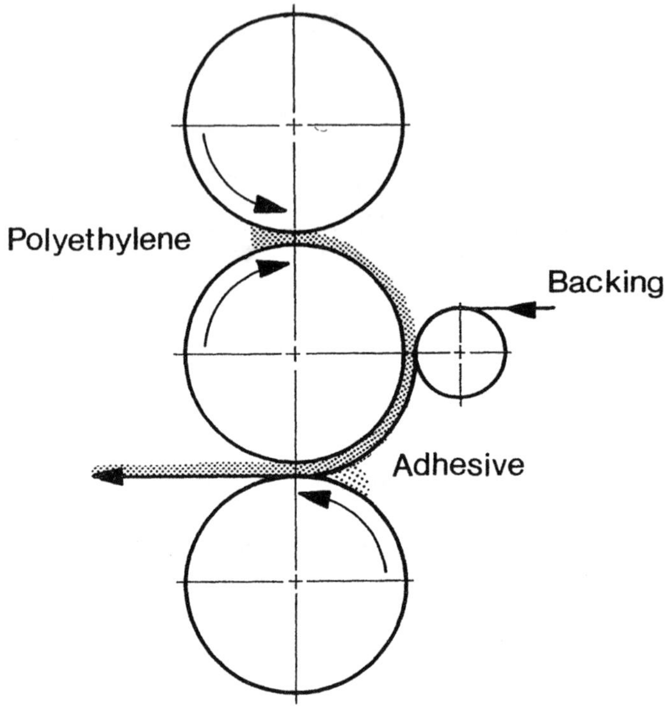

Fig. 34-44. Calendering of polyethylene coated fabric tape.

Doctor's knives have been used to separate the adhesive from the roll in case of fragile backings.

Calendering is suitable for application of heavy adhesive coatings. While coatings as thin as 0.05 mm could be produced by calendering, the uniformity of the coatings below 0.1 mm might not be acceptable. The process is susceptible to weight variation across the width because of roll bending. The weight variations because of bending are compensated by crowning of the rolls. This decreases the gap in the center and compensates for bending when coating. Roll crossing and roll bending are continuous methods to compensate for the increase in the gap in the calender center. Roll crossing is accomplished by moving the bearings horizontally, thus increasing the gap at the roll ends. Roll bending is accomplished by applying hydraulic pressure at the bearings to counteract the shear forces causing the bending.

Calendering equipment is expensive and it is best suited for large-volume manufacturing of products that require a heavy adhesive coating. Calendering can convey a soft material at high rates at a much lower mechanical energy input than extrusion. The largest volume product made by calendering is corrosion protective tape.

The analysis of the calendering process requires simplifying assumptions regarding polymer behavior. Middleman[10] presents a mathematical analysis of several cases of the calendering process assuming Newtonian and power law models of polymer behavior during calendering. Calendering analysis by the finite element method is reviewed by Tadmor and Gogos.[35]

REFERENCES

1. Satas, D. *Web Processing and Converting Technology and Equipment*. Van Nostrand Reinhold, New York, 1984.
2. Weiss, H. L. *Coating and Laminating Machines*. Converting Technology Co., Milwaukee, Wis., 1977.
3. Booth, G. L. *Coating Equipment and Processes*. Lockwood, New York, 1970.
4. Zink, S. C. Coating Processes. *Kirk Othmer Encyclopedia of Chemical Technology*. Third edition, Vol. 6, pp. 386–426. Wiley–Interscience, New York, 1979.
5. Weiss, H. L. Coating Systems for Emulsion PSA's. In *Proc. Technical Seminar*, Pressure-Sensitive Tape Council, Itasca, Ill., May 9–11, 1984.
6. Meyers, R. R. *J. Polym. Sci., Part C*, **35:** 3–21 (1971).
7. Matsuda, T., and Brendley, W. H., Jr. *J. Coatings Technol.* **51**(658): 46–60 (1979).
8. Freeston, W. D., Jr. *Coated Fabrics Technology*, pp. 25–41. Technomic Publishing Co., 1973.
9. Hwang, S. S. *Chem. Eng. Sci.* **34:** 181–189 (1979).
10. Middleman, S. *Fundamentals of Polymer Processing*. McGraw-Hill, New York, 1977.
11. Eklund, D. Blade and Air Knife Coating. *Web Processing and Converting Technology and Equipment*, pp. 34–59 (D. Satas, ed.). Van Nostrand Reinhold, New York, 1984.
12. Grant, O. and Satas, D. Other Knife and Roll Coaters. *Web Processing and Converting Technology and Equipment*, pp. 60–80 (D. Satas, ed.). Van Nostrand Reinhold New York, 1984.
13. Gargatanis, P. A., Cleland, A. J., and Wairegi, T. *TAPPI* **61**(4): 77–81 (1978).
14. Bald, T. J. Solid Silicone Coating System. In *Proc. Technical Seminar*, pp. 101–105. Pressure-Sensitive Tape Council, Itasca, Ill. May 9–11, 1984.
15. Taylor, P. C. *Paper Technol. Ind.* **19**(9): 305 (1978).
16. Mushel, L. A. *Package Print.* (June 1979).
17. Miller, J. C., and Meyers, R. R. *Trans. Soc. Rheol.* **II:**77–93 (1958).
18. Schneider, G. B. *Trans. Soc. Rheol.* **VI:** 209–221 (1962).
19. Swain, R. H. *Paper Film Foil Convert.* **59**(6 and 7): (1985).
20. Kosta, G. U.S. Patent 4,029,833 (1977) (assigned to Midland Ross Corp.).
21. Aurin, W. Kroenert's Developments in Coating Equipment for Pressure Sensitive Tapes. In *Proc. Technical Seminar*, pp. 15–34. Pressure-Sensitive Tape Council, Itasca, Ill., May 8–10, 1985.
22. Larkin, M. J. Gravure Coaters. *Web Processing and Converting Technology and Equipment*, pp.15–33 (D. Satas, ed.). Van Nostrand Reinhold, New York, 1984.
23. Lankford, A. P. Precision Coating with Offset Gravure. In *Proc. 1987 TAPPI Polymers, Laminations and Coatings Conference*, pp. 189–192.
24. Lippert, H. G. Slot Die Coating—A Technical Update on Recent Techniques and Design

Developments. In *Proc. 1987 TAPPI Polymers, Laminations and Coatings Conference*, pp. 193–203.

25. Satas, D. Hot Melt Coaters. *Web Processing and Converting Technology and Equipment*, pp. 147–172 (D. Satas, ed.). Van Nostrand Reinhold, New York, 1984.
26. Park, G. C. One Hot Melt and Water-borne Application System. In *Proc. Technical Seminar*, pp. 156–160. Pressure-Sensitive Tape Council, Rosemont, Ill., June 23–24, 1981.
27. Mainstone, K. A. *1981 Hot Melt Adhesives and Coatings Short Course Notes of the Technical Association of the Pulp and Paper Industry*, pp. 75–79.
28. Urquhart, G. New GID System for Emulsions, Hot Melts and Solvents. In *Proc. Technical Seminar*, pp. 96–100. Pressure-Sensitive Tape Council, Itasca, Ill., May 7–9, 1986.
29. Houseman, G. *Course Notes 1978 International Hot Melt Short Course*, pp. 85–87. TAPPI, Frankfurt, West Germany, May 8–11, 1978.
30. Womack, H. G., and Wallace, R. R. *Course Notes 1978 International Hot Melt Short Course*, pp. 94–97. TAPPI, Frankfurt, West Germany, May 8–11, 1978.
31. Plasschaert, J. *Course Notes 1978 International Hot Melt Short Course*, pp. 68–84. TAPPI, Frankfurt, West Germany, May 8–11, 1978.
32. Palermo, F. C., and Korpman, R. *Course Notes 1978 International Hot Melt Short Course*, pp. 115–118. TAPPI, Frankfurt, West Germany, May 8–11, 1978.
33. Smith, D. J. U.S. Patent 3,929,323 (1975) (assigned to Egan Machinery Co.).
34. Morris, J. F. U.S. Patent 2,879,547 (1956) (assigned to The Kendall Co.).
35. Tadmor, Z., and Gogos, C. G. *Principles of Polymer Processing*. Wiley, New York, 1979.

35. Coating Additives

Donatas Satas
Satas & Associates
Warwick, Rhode Island

Pressure sensitive adhesives are used as thin coatings applied over various substrates. Various specialized equipment is used to produce such products. The adhesive coating must be of sufficiently low viscosity to flow and uniformly cover the substrate. The coating, therefore, must be dissolved in organic solvents, applied as an aqueous dispersion, a hot melt, or forced into a thin sheet by high-pressure forces developed on a calender or an extruder.

Two physical properties, viscosity and surface tension, are most important in the coating's machinability—its ability to produce a thin film and to establish a good contact with the substrate.

VISCOSITY

Figure 35-1 is helpful in defining the viscosity, that is, the resistance exhibited by a fluid to flow. The force F acting on a parallel plate arrangement that has the fluid confined between the plates, causes the plates to slide at a velocity v. The ratio of the incremental velocity (dv) and incremental thickness (dx) is the shear rate.

$$D = \frac{dv}{dx} \tag{1}$$

The units of shear rate are reciprocal seconds. The shear stress (τ) is defined as shear force (F) per unit area (A)

$$\tau = \frac{F}{A} \tag{2}$$

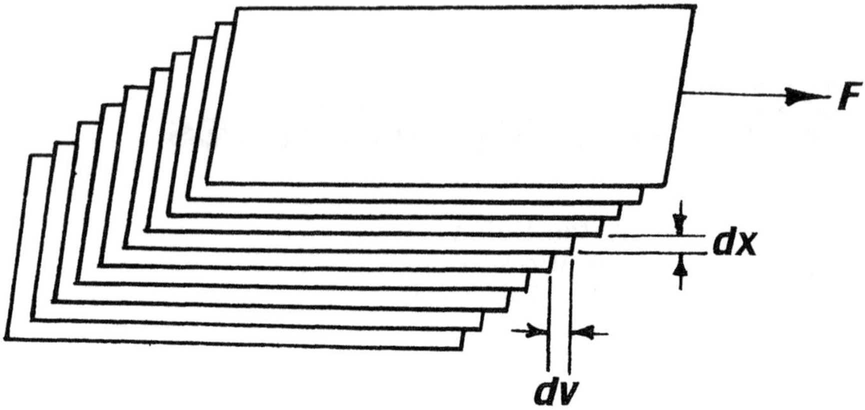

Fig. 35-1. Schematic diagram of a parallel plate arrangement illustrating the Newtonian flow.

The shear force is expressed in force units, such as dynes/cm^2. The viscosity is defined as the ratio of shear stress over shear rate.

$$\eta = \frac{\tau}{D} \tag{3}$$

The units of viscosity are dynes sec/cm^2 or g/cm·sec.

In an ideal system, the shear stress–shear rate relationship is represented by a straight line (Fig. 35-2), and the viscosity is independent of the shear rate (Fig. 35-3). Such ideal or Newtonian flow behavior is quite common in real coatings: polymer solutions in organic coatings exhibit a largely Newtonian flow. Some coatings, however, exhibit deviations from ideal behavior and may exhibit plastic, pseudoplastic (shear thinning), or dilatant (shear thickening) flow (see Figs. 35-2 and 35-3). Pseudoplastic behavior is quite common in pressure sensitive adhesive coatings. It exhibits a decrease of viscosity on application of shear, which levels off at higher shear rates. Plastic and pseudoplastic flow behavior exhibit a yield value: a minimum shear stress to induce the flow. Many latex coatings exhibit pseudoplastic flow, especially at low shear rates. This does not have much effect on the coating process, but is very important for the leveling of the coating surface.

Dilatant flow is characterized by the increase of viscosity with increasing shear rate. It is less common than pseudoplastic flow, but some highly filled coatings and plastisols may exhibit a pronounced dilatant flow that may cause serious problems in the coating process at higher shear rates. Shear rates associated with various processes are shown in Table 35-1.

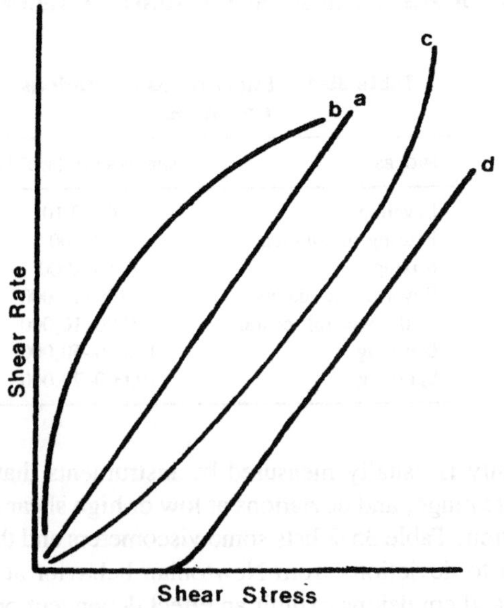

Fig. 35-2. Shear stress–shear rate relationship for various coatings: (a) Newtonian, (b) dilatant, (c) pseudoplastic, (d) plastic.

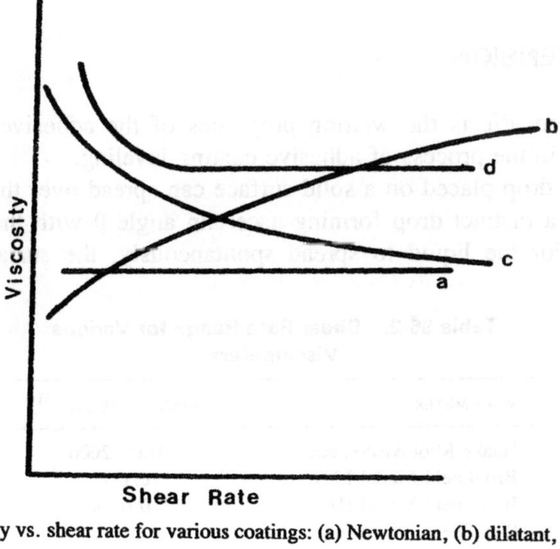

Fig. 35-3. Viscosity vs. shear rate for various coatings: (a) Newtonian, (b) dilatant, (c) pseudoplastic (d) plastic.

Table 35-1. Shear Rates for Various Processes

PROCESS	SHEAR RATE (sec^{-1})
Leveling	0.01–0.10
Pouring of solution	10–100
Mixing	50–1000
Reverse roll coating	100–10,000
Knife-over-roll coating	1000–10,000
Brushing	10,000–50,000
Spraying	10,000–70,000

The viscosity is usually measured by instruments that perform in a fairly narrow shear rate range, and deviations at low or high shear rates might therefore escape observation. Table 35-2 lists some viscometers and their shear rate range.

In addition to deviations from Newtonian behavior at various shear rates, some solutions and emulsions exhibit an effect dependent on the shear duration. Thixotropy is the decrease of viscosity as a result of mixing over a period of time. The viscosity decreases with increasing mixing time (Fig. 35-4). When mixing is stopped, the viscosity may gradually increase to its original value over a period ranging from a few seconds to days or weeks. Rheopectic is opposite to thixotropic behavior, but it is rarely observed in adhesive coatings.

SURFACE TENSION

Surface tension affects the wetting properties of the adhesive. It is also the driving force in the process of adhesive coating leveling.

A liquid drop placed on a solid surface can spread over the entire surface or remain as a distinct drop forming a certain angle θ with the substrate (see Fig. 35-5). For the liquid to spread spontaneously, the surface free energy

Table 35-2. Shear Rate Range for Various Viscometers

VISCOMETER	SHEAR RATE (sec^{-1})
Haake Rhotoviscometer	0.02–2000
Brookfield Model LVT	0.1–1
Brookfield Model HAT	0.6–24
Sun Chemical viscometer	24–480
Severs extrusion rheometer	50–1000
Instron extrusion rheometer	500–46,000
ICI cone and plate viscometer	10,000

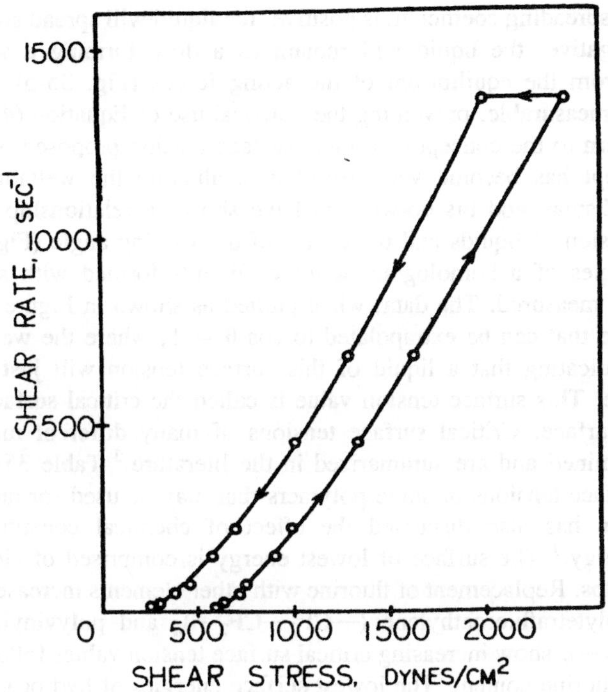

Fig. 35-4. Thixotropic loop.

must decrease as a result of spreading. The driving force for this spreading process is called the spreading coefficient S and may be expressed as follows:

$$S = \gamma_{SA} - (\gamma_{SL} + \gamma_{LA}) \quad (4)$$

where

γ_{SA} = interfacial free energy at the solid/air interface
γ_{SL} = interfacial free energy at the solid/liquid interface
γ_{LA} = surface tension of the liquid

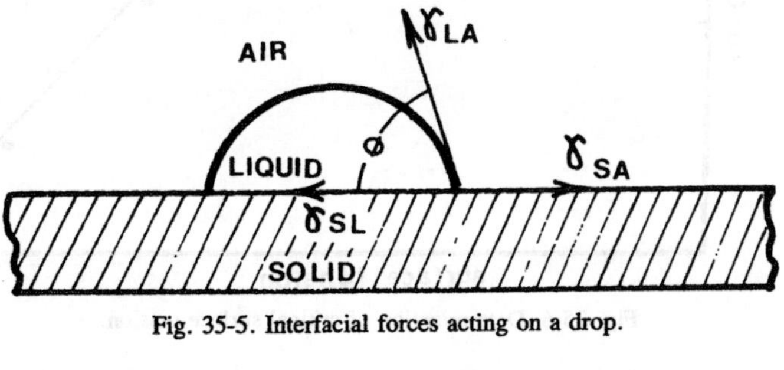

Fig. 35-5. Interfacial forces acting on a drop.

If the spreading coefficient is positive, the liquid will spread spontaneously. If it is negative, the liquid will remain as a drop forming a wetting angle resulting from the equilibrium of the acting forces (Fig. 35-5). Neither γ_{SA} nor γ_{SL} is measurable, preventing the practical use of Equation (4). Therefore, we must turn to the concept of critical surface tension proposed by Zisman.[1,2] This concept has become very useful in evaluating the wettability of solid surfaces. Zisman and his co-workers have shown a relationship between the surface tension of liquids and the cosine of the wetting angle (Fig. 35-6). The contact angles of a homologous series of liquids formed with a given solid surface are measured. The data, when plotted as shown in Figure 35-6, give a straight line that can be extrapolated to $\cos \theta = 1$, where the wetting angle θ is zero, indicating that a liquid of this surface tension will just spread over that surface. This surface tension value is called the critical surface tension of the solid surface. Critical surface tensions of many different materials have been determined and are summarized in the literature.[3] Table 35-3 shows the critical surface tensions of some polymers that may be used for tape backings.

Zisman has also discussed the effect of chemical constituents on the surface energy.[1] The surface of lowest energy is comprised of closely packed $-CF_3$ groups. Replacement of fluorine with other elements increases the surface energy. Polytetrafluoroethylene $(-CF_2-CF_2-)_n$ and polyvinylidenefluoride $(CF_2-CH_2-)_n$ show increasing critical surface tension values following the decrease of fluorine content. The lowest surface energies of hydrocarbon surfaces are exhibited by closely packed methyl groups. The presence of aromatic groups in the surface increases the critical surface tension. Gardon[4] has related the

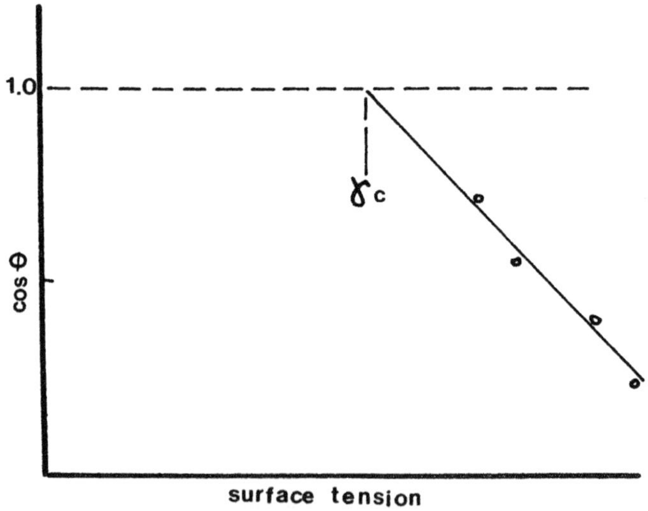

Fig. 35-6. Determination of critical surface tension.

Table 35-3. Critical Surface Tension of Some Plastic Film Surfaces

MATERIAL	CRITICAL SURFACE TENSION (dynes/cm at 20°C)
Polytetrafluoroethylene	22
Silicone rubber	22
Polyvinylidene fluoride	25
Butyl rubber	27
Polyvinyl fluoride	28
Polybutyl acrylate	28
Polyurethane	29
Polyethylene	31
Polychlorotrifluoroethylene	31
Polypropylene	31
Polystyrene	33–35
Polyethyl methacrylate	33
Polyvinyl chloride	33–38
Cellulose	35.5–41.5
Polyvinyl acetate	37
Nylon 6	38–42
Cellulose acetate	39
Polyethylene terephthalate	43

critical surface tension (γ_c) to the solubility parameter of a polymer (δ) as follows:

$$\gamma_c = 3.61\delta \qquad (5)$$

The critical surface tension can be approximately determined by employing liquids of known surface tensions. A liquid that just wets the surface has a wetting angle of zero, and its surface tension is equal to the critical surface tension of the solid surface.[5] A series of liquids of varying surface tensions can be prepared by blending two (or more) solvents as shown in Table 35-4. Such liquid blends can also be purchased as ready-to-use-kits.

LEVELING

Leveling is the capability of a liquid surface to eliminate minor nonuniformities and scratches introduced during coating. Most coating processes leave a nonuniform surface. Such coating marks are expected to level before the coating solidifies. The driving force of the leveling process is the surface tension that must overcome the resistance to flow offered by viscosity. Leveling takes place at low shear rates of 0.01–0.1 sec^{-1}, and the viscosity at low shear rates is

Table 35-4. Surface Tension of Various Formamide-Cellosolve Mixtures

FORMAMIDE (vol %)	CELLOSOLVE (vol %)	SURFACE TENSION (dynes/cm)
99.0	1.0	56
96.5	3.5	54
93.7	6.3	52
90.7	9.3	50
87.0	13.0	48
83.0	17.0	46
80.3	19.7	45
78.0	22.0	44
74.7	25.3	43
71.5	28.5	42
67.5	32.5	41
63.5	36.5	40
59.0	41.0	39
48.5	51.5	37
42.5	57.5	36
35.0	65.0	35
26.5	73.5	34
19.0	81.0	33
10.5	89.5	32
2.5	97.5	31
0	100.0	30

Note: This test is a part of ASTM Standard D 2578-67. Formamide–water blends for surface tension in the range of 56 to 70 dynes/cm are given by Markgraf.[6] For other blends, see reference 7.

therefore of utmost importance.[7] It is obvious that the most desirable coating properties for best leveling are high surface tension and low viscosity at low shear rates. These conditions are not easily met: the surface tension must often be decreased in order to wet the substrate and often the viscosity must be increased in order to obtain sufficiently thick coatings so as to be able to process them on coating equipment.

Most of the solventborne coatings, unless they are of a very high viscosity, level well. Aqueous emulsions, however, often exhibit insufficiently good leveling. Often they are pseudoplastic and may have a yield point, that is, the viscosity is high at low shear rates at which the leveling takes place (see Fig. 35-7).

Figure 35-8 shows viscosity shear rate curves for several emulsion coatings formulated with thickeners.[23] Emulsions 1 and 2 have leveled well, while emulsions 3 and 4 did not.

The Rheology Index is a useful measure of the deviation from Newtonian flow. It is defined as the ratio of high shear rate viscosity to low shear rate

viscosity. If the Rheology Index is equal to one, the
flow characteristics; if it is larger than one, the
smaller than one, it is pseudoplastic. A
leveling. Aqueous adhesive emulsions should
0.25 for acceptable leveling.

The leveling process has been extensively stud
brush application of paint, especially of paint
and with the leveling of emulsion coatings
methods. Reverse roll coating produces longitudinal
splitting between the rolls. The geometry of such
35-9.

Orchard[9] has proposed a mathematical model
the leveling half-time is expressed as follows:

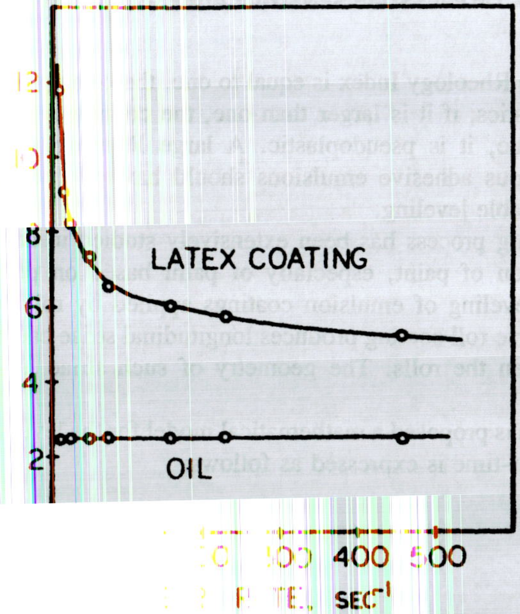

where

η_L = viscosity at low

lsion coating as a function of shear rate compared to

viscosity. If the Rheology Index is equal to one, the coating exhibits Newtonian flow characteristics; if it is larger than one, the coating is dilatant; and if it is smaller than one, it is pseudoplastic. A large Rheology Index favors good leveling. Aqueous adhesive emulsions should have a Rheology Index above 0.25 for acceptable leveling.

The leveling process has been extensively studied in connection with the brush application of paint, especially of paint based on polymer emulsions, and with the leveling of emulsion coatings applied by roll coating and other methods. Reverse roll coating produces longitudinal striae caused by the coating splitting between the rolls. The geometry of such striae is shown in Figure 35-9.

Orchard[9] has proposed a mathematical model for the leveling process where the leveling half-time is expressed as follows:

$$T^{1/2} \propto \frac{\eta_L}{\gamma} \; \frac{\lambda^4}{h^3} \tag{6}$$

where

γ = surface tension

η_L = viscosity at low shear rate

Leveling is favored by the decrease of viscosity, increase in surface tension, decrease of the distance between striation marks, and increase of coating thickness. Leveling has also been reviewed by Mahoney and Hagan.[10]

Some emulsions exhibit viscoelastic properties. The elastic component affects the coating process and causes the production of more pronounced striation marks, even though such emulsions might have a high (0.5) Rheology Index.[8] Elasticity also has a retarding effect on leveling, although less important than that of viscosity.[11]

The compounding of adhesives for the best coating and leveling properties requires decreasing the pseudoplastic behavior as much as possible, and removing the yield value at low shear rates, which does not allow proper leveling. Figure

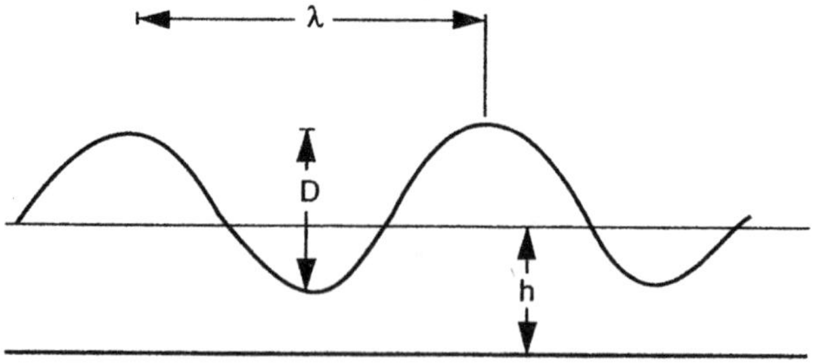

Fig. 35-9. Profile of coating striation marks.

Adhesives that are applied as hot melts, or that are calendered or extruded, have a high viscosity. Dewetting is not a problem after the adhesive has been applied as a thin coating.

Aqueous adhesive emulsions, on the other hand, often require adustments of viscosity and surface tension for acceptable machinability during the coating process. Emulsions are not usually sufficiently viscous and the viscosity must be adjusted. Aqueous emulsions might also require an improvement in surface wetting, a decrease of foaming, and an improvement in mechanical stability.

THICKENERS

The viscosity of many carboxylated polymer emulsions is pH dependent and it can be increased simply by the addition of a base, such as ammonium hydroxide. A typical pH-viscosity curve for such a polymer is shown in Figure 35-11.[12]

The adhesive properties of self-thickening latexes may be slightly affected by the neutralization of carboxylic groups. Table 35-5 shows the effect of neutralization on tack and peel adhesion.

Addition of solvent or plasticizer may also increase the viscosity. Solvent swells the emulsion particles and increases interaction between particles, thus causing a viscosity increase. Solvent evaporates on drying and does not leave any permanent effect on the adhesive properties; plasticizer, of course, softens

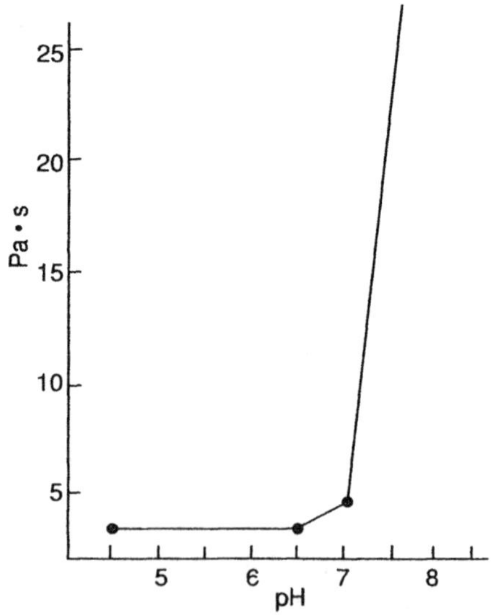

Fig. 35-11. Viscosity dependence on pH for an acid terminated acrylic pressure sensitive adhesive.

Table 35-6. Thickening Efficiency of Alkali-Soluble Acrylic Emulsion Thickeners for Various Polymer Emulsions

EMULSION TYPE	PERCENT THICKENER ON TOTAL SOLIDS REQUIRED TO RAISE THE VISCOSITY TO 2 PA·S[a] A	B
Nitrile latex (20%)	4.0	6.2
Styrene copolymer latex (20%)	4.2	9.0
Natural rubber latex (50%)	0.8	0.6
SBR latex (20%)	2.5	1.0
Acrylic latex (20%)	2.5	>6.0

[a] A. Acrysol ASE-60. B. Acrysol ASE-75.

Table 35-7. Thickening Efficiency of Various Thickeners in an Acrylic Adhesive Latex (Viscosities Pa·s)

THICKENER TYPE	PERCENT THICKENER ON DRY BASIS 0	0.4	1.0	2.0
Acrylic emulsion	0.16	1.55	6.75	10
Cellulosics	0.16	1.2	2.85	8.9
Polyvinyl alcohol	0.16	3.8	7.1	7.5

tics as close to Newtonian as possible. There are rheology modifiers, such as Acrysol RM-3, that reduce low shear viscosity without much effect on the high shear viscosity. The yield value required to obtain the smallest visually discernible brush marks is considered to be 0.5 dyne/cm^2. The yield value of 20 dyne/cm^2 gives pronounced brush marks in a paint coating. The effect of various thickeners on the rheology of styrene-butadiene copolymer/tackifying resin emulsion was investigated by Hickman and Kokke-Hall,[18] and the results are described in Chapter 12. Polyurethane and cellulosic thickeners had no yield stress and therefore leveled well.

Polyacrylate Thickeners. These thickeners are available in two forms: as alkaline viscous solutions of acrylic acid and as alkali-soluble polyacrylic copolymer emulsions. The latter are of low viscosity and are therefore easily added to the adhesive. The viscosity is increased by the addition of ammonia

Table 35-10. Effect of Polyvinyl Methyl Ether on the Adhesive Properties of an Acrylic Adhesive

AMOUNT OF THICKENER (phr)	180-DEGREE PEEL ADHESION[a] (N/m)(oz/in.)	QUICK TACK (kPa)	POLYKEN TACK (g/cm^2)	ROLLING BALL TACK (cm)	STATIC SHEAR[b] (min)
0	235 (21.5)	16.8	742	18	2.6
0.5	291 (26.6)	17.2	570	19	2.0
1.0	235 (21.5)	15.9	707	20	1.6
2.0	290 (26.5)	15.9	765	15	1.7

[a] Coating thickness is 0.035 mm.
[b] Weight on a 1.3-cm-by-1.3-cm sample is 400 g.

Polymer emulsions contain surface active agents added before polymerization, but that might not be sufficient to assure adequate wetting and emulsion stability, so that additional surfactants and stabilizers might be required. Solvents and plasticizers might also be used to decrease the surface tension.

Surface active agents may affect the properties of a pressure sensitive adhesive. In addition to increased water sensitivity, the surfactant may migrate, if incompatible, to the adhesive surface, causing a decrease of tack. Surfactants that are compatible with the adhesive act as plasticizers; they may increase the tack and peel adhesion and decrease the shear resistance. Fatty acid soaps, such as oleates, lauryl sulfates, and stearates migrate rapidly even at low concentration levels and cause loss of tack.[21] However, when used as an additive to styrene butadiene latex, potassium rosinate has even caused an improvement of tack and peel adhesion after aging. Ethoxylated tetramethyldecynediol and octyl phenoxy polyethoxyl ethanol also show similar results. Table 35–12 shows the effect of some surfactants.

Nonionic surfactants based on acetylenic glycol have been proposed for use with pressure sensitive adhesive latexes. These surfactants are nonfoaming, effectively reduce the surface tension, and do not increase water sensitivity because of the presence of a hydrophobic acetylenic glycol group. Peel adhesion remains constant on aging and after immersion in water.[22]

It has also been reported that surfactants in some cases cause a large change in the adhesive's shear strength, as shown in Table 35-13.[13] Similar effect with several surfactants has been observed independently by the author. Surfactants may act as plasticizers and in small quantities could conceivably soften the adhesive slightly, resulting in a slight change in peel adhesion and in a slight decrease in shear resistance. The drastic decrease of shear resistance with some adhesive–surfactant combinations is much greater than can be expected from the plasticizing effect. It is conceivable that even a small amount of surfactant may cover the surface of latex particles and prevent complete fusion, causing this change in shear resistance. This explanation is also consistent with the fact that the effect is not observed with all pressure sensitive adhesives; the shear resistance of some adhesives does not depend on the secondary bonds, but on molecular entanglement and covalent bonding.

Emulsion Stabilization

Natural rubber latex historically has been stabilized by the addition of potassium hydroxide. Several surfactants in low concentrations of 0.5 phr are effective in increasing the mechanical and chemical stabilities of natural rubber latex. Ethylene oxide condensates of alkyl phenols are low foamers and impart mechanical stability, especially when an octylphenol or a nonylphenol condensation

Table 35-13. Effect of Surfactants on Adhesive Properties

PROPERTY	LATEX ALONE	TRITON X-100[a] (5 phr)	DOWFAX 2A-1[b] (5 phr)
Thickness (μm)	28	30	36
Williams plasticity	4.06	2.38	4.74
Static shear (hr) 70°C, 1000 g/in²	>95	1–11	>95
Static shear (hr) room temperature 1000 g/in²	>199	8.23	>199
180-degree peel adhesion (N/m)	550	525	370

Note: Hycar 2600x207 by BF Goodrich Chemical Co.
[a] Surfactant by Rohm & Haas Co.
[b] Surfactant by Dow Chemical Co.

product with 30–40 moles of ethylene oxide is used.[23] Among the anionic surface active agents, sodium alkyl naphthalene sulfonates produced excellent mechanical and chemical stability. Anionic sulfonates, however, caused foaming. Octylphenoxy ethanol-based surfactants did not cause much change in the adhesive properties of natural rubber/tackifying resin emulsion adhesives.

Adhesive emulsions may be acidic or alkaline. For alkaline emulsions a dilute solution of ammonium hydroxide may be added as a stabilizer. Sodium tripolyphosphonate and tetrasodium pyrophosphate are buffers used with both acidic and alkaline emulsions as stabilizers.

Insufficiently stable adhesive latexes may cause polymer deposition on rolls and other equipment parts, requiring frequent cleaning. Stabilization of the latex by either adjustment of the pH or by the addition of surfactants, glycols, or colloidal stabilizers may be required. Keeping the rolls wet by spraying, or by a moistened felt, or by coating the rolls with a fluorocarbon coating may be helpful in preventing the adhesion of coagulum.

Prevention of Foaming

Foaming of aqueous emulsions interferes with mixing and coating operations, but foaming is not limited to emulsions only: adhesive solutions can also foam.

The presence of surfactant lowers the surface tension and produces an elastic membrane around the air bubble, thus enhancing foaming and foam stability. Foaming ability and foam stability can be evaluated by several methods. Generally, the height of the foam is measured immediately after shaking and

8. Matsuda, T., and Brendley, W. H., Jr. *J. Coating Technol.* **51**(658): 46–60 (November 1979).
9. Orchard, S. *Appl. Sci. Res.* **A11:** 451 (1962).
10. Mahoney, D. M., and Hagan, J. W. Technical Considerations in Converting to Water-Based PSA's. In *Proc. Technical Seminar,* pp. 78–101. Pressure-Sensitive Tape Council, Cherry Hill, N.J., June 22–23, 1982.
11. Bierman, M. *Rheol. Acta* **7:** 138 (1968).
12. Sanderson, F. T. Water Based Acrylic Pressure Sensitive Adhesives. In *Proc. Technical Meeting on Water Based Systems,* pp. 13–27. Pressure-Sensitive Tape Council, Chicago, Ill., June 21–22, 1978.
13. *Latexes in Adhesive Systems.* Bulletin L-14, B.F. Goodrich Chemical Co.
14. Blake, D. M., Jr. Thickeners for Waterborne Coatings. *Handbook of Coating Additives,* pp. 43–117 (L. J. Calbo, ed.). Dekker, New York and Basel, 1986.
15. Greenwald, H. L., and Luskin, L. S. Poly(Acrylic Acid) and Its Homologs. *Handbook of Water-Soluble Gums and Resins* (R. L. Davidson, ed.). McGraw-Hill, New York, 1980.
16. Garret and Brown. *J. Appl. Polym. Sci.* **1:** 287–295 (1959).
17. Sanderson, F. T., and Gehman, D. R. Acrylic Thickeners for Latex Adhesives. In *Proc. Technical Meeting on Water Based Systems,* pp. 91–108. Pressure Sensitive Tape Council, Chicago, Ill., June 13–14, 1979.
18. Hickman, A. D., and Kokke-Hall, S. The Versatility of S/B Latex for PSA Applications. In *Proc. Technical Seminar,* pp. 136–148. Pressure Sensitive Tape Council, Itasca, Ill., May 8–10, 1985.
19. Powell, G. M. III. Hydroxyethylcellulose. *Handbook of Water-Soluble Gums and Resins* (R. L. Davidson, ed.). McGraw-Hill, New York, 1980.
20. Rifi, M. R. Water-Based Pressure Sensitive Adhesives Structure vs. Performance. In *Proc. Technical Meeting on Water-Based Systems,* pp. 28–39. Pressure Sensitive Tape Council, Chicago, Ill., June 21–22, 1978.
21. French, J. A., and Lee, I.S.P. Factors Affecting the Performance of Carboxylated Styrene Butadiene Latices in Pressure Sensitive Adhesives. In *Proc. Technical Meeting on Adhesive Coating Technology,* pp. 48–60. Pressure Sensitive Tape Council, Rosemont, Ill., June 23–24, 1981.
22. Morrell, S. P. The Role of Acetylenic Glycols in Water-Based Pressure Sensitive Adhesives. In *Proc. Technical Meeting on Adhesive Coating Technology,* pp. 55–75. Pressure Sensitive Tape Council, Rosemont, Ill., June 18–19, 1980.
23. Oldack, R. C. PSA's from Natural and Carboxylated SBR Latexes. In *Proc. Technical Meeting on Water-Based Systems,* pp. 38–63. Pressure Sensitive Tape Council, Chicago, Ill., June 13–14, 1979.
24. Rosen, M. J. *Surfactants and Interfacial Phenomena,* pp. 201–206. Wiley, New York, 1978.
25. Nielsen, A. C. Antifoam Selection: Case Studies Based on Several Resin Systems. In *Proc. Technical Meeting on Water-Based Systems,* pp. 40–59. Pressure Sensitive Tape Council, Chicago, Ill., June 21–22, 1978.

increasing the rate of the slowest step. The study of the drying process by varying the external factors requires the observation of the effect of temperature in various zones, web speed, air velocity, and other process variables. Based on these observations, the dryer conditions are set to obtain optimum results in terms of drying rate and coating appearance.

DRYING PROCESS

The drying of a pressure sensitive adhesive coating is a complex process, because heat and mass transfer take place simultaneously by several different mechanisms that may change as drying proceeds.

During the initial drying stages the liquid evaporates from the saturated surface. The concentration of the liquid in the adhesive is high and its movement to the surface is sufficiently rapid to keep the surface saturated with liquid. During this stage, called constant rate drying, the rate of mass transfer is determined by the rate of heat transfer into the adhesive. An increase in drying rate can be realized by increasing the heat transfer rate.

At the conditions of dynamic equilibrium between the heat transfer into the adhesive film and the vapor removal from the surface, the drying rate is expressed by the following equation:

$$\frac{dw}{d\theta} = \frac{hA\Delta t}{\lambda} = k_g A\Delta p, \tag{1}$$

where

$\frac{dw}{d\theta}$ = drying rate
k_g = mass transfer coefficient
h = heat transfer coefficient
A = area for heat transfer and evaporation
λ = latent heat of evaporation at t_s
$\Delta t = t - t_s$, where t = gas temperature and t_s = temperature of evaporation surface
$\Delta p = p_s - p$, where p_s = vapor pressure of liquid at the surface temperature t_s and p = partial pressure of liquid in the gas

As soon as the liquid concentration on the surface decreases below saturation, the drying enters into the falling rate period. The rate of solvent removal decreases with drying time during this period. The initial portion of the falling rate period is sometimes called the warming-up period, because the adhesive surface temperature starts to increase at a fast rate during this short period. The adhesive

the thickness. Such data are very difficult to obtain in thick slabs; they are practically impossible to obtain in thin adhesive films.

The deviation of the theoretical diffusion equations from the experimental results is due to the assumption that the diffusion coefficient is constant. In fact, it is affected by many variables. It decreases with decreasing liquid concentration, with decreasing temperature, with increasing pressure, and with increasing density. Van Arsdell[3] has elaborated on the diffusion calculations, assuming a variable diffusion coefficient in the falling rate period. Hansen[4] points out that the diffusion coefficient for solvents in polymers is highly dependent upon the solvent concentration. This dependence is exponential. In the case of drying a paint film, a somewhat different mechanism has been suggested. The rate is limited by boundary conditions, and diffusion through the stagnant air layer at the boundary determines the drying rate.[5]

The drying process is also affected by the heat transfer mechanism. In the case of one-sided convection drying, heat is transferred from the top surface down. The formation of a surface skin may take place sooner than in ovens where heat is introduced from both sides. Drying by radiation may help to extend the constant rate drying period, since heating takes place from the inside of the coating; the radiation penetrates somewhat below the surface.

DRYING CURVES

Drying data are best represented graphically in one of several ways as shown in Figures 36-1 through 36-3. These curves have been constructed from the drying data of a typical pressure sensitive rubber-resin adhesive solution containing 25% solids, the solvent being heptane-toluene blend (80/20), with a coating weight of 105 g/m^2, dried by convection at 66°C.

Figure 36-1 shows the remaining solvent in the adhesive plotted against drying time. While it is simple to construct this type of a drying curve, the information gained is minimal. The curve shows that the drying rate decreases with drying time, but does not show drying rate values at various points in time. The change in the drying rate is better illustrated in Figure 36-2, constructed from the same data, which shows a plot of drying rate versus drying time. Various drying periods are clearly shown in this plot. At the beginning of drying, at the point immediately after the coating head, rapid evaporation takes place accompanied by the lowering of film surface temperature. This cooling-down period is represented by the segment A–B and is commonly observed with most coatings containing rapidly evaporating solvents. The B–C segment represents the constant rate drying period, which is usually of short duration for solution coatings. Obviously, it is advantageous to increase the length of this period, because the evaporation rate starts dropping rapidly at the end of

up period. Segment D–E represents the falling rate drying period. This segment clearly dominates the drying process and corresponds to the onset of diffusion as the dominating mass transfer mechanism. The drying rate during this period can be accelerated by an increase in temperature (mass diffusion coefficient increases with increasing temperature). However, the increase is limited by the formation of blisters in the adhesive surface caused by expanding solvent vapor trapped below the dry film surface. Surface disturbances occur because of surface skin formation before the adhesive coating is completely dried and can manifest in various ways as blisters, surface craters, or pinholes.[6]

It has sometimes been advocated that the addition of a higher boiling solvent, which would leave the coating last, might allow the use of a higher drying temperature in the last zones, thereby decreasing drying time. It should be kept in mind, however, that bulky molecules have lower diffusion coefficients and therefore, diffuse at a slower rate, counteracting the effect of the increased temperature. It has been observed that the addition of a liquid plasticizer to the adhesive formulation increases the drying rate. Apparently, the liquid plasticizer provides a path of lower resistance for the diffusion of solvent molecules.

The drying data may also be represented by plotting the drying rate versus the solvent content of the adhesive, as shown in Figure 36-3.

Galbraith[7] has presented the drying data on a special graph paper that allows for an easier analysis of the effects of various drying condition changes. Such graphs are constructed by plotting the percent of total volatile remaining (percent of the original solvent still remaining) against the total accumulated time in the oven. The abscissa is plotted on log-log scale and the ordinate on

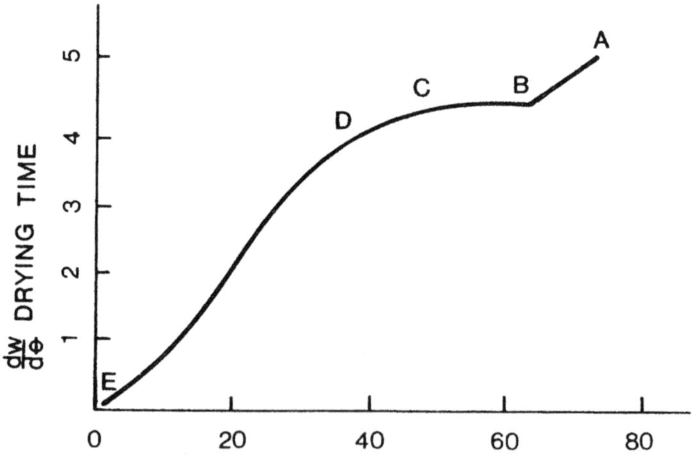

Fig. 36-3. Drying rate as a function of remaining solvent.

log probability scale. Figure 36-4 shows such a plot. The data points fall on three straight lines, the first of which is the constant rate drying period, the second the warming-up period, and the third the falling rate period. The changes in the drying process resulting from changes in drying conditions can be easily analyzed if the data are presented in this manner. Figure 36-5 illustrates a change that affects the constant rate drying period. A series of parallel lines represent drying processes that differ in the length of constant rate drying period. Figure 36-6 shows the curve expected when the change in drying conditions affects the falling rate period only. The converging lines represent different drying conditions during the falling rate period, while the upper portion of the curve representing constant rate and warming-up periods remains unchanged.

Figures 36-7 through 36-10 show the results of various changes in the

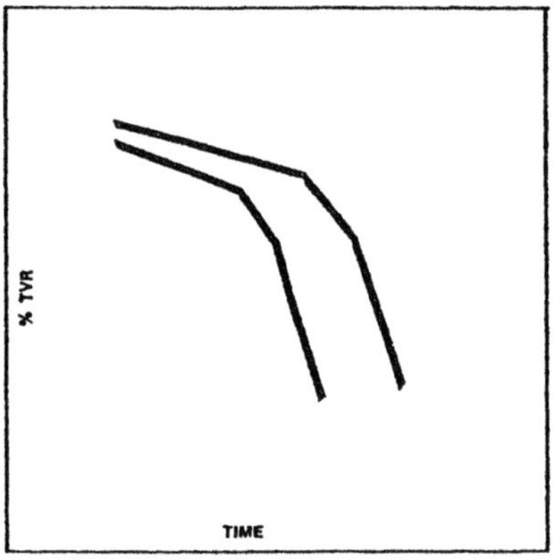

Fig. 36-5. Drying curve change affected by a change in the constant rate drying period.[7]

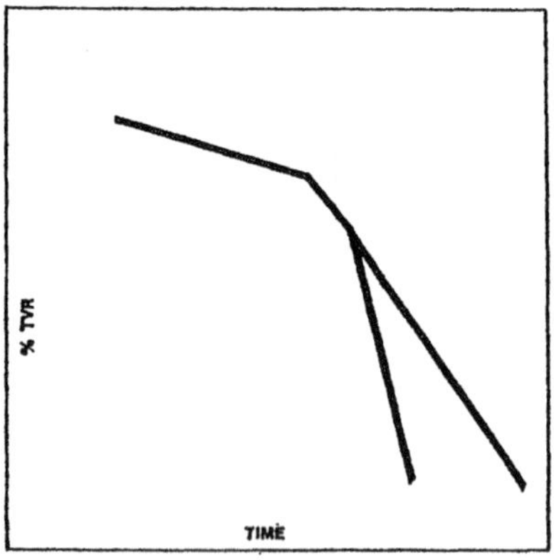

Fig. 36-6. Drying curve change affected by a change in the falling rate drying period.[7]

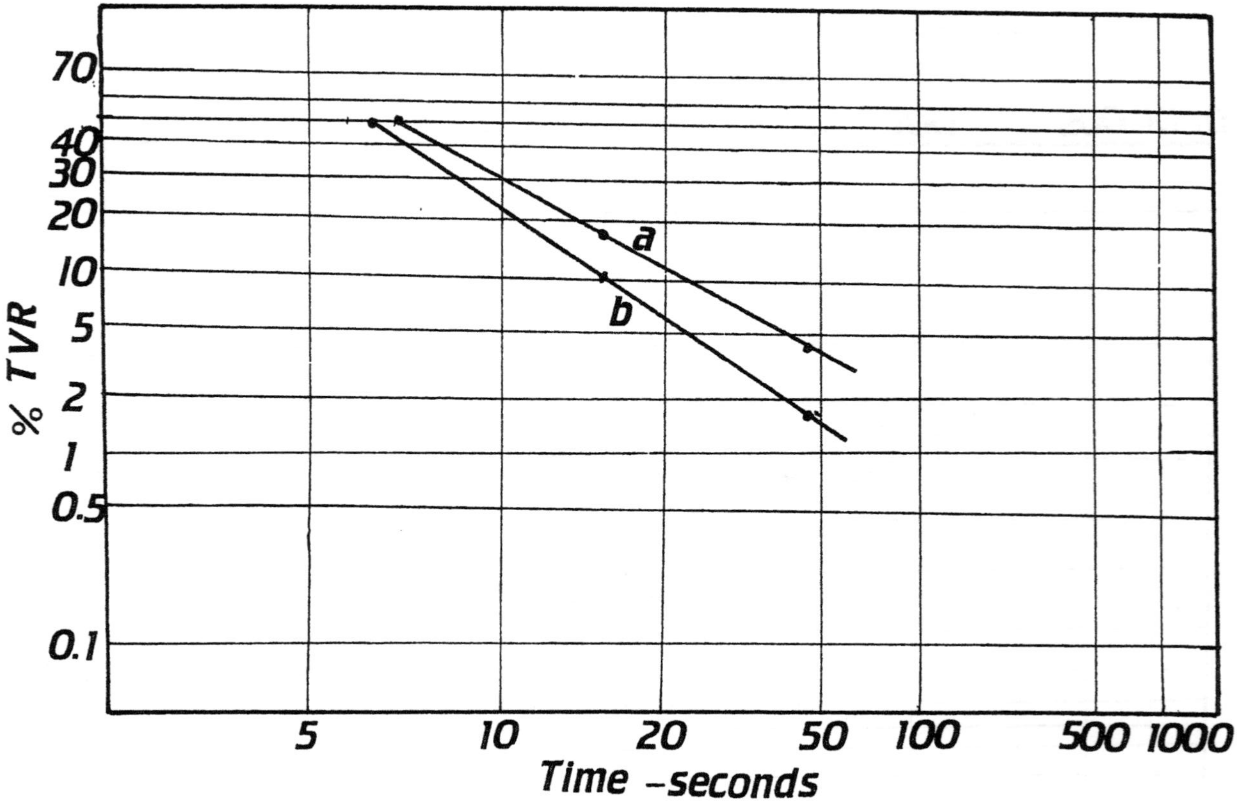

Fig. 36-8. The effect of temperature on drying of a solution adhesive.[7] (a) 66°C (150°F); (b) 121°C (250°F).

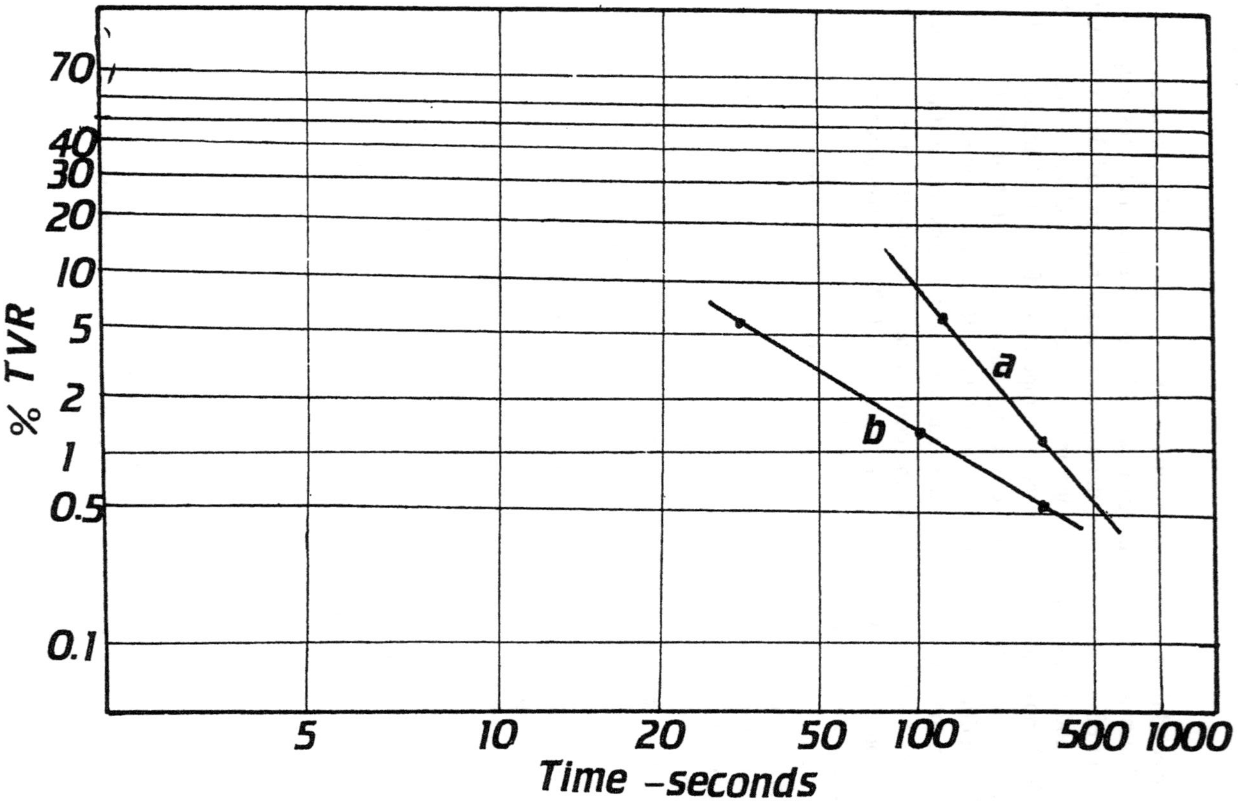

Fig. 36-9. The effect of temperature on drying of an emulsion adhesive.[7] 2-mil-thick coating. (a) 66°C (150°F); (b) 121°C (250°F).

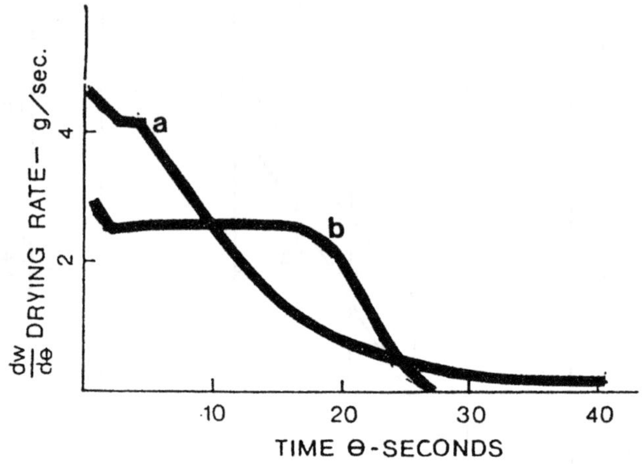

Fig. 36-11. Drying curves with different constant rate drying periods.[8]

resulting in a gain in overall production rates. Figure 36-12 compares the drying curve for an adhesive coating 50 μm thick (Curve a), with the coating and drying performed in two stages on two consecutive coatings each 25 μm thick (Curves b_1 and b_2). Drying time was decreased from 40 to 20 sec. Obviously, the equipment costs for such a multiple-stage coater-dryer are much higher.

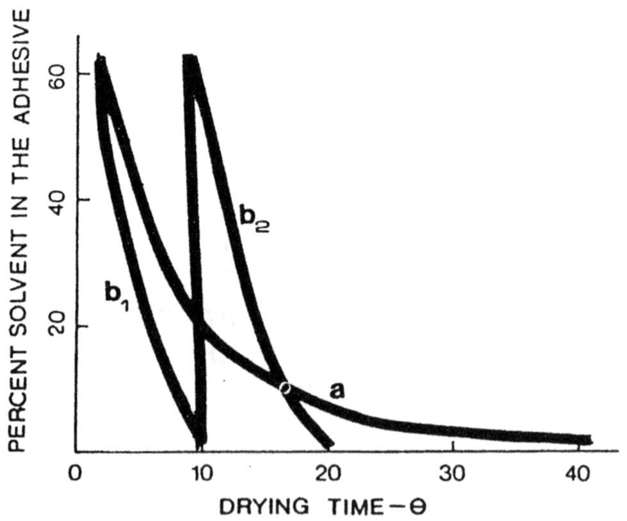

Fig. 36-12. Drying of a 50-μm-thick adhesive coating; (a) drying in one stage; (b_1, b_2) drying of two consecutive coatings.[8]

Table 36-1. Solvents in Order of Increasing Retention

SOLVENT	EVAPORATION RATE (*n*-butyl acetate = 1.0)	MOLAR VOLUME (cm^3/mole)
Methanol	4.1	40
Acetone	10.2	73
2-Methoxyethanol	0.51	79
MEK	4.5	90
Ethyl acetate	4.8	97
2-Ethoxyethanol	0.35	97
n-Heptane	3.3	146
2-Butoxyethanol	0.076	130
n-Butyl acetate	1.0	132
Benzene	5.4	88
2-Methoxyethyl acetate	0.35	117
2-Ethoxyethyl acetate	0.23	135
Dioxan	—	85
Toluene	2.3	106
Chlorobenzene	—	101
2-Nitropropane	1.5	90
m-Xylene	0.75	122
MIBK	1.4	124
Isobutyl acetate	1.7	133
2,4-Dimethyl pentane	5.6	148
Cyclohexane	5.9	108
Diacetone alcohol	0.095	123
Pent-Oxone	0.26	143
Methyl cyclohexane	3.5	126
Cyclohexanone	0.25	103
Methyl cyclohexanone	0.18	122
Cyclohexyl chloride	—	118

is poor, the resin can precipitate during drying and give a heterogeneous adhesive of decreased peel adhesion. Engel and Fitzwater[15] report that the adhesion of methyl methacrylate/lauryl methacrylate (90/10) copolymer films depends on the solvent composition used for the casting of these films. The polymer chains are extended in good solvents. The probability of functional groups contacting the substrate is higher than in the case of poor solvents where the polymer chains are coiled.

DRYING OF LATEXES

In the drying of latex adhesive films, the mass transfer mechanism during the falling rate period is different. It is believed that, at least during part of the

evaporation and film formation take place simultaneously. In dry sintering, film formation would start only after the water has evaporated.

Sheetz[19] reviewed the other theories and proposed a diffusion mechanism. According to this theory, capillarity, wet sintering, and diffusion are important at the early stages of latex coating drying. In later stages, the diffusion of water through the polymer particles to the film surface becomes the dominant mass transfer mechanism. Sheetz points out that the rate of water evaporation would become very slow at the end of drying if the water were supplied to the surface by capillary forces. Experimental data show that the rate of water evaporation becomes slower at about 70% solids, but not as slow as would be expected from the capillary theory. Furthermore, the drying rate depends on the type of polymer used, which also suggests a diffusional process.

Voyutskii[20] has proposed a theory that also considers the presence of surfactants in latexes. Latex drying consists of three stages. In the first stage, water is evaporated and a fairly well packed coating is formed. In the second stage, the spherical latex particles are deformed and accept a polyhedron form as contact between the latex particles increases. During the third stage, the surfactant either dissolves in the polymer or, if it is not soluble, accumulates between the latex particles and forms separate aggregates. In the first case, the surfactant acts as a plasticizer and may improve the cohesion between the particles; in the second stage, the adhesive film represents mutually penetrating networks of surfactant and adhesive. Such films may exhibit a lower tack and adhesion and have poor water resistance.

The mechanism of latex drying was reviewed by Imoto,[21] who pointed out that none of the proposed theories are completely satisfactory.

Adhesive coatings from latexes can be dried at higher temperatures and at higher heat transfer rates than solution coatings, although the temperature should be kept below 100°C at the early stages of drying. The fast decrease of drying times with increasing temperature for various thicknesses of acrylic coatings is shown in Figure 36-14.[22] The same data are presented in Figure 36-15.[8]

DeSanti has compared the single-sided and two-sided drying of pressure sensitive adhesive latex coatings.[23] Laboratory data were used to verify a computer model used to simulate web drying at various positions throughout the oven. For latex coatings, heat transfer is much more important, because a larger amount of heat must be transferred into the coating in order to evaporate water from a latex than solvent from an adhesive solution. The dryer should have at least two temperature zones in order to keep its length shorter. No skinning takes place if the average evaporation rate in the first zone is maintained at a sufficiently low level. The drying rate can be much higher in localized drying areas. The evaporation rate must be restricted until the adhesive coating is set at least 70% dry. After that, blistering does not take place and more

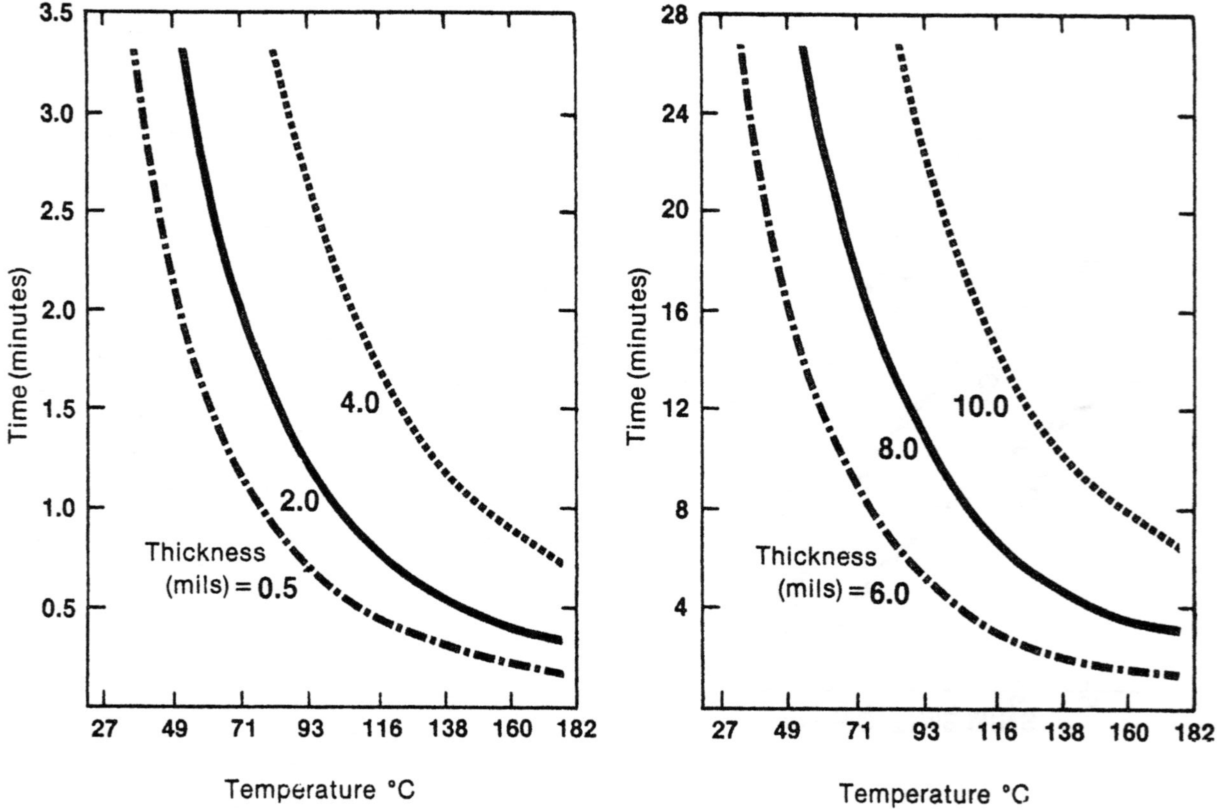

Fig. 36-15. Drying time vs. oven temperature for several film thicknesses of an acrylic polymer. (*Courtesy of B.F. Goodrich Chemical Co.*)[22]

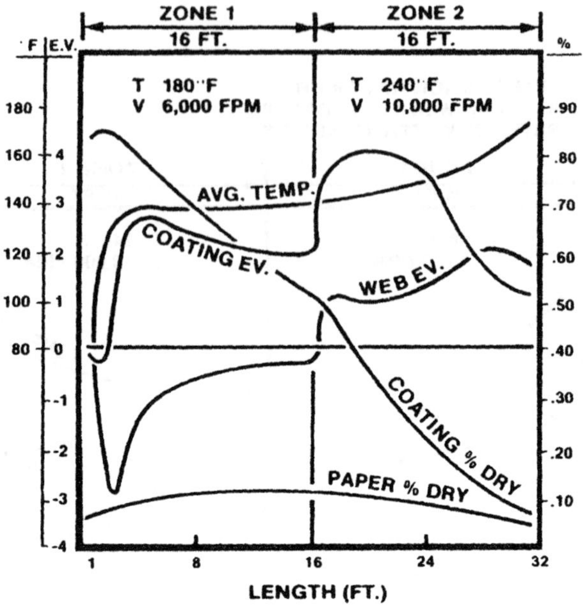

Fig. 36-17. Two-sided drying evaporation curves for an aqueous adhesive coating.[23]

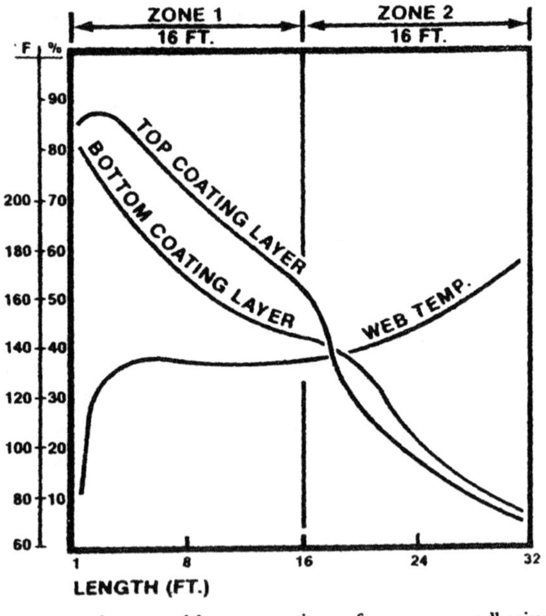

Fig. 36-18. Moisture content in top and bottom portions of an aqueous adhesive coating in a two-sided coater.[23]

higher temperature to the adhesive surface. In two-side convection ovens, some heat is also transferred by conduction from the backing to the adhesive. In infrared ovens, heat is transferred by radiation and also by convection. Some heat within the adhesive film is always transferred by conduction.

In the case of a turbulent air stream passing over the adhesive coating, the heat must be transferred through a boundary layer of stagnant air, which offers resistance to heat flow and determines the numerical value of the heat transfer coefficient. Most convection ovens operate well in the turbulent flow region at high Reynolds numbers [$LV\ (\rho/\mu)$]. At Reynolds numbers above 500,000, the Colburn relation can be used, which in its simplified form for air is shown below:[24]

$$h_m = 0.0128G^{0.8} \tag{4}$$

where

h_m =heat transfer coefficient based on length mean Δt
G = mass flow

The heat transfer is increased by the 0.8 power of the air velocity. An increase in air velocity to improve the drying rate has its limitations, for example, power consumption is increased at a faster rate than the heat transfer coefficient and high velocity air may disturb the adhesive coating.

At Reynolds numbers below 80,000, the laminar boundary layer determines the heat transfer, and the Pohlhausen equation becomes valid, as shown in its simplified form:[25,26]

$$h = n(V/L)^{1/2} \tag{5}$$

where

n = a constant dependent on air properties
V = average air velocity
L = heated length

Heat transfer is increased as the square root of air velocity and as the reciprocal square root of L, which depends on the discharge nozzle spacing. The closer the nozzles are spaced together, the smaller L is.

While most convection dryers operate in the turbulent flow region, some have been developed to operate at low Reynolds numbers. In such cases, the close spacing of nozzles becomes very important. The theory behind the construction of such dryers has been discussed by Gardner.[27]

Infrared radiation in combination with convective drying is used quite often for pressure sensitive adhesives. The infrared energy band lies just outside the visible light band in the electromagnetic spectrum. The useful wavelength band for drying is in the region of 0.7–11 μm. This band is subdivided into

ing vehicle below the surface skin. Coating of impervious webs is especially susceptible to the formation of blisters. This can be remedied by lowering the rate of evaporation in the early zones in order to delay the formation of surface skin. A decrease in temperature, airflow, or line speed helps in such cases. Figure 36-21 shows a typical cratered coating surface. Various other surface defects in coatings have been discussed in an earlier publication.[6] The presence of a high boiling diluent is especially conducive to the formation of blisters. The extreme case is the presence of water in the solvent solution. Water, unless it forms an azeotrope, would tend to remain behind. It does not diffuse easily through the polymer film and later causes blow holes in the coating, when subjected to temperatures above its boiling point. Smaller quantities of trapped solvent, including water, may be insufficient to form blow holes, but sufficient to form a foamy layer within the adhesive coating, usually close to the substrate. The coating is weak in this plane and fails more easily in shear than the rest of the adhesive.

The substrate, especially paper, may contain water that can be released

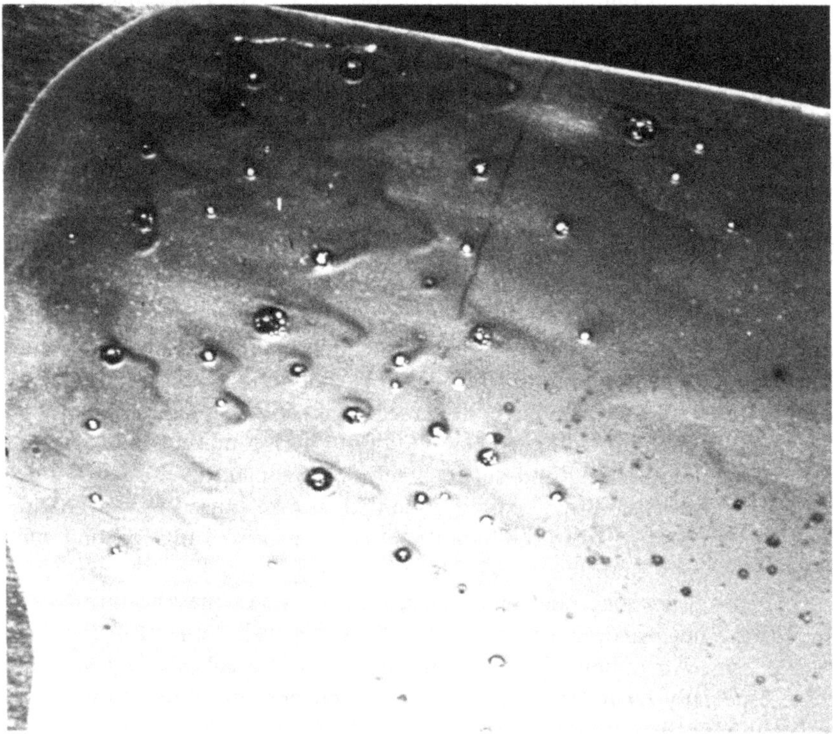

Fig. 36-21. Craters in a dried coating. (*Courtesy of Nalco Chemical Co.*)

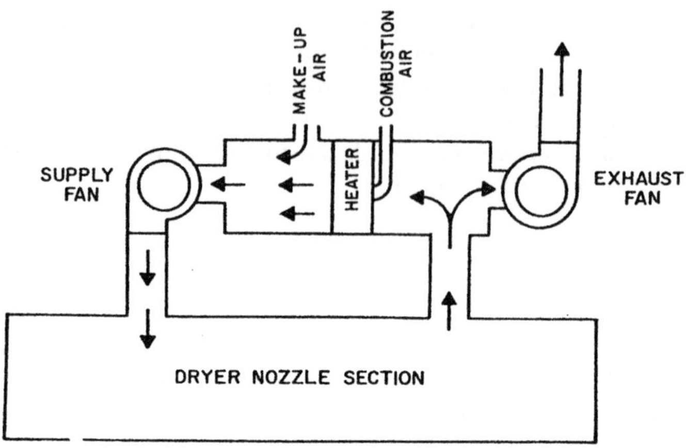

Fig. 36-22. Schematic diagram of the air flow in a convection oven. (*Courtesy of Overly, Inc.*[28])

is about 1.1–1.3 volume percent. Solvent drying operations computerized for maximum safety have been proposed.[29] Most of the solvent evaporates in the first zone, and therefore less recycling is possible in that zone. To conserve energy, cascading may be employed. Cascading is the reusing of exhaust gases of one zone as supply air for another zone. Cascading can be used effectively in pressure sensitive adhesive drying, because the air in the last zones does not have much vapor and can be gainfully employed in the early zones as the supply air. Heat recovery units may be used to extract heat from the exhaust and to utilize it for the preheating of makeup air.

Air recirculation in ovens used to dry aqueous adhesives is much higher, resulting in lower overall energy requirements, even though the heat of evaporation of water is higher than that of solvents. In this case, the extent of air recirculation is not limited by safety considerations, but by the effect of increasing vapor concentration on the drying rate. In convection ovens, most of the energy is expended for heating a large volume of air and not for the evaporation of the vehicle. The exceptions are dryers operating in an inert gas atmosphere.

Gardner[30] gives heat transfer coefficients of various drying oven constructions for comparison purposes.

Quiet air, 10 kcal/(hr) (m^2) (°C)
15 mph wind, 29 kcal/(hr) (m^2) (°C)
Early dryers, 50–120 kcal/(hr) (m^2) (°C)
High velocity dryers, 170–370 kcal/(hr) (m^2) (°C)

Drying rates vary depending on dryer design and drying conditions. The approximate range of water drying rates in various dryers is given below.

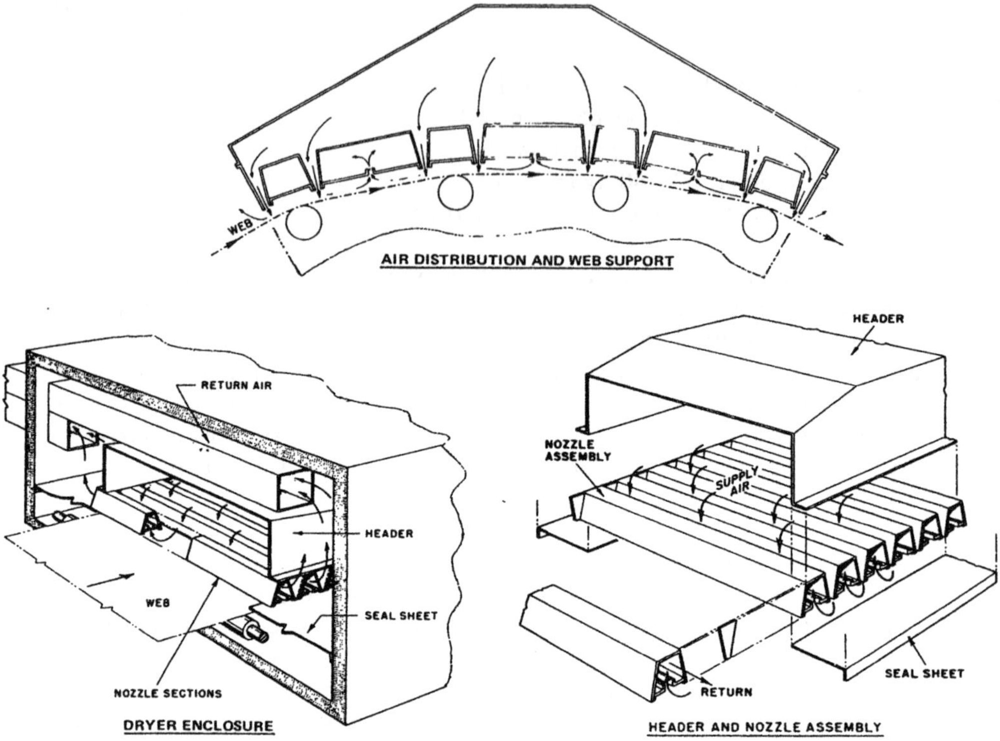

Fig. 36-23. Air distribution in a slotted nozzle convection oven. (*Courtesy of Black Clawson Co.*)

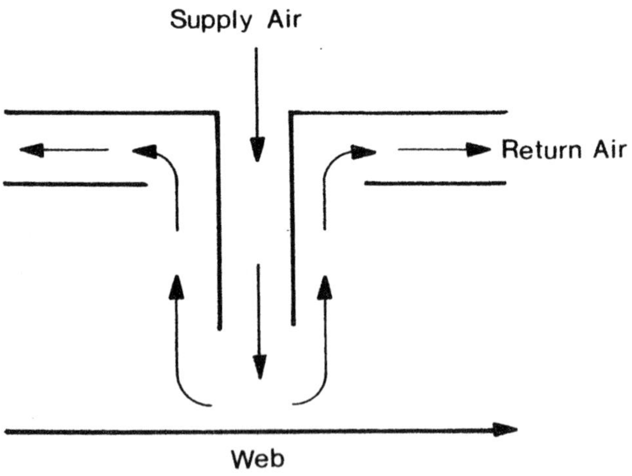

Fig. 36-24. Air tube nozzle.

Fig. 36-27. Floater oven, nozzles, and web positioning. (*Courtesy of Pagendarm KG.*)

dryer has been discussed by several authors,[35–37] as has drying by airfoil compared to high velocity air nozzles.[38]

Infrared Radiation Drying

Infrared dryers, usually in combination with convection dryers (since airflow is needed to remove the vapor from the adhesive surface), are used for drying pressure sensitive adhesives.

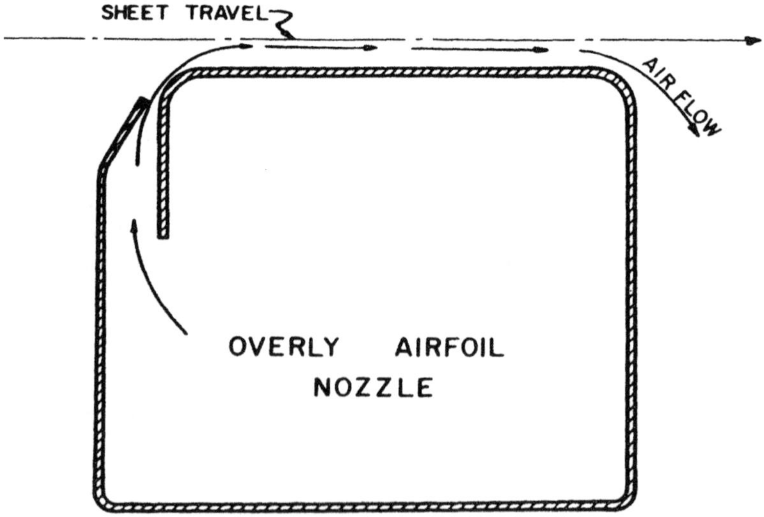

Fig. 36-28. Airfoil nozzle.[38]

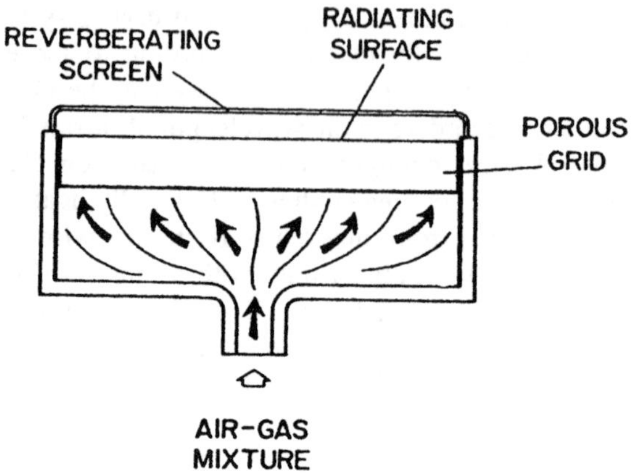

Fig. 36-30. Surface combustion burner constructed from porous ceramic. (*Courtesy of Eclipse, Inc.*)

surface of a porous ceramic tile, metallic wire screen, or a similar material (Fig. 36-30).

4. *Direct impingement burner*. The hot gas flame impinges upon refractory.
5. *Catalytic burner*. The catalyst bed allows gas combustion at lower temperatures (315–425°C).

Infrared heaters are controlled by periodically interrupting the flow of electrical current. Gas-heated emitters are controlled by varying gas flow. Reflectors are employed with radiation heaters to direct the radiation toward the web surface and are usually nonfocusing. Some air circulation is provided to carry away vapors. Infrared radiation is often combined with convection-type heating. A schematic diagram of a heater of this type is shown in Figure 36-31. Instead of heated air, ambient temperature air may be used, especially with high radiation intensity shortwave infrared heaters. Shortwave infrared radiation is not easily absorbed by the adhesive coating, and most of the energy is absorbed at the adhesive–substrate interface. Thus heating takes place from the inside, which helps to avoid the formation of a surface film.[39,40] Ambient temperature air may be impinged upon the adhesive surface in order to remove water (or solvent) from the adhesive surface. An adhesive coating as heavy as 0.5 mm was dried without blistering by the combination of shortwave infrared radiation and ambient temperature air impingement.[41]

Dryer Construction

Many different dryer designs are available, but the design types used for pressure sensitive adhesive products are limited to only a few types.

Fig. 36-32. Pressure sensitive adhesive coating line. (*Courtesy of Wolverine Corp.*)

must be exhausted because of a low allowable recirculation ratio. Heat recovery is therefore important in conserving energy. Several types of heat exchange equipment are used. The power required to move the air and exhaust gases in the ovens is provided by blowers. For a more detailed discussion of heat exchangers and of air blowers, see reference 31.

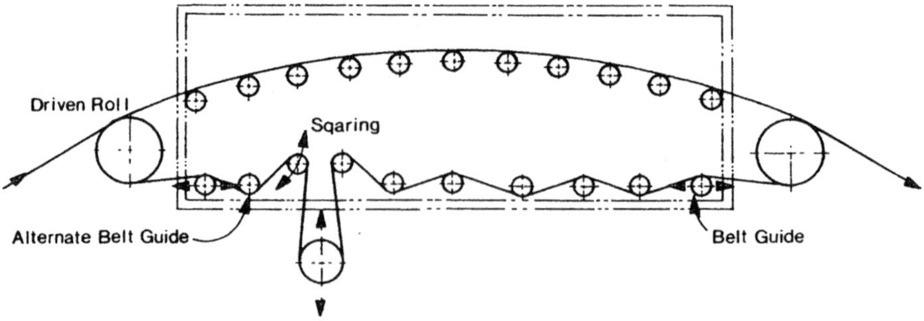

Fig. 36-33. Conveyor-belt dryer.

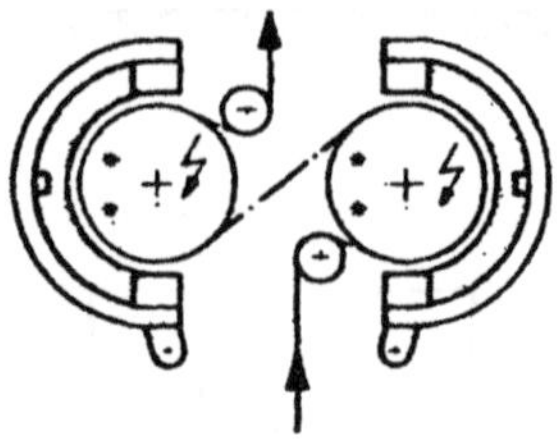

Fig. 36-35. Moisture addition by steaming the web supported on chilled rolls. (*Courtesy of Pagendarm Apparatebau GmbH.*)

order to minimize curling in such laminates. It is difficult to add the exact amount of moisture that was lost during drying, but the addition of moisture slightly in excess is not harmful: moisture plasticizes paper fibers and helps to relieve internal stresses.

Curling and its prevention has been discussed by several authors. Spitz and Blickensderfer[42] have suggested that a property called curl tendency is more fundamental than the extent of curl. In the case of coated and uncoated papers, this property can be measured and quantitatively defined. Haagen[43] has discussed the causes of stresses in coatings. Problems of curl in extrusion coated papers have been discussed by Bates and Marsella.[44] Measurements of internal stresses have been described by Gussman.[45]

Moisture lost during procesesing is often added at the dryer's end. Moisture can be added by wetting the back of a roll coater, by mechanical spraying, by steaming, or by employing electrically charged water particles. Steaming may be carried out by passing the web through a steam chamber where the steam condenses on the surface and inside the fibrous structure. The amount of steam condensed depends on the entering web temperature. The lower the web temperature, the more steam will condense. Figure 36-35 shows a unit for moisture addition where the web is supported by cooling rolls when passing through a condensation chamber.[46] Curl produced by roll set is eliminated by breaking the web over a sharp edge. A decurling unit consisting of pressure rolls has been described by Manfredi.[47]

SOLVENT RECOVERY AND INCINERATION*

The removal of solvent vapor from exhaust may be required by environmental regulations, or it may be economically worthwhile to recover the solvent for reuse. Solvent recovery requires a large-volume operation, or the use of a

* Mr. Richard K. Rathmell contributed to this section.

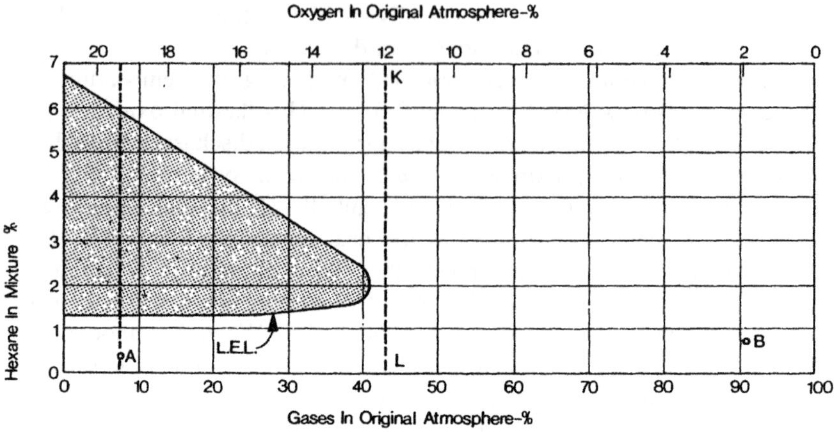

Fig. 36-36. Relation between oxygen, nitrogen, and heptane concentration in forming a flammable mixture. (*According to Janett.*[51] *Courtesy of Ross-Waldron Division, Midland Ross Corp.*)

the normal solvent concentration in drying ovens. Addition of solvent to this mixture raises the solvent concentration, as represented by the dotted line, and the mixture soon enters the flammable area. If the oxygen concentration is decreased below 12%, the addition of solvent to such a gas does not form a flammable mixture, as shown by line KL. Usually inert gas drying takes place in a gas containing only 2% oxygen, as represented by Point B. Such a mixture is safe and cannot be ignited.

Such a dryer can be operated at a very low emission level by repeatedly recirculating the same volume of gases within the drying chamber and associated vapor condenser. If a perfectly airtight dryer were used, the same volume of noncondensible gases could be maintained and recirculated within it indefinitely. In actual practice, a relatively small flow of low oxygen content gases is required to sweep air off the wet web entering the dryer and to maintain low oxygen concentration within the dryer. Liquid solvent in the wet coating is evaporated by heat, and the vapor is condensed at the same rate in a chilled condenser. Noncondensible gases and noncondensed vapors discharged from the chilled condenser are reheated and returned to the heated dryer chamber where new vapors are forming.

Closed-loop recirculation of gases from a dryer to a condenser and back to the heated dryer was demonstrated as early as 1911 in Rhode Island; however, the concept for maintaining low oxygen concentration for safety assurance was not recognized at the time. Approximately 10 years later a process to maintain vapor concentration too rich to be combustible was described, but it did not find general acceptance. Obviously, start-ups and shutdowns would require passing through a hazardous condition.

In 1972, the Midland Ross Corporation installed a drying line designed to recirculate safe low oxygen concentrations within the dryer, with controlled air inflow into a vapor burner, recirculation of the hot combustion gases within the dryer, and exhaust of the excess volume. This system likewise did not find general acceptance; insufficient oxygen in the vapor burner could generate black soot in the recirculated gases.

Airco, Inc. developed a vapor recovery process wherein liquid nitrogen was used both as a refrigerant and a source of inert gas.[52] Although liquid nitrogen is a relatively expensive refrigerant, the energy saved relative to a carbon adsorption/desorption process can justify the expense.

The Wolverine Corporation has developed further improvements by supplying integrated systems, including airtight dryers and condensers that use inexpensive mechanical refrigeration. The requirement for makeup inert (low oxygen) gases is relatively low, and by-product low oxygen flue gases taken from a steam boiler may be used or any inert industrial gas, such as nitrogen or carbon dioxide, may be purchased. A schematic diagram of the system is shown in

Fig. 36-38. Condensation solvent recovery unit. (*Courtesy of Wolverine Corp.*)

31. Satas, D. Drying. *Web Processing and Converting Technology and Equipment,* pp. 250–318 (D. Satas, ed.). Van Nostrand Reinhold, New York, 1984.
32. Wedel, G. L. *TAPPI* **43**(8): 89–92 (1980).
33. Janett, L. G. Update on Drying Systems, Coated Products, Systems and Markets. In *Proc. Institute of Graphic Communications Conference.* Andover, Mass., August 24–26, 1980.
34. Overly, W. F., and Pagel, K. J. U.S. Patent 3,629,952 (1971) (assigned to Overly, Inc.)
35. Richardson, C. A., and Lawton, D. W. *TAPPI* **56**(4): 86–89 (1973).
36. Lawton, D. W. *TAPPI* **57**(6): 105–107 (1974).
37. Fraser, W. A. R. *Paper, Film Foil Converter* **57**(5): 162–165 (1983).
38. Johns, R. E. *TAPPI* **61**(2): 41–44 (1978).
39. De Yoe, J. Dry/Cure of Water Based Pressure Sensitive Adhesive Using High Density Electric Infrared/Convection Drying. In *Proc. Technical Meeting on Water Based Systems,* pp. 109–116. Pressure Sensitive Tape Council, Chicago, Ill., June 13–14, 1979.
40. Stephanson, E. W. *TAPPI* **69**(2): 42–44 (1986).
41. Pitt, S., and Terpstra. Drying and Curing Coatings on Webs with Radiant Energy Systems. In *Proc. Technical Seminar,* pp. 9–17. Pressure Sensitive Tape Council, Cherry Hill, N.J., June 22–23, 1982.
42. Spitz, D. A., and Blickensderfer, P. S. *TAPPI* **46**(11): 676–689 (1963).
43. Haagen, H. *Farbe Lack* **85**(2): 94–100 (1979).
44. Bates, A. and Marsella, L. J. *TAPPI* **47**(7): 133A–135A, 168A–170A (1964).
45. Gussman, S. *Off. Dig.* **34**(2): 895 (1962).
46. Pagendarm, R. *Wochenbl. Papierfabr.* **17:** 670–673 (1979).
47. Manfredi J. L. *Paper, Film Foil Converter* **52**(11): 78–80 (1978).
48. Fielding, H. W. Positive Results for Meeting VOC Control Regulations. In *Proc. Technical Seminar,* pp. 29–35. Pressure Sensitive Tape Council, Cherry Hill, N.J., June 22–23, 1982.
49. Bethea, R. M. *Air Pollution Control Technology.* Van Nostrand Reinhold, New York, 1978.
50. Rothman, H. V. Solvent Recovery on a Fixed Carbon Bed with Inert Gas Regeneration. In *Proc. Technical Seminar,* pp. 184–190. Pressure Sensitive Tape Council, Itasca, Ill., May 8–10, 1985.
51. Janett, L. G. Update on Drying Systems, Coated Products, Systems and Markets. In *Proc. Institute of Graphic Communications Conference.* Andover, Mass., August 24–26, 1980.
52. Rothchild. R. D. U.S. Patent 4,150,494 (1979) (assigned to Airco, Inc.).
53. Greenfell, T. N. What's New in Oven Design? Interair Systems. In *Proc. Technical Meeting on Water-Based Systems,* pp. 138–149. Pressure Sensitive Tape Council, Chicago, Ill., June 21–22, 1978.
54. Wildenberg, H. N., DeSanti, E. A. and Rennes, J. M. An Alternative to Carbon Bed Adsorption. In *Proc. Technical Seminar,* pp. 18–28. Pressure Sensitive Tape Council, Cherry Hill, N.J., June 22–23, 1982.
55. Nikityn, J. Energy Savings through Solvent Recovery. In *Proc. Technical Seminar,* pp. 101–109. Pressure Sensitive Tape Council, Itasca, Ill, April 5–7, 1983.
56. Nikityn, J. *J. Ind. Fabr.* **1**(4) (1983).

Fig. 37-1. Pair of draw shear slitters.

knife is cup shaped, providing a surface for web support (see Fig. 37-1).[1] The top knife blade is usually beveled and overlaps the bottom by 0.25 to 0.75 mm, and it penetrates the web slightly. The bottom knife is directly driven, and it drives the top knife by friction.

An enlarged shear slitting section is shown in Figure 37-2. The edge of the web supported by the female knife is cut clean and square. The other edge, however, is deformed because it is pushed down by the male knife. Some webs, such as foil tapes, do not recover by themselves from this deformation and must be ironed during winding.

One of two types of slitting arrangements may be used. A web may pass tangentially between the slitter knives—this system is commonly known as

Fig. 37-4. Wraparound shear slitters.

slit width. There is no need for supporting idler rolls in this type of slitting. The web path is illustrated schematically in Figure 37-5.

In the draw slitting arrangement, the bottom (female) knives are driven at synchronous web speed, while the driven roll just prior to the female knives is driven slightly slower. The driven roll immediately after the knives is driven slightly faster. This avoids the formation of slack web areas in the slitting station.

Figure 37-6 shows a schematic diagram of web and slitting knife configuration. The arc of knife contact should be much smaller than the arc of web

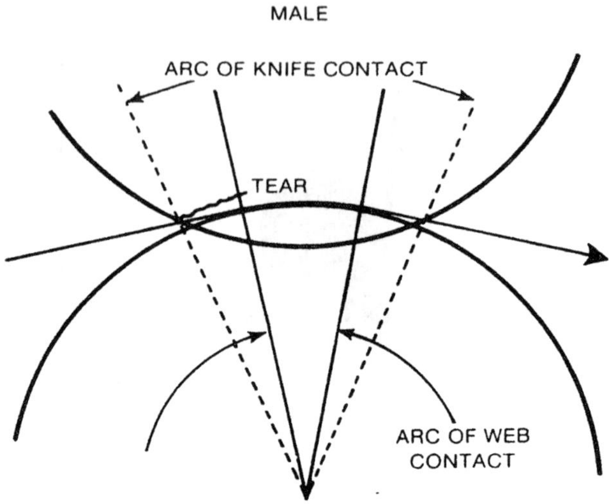

Fig. 37-7. Web tear in a shear slitter.

the male knife and the web is less and the burnishing of web edges is less likely. Wraparound shear slitting also allows the maintenance of a tighter width tolerance, because the web is supported across its entire width when slit, while in the kiss method, the web may sag between the knives. Serrated knives may be used to obtain an easy-to-tear tape edge, which is a desirable feature for some products.

The advantage of draw shear slitting is fast set up. Female knives can be easily repositioned on the shaft, unlike wrapround slitting, where the shaft must be removed from the machine and the knives and spacers rearranged in the required order. Shear slitting, however, leaves a tacky edge and some adhesive may accumulate on the knife.

SCORE CUT

Score (crush) cut is the oldest slitting method used for pressure sensitive tapes. It is the preferred method for heavier tapes, such as fabric or paper-backed products. Score slitting is easy to set up. The slitter consists of hardened-steel round-edge knives set against a hardened sleeved roll, called a platen roll. The knives are pressed against the roll by pneumatically or hydraulically actuated cylinders.

A score cut is rough, and it tends to push the adhesive away from the edge, which helps to obtain dry edges of the adhesive tape roll. It also can

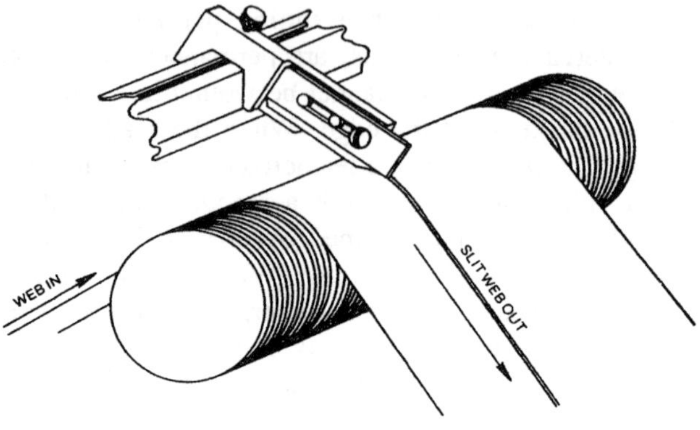

Fig. 37-9. Web slitting by a razor blade in a grooved roll.

Mounting the blades is made easier by the use of a comb. The comb provides many narrowly spaced slots for insertion of standard razor blades. Figure 37-10 shows such a comb in detail.

At high slitting speeds (over 500 m/min), blades burnish the film edge. Sufficient heat may be generated by friction to cause the film edge to melt, forming a bead that may interfere with winding and roll appearance.

A different razor blade setup is shown in Figure 37-11. The web contacts only the finely honed part of the blade, thereby keeping the friction between the film and the blade to a minimum.

SLICING

Instead of slitting when the tape is being rewound, wound tape rolls can be sliced directly without rewinding. A rotary blade slices through the wound material and through the core, leaving a clean, smooth edge.

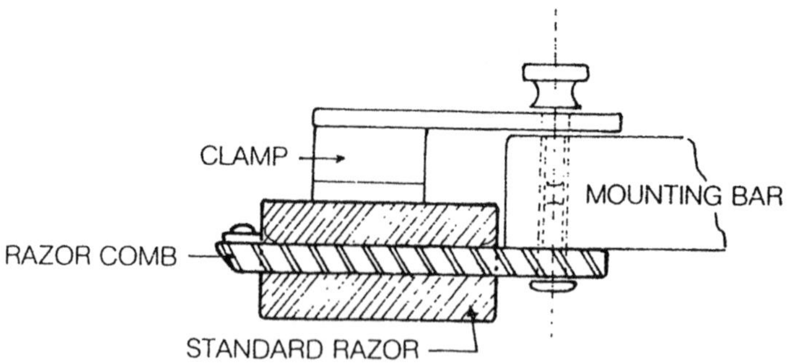

Fig. 37-10. Detail of a razor comb.

for the displacement type of slicing. The displacement slicing method may be used to cut very narrow (down to 0.4 mm) charting tapes.

Another method is compression slicing, where the material is compressed instead of being displaced. A bayonet-type knife is used for compression cutting. The method is suitable for the slicing of pressure sensitive adhesive tapes, especially those made from a compressible web, such as foam.

Removal cutting is another slicing method. This method entails the removal of a narrow strip of material between the cuts. A double-edged circular knife cuts into the roll, making two parallel incisions. The knife is followed by a narrow tool that removes the material between the incisions.

Slicing is an accurate method in the production of slit rolls. Cut widths may be produced to a tolerance of ±0.005 mm. The edges are perfectly aligned, since the material is not rewound.

CORE CUTTING

Paper or plastic cores used for tape winding must be precut to the required width. Some special products may employ injection-molded plastic cores, but usually a paper or plastic tube is used. Core cutters are simple score cutting devices. A detail of a core cutter is shown in Figure 37-13. In differential winding, it is important to have well-cut cores with clean edges in order to maintain a uniform friction between the cores and spacers.

Fig. 37-13. Core cutter.

Fig. 37-15. Duplex slitter rewinder. (*Courtesy of John Dusenbery Co.*)

on a duplex machine by placing cores and spacers on the mandrel. The spacers are keyed and restrained from rotation, but they can slide axially on the mandrel, while cores can slip around the mandrel by overcoming the frictional force between the core and spacer edges. The friction level between cores and spacers is determined by the pressure applied at the mandrel ends and is transmitted to the cores along the mandrel. Thus, the tension at which the slit tape is wound depends on the lateral pressure applied. This is usually accomplished through air cylinders. This arrangement allows the winding of materials of nonuniform thickness at the same tension. It is not uncommon for tape thickness to vary across the web width, either because of variation in the backing thickness, or because of the variation in the thickness of adhesive applied. This arrangment produces rolls that may be of a somewhat different diameter, but they will be wound at the same tension and therefore will be of much better quality.

Various accessories are used with differential winding. Core adapters might be used to lower the friction between the mandrel spacer and the core for light tension rewinding.[4] The controlling frictional surface becomes the surface between the core adapter and spacer. It can be adjusted at a very low level to allow winding at low tension, as might be required when slitting lightweight, low modulus film tapes. Core adapters are also used for winding on cores of a larger diameter than the mandrel. Several types of core adapters are available,

such as a recessed adapter and a flush-sided adapter. Adapters allow the use of different-width cores.

Core drivers may be used where the winding tension must be higher than that generated by frictional forces between the core and spacer edges—generally in the range of 50 kg. Core drivers provide a high friction surface against which the core slips. Core drivers prevent core collapse and locking on the mandrel, as occurs when wound on mandrels of the same diameter as core ID.

Top-riding rolls may be used to smooth out upturned edges or to remove wrinkles. This is especially needed when slitting foil tapes. Top-riding rolls also help to force out the air layer that is carried by the web when wound at high speeds. Surface air is carried into the roll by all materials being rewound. At low speeds, the air can escape at the roll edges, but at high winding speeds, it may get trapped between the tape layers, forming bubbles or balloons. The top-riding roll should be positioned at the point where the slit web contacts the rewinding roll. A top-riding roll also helps to eliminate the force concentration in the area of a thicker band. The web can be wrapped around the top-riding roll for maximum effect (see Fig. 37-17). Individual top riding rolls for each tape roll may be used to accommodate variation in roll diameters.

Surface winding may be used for materials that do not have an exposed pressure sensitive adhesive. In surface winding, the force is transmitted through one or two driven drums. Two threading patterns of a two-drum surface winder are shown in Figure 37-18. A rider roll on top may be used to control the nip pressure between the drums and the roll surface.[1] Duplex surface-driven winders are also available. Staggered rolls may be wound in two positions against a rotating drum.

Pressure sensitive label stock is usually wound by using a combination of center–surface winding. For the most part pressure sensitive label stock is not uniform enough in caliper to run on a two-drum surface winder, hence duplex

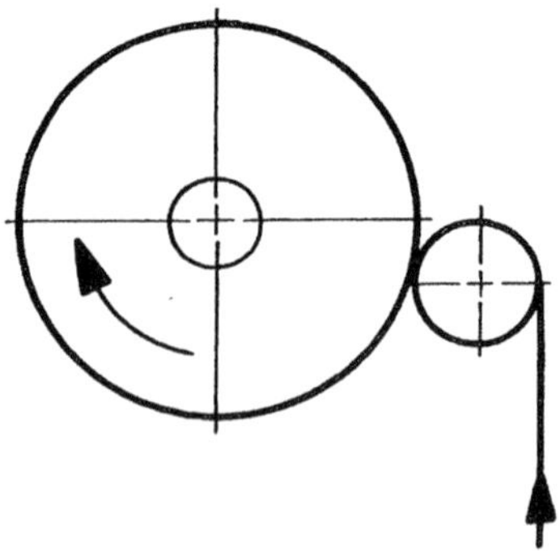

Fig. 37-17. Top-riding roll for rewind.

Fig. 37-20. Individual-arm center–surface slitter. (*Courtesy of John Dusenbery Co.*)

Slitters are designed to run mechanically at speeds as high as 1000 m/min, but the usual fast speed is about 500 m/min. But even at those speeds, the production of a slitter may be low. Tape-slitting machines are very inefficient because of long setup times. They are utilized for tape slitting only 10–15% of the time. The slitter might be stopped often to take out defective material. It is quite convenient to use the slitter as an inspection machine. Inspection requires much lower speeds, perhaps up to only 100 m/min. Therefore, the mechanization and automation of core loading and finished tape removal is very important in order to achieve a higher productivity rate. This problem has been addressed by simply providing multiple shafts that can be prepared ahead of time or by modern, fully automatic slitters with automatic core loading, tape removal, and packaging.

REFERENCES

1. Greiner, T. S. Winding, Slitting and Splicing. *Web Processing and Converting Technology and Equipment*, pp. 364–393 (D. Satas, ed.). Van Nostrand Reinhold, New York, 1984.
2. Rienau, J. H. *Paper, Film Foil Converter* 53(4): 61–66 (1979)
3. Satas, D. Web Handling. *Web Processing and Converting Technology and Equipment*, pp 394–433 (D. Satas, ed.). Van Nostrand Reinhold, New York, 1984.
4. Rienau, J. H. *Paper, Film Foil Converter* 53(5): 172–178 (1979).

38.

The addition
comprehensiv
time and pr
be ignored

The c
is developi
constantly.

Very i
microelectr
tage it giv
are as impr

Sinc
sensitive

1. M
2. M
3. In

Any
methods,
later. F
to clos
loop co

Pr
ential (
comput
in respo

is often underestimated. In any event, information collection is the first step in sophisticated control, and it deserves careful planning.

MACHINE AND PROCESS LOGIC

The types of control logic used on converting machines must be defined. It is especially important to differentiate between "machine logic" and "process logic." Machine logic refers to the electrical apparatus used to turn things on and off: relays, timers, and counting devices are typical examples of the hardware. The best example of an application is the operation of an automatic turret-type unwind or rewind with splicer or transfer assembly.

A common but by no means exclusive characteristic of machine logic is its "on–off" or digital character: the turret rotates, the paster roll swings to position, the new core comes to speed, a transfer is made, and the knife fires. With the exception of changing roll diameters or speed match variability, all of the actions are positive. There is no room for interpretation. Conditions are established and confirmed, and positive on–off activity is executed.

Process logic applies to the sophisticated measurement and gauging systems employed to analyze the viscosity, thickness, temperature, moisture content, and the other parameters that are of importance in maintaining the converting process.

Although the actual setpoint value—the ideal—is important, be prepared to deal with tolerances and even the unexpected. Clearly, these are not questions easily resolved by on–off decisions. Instead, tempered response is called for. In the past, operator judgment or relatively unsophisticated process controllers have been relied on to cope with a wide variety of factors that tend to push a process out of the range of acceptability.

For a variety of reasons, the limits of acceptability are narrowing. A control system must respond to process stress in a timely manner, anticipating factors contributed not only by the process but by the controller itself, and associated sensors and actuators as well. This is now done by the modern PID controller which is a small box, usually with its own microprocessor, that takes a voltage or current from some instrument, processes it, and produces a command to some valve, actuator, or switch that actually varies a process parameter. Some of these controllers are even capable of self-adjustment. If the process itself is not too demanding, the PID controller is a fine solution to many ordinary process logic problems. Unfortunately, there is a tendency to try to control too much with it.

Pressure sensitive products converting is now so complex that it is frequently desirable to automatically change the set points themselves in response to still other variables. This activity is clearly beyond the capability of the PID controller.

The need to mathematically analyze collected information is not always obvious. But the increased need to interpret the new information is a logical result of our increased ability to measure. Indeed, it is easily arguable that the human ability to respond in a timely fashion to the quantity of data pouring from the gauges has been exceeded. But the sheer mass of information is not the most serious problem. Of greatest concern should be appropriate processing and correct mathematical interpretation, so that the operator or controlling computer receives concise and reliable guidance.

Consider the problem of a gauged cross-web profile of adhesive thickness. Also, assume that attention has been paid to those details needed for accurate measurement, such as careful calibration and selection of measurement and reference frequencies for infrared instruments, and differential and point-for-point provisions for particulate gauges.

For the moment, consider only problems of interpretation. Figure 38-1 shows a typical video display in the popular hanging format.[1]

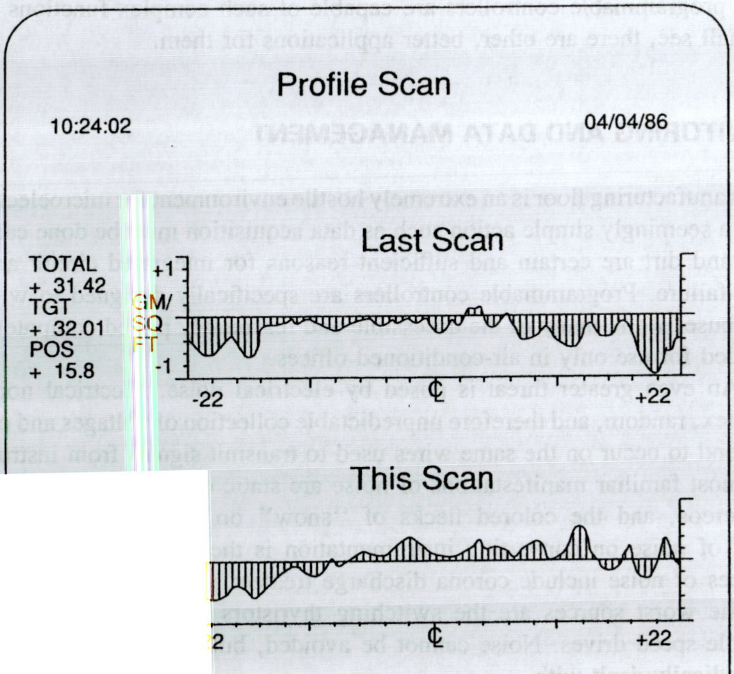

anging format. While used most often for display of
ay any parameter, such as viscosity or temperature.

The setpoint value is a single horizontal line with deviations above and below, and a vertical dimension proportional to actual gauged thickness. If the coating is being applied with a fully adjustable slot die, manual correction should be possible. Assuming a fairly static process, skill and experience should allow the operator to achieve good coating accuracy.

But if the coating method used is reverse roll, and unless the coater is equipped with a calender-type controlled-temperature air manifold for full cross-web compensation of mechanical variables, the operator can only make drive-side and operator-side adjustments. He has no way of knowing how to optimize a complex thickness profile, nor has he control over local variations, such as those caused by pan circulation thinning and settling, or metering and casting roll irregularities. Still, there must be optimum adjustment. This adjustment must consider as many factors as possible: setpoint, tolerances, trends, and customer preference.

This requirement can be satisfied best through a well-written control program executed by a process control computer. It is certainly true that the more sophisticated programmable controllers are capable of such complex functions but, as we shall see, there are other, better applications for them.

MONITORING AND DATA MANAGEMENT

The manufacturing floor is an extremely hostile environment for microelectronics. Even a seemingly simple action such as data acquisition must be done carefully. Heat and dirt are certain and sufficient reasons for integrated circuit and disk drive failure. Programmable controllers are specifically designed to withstand this abuse, while many of the accessible and reasonably priced computers were intended for use only in air-conditioned offices.

An even greater threat is posed by electrical noise. Electrical noise is a complex, random, and therefore unpredictable collection of voltages and currents that tend to occur on the same wires used to transmit signals from instruments. The most familiar manifestations of noise are static on radios, clicks and pops on stereos, and the colored flecks of "snow" on television receivers. The effect of noise on converting instrumentation is the same: distortion of data. Sources of noise include corona discharge treaters, motor starters, and relays. But the worst sources are the switching thyristors that operate constantly in variable-speed drives. Noise cannot be avoided, but it must be controlled and methodically dealt with.

Severe noise can actually destroy microelectronic components. But it can be managed by careful attention to the principles of shielding and grounding, and through the use of signal isolators.

Gauging instruments will nearly always offer a current output that is inher-

circuitry that protects the electronics from machine-generated overstress. Asking the controller or computer to perform arithmetic and logic operations only serve to worsen the situation.[3] Finally, serial communications that are inherently slow are often used. When added together, these delays can result in a poorly operating system—one that simply cannot respond to the needs of the machine and process in a timely fashion. Fortunately, a quite simple solution exists for this problem: distributed control, which is defined as using as many controllers, microprocessors, and computers as needed and tying them all together with some sort of common communications link, or bus. Many instrument and gauge manufacturers have therefore equipped their instruments with microprocessors. But unless the device must perform sophisticated functions, such as signal linearization or high-speed PID functions, such a capability may be more of a matter of gadgetry than actual control need.

The drive system is an excellent example of a situation where the use of dedicated multiple processors is the only reasonable control scheme. To maintain stability and accuracy of speed control, much work must be done in a very short time. These requirements are beyond the capabilities of currently available microprocessors, but given the current rate of technological development, that situation can be expected to change in a few years.[3,4] Until that happens, this work must be distributed.

A separate microprocessor is used in each digital drive section, with all sections tied together by a communications bus, and with all sections following the lead section. Primary concern lies in the existence of the communications bus. This means that monitoring and control can be achieved remotely with relatively little effort. The communications bus allows the examination of many points, twenty or more in each of the drives, for both control and troubleshooting. Best of all, this examination can be done with very little effort or expense, since a bus is commonly a single coaxial cable or twisted pair of wires.

This is probably an appropriate time to emphatically state that motor control is not an elementary task and that the temptation to independently develop a digital drive controller is best resisted. Drive systems available today represent years of specialized development. Independent effort at system development will be more profitable if it is directed at solving specific and unique process problems that the designer is familiar with and to which he can bring his own expertise.

The high-speed update of monitored values is not always needed for satisfactory process or system operation. Viscosity, for example, will change very slowly. Little variation will occur over a period of several minutes in a typical multigallon reservoir and pan. This contrasts dramatically to a DC drive system, where an error of several revolutions in position can develop in less than a second. Therefore, it is reasonable to treat these factors quite differently. Unjustified insistence upon high-speed reaction will do nothing but add to system cost without providing discernible benefit.

Sophisticated features and advanced capabilities are expensive, so do not set specifications that are unnecessary from the standpoint of process and need. Planning for future expansion must be done with special care. Given the very rapid rate of development of control technology, it probably is a waste of time to attempt to predict the future. Accommodate future need only if doing so does not add to the project cost.

It is now necessary to examine a very significant point: resolution. Digital-to-analogue and analogue-to-digital converters translate in discrete stepwise jumps.

For example, consider the drive control signal. Typically, a DC regenerative drive will use a speed setting voltage of −10 VDC to +10 VDC. To cover the full speed range from the base speed of a motor with counterclockwise rotation to the base speed of a motor with clockwise rotation, a digital-to-analogue converter would have to produce a total signal range of 20 volts. An inexpensive converter would be able to divide this voltage range into 256 discrete steps (two to the eighth power; 20/256 = 0.078). The voltage produced by the digital-to-analogue converter will vary in discrete jumps of 78 millivolts. A change of 149 or 22 millivolts cannot be specified—the device simply cannot produce it directly. Now calculate the significance of this in motor speed. Assuming a base speed (in each direction) of 1800 rpm, the total speed range commanded by the 20-volt range is 3600 rpm (3600/256 = 14 rpm per 78-millivolt increment). The motor speed cannot be changed in steps smaller than 14 rpm. Clearly, this is hardly precise control—but many drives successfully use electronics with no greater precision. Obviously, other factors are of even greater importance.

An A/D converter with 10-bit resolution would have 1024 increments or 3.52 revolutions per step (0.06 revolution/sec)—a vast improvement, if necessary. But as resolution and, perhaps, accuracy improve, two other factors deteriorate. That is, both cost and the time required for conversion rise sharply.

Consider an input example of considerable interest: adhesive weight or thickness. Since a relatively narrow range of values is being considered, extreme precision is unnecessary. The range is 0–28 lb./3000 ft^2 rm (0–17.4 g/m^2). Dividing 28 by 256 yields a resolution of little more than 0.11-lb (0.07-g) increments, which is perfectly adequate for ordinary pressure sensitive work. And it can be achieved with inexpensive analogue-to-digital converters. The input/output unit cost can also be reduced without detriment, thus protecting the project and producing possible additional capabilities that, in turn, make justification easier.

ASSEMBLING THE ACTUAL CONTROL SYSTEM

Some of the most significant concepts involved in the basic monitor-only system are now understood, though a system with closed-loop circuits is beyond the scope of this discussion.

Fig. 38-3. A typical programmable logic controller, capable of two-way communications with a personal computer. (*Courtesy of Cutler-Hammer.*)

and so he must have practical hands-on experience with converting. It is only in this way that he gains the insights necessary to manipulate process real-time problems inside the limits imposed by the microelectronics.

SOFTWARE AND PROGRAMMING

As long as the system is intended to monitor or gather data only, the task of finding or creating computer operating instructions (software) is quite simple. There are two options: "configure" or modify software written by others to satisfy the present job requirements; or write an original program.

A large number of so-called third-party general-purpose software is available. Much of it is offered by manufacturers of complete hardware/software systems. A box with screw terminals on the back, ready to receive signals

Graphic ASCII Character Codes

ASCII CODE	CHARACTER	ASCII CODE	CHARACTER	ASCII CODE	CHARACTER
176	░	192	└	208	╨
177	▒	193	┴	209	╤
178	▓	194	┬	210	╥
179	│	195	├	211	╙
180	┤	196	─	212	╘
181	╡	197	┼	213	╒
182	╢	198	╞	214	╓
183	╖	199	╟	215	╫
184	╕	200	╚	216	╪
185	╣	201	╔	217	┘
186	║	202	╩	218	┌
187	╗	203	╦	219	█
188	╝	204	╠	220	▄
189	╜	205	═	221	▌
190	╛	206	╬	222	▐
191	┐	207	╧	223	▀

Fig. 38-4. Special graphic programming is not absolutely necessary for the creation of highly effective video screen displays. The graphic ASCII commands shown here can be used for many applications.

of a so-called graphic mode. Refer to Figure 38-4, which shows a segment of the graphic ASCII character codes that can be used to generate graphic images.

THE PROCESS CONTROL COMPUTER

The most interesting aspect of hardware selection must surely be selecting the process control computer. Specially hardened industrial computers, such as the example shown in Figure 38-5, can be used. Journals and catalogs are filled with descriptions of perfectly adequate computers. But, if data acquisition is the primary goal, with actual control being limited to processes with long time constants, there is no reason for not using the ubiquitous personal computer.

It is true that care and thought must be given in selecting a location that will offer protection from environmental hazards, but bus communications and suitable interface boxes will simplify that problem. The programmable controller is used to acquire machine logic status, with process logic coming from any other convenient source. The personal computer can then be used for generating

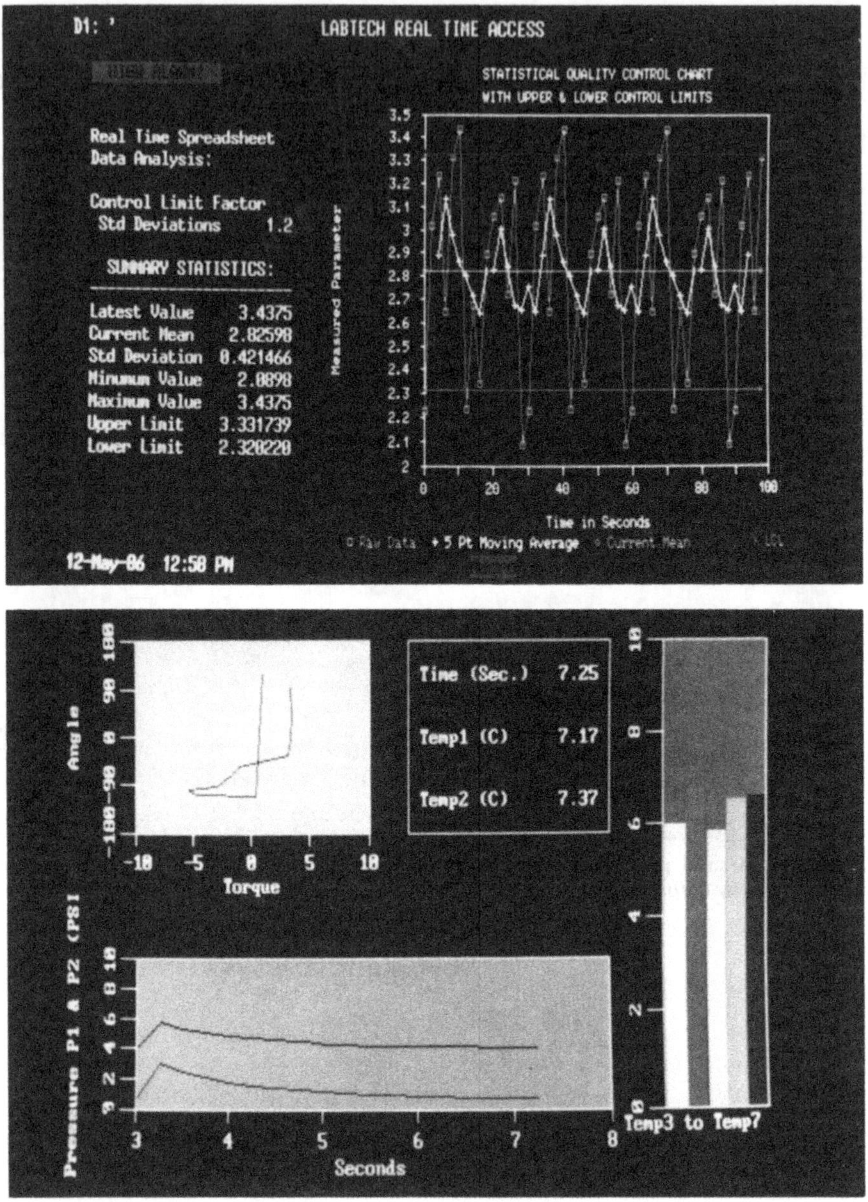

Fig. 38-6. Typical displays developed by commercial third-party software. The use of such software eliminates the need for programming in many data acquisition situations. (*Courtesy of Laboratory Technology Corp.*) (a) This screen shot shows some of the display capabilities of LABTECH NOTEBOOK, including windows, traces, plots, and bar graphs, that can be dynamically updated in real time with newly acquired information. (b) *Lotus Symphony* and other applications can now analyze and chart scientific and industrial process data in real time, with LABTECH's *Real Time Access*. This pipeline exchanges measurements and control commands between the LABTECH NOTEBOOK data acquisition package, and MS/PL-DOS application programs operating concurrently.

for the IBM PC. Osborne/McGraw-Hill, Berkeley,

fann, R. T. *Design of Feedback Control Systems*. CBS
inston, New York, 1982.
Prentice-Hall, New York, 1986.
t Press Division of Microsoft Corp., Redmond, Wash.,

39. Radiation Curing

R. Dowbenko
PPG Industries, Inc.
Allison Park, Pennsylvania

Pressure sensitive adhesives are made by irradiating an uncured mixture of polymerizable compounds either by ultraviolet light (UV) or by electron beam (EB) to form a polymeric film of an adhesive. For short, they are called radiation adhesives or UV (EB) adhesives. They are intended to be formed directly on the substrate on which they will be used. Many intermediate steps employed in the manufacture of conventional pressure sensitive adhesives are eliminated. In some cases, the radiation curable adhesives are formed on release paper. This way of preparation still retains the advantages of these adhesives over the conventional pressure sensitive adhesives.

The interest in radiation pressure sensitive adhesives derives from several intrinsic advantages of these materials. The most important properties of these materials involve the abatement of air pollution because no solvents (or very reduced amounts of them) are used, savings in manufacturing cost because of elimination of solvents, as well as simplification of the manufacturing process. Moreover, the performance characteristics are improved, which is an additional reason for a high degree of interest in these materials. Other special features of these UV and EB adhesives will become apparent later on in this chapter.

Preparation of radiation pressure sensitive adhesives is very simple in concept. As mentioned, the formation of the adhesive occurs directly on the substrate to be used or on the release paper. In a most direct and simple form, a mixture of monomers is applied in a film thickness that is used with conventional adhesives and is then irradiated by either ultraviolet light or by an electron beam. The adhesive is formed directly on the substrate to be used later as desired. The operation is usually a continuous one on a web of a release paper or another substrate, or by coating and curing discrete objects on a conveyor belt, such as adhesive-backed floor tiles. This is the essential difference in manufacturing

FREE RADICAL POLYMERIZATION

Free radical initiated polymerization occurs when a monomer, such as 2-ethylhexyl acrylate, is subjected to the action of free radicals that are derived from a free radical initiator, such as benzoyl peroxide or azobis(isobutyronitrile). There are several key steps in the chemical mechanism of this process, the first being the decomposition of the initiator, benzoyl peroxide, to form free radicals (1). Next, the free radical, which has been formed as a result of the decomposition of the initiator, reacts with a molecule of a monomer, such as 2-ethylhexyl acrylate or methyl methacrylate, to form an intermediate (2), which reacts further (3) with additional monomer molecules in a propagation reaction. This propagation or polymer growth reaction continues with numerous monomer molecules until the polymeric free radical encounters a molecule that has a hydrogen atom that is able to react with the polymeric radical. When that happens, a termination occurs (4), and an inactive polymer molecule is formed that theoretically does not continue to undergo polymerization. These four reactions, roughly speaking, describe the intimate steps in the free radical additional polymerization process.

$$R'\text{—}R' \longrightarrow R'\cdot + R'\cdot \qquad (1)$$

$$R'\cdot + CH_2 = CHR \longrightarrow R'\text{—}CH_2\text{—}\dot{C}HR \qquad (2)$$

$$R'\text{—}CH_2\text{—}\dot{C}HR + n(CH_2 = CHR) \longrightarrow R'\text{—}(CH_2\text{—}CHR)_n\text{—}CH_2\text{—}\dot{C}HR \qquad (3)$$

$$R'\text{—}(CH_2\text{—}CHR)_n\text{—}CH_2\text{—}\dot{C}HR \xrightarrow{R''H} R'\text{—}(CH_2\text{—}CHR)_n\text{—}CH_2CH_2R + R''\cdot \qquad (4)$$

Radiation polymerization is in principle completely analogous to the conventional addition polymerization with one important difference: a different compound is used to initiate the polymerization in UV and in EB.

In the case of UV, a photoinitiator is used. Its function is completely analogous to the free radical initiator. Both furnish free radicals that are necessary for initiation of polymerization, but with a photoinitiator, the process of radical formation is different. In contrast to the conventional initiators that possess a thermally labile bond that is cleaved to form free radical species, the photoinitiator has a bond that breaks upon absorption of radiant energy. It forms free radical species that initiate polymerization by an addition reaction to a monomer molecule and then continues by propagation and eventual termination. The photoinitiator is therefore one of the key components in a UV polymerization, and the outcome of such a polymerization is critically dependent on the choice of the photoinitiator, including its chemical nature and the amount.

As mentioned earlier, both the UV and EB polymerizations occur essentially by free radical processes. In the case of UV polymerization, the energy required to break the bond is associated with a wavelength of about 300 nm and is

storage stability and is almost never used in practical systems where storage stability is of importance.

Benzyl Dialkyl Ketals

The well-known 2,2-dimethoxy-2-phenylacetophenone is a very efficient photoinitiator and decomposes on UV irradiation according to the following scheme:

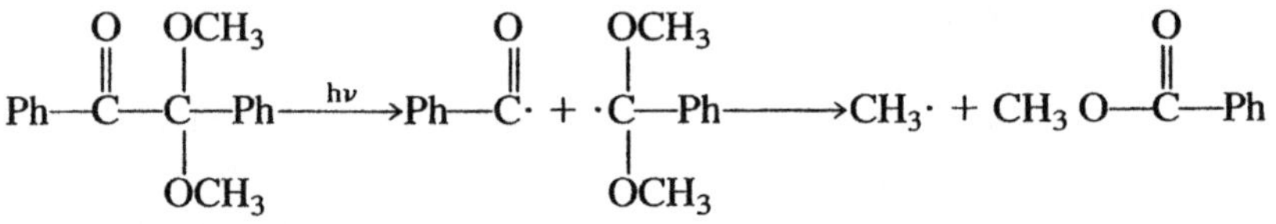

with the dimethoxybenzyl radical further fragmenting to methyl benzoate and the methyl radical.[9] Compositions containing this photoinitiator have fast cure and good storage stability.

Benzophenone

On irradiation, this compound undergoes a conversion to the singlet benzophenone, followed by a conversion to the triplet excited state. The triplet then abstracts a hydrogen atom from a different compound to form a benzhydryl free radical, which is an active initiating species.

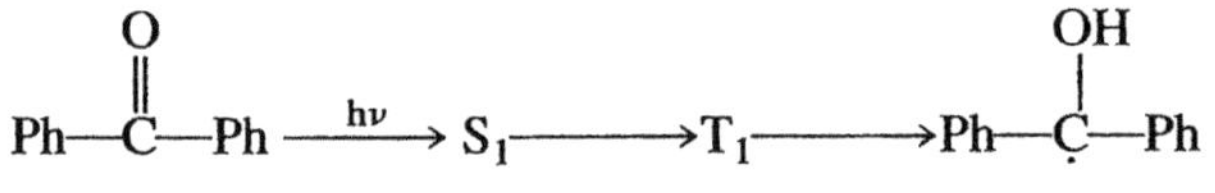

The activity of benzophenone can in general be enhanced by addition of small amounts of amines.[10] Benzophenone is an inexpensive photoinitiator and can be used in mixtures with the benzoin ethers and with 2,2-dimethoxy-2-phenylacetophenone.

Michler's ketone is a substituted benzophenone that initiates polymerization in a similar manner. It can also be used together with the photoinitiators previously discussed, but its main disadvantage is its dark color.

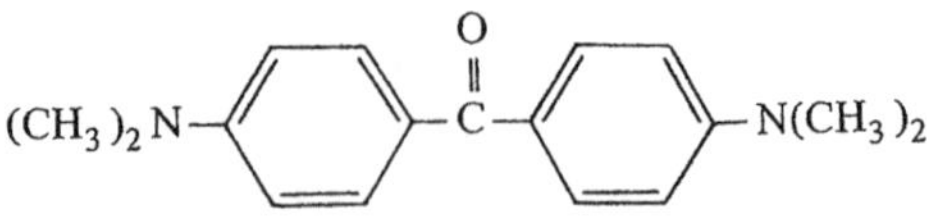

stream of electrons that are typically accelerated through 300–500 kV in a vacuum. The beam issues through a thin metal window and is magnetically scanned through the length of the line. This type of equipment requires thick shielding and constant caution to prevent exposure to people.

The linear accelerator is highly efficient and compact, requires much less shielding, and comes in small and compact units[14–17] with an effective length of as little as 1.7 m. The linear filament system is of lower energy than the scanned beam system, but it is quite adequate for curing thin coatings of organic materials.

Measurement of UV Intensity and Dose Rate

Among the many variables that can affect the outcome of radiation curing is the intensity and the rate of deposition of radiant energy into the adhesive composition to be cured. There are some commercially available on-line devices to measure the intensity of UV radiation.[18] The system should also be periodically checked by curing a known composition and measuring its properties.

In electron-beam curing, the dose rate is measured by megarads (Mrad) per second (one Mrad is energy absorbed equal to 10 joules). It can also be estimated by similar devices measuring the amount of color increase in radiation-sensitive films. Here also it is advisable to check the equipment by using a composition with a known rate.

Atmosphere in Radiation Curing

There are several types of compositions known, especially in the area of coatings, that can be effectively cured in air atmosphere. This is especially true of compositions polymerizing by ionic mechanisms (which will not be discussed here since such pressure sensitive adhesives apparently are unknown). Most free radical polymerizations are more or less strongly inhibited by molecular oxygen. Propagating free radicals react with oxygen to form a peroxy radical that is useless for further propagation.

$$\text{--}CH_2\cdot + O^2 \longrightarrow \text{--}CH_2\text{—}O\text{—}O\cdot$$

The effect is to produce considerably lower molecular weight species, especially on the surface where subsequently the adhesive bond is formed. Because of the lower molecular weight, such species contribute much poorer adhesive properties. For this reason, many of the radiation-cured (both UV and EB) adhesives are made in an inert gas atmosphere, most commonly in relatively inexpensive

tuted styrenes fall in the same category. Another important factor is that UV polymerization for pressure sensitive adhesives nearly always occurs in films of less than about 1 mil. Therefore, volatile monomers, such as vinyl acetate or ethyl acrylate, are used only rarely, because they usually evaporate faster than they can be polymerized. For this reason, volatile compounds are in general unsuitable for preparation of UV pressure sensitive adhesives.

In addition to the simple acrylic and vinyl esters that contain only a polymerizable group and an ester group, other monomers with different functionality can be used. Thus, ethers, amides, carbamates, hydroxyacrylates, acids, and other functional compounds can be used in UV pressure sensitive adhesives where special properties are desired.

Compounds containing two or more allyl groups present a special case in that they undergo an addition with di- or trithiols as shown, to form

$$CH_2{=}CHCH_2{\sim}CH_2CH{=}CH_2 + HS{\sim}SH \rightarrow [SCH_2CH_2CH_2{\sim}CH_2CH_2CH_2S]_n$$

polymers which, with the proper choice of backbones that carry the allyl and the mercaptan groups, can give useful pressure sensitive adhesives.[19] The processing of such compositions in UV polymerization is completely analogous to polymerizations with acrylic compounds, including the necessity of a photoinitiator.

The possibilities of raw materials to prepare UV pressure sensitive adhesives are indeed numerous and can encompass nearly every type of compound. In contrast to the solution or emulsion pressure sensitive adhesive polymers to which additives (such as tackifiers, UV, and heat stabilizers) are usually added in a separate formulating step, the composition that will form a pressure sensitive adhesive by radiation polymerization must contain all the additives before polymerization. This is obvious since little can be done to add another component to a thin film of a pressure sensitive adhesive once it has been formed. Therefore, all formulation to produce a complete final product must be done before the composition is polymerized by radiation.

Functional monomers of all kinds may be used in radiation pressure sensitive adhesives. In addition to this, it is often useful, as it is in UV-cured coatings, to use compounds containing two acrylic (or vinyl) double bonds. Although this practice must be used carefully in conventional solution adhesive polymers since it would lead to unwanted gelation, in UV-cured pressure sensitive adhesives, this approach is often used to counteract the generally lower molecular weight obtained by radiation polymerization and to promote the formation of higher molecular weight polymers. In radiation-cured adhesives, the finished product is often formed in a fraction of a second, favoring low molecular weight compounds. Other components (additives) in the composition may not favor the formation of high molecular weight polymers.

Table 39-1. Performance of an EB Syrup Adhesive[a]

			EB ADHESIVE	CONVENTIONAL ADHESIVE
Peel adhesion, stainless steel (g/cm)	Initial	20 min	396	318
		24 hr	735	643
	7 days roll storage @ 70°C	20 min	299	404
		24 hr	648	579
Quickstick (grams, triplicate values)	Initial		780	480
	7 days roll storage @ 70′		360	370
Dead load strength (hr, 0.91 kg weight)	Initial		9.5	4
	7 days roll storage @ 70°C		9	4

[a] Conditions of cure: 10 Mrads, nitrogen atmosphere containing 250–300 ppm of oxygen cured on 2 mil Mylar film.

materials as compared with the properties of conventional solvent adhesives, it is of interest to quote them here (Tables 39-1 and 39-2).

In the adhesive cured by EB, high peel and high tack combined with high shear strength are especially noteworthy as such properties are difficult to achieve in conventional pressure sensitive adhesives.

Other work is based on the use of compositions containing polyvinyl ethers and acrylic monomers, both mono- and polyfunctional, together with a light initiator.[24] Even though the main claim of the patent encompasses compositions containing both mono- and polyfunctional acrylates, one of the best adhesives prepared (peel adhesion 570 g/cm, shear time 60 hr) used neopentyl glycol

Table 39-2. Performance of a UV Syrup Adhesive[a]

			UV ADHESIVE	CONVENTIONAL ADHESIVE
Peel adhesion, stainless steel (g/cm)	Initial	20 min	401	362
		24 hr	541	386
	7 days roll storage @ 70°C	20 min	145	75
		24 hr	240	132
Percent shrinkage, 13 × 13 cm vinyl/steel, 7 days @ 70°C	Machine direction		0.5%	1.05%
	Cross direction		0.0%	0.55%

[a] Conditions of cure: N_2 flow 1 m^3/min; lamp intensity 333 W/in., conveyor speed 30 m/min, 12 passes.

propionic acid with a polyol to form a β-mercaptopropionate, and the polyallyl compounds, the second constituent in the syrup, from an isocyanate-terminated prepolymer, and allyl alcohol. Then, to form an adhesive, a diallyl compound, prepared from a mixture of polypropylene and polyethylene glycols, toluene diisocyanate, and allyl alcohol, is mixed with pentaerythritol tetra(β-thiopropionate) and a small amount of benzophenone. After curing with a sun lamp for 20 sec, the resulting adhesive had a peel adhesion between 187 and 464 g/cm. Another adhesive is reported to have a peel adhesion of 630 g/cm.[19]

ADHESIVE POLYMERS FOR RADIATION CURE

The volume of work that has been recorded in the literature with respect to radiation cure of essentially polymeric materials for the purpose of improving their properties is much larger than the volume reported on adhesive syrups.

One of the earliest references in radiation cure[31] describes EB irradiation at 1–24 Mrads of polymers such as polyacrylates, polyethers, and rubber–tackifier blends to obtain pressure sensitive adhesive tapes with improved high-temperature shear adhesion. Similar work is described in another patent[32] where solutions containing a photoinitiator and elastomers, such as crepe rubber, styrene/butadiene rubber, 2-ethylhexyl acrylate/vinyl acetate copolymer, and polyvinyl ethyl ether are irradiated in UV, after evaporation of the solvent, to improve cohesive and shear strengths of adhesives.

A different approach, although still working with essentially "nonfunctional" polymers, is exemplified by another patent[33] in which a conceptual departure is made from previous work in that solvent-free, 100% solids hot melt adhesives are irradiated to produce pressure sensitive adhesives with substantially improved adhesive properties. Predominantly acrylic polymers, with a Williams' plasticity number of up to about 1.5 mm and a melt viscosity (at 75°C) between 500 and 100,000 cps, are applied as hot melts and irradiated at doses between 0.5 and 12 Mrads. Compositionally, most of them are 2-ethylhexyl acrylate and vinyl acetate copolymers with small amounts of *N,N*-dimethylaminoethyl methacrylate, diacetone and acrylamide, *N*-vinylpyrrolidone, or acrylic acid. Polyfunctional acrylates, such as trimethylolpropane trimethacrylate, and thermal stabilizers can also be used.

Since the data are extensive,[23] it will be instructive to quote them completely for both the EB and UV adhesives (Tables 39-3 and 39-4).

It can readily be seen that remarkable improvements in the properties of hot melt pressure sensitive adhesives can be made by irradiating them with EB and UV, but especially with EB.

Work very similar to the preceding irradiation of hot melt adhesives is described in a patent by Skoultchi and Davis.[34] Compositions, such as copolymers

acrylate. Improvements in adhesive properties obtained are similar to those given in reference 34. Thus, peel adhesion of 1390 g/cm and 12-hr shear adhesion are obtained after irradiation of 8 min with a UV lamp (275 watts) of a polymer that before irradiation had peel adhesion of 1474 g/cm and a shear adhesion of 4.5 min.

A further variation of the copolymerized photoinitiator approach is presented in the use of the reaction products of polychlorophenyl glycidyl ethers (such as the pentachlorophenyl ether) and acrylic or methacrylic acids.[36] Such polymerizable photoinitiators are copolymerized with the usual adhesive-producing monomers, such as methyl acrylate/2-ethylhexyl acrylate and vinyl acetate, to produce polymers that are applied as hot melts and irradiated by UV. One of the advantages of these compositions is said to be the constancy of adhesive properties over a wide range of cure conditions. In other words, they are difficult to overcure with the resulting loss of adhesive properties. That the chlorophenyl compounds are indeed active photoinitiators is shown by omitting them from a polymer (in which case, the adhesive properties are essentially unchanged on irradiation).

Other work with polymerizable photosensitizers[37] includes the use of benzoin acrylate, which is copolymerized with a variety of monomers, such as 2-ethylhexyl acrylate, butyl acrylate, tetrahydrofurfuryl acrylate, acrylamide, and acrylic acid, and then irradiated for varying periods of time. Voluminous adhesive data are given to show improvements in adhesive properties on irradiation. An earlier patent assigned to the same company also uses benzoin acrylate as a copolymerized constituent in the copolymer.[38]

Other work to improve the adhesive properties of acrylic pressure sensitive adhesive polymers includes cocuring of an acrylic copolymer by EB, together with the usual additives, with polyfunctional acrylates or methacrylates.[39]

Similar improvements can be obtained in an ethylene/vinyl acetate copolymer on cocuring it by EB with polyol acrylates or methacrylates and with other additives.[40,41]

A departure from the work described is to introduce a reactive acrylic bond into an acrylic copolymer by means of the reaction of glycidyl methacrylate with the carboxyl groups on the polymer backbone and to cure the resulting "activated" copolymer.[42] Thus a copolymer from 2-ethylhexyl acrylate, ethyl acrylate, and methacrylic acid, postreacted with glycidyl methacrylate and cured by UV light gave the 180-degree peel adhesion of 846–984 g/cm.

Improvement of pressure sensitive adhesive properties can also be obtained with compositions based on polybutadienes and modified polybutadienes. Thus, a patent[43] describes EB curing (at about 2 Mrads) of a mixture of an acrylic acid modified polybutadiene and nitrile rubber, together with 2-ethylhexyl acrylate and acrylic acid, as well as other additives. The properties of the improved adhesive included peel adhesion of 412 g/cm and shear adhesion of 40 degrees for more than 3 hr. Two other patents[44,45] describe similar compositions from

coatable at elevated temperature (100°C), are suitable for pressure sensitive adhesives.[53] The low level of irradiation by uncured materials is an important advantage of this product.

Formulations including acrylic monomers and tackifying resins for ultraviolet radiation curable pressure sensitive adhesives has been discussed by Johnson.[54] The effect of a photoinitiator system on adhesive properties has been discussed by Herze *et al.*[55] Curing under inert atmospohere helps to eliminate the undercured upper layer that may be formed because of oxygen inhibition.

CONCLUSIONS

In spite of the considerable amount of work that has been published and in spite of the interest this evolving technology has engendered, there is no published information that radiation-cured pressure sensitive adhesives are commercial. It is probable, however, that radiation curing processes are being used by companies that manufacture adhesives without any public announcements. At any rate, this developing field is far from being exhausted. Nevertheless, many problems still await solutions before these adhesives will achieve full significance.

The problem of application of these new materials, especially of the polymeric type, is far from being a trivial one. An analysis of the cost of these materials compared with the conventional ones must be made, and it must, of course, include the cost of the energy required for curing and capital expenditures for equipment. The future availability of solvents for conventional adhesives and possibly their cost increase must be taken into account, as well as the present (and future) pollution legislation. However, the most important factor in commercialization of these materials may likely be to adapt them to a specialty area where their performance advantages over those of the conventional adhesives will be so desirable that the user will be willing to pay the increased price.

REFERENCES

1. Steuben, K. C. *Adhesives Age* **20**(6): 16 (June 1977).
2. Pappas, S. P. (ed.). *UV Curing: Science and Technology*. Technology Marketing Corp., Stanford, Conn., 1978.
3. Oster, G., and Yang, N. L. Photopolymerization of Vinyl Monomers. *Chem Rev.* **68:** 125 (1968).
4. Calvert, J. A. and Pitts, J. N. *Photochemistry*. Wiley, New York, 1966.
5. Charlesby, A. *Atomic Radiation and Polymers*. Pergamon, Oxford, 1960.
6. Tawn, A. R. H. *J. Oil Col. Chem. Assoc.* **51:** 782 (1968).
7. Ledwith, A., Russell, P. J., and Sultcliffe, L. H. *J. Chem. Soc., Perkin Transact.* **2:** 1925 (1972).
8. Ledwith, A. *J. Oil Col. Chem. Assoc.* **59:** 157 (1976).

38. Guse, G., Lukat, E., Jaucken, P., Lende, W., and Pietsch, H. Ger. 2,443,414 (March 25, 1976) (assigned to Beiersdorf AG).
39. Gleichenhagen, P., and Karmann, W. Ger. 2,455,133 (May 26, 1976) (assigned to Beiersdorf AG).
40. Karmann, W., and Gleichenhagen, P. Ger. 2,350,030 (April 17, 1975) (assigned to Beiersdorf AG).
41. Yoshikawa, N., Kamano, T., and Sunagawa, M. Jap. Kokai 51 046,331 (April 20, 1976) (assigned to Nitto Electric Industrial Co.).
42. Uehara, K., Murakoshi, S., and Hisamtsu, H. Jap. Kokai 74 05,145 (January 17, 1974) (assigned to Dainippon Ink and Chemicals, Inc.).
43. Sasaki, T., Araki, K., Kawaguchi, T., and Ishiyama, H. Jap. Kokai 76 26,940 (March 5, 1976) (assigned to Japan Atomic Energy Research Institute).
44. Sasaki, T., Araki, K., Kawaguchi, T., and Ishiyama, H. Jap. Kokai 76 26,941 (March 5, 1976) (assigned to Japan Atomic Energy Research Institute).
45. Sasaki, T., Araki, K., Kawaguchi, T., and Ishiyama, H. Jap. Kokai 76 26,942 (March 5, 1976) (assigned to Japan Atomic Energy Research Institute).
46. Fukukawa, T., Shimomura, T., Yoshikawa, S., and Sunagawa, M. Jap. Kokai 74 037,692 (December 4, 1975) (assigned to Nitto Electric Industrial Co.).
47. Nakata, S., Honi, H., and Ohnishi, K. Jap.74 029,613 (August 6, 1974) (assigned to Sekisui Chemical Co., Ltd.).
48. Kasper, A. A., U.S. Patent 3,328,914 (June 27, 1967) (assigned to Kendall Co.).
49. Hansen, D. R., and St. Clair, D. J. U.S. Patent 4,133,731 (January 9, 1979) (assigned to Shell Chemical Co.).
50. Hanrahan, M. J., and Meshirer, M. Radiation Curable Pressure Sensitive Adhesives. In *Proc. Radcure 86 Conference,* Association for Finishing Processes of the Society of Manufacturing Engineers, Baltimore, Md., September 8–11, 1986.
51. Kardashian, R. Applications of Electron Beam Technology for Pressure Sensitive Adhesive Products. In *Proc. Technical Seminar,* Pressure Sensitive Tape Council, Itasca, Ill., May 7–9, 1986.
52. Lin, J., Sun, B., and Wen, W. UV Curable, Screen Printable Pressure-Sensitive Adhesive. Technical Paper FC84-1019, Society of Manufacturing Engineers, Dearborn, Mich., 1984.
53. Huber, H. F., and Mueller, H. Pressure-Sensitive Adhesives Based on Radiation Curable Polyesters. In *Proc. Radcure 86 Conference,* Association for Finishing Processes of the Society of Manufacturing Engineers, Baltimore, Md., Sept. 8–11, 1986.
54. Johnson, M. Some Recent Approaches to the Development of Acrylic UV-Curable Pressure Sensitive Adhesives. In *Proc. Radcure 86 Conference,* Association for Finishing Processes of the Society of Manufacturing Engineers, Baltimore, Md., September 8–11, 1986.
55. Herze, P. Y., Philips, M., and Loutz, J. M. Pressure Sensitive Adhesives Obtained by Irradiation. In *Proc. Radcure 86 Conference,* Association for Finishing Processes of the Society of Manufacturing Engineers, Baltimore, Md., September 8–11, 1986.

9781475708684